Molecular Modeling *for the* Design *of* Novel Performance Chemicals *and* Materials

T0179219

Molecular Modeling *for the* Design *of* Novel Performance Chemicals *and* Materials

Edited By
Beena Rai

CRC Press
Taylor & Francis Group
Boca Raton London New York

CRC Press is an imprint of the
Taylor & Francis Group, an **informa** business

CRC Press
Taylor & Francis Group
6000 Broken Sound Parkway NW, Suite 300
Boca Raton, FL 33487-2742

First issued in paperback 2019

© 2012 by Taylor & Francis Group, LLC
CRC Press is an imprint of Taylor & Francis Group, an Informa business

No claim to original U.S. Government works

ISBN-13: 978-1-4398-4078-8 (hbk)
ISBN-13: 978-0-367-38157-8 (pbk)

**Visit the Taylor & Francis Web site at
http://www.taylorandfrancis.com**

**and the CRC Press Web site at
http://www.crcpress.com**

Anor aniyan mahaan mahiyan
[The greatest of the great is hidden in the smallest of the small like atomic energy.]

The Vedic dictum

Contents

Foreword

I am extremely pleased to write the Foreword for *Molecular Modeling for the Design of Novel Performance Chemicals and Materials*, edited by Dr. Beena Rai. The techniques of modeling and simulation have been developed extensively over the last few years. The approximations involving *ab initio* and density functional theory at quantum length scales, as well as the classical atomistic simulation and coarse-grained simulation at larger length scales, have made the computational tools potent in the modeling of materials. I expect the multiscale approach to simulation of materials to be further refined in future years, such that the modeling of designer materials may be attempted in a serious manner. At the same time, there has been tremendous growth in computational hardware in terms of speed, architecture, and affordability. This has been coupled with developments in numerical algorithms and efficient strategies of coding. All these together make the future of materials modeling extremely promising.

It is thus fitting that at this stage it is necessary to take stock of the current status of the modeling of materials and the applications of such techniques to meaningful problems. This book brings together topics on various facets of such modeling techniques and applications in diverse ranges from pharmaceuticals to cement. The interesting point to note is that even though the case studies presented are very diverse, the chapters are organized according to certain underlying themes. The book starts with a review of molecular modeling tools in Chapter 1, subsequently describing several applications in succeeding chapters. Chapters 2 to 4 deal with the modeling of mineral–reagent interactions. The adsorption of surfactants at air–liquid and solid–liquid interfaces, as studied through various molecular modeling techniques, and their effect on the macroscopic behavior of the system under study has been illustrated with various case studies drawn from the work of the authors. Chapter 5 deals with the application of molecular modeling tools in the crystallization and design of pharmaceutical products of desired properties. Chapters 6 and 7 deal with studies on the microstructure in water—surfactant systems using atomistic and mesoscale applications. Chapter 8 emphasizes the study of phase transitions in porous media and their role in the design of novel porous materials for improved industrial processes. The study of wettability on solid surfaces is examined in detail in Chapter 9, with a focus on the advantages and disadvantages of various methodologies and relevant case studies. Chapter 10 illustrates applications in the area of transport properties of materials, with a special focus on zeolite and nanofluids—novel heat-transfer liquids that possess higher thermal conductivity. Chapters 11 to 14 describe the use of density functional theory in hydrogen storage materials, semiconductor alloys, cement clinker compounds, and catalysis, each of which is an important area. The authors are all accomplished in their areas, and the book brings together a suite of competencies on molecular modeling in different parts of material research.

I think it is an extremely important addition to the growing field and will serve as an important reference book.

Sourav Pal, FNA, FASc, FNASc
Director
National Chemical Laboratory
Adjunct Professor
Indian Institute of Science Education and Research
Pune, India

Preface

Beginning with the 1980s, molecular modeling (MM) has become an important tool in many academic institutions and industrial laboratories. Although the role of MM in biological fields—especially in the design and development of novel drug molecules or formulations—is well established and acknowledged, its direct role in the design and development of performance chemicals and novel materials is still not well known. Questions such as which new products have resulted from an MM-based approach are often still asked. Although MM may be playing an important role in product development, quite often it becomes difficult to predict its direct impact because most of the time the problem being addressed involves a multidisciplinary approach. Further, the assumption that the fundamental phenomena being modeled through MM will have direct impact on the macroscopic and functional properties of a product makes the situation more complicated. In most cases, MM actually works as an enabler toward the development of novel products and materials rather than directly coming up with these new products and materials, such as novel drug molecules in biological application. This is precisely the reason that, despite seeing value in MM tools, most engineers and practitioners often focus on the question of how to leverage these tools to design and develop novel materials or chemicals for the industry the person is working in? Unfortunately, there is no simple answer to this question. Excellent books and research publications highlight the most intricate, fundamental, and theoretical details about MM techniques and tools. However, the use of highly specialized terminologies and a lack of practical case studies make it almost impossible for those who do not specialize in MM to assimilate the knowledge and use it for their benefit. What is required from the user's perspective is a simple guide that highlights the utility and power of MM tools.

The goal in editing this book is to provide answers to these important questions. The attempt here is to provide a simple and practical approach to the MM paradigm. The focus of this book is to highlight the importance and usability of MM tools and techniques in various industrial applications as illustrated by several case studies. The contributing authors have provided numerous examples in which MM tools have been used to understand the properties of the materials and rationally design novel chemicals and materials.

The book is divided into 14 chapters that cover a diverse range of industrial applications from pharmaceuticals to cement. The interesting point to note is that even though the case studies presented are wide ranging, the chapters are organized according to certain underlying themes. Chapter 1 provides a brief overview of MM techniques and tools as may be appropriate for a beginner. Adequate references are provided for more expert learning.

Chapters 2 to 4 deal with the modeling of mineral–reagent interactions. The adsorption of surfactants at air–liquid and solid–liquid interfaces and its effect on their macroscopic behavior is illustrated through various case studies. The applications range from mineral processing to pharmaceuticals. Concepts like the wetting of bare mineral surfaces and those coated with a monolayer of surfactants are also addressed. These chapters clearly provide an insight into the utility and power of MM in the design and development of performance chemicals.

The application of MM tools in the crystallization and design of pharmaceutical products is described in Chapter 5.

Chapters 6 deals with some of the fundamental aspects of surfactant assembly, while in Chapter 7 the behavior of liquid in confinement and its implications in industrial applications are described. Phase transitions in porous media and their role in the design of novel porous materials for improved industrial processes are illustrated in Chapter 8. Chapter 9 deals with the advantages and disadvantages of various methodologies employed in the studies of wettability on solid surfaces, as illustrated through relevant case studies. The application of MM in the study of transport properties of materials—with a special focus on zeolite and nanofluids—is described in Chapter 10.

Chapters 11 to 14 demonstrate applications of density functional theory and simulations in the design and development of novel materials for hydrogen storage, new semiconductor alloys, cement, and catalysis.

This book has been compiled with two types of audiences in mind—practicing engineers and practicing chemists, each approaching MM from different perspectives but sharing a common desire to solve the practical problems in the most effective manner. I hope that this book serves both communities and provides a fresh perspective on the application of MM tools and techniques in various industries. I am sure that researchers entering the fascinating field of MM will find it useful as a reference guide.

I thank all the authors for their contributions and encouragement in the creation of this book. I also thank all the reviewers for their valuable comments and suggestions. My thanks to the members of the technical communication team at Tata Research Development and Design Centre Pune and to Deenaz Bulsara for their editorial help. I thank all my previous and present colleagues for their direct or indirect contributions to the work presented here. I am extremely thankful to Dr. Sourav Pal, Director, National Chemical Laboratory, Pune, India for writing a befitting Foreword for the book. Special thanks are due to Dr. Pradip, who has been a constant source of encouragement and a guide throughout the compilation of this book. I also thank K. Ananth Krishnan, Corporate Technology Officer, Tata Consultancy Services, Ltd., for his encouragement and support. Finally, my sincere thanks are due to my family—Kulbhushan, Kritika, and Kaustubh—who have silently supported me in this endeavor.

Beena Rai
Pune, India

Editor

Beena Rai earned her PhD in 1995 from the National Chemical Laboratory, Pune, India. Currently, she is working as senior scientist at the Process Engineering Innovation Lab, Tata Research Development & Design Centre (TRDDC) (a division of Tata Consultancy Services Ltd.), Pune, India. At TRDDC, she leads the research program on the rational design of industrial performance chemicals using molecular modeling techniques. She, along with her coworkers, has successfully applied these techniques to the design and development of performance chemicals for various industrial applications. Her research interests include mineral processing, rational design, synthesis and testing of performance chemicals, colloids and interfacial science, ultrafine grinding, waste recycling, cements, nanofluids, drug delivery, and biosensors. She has several industrial projects to her credit and has authored 50 research publications, 7 patents, and several reports. She has been a visiting scientist at Laboratoire Environnment et Mineralogie, INPL-CNRS, Nancy, France.

List of Contributors

Jhumpa Adhikari
Department of Chemical Engineering
Indian Institute of Technology Bombay
Mumbai, India

K. Ganapathy Ayappa
Department of Chemical Engineering
Indian Institute of Science
Bangalore, India

Pui Shan Chow
Department of Crystallization and
    Particle Science
Institute of Chemical and Engineering Sciences
A*STAR (Agency for Science, Technology, and
    Research)
Jurong Island, Singapore

Hao Du
National Engineering Laboratory for
    Hydrometallurgical Cleaner Production
    Technology
Institute of Process Engineering
Chinese Academy of Sciences
Beijing, People's Republic of China

Ricardo Grau-Crespo
Department of Chemistry
University College London
London, United Kingdom

Yoshiyuki Kawazoe
Institute for Materials Research
Tohoku University
Sendai, Japan

Sandip Khan
Department of Chemical Engineering
Indian Institute of Technology Kanpur
Kanpur, India

Vijay Kumar
Dr. Vijay Kumar Foundation
Gurgaon, India

T. K. Kundu
Department of Metallurgical and Materials
    Engineering
Indian Institute of Technology
Kharagpur, India

Sang Kyu Kwak
Division of Chemical and Biomolecular
    Engineering
School of Chemical and Biomedical Engineering
Nanyang Technological University
Singapore

Jin Liu
Department of Metallurgical Engineering
College of Mines and Earth Sciences
University of Utah
Salt Lake City, Utah

Jan D. Miller
Department of Metallurgical Engineering
College of Mines and Earth Sciences
University of Utah
Salt Lake City, Utah

Palanichamy Murugan
Functional Materials Division
Central Electrochemical Research Institute
Karaikudi, Tamil Nadu, India

Ajay Nandgaonkar
Research and Innovation Group
Computational Research Laboratories Limited
Pune, India

Orhan Ozdemir
Department of Mining Engineering
Faculty of Engineering
Istanbul University
Istanbul, Turkey

Venkata Gopala Rao Palla
Research and Innovation Group
Computational Research Laboratories Limited
Pune, India

S. C. Parker
Computational Chemistry Group
Department of Chemistry
University of Bath
Bath, United Kingdom

Mario Pinto
Research and Innovation Group
Computational Research Laboratories
    Limited
Pune, India

Sendhil K. Poornachary
Department of Crystallization and
    Particle Science
Institute of Chemical and Engineering Sciences
A*STAR (Agency for Science, Technology, and
    Research)
Jurong Island, Singapore

Pradip
Tata Research Development & Design Centre
Pune, India

Beena Rai
Tata Research Development & Design Centre
Pune, India

K. Hanumantha Rao
Mineral Processing Group
Division of Sustainable Process Engineering
Luleå University of Technology
Luleå, Sweden

Ryoji Sakurada
Department of Civil Engineering
Akita National College of Technology
Akita, Japan

Abhishek Kumar Singh
Materials Research Centre
Indian Institute of Science
Bangalore, India

Jayant K. Singh
Department of Chemical Engineering
Indian Institute of Technology Kanpur
Kanpur, India

Sudhir K. Singh
Department of Chemical Engineering
Indian Institute of Technology Kanpur
Kanpur, India

Reginald B. H. Tan
Department of Crystallization and
    Particle Science
Institute of Chemical and Engineering Sciences
A*STAR (Agency for Science, Technology, and
    Research)
and
Department of Chemical and Biomolecular
    Engineering
National University of Singapore
Singapore

Foram M. Thakkar
Department of Chemical Engineering
Indian Institute of Science
Bangalore, India

Umesh V. Waghmare
Theoretical Sciences Unit
Jawaharlal Nehru Centre for Advanced
    Scientific Research
Bangalore, India

Xuming Wang
Department of Metallurgical Engineering
College of Mines and Earth Sciences
University of Utah
Salt Lake City, Utah

Xihui Yin
Department of Metallurgical Engineering
College of Mines and Earth Sciences
University of Utah
Salt Lake City, Utah

Shili Zheng
National Engineering Laboratory for
    Hydrometallurgical Cleaner Production
    Technology
Institute of Process Engineering
Chinese Academy of Sciences
Beijing, People's Republic of China

1 Basic Concepts in Molecular Modeling

Beena Rai

CONTENTS

1.1 INTRODUCTION

With ever-changing needs for innovative materials and improved performance chemicals that exhibit unique functional properties, researchers and engineers have become increasingly involved in designing materials at the molecular level. The study of substances at the molecular level is not new, but using the knowledge about the interactions between atoms and molecules to design materials and chemicals with the desired functional properties is new. A computational chemistry approach leading to a rigorous systematic scientific approach toward the design and development of novel products seems to be the most logical option. In fact, with the high costs associated with experiments, computational chemistry definitely offers a relatively less expensive alternative for the purpose of design of materials. The popularity of using the computational approach to investigate structure–property relationships for both macroscopic and nanoscaled systems is reflected by the increasing number of computational chemistry software vendors at various chemical engineering conferences worldwide.

Molecular modeling can be defined as the application of computational techniques, grounded in theory, to predict or explain observable biological, physical, or chemical properties of molecules. The most accepted definition, as stated by Pensak (1989), is "Molecular modeling (MM) is anything that requires the use of a computer to paint, describe, or evaluate any aspect of the properties of the structure of a molecule." Most of the methods applied in MM can be broadly classified as computational methods based on energetic models of the systems and other concepts such as structure matching, chemical similarity, quantitative structure–activity relationships (QSAR), molecular shape compatibility, and materials informatics. Among the energetic models, techniques based on electronic structure calculation (molecular orbital, density functional, semiempirical methods, *ab initio*) can be differentiated from the empirical energy function (force field) methods. Force field methods do not treat electrons explicitly but model their effects in terms of analytical functions expressing energy contributions from bond stretching, angle bending, torsion and nonbonded interactions. As compared with the energy-based methods, the nonenergy-based methods are generally much more qualitative in nature, deriving benefits from some of the well-known concepts in statistics, mathematics, and computer science. In the following sections, a brief summary of these molecular modeling concepts is presented. The objective here is to introduce the reader to different paradigms without exploring the theoretical details.

1.2 MOLECULAR MECHANICS OR ATOMISTIC SIMULATIONS

Molecular mechanics or *atomistic simulations,* consider the atomic composition of a molecule to be a collection of masses that interact with each other through harmonic forces. Because of this simplification, it is a relatively faster method applicable to large systems, as compared with quantum chemical methods. Furthermore, with the recent developments of accurate force field parameters for diverse kinds of molecules, it is possible to treat almost all types of molecular systems, including transition metals. Atomistic simulation provides a set of tools for predicting many useful functional properties of systems of industrial relevance. These properties include thermodynamic properties (e.g., equations of state, phase equilibria, and critical constants), mechanical properties (e.g., stress–strain relationships and elastic moduli), transport properties (e.g., viscosity, diffusion, and thermal conductivity), and morphological information (e.g., location and shape of binding sites on a bio-molecule and crystal structure). This list is by no means exhaustive and continues to grow as algo-rithmic and computer hardware advances make it possible to access additional properties. Some of the most relevant textbooks of molecular simulation provide excellent reviews (Allen and Tildesley 1987; Frenkel and Smit 1996; Sadus 1999).

In the framework of molecular mechanics, the atoms in molecules are treated as rubber balls of different sizes (atom types) joined by springs of varying length (bonds). The potential energy of a system can be expressed as a sum of valence (or bond), cross-term, and nonbond interactions, which together are commonly referred to as force fields:

$$E_{\text{total}} = E_{\text{valence}} + E_{\text{crossterm}} + E_{\text{nonbond}} \tag{1.1}$$

1.2.1 VALENCE INTERACTIONS

The energy of *valence interactions* is generally accounted for by diagonal terms such as bond-stretching, angle-bending, dihedral-angle torsion, and inversion or out-of-plane interactions. Generally, the interatomic forces are assumed to be harmonic and the bond-stretch term is repre-sented by a simple quadratic function:

$$E_{\text{str}} = \frac{1}{2} k_b \left(b - b_0 \right)^2 \tag{1.2}$$

where k_b is the stretching force constant, b_0 is the equilibrium bond length, and b is the actual bond length. For angle bending, a simple harmonic expression is used:

$$E_{\text{bend}} = \frac{1}{2} k_\theta \left(\theta - \theta_0 \right)^2 \tag{1.3}$$

where k_θ is the angle-bending force constant, θ_0 is the equilibrium bond angle, and θ is the actual bond angle. A cosine expression is commonly used for the dihedral potential energy, as represented in Equation 1.4:

$$E_{\text{tors}} = \frac{1}{2} k_\varphi \left(1 + \cos(n\varphi - \varphi_0) \right) \tag{1.4}$$

where k_φ is the torsional barrier, φ is the actual torsion angle, n is the periodicity, and φ_0 is the refer-ence torsional angle.

1.2.2 Valence Cross-Terms

Some of the modern force fields also include *cross-terms* to account for bond or angle distortions caused by nearby atoms. These terms are required to accurately reproduce experimental vibrational frequencies of molecules. Cross-terms may include stretch–stretch, stretch–bend–stretch, bend–bend, torsion–stretch, torsion–bend–bend, bend–torsion–bend, stretch–torsion–stretch terms.

1.2.3 Nonbond Interactions

The energy of interactions between nonbonded atoms is accounted for by the following:
- van der Waals interactions between nonbonded atoms that are usually represented by Lennard–Jones potential:

$$E_{vdw} = \sum \frac{A_{ij}}{r_{ij}^{12}} - \frac{B_{ij}}{r_{ij}^{6}}$$

(1.5)

where A_{ij} and B_{ij} are the repulsive and attractive term coefficients, respectively, and r_{ij} is the distance between the two atoms.
- Electrostatic interactions are described by a coulomb term:

$$E_{elec} = \frac{1}{\varepsilon} \frac{Q_1 Q_2}{r_{ij}}$$

(1.6)

where Q_1 and Q_2 are the charges on the interacting atoms, ε is the dielectric constant, and r_{ij} is the interatomic distance.

1.2.4 Constraints

Constraints that can be added to an energy expression include distance, angle, torsion, and inversion constraints. Constraints are useful for examining only part of a structure.

The equilibrium values of bond lengths, angles, torsional angles, and various constants associated with the functions used to describe the potential energy surface in a force field are referred to as *force field parameters*. These parameters are derived empirically from various experimental techniques or from more accurate quantum chemical calculations. The purpose of a force field is to describe the potential energy surface of entire classes of molecules with reasonable accuracy. Accordingly, there are some generic force fields offering broadest possible coverage of the periodic table, with necessarily lower accuracy (Casewit et al. 1992a, 1992b; Mayo et al. 1990; Rappe' et al. 1992, 1993). However, there are also some force fields that aim for high accuracy for a limited set of elements, thus enabling good predictions of many molecular properties (Sun 1998).

A typical force field contains all the necessary elements for calculations of energy and force for a given molecular system, and these elements include

- A list of force field types
- A list of partial charges
- Force field–typing rules
- Functional forms for the components of the energy expression
- Parameters for the function terms
- If applicable, rules for generating parameters that have not been explicitly defined
- In some cases, a way of assigning functional forms and parameters

In general, many experimental properties such as vibrational frequencies, sublimation energies, and crystal structures can be reproduced with a force field, even though it includes most of the quantum effects empirically. However, it is important to appreciate the fundamental limitations of force field approach. Applications beyond the capability of most force field methods include

- Electronic transitions (photon absorption)
- Electron transport phenomena
- Proton transfer (acid–base reactions)

It is interesting to note that the true power of the atomistic simulations lies in following three major areas:

1. First, these simulations can handle large systems, are several orders of magnitude faster (and cheaper) than quantum-based calculations, and can be used for studying condensed-phase molecules, macromolecules, crystal morphology, inorganic and organic interfaces, and so on.
2. Second, the energy contributions can be analyzed at the level of individual or classes of interactions. One can decompose the energy into bond energies, angle energies, nonbond energies, and so on—or even to the level of a specific hydrogen bond or van der Waals contact—to understand the observed physical properties or to make a prediction.
3. Third, the energy expression can be modified to bias the calculation by imposing constraints.

1.3 ENERGY MINIMIZATION

For any given molecular system, deviations from equilibrium values, such as those in bond length, angle, and torsions, result in increased total energy. In a true sense, the total energy of a molecule is a measure of intramolecular strain relative to a hypothetical molecule with an ideal geometry at equilibrium. Various *energy minimization* techniques are used to obtain the most stable conformation of a given molecule that lies as close as possible to this hypothetical equilibrium structure. The optimization can be regarded in mathematical terms as an optimization in a multidimensional space. The minimization methods are broadly classified into two categories based on the order of the derivatives used to locate a minimum on the potential energy surface. Zero-order methods such as SIMPLEX use only the energy function to locate regions of low energy through a grid search procedure. The higher-order methods use a gradient of the function or hessian to locate a minimum. Some of the commonly used standard methods are briefly described next.

1.3.1 STEEPEST DESCENT

The first derivative of an energy function is computed to find the minimum. The energy is computed for the given confirmation and then again after one of the atoms is moved in small increments in one of the directions of the coordinate system. This process is followed for all the atoms and finally whole molecule is moved to a position downhill on the potential energy surface. The process is repeated until the desired threshold condition is fulfilled. The process is more suitable for the optimization far away from the minimum as a rough and fast method to reach closer to the minimum.

1.3.2 NEWTON–RAPHSON METHODS

As a rule, N^2 independent data points are required to solve a harmonic function with N variables numerically. Because a gradient is a vector N long, the best one can hope for in a gradient-based minimizer is to converge in N steps. However, if one can exploit second-derivative information, an optimization could converge in one step, because each second derivative is an $N \times N$ matrix. This is the principle behind the variable metric optimization algorithms such as *Newton–Raphson method*.

Here, in addition to using the gradient to identify a search direction, the curvature of the function (the second derivative) is also used to predict where the function passes through a minimum along that direction. Because the complete second-derivative matrix defines the curvature in each gradient direction, the inverse of the second-derivative matrix can be multiplied by the gradient to obtain a vector that translates directly to the nearest minimum. However, it is important to note that the method is computationally intensive and becomes unstable when the structure is far from the minimum. Also, calculating, inverting, and storing an $N \times N$ matrix for a large system may become impractical. Therefore, pure Newton–Raphson method is applied primarily for calculations in which rapid convergence to an extremely precise minimum is required while performing vibrational normal-mode analysis. In addition to the iterative Newton–Raphson method, variants of the Newton method, such as quasi-Newton and truncated Newton methods, are also used.

1.3.3   CONJUGATE GRADIENT

This *conjugate gradient method* uses an algorithm that produces a complete basis set of mutually conjugate directions such that each successive step continually refines the direction toward the minimum. If these conjugate directions truly span the space of the energy surface, then, in turn, minimization along each direction must arrive at a minimum. This method is more time consuming than steepest descent but is more accurate. It is the method of choice for large models because, in contrast to Newton–Raphson methods, in which storage of a second-derivative matrix [$N(N + 1)/2$] is required, only the previous $3N$ gradients and directions have to be stored. However, it is important to note that the conjugate gradient method can be unstable if the conformation is so far away from a local minimum that the potential energy surface is not close to being quadratic.

The choice of minimization method depends on two factors, the size of the system and its conformational state. As a guiding principal, if the structure is far from the minimum then the steepest descent should be used for the first few steps, followed by a more precise conjugate gradient or Newton–Raphson methods. For detailed information, the reader is requested to refer to some of the standard textbooks (Allen and Tildesley 1987; Frenkel and Smit 1996).

1.4   HANDLING CHARGES IN MOLECULAR SIMULATIONS

A force field charge is nothing but the partial charge of an atom. It is important to assign the correct charge because electrostatic interactions play a critical role in determining the structures of inorganic systems and the packing of organic molecules. Some of the force fields support charges, and they are assigned automatically when the force field types are assigned (Mayo et al. 1990; Sun 1998). However, in some cases, charges need to be supplied externally (Casewit et al. 1992a, 1992b; Rappe' et al. 1992, 1993). Different methods for the calculation of partial atomic are based on *ab initio*, semiempirical, and empirical procedures. Comprehensive reviews describing different approaches to the evaluation of partial charges have been reported (Bachrach 1993; Willium 1991). Charges can be computed from quantum chemical wave functions such as the Mulliken population analysis (Mulliken 1955) or using topological procedures such as Gasteiger (Gasteiger and Marsili 1980) and QEq (Rappe and Goddard 1991) methods.

1.5   PERIODIC BOUNDARY CONDITIONS

Periodic boundary conditions refer to the simulation of structures consisting of a periodic lattice of identical subunits. Periodic boundaries help simulate bulk-material, solvent, and crystalline systems. Ideally, a periodic system infinitely replicates in all dimensions to form a periodic lattice. However, in practice, all periodic boundary algorithms imply a cutoff criterion for computational efficiency (Figure 1.1). In these cutoff schemes, each atom interacts with the nearest images of other $N - 1$ atoms (minimum-image convention) or only with the explicit images contained in a sphere

(a)
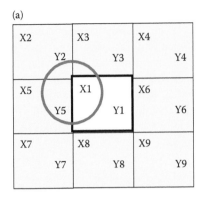

(b)
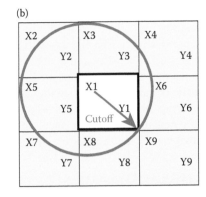

FIGURE 1.1   (a) Minimum-image structure and (b) explicit-image structure showing cutoff radius.

of radius R_{cutoff} centered at the atom. Minimum-image convention implies that the maximum cutoff distance should not be more than half the cell dimensions. The use of cutoff methods has been known to introduce errors in simulations (Bader and Chandler 1992; Schreiber and Steinhauser 1992; York et al. 1994).

The Ewald summation method is a more robust technique (Ewald 1921; Tosi 1964) for periodic systems. Crystalline solids are the most appropriate candidates for Ewald summation, because the error associated with using cutoff methods is much greater in an infinite lattice. However, the technique can also be applied to amorphous solids and solutions. For a detailed description of this method, the reader is requested to refer to relevant research papers (Deem et al. 1990; Karasawa and Goddard 1989). As opposed to traditional long-range interactions that converge extremely slowly, the Ewald method converts the summation of potential energy into two series, each of which converges much more rapidly, and a constant term:

$$U_{Ewald} = U^r + U^m + U^0 \tag{1.7}$$

This is achieved by considering each charge to be surrounded by a neutralizing charge distribution (Gaussian distribution) of equal magnitude and opposite sign. The sum-over-point charges are thus converted to a sum of interactions between the charges plus the neutralizing distributions. This is referred to as the *real space* or *direct sum* U^r. A second charge distribution, which counteracts the first, is performed in the reciprocal space and termed as U^m. The self-term U^0 is a correction term that cancels the interaction of each of the Gaussian countercharges with itself. The direct sum is primarily evaluated using cutoff methods, while the reciprocal sum is approximated using Fast Fourier transform (FFT) with convolutions on a grid on which charges are interpolated in the grid points (Darden et al. 1993).

1.6   NEIGHBOR LIST

MM simulations, being many-body systems, require dealing with the particles that are coupled with short-range potentials. Thus a key part of the simulation is to find which particles {j} interact with particle {i}. The set of such particles is known as the *neighbor list* of particle {i}. All the simulations using limited-range interactions require a means of tracking which particles neighbor which, a task handled by a neighbor list generation algorithm. Two of the most common algorithms used to construct a neighbor list are the Verlet neighbor list and the linked-cell method. The Verlet neighbor list (Allen and Tildesley 1987; Frenkel and Smit 1996) stores for each particle {i} an array of those particles {j} within some fixed radius. However, this method is suitable for modeling smaller systems only because with increasing system size the memory requirement also increases. This neighbor list requires occasional updating if particles in the system are allowed to move far from their original positions. In absence of any means of finding the particles just outside the neighbor list, this update requires a lengthy search. The linked-cell method (Allen and Tildesley 1987; Frenkel and Smit 1996) for neighbor list construction divides the supercell into a number of identical smaller cells, axis-aligned with the supercell. Associated with each smaller cell is a linked list of those particles within. If particles move from one smaller cell to another, it is then a simple matter to determine which neighboring cell the particle has moved to and place it at the head of that cell's linked list. It is important to note that the performance of these neighbor list techniques depends on a variety of parameters that may be adjusted for maximum efficiency.

1.7 MOLECULAR DYNAMICS

Once the mathematical models (i.e., force fields) for the internal structure of each molecule (i.e., the intramolecular potential) and the interaction between molecules (i.e., the intermolecular potential) are known through classical statistical mechanics, one can predict the properties of a macroscopic sample of such molecules based on statistical averaging over the possible microscopic states of the system as it evolves under the rules of classical mechanics (Chandler 1987; McQuarrie 1976; Wilde and Singh 1998). Two of the most common techniques used are molecular dynamics (MD) (Alder and Wainwright 1957, 1959; McCammon et al. 1977; Rahman 1964; Stillinger and Rahman 1974) and Monte Carlo (MC) (Milik and Skolnick 1993; Ojeda et al. 2009) simulations.

In its simplest form, MD solves Newton's familiar equation of motion:

$$F_i = m_i a_i(t) \tag{1.8}$$

where F_i is the force, m_i is the mass, and a_i is the acceleration of atom i. The force on atom i can be computed directly from the derivative of the potential energy V with respect to the coordinate r_i:

$$-\frac{\partial V}{\partial r_i} = m_i \frac{\partial r_i^2}{\partial t^2} \tag{1.9}$$

The classical equations of motion are deterministic; therefore once the initial coordinates and velocities are known, the coordinates and velocities can be determined later. The coordinates and velocities for a complete dynamics run are called the *trajectory*. Thus solving the classical equations of motion as a function of time (typically over a period limited to tens of nanoseconds) generates the microscopic states of the system. The system may thus relax to equilibrium (provided the time for the relaxation falls within the time accessible to MD simulation), leading to the extraction of transport properties, which at the macroscopic scale describe the relaxation of the system in response to inhomogeneities.

MC simulations generate equilibrium configurations stochastically according to the probabilities rigorously known from statistical mechanics. Because it generates equilibrium states directly, one can use it to study the equilibrium configurations of systems that may be expensive or impossible to access via MD. The drawback of MC is that it cannot yield the kind of dynamic response information that leads directly to transport properties.

1.7.1 INTEGRATORS IN MOLECULAR DYNAMICS

Potential energy is a function of the atomic positions ($3N$) of all the atoms in the system. Due to the complicated nature of this function, there is no analytical solution to the equations of motion; they must be solved numerically. Numerous numerical algorithms have been developed for integrating the equations of motion. When choosing which algorithm to use, one should consider the following criteria:

- The algorithm should conserve energy and momentum.
- It should be computationally efficient.
- It should permit a long time step for integration.

All the integration algorithms assume that the positions, velocities, and accelerations can be approximated by a Taylor series expansion:

$$r(t + \delta t) = r(t) + v(t)\delta t + \frac{1}{2}a(t)\delta t^2 + \ldots$$

$$v(t + \delta t) = v(t) + a(t)\delta t + \frac{1}{2}b(t)\delta t^2 + \ldots$$

$$a(t + \delta t) = a(t) + b(t)\delta t + \ldots \tag{1.10}$$

where r is the position, v is the velocity (the first derivative with respect to time), a is the acceleration (the second derivative with respect to time), and so on.

1.7.1.1 Verlet Algorithm

The Verlet algorithm (Verlet 1967) is derived as follows:

$$r(t + \delta t) = r(t) + v(t)\delta t + \frac{1}{2}a(t)\delta t^2 \tag{1.11}$$

$$r(t - \delta t) = r(t) - v(t)\delta t + \frac{1}{2}a(t)\delta t^2 \tag{1.12}$$

Addition of Equations 1.11 and 1.12 gives:

$$r(t + \delta t) = 2r(t) - r(t - \delta t) + a(t)\delta t^2 \tag{1.13}$$

The Verlet algorithm uses positions and accelerations at time t and the positions from time $t - \delta t$ to calculate new positions at time $t + \delta t$. No explicit velocities are used. It is straightforward and modest on the storage requirements, but the precision is moderate.

1.7.1.2 Leapfrog Algorithm

The velocities are first calculated at time $t + (1/2)\,\delta t$ and then are used to calculate the positions r at time $t + \delta t$. In this way, the velocities leap over the positions and then the positions leap over the velocities. The advantage of this algorithm is that the velocities are explicitly calculated; however, they are not calculated at the same time as the positions:

$$r(t + \delta t) = r(t) + v\left(t + \frac{1}{2}\delta t\right)\delta t \tag{1.14}$$

$$v\left(t + \frac{1}{2}\delta t\right) = v\left(t - \frac{1}{2}\delta t\right) + a(t)\delta t \tag{1.15}$$

1.7.1.3 Verlet Velocity Algorithm

The Verlet velocity algorithm overcomes the out-of-synchrony shortcoming of the Verlet leap-frog method. The advantage here is that the positions, velocities, and accelerations are computed at the same time t. There is no compromise on precision. The Verlet velocity algorithm is as follows:

$$r(t+\delta t) = r(t) + v(t)\delta t + \frac{1}{2}a(t)\delta t^2 \qquad (1.16)$$

$$v(t+\delta t) = v(t) + \frac{1}{2}[a(t) + a(t+\delta t)]\delta t \qquad (1.17)$$

1.7.1.4 Beeman's Algorithm

This algorithm is closely related to the Verlet algorithm:

$$r(t+\delta t) = r(t) + v(t)\delta t + \frac{2}{3}a(t)\delta t^2 - \frac{1}{6}(t-\delta t)\delta t^2 \qquad (1.18)$$

$$v(t+\delta t) = v(t) + v(t)\delta t + \frac{1}{3}a(t)\delta t + \frac{5}{6}a(t)\delta t - \frac{1}{6}a(t-\delta t)\delta t \qquad (1.19)$$

It provides a more accurate expression for the velocities and better energy conservation but is more expensive.

1.7.1.5 Runge–Kutta-4 Integrator

The fourth-order Runge–Kutta method (Press et al. 1986), one of the oldest numerical methods for solving ordinary differential equations, is self-starting but requires four energy evaluations per step and thus is very slow. However, the method is very robust and can deal with almost any kind of equations, including stiff ones.

1.7.1.6 Predictor–Corrector Algorithm

Predictor–corrector algorithms (Allen and Tildesley 1987) constitute another commonly used class of method to integrate the equations of motion. The method consists of three steps:

1. Predictor: From the positions and their time derivatives up to a certain order q, all known at time t, one "predicts" the same quantities, such as accelerations, at time $t + \Delta t$, by means of a Taylor expansion.
2. Force evaluation: The force is computed, taking the gradient of the potential at the predicted positions. The resulting acceleration is in general different from the predicted acceleration. The difference between the two constitutes an error signal.
3. Corrector: This error signal is used to correct positions and their derivatives. All the corrections are proportional to the error signal, the coefficient of proportionality being a "magic number" determined to maximize the stability of the algorithm.

1.7.2 STATISTICAL ENSEMBLES

Molecular dynamics allows us to explore the constant-energy surface of a system. However, most natural phenomena occur under conditions in which the system is exposed to external pressure and/or exchanges heat with the environment. Under these conditions, the total energy of the system is no longer conserved and hence different forms (i.e., ensembles) of molecular dynamics are required. An *ensemble* is a collection of all possible systems that have different microscopic states but have identical macroscopic or thermodynamic states. Depending on which state variables (energy, E; enthalpy, H; number of particles, N; pressure, P; stress, S; temperature, T; and volume, V) are kept fixed, different statistical ensembles can be generated. A variety of structural, energetic, and dynamic properties can be calculated from the averages or the fluctuations of these quantities over the ensemble generated. There exist different ensembles with different characteristics.

1.7.2.1   Microcanonical Ensemble (NVE)

The fixed number of atoms (N), constant-energy (E), and constant-volume (V) ensemble are obtained by solving Newton's standard equation of motion without any temperature and pressure control. Energy should be ideally conserved in this ensemble, but due to rounding and truncation errors during the integration process a slight fluctuation or drift in energy is often observed. It is not recommended for equilibration phase. However, during the data collection phase, if one is interested in exploring the constant-energy surface of the conformational space, this is a useful ensemble. The results can be used to calculate the thermodynamic response function (Ray 1988).

1.7.2.2   Canonical Ensemble (NVT)

The *canonical ensemble* is a collection of all systems whose thermodynamic state is characterized by a fixed number of atoms (N), a fixed volume (V) and a fixed temperature (T). Temperature is controlled by a direct velocity scaling method (Hermansson et al. 1988) or by exchanging heat with a heat bath (Andersen 1980; Berendsen et al. 1984; Hoover 1985; Nosé 1984a, 1984b, 1991). Direct velocity is generally used only during the initialization stage, followed by any of the other temperature-control methods during the data collection phase. It can be used to compute the linear response transport properties such as diffusivity (D), thermal conductivity (λ), and viscosity (η).

1.7.2.3   Isobaric–Isothermal Ensemble (NPT)

The *isobaric–isothermic ensemble* is characterized by a fixed number of atoms (N), a fixed pressure (P), and a fixed temperature (T). This method is applicable to periodic systems only. The unit cell vectors are allowed to change, and the pressure is adjusted by adjusting the volume (the size and shape of the unit cell). Several methods are available to control pressure. Those of Berendsen et al. (1984) and Anderson (1980) only vary the size of the unit cell, whereas that of Parrinello and Rahman (1982) allows both the cell volume and its shape to change. NPT is the ensemble of choice when the correct pressure, volume, and densities are important in the simulation. This ensemble can also be used during equilibration to achieve the desired temperature and pressure before changing to the constant-volume or constant-energy ensemble when data collection starts.

1.7.2.4   NPH Ensemble

The constant-pressure, constant-enthalpy ensemble (Andersen 1980) is the analogue of constant-volume (V), constant-energy (E) ensemble, in which the size of the unit cell is allowed to vary. Enthalpy (H) remains constant when the pressure is kept fixed without any temperature control. The natural response functions (specific heat at constant pressure, thermal expansion, adiabatic compressibility, and adiabatic compliance tensor) are obtained from the proper statistical fluctuation expressions of kinetic energy, volume, and strain (Ray 1988).

1.7.2.5 Grand Canonical Ensemble (μVT)

The thermodynamic state for this ensemble is characterized by a fixed chemical potential (μ), a fixed volume (V), and a fixed temperature (T).

1.7.3 Equilibrium Thermodynamic Properties

The fluctuations in different ensembles are related to thermodynamic derivatives, such as the specific heat or the isothermal compressibility. The transformation and relation between different ensembles has been discussed in detail by Allen and Tildesley (1987). To obtain the equilibrium thermodynamic properties of a structure, the time average of a variable, A, (Equation 1.20) yields the thermodynamic value for the selected variable:

$$A = \lim_{T} \to \infty \int_{0}^{T} A(T)\, dt \tag{1.20}$$

This dynamic variable can be any function of the coordinates and momentum of the particles in the structure. Through time averaging, the first-order properties of a system such as internal energy, kinetic energy, pressure, and virial can be computed. The microscopic expressions in the form of fluctuations of these first-order properties can be used to calculate thermodynamic properties such as specific heat, thermal expansion, and bulk modulus. In the thermodynamic limit, the first-order properties obtained in one ensemble are equivalent to those obtained in other ensembles. However, second-order properties differ among ensembles. Therefore, it is important to use the appropriate ensemble to obtain the desired properties.

1.7.4 Constraints during Molecular Dynamics

Bond vibrations constitute the highest frequencies in the system and thus determine the largest time step that can be used during dynamics. If the bonds are constrained, longer time steps can be used during dynamics, leading to considerable savings in computational time. Constraints and restraints can be applied accordingly during a dynamics run to increase computational efficiency or to focus the simulation on more interesting parts of the structure. Two of the most popular methods are the RATTLE (Andersen 1983) and the SHAKE algorithms (Ryckaert et al. 1977). The RATTLE procedure involves adjusting the coordinates to satisfy each of the constraints individually. The procedure is reiterated until all the constraints are satisfied within a given tolerance. This method is more suitable for use with the velocity version of Verlet integrators. RATTLE can be used to constrain bonds and angles spanned by two constrained bonds, as well as the distance between any pair of atoms in periodic and nonperiodic systems. It can be used with the constant-volume ensemble but not with constant-pressure dynamics.

1.8 NONEQUILIBRIUM MOLECULAR DYNAMICS

While equilibrium MD is used extensively for calculating transport properties (Cummings and Evans 1992), in recent years nonequilibrium MD (NEMD) methods have become popular for studying transport processes such as the rheological properties. While equilibrium MD methods can only access the linear regime of transport properties, NEMD provides access to transport properties in the nonlinear regime, as well. The basic philosophy of the NEMD method (Cummings and Evans 1992; Evans and Morriss 1990; Sarman et al. 1998) is to introduce a (fictitious or real) field X into the equations of motion of the system, which drives the conjugate thermodynamic flux J (momentum flux or the heat current). The first requirement for this applied field is that it be consistent with periodic boundary conditions to ensure sample homogeneity. The second requirement is that the transport property δ of interest can be calculated from the constitutive relation:

$$\delta = \lim_{x \to 0} \lim_{x \to \infty} \frac{J}{X} \tag{1.21}$$

Though NEMD methods have been known since the 1980s, only recent theoretical and algorithmic developments (Evans and Morriss 1990; Sarman et al. 1998) have made it possible to use it in many applications (McCabe et al. 2001; Moore et al. 1997, 1999, 2000a, 2000b, 2000c).

1.9 MESOSCALE MODELING

Even though MD is the most accurate method for calculating the dynamics of an atomistic system by integrating the equations of motion of all the atoms in the system, it suffers from large memory demand and speed required to simulate realistic length and time scales. To understand many physical processes occurring on a microsecond or longer timescales, coarse-grained methods, or *mesoscale modeling*, is often applied. Mesoscale methods apply classical simulation techniques to coarse-grained systems without compromising the underlying molecular level interactions. These methods are primarily divided into two categories: particle based and grid based. The grid-based methods include the lattice Boltzmann method (LBM) (Succi 2001) and the lattice gas automata (LGA) (Rivet and Boon 2001). The particle-based methods include Brownian dynamics (Ermak and McCammon 1978), Stokesian dynamics (Brady and Bossis 1988), dissipative particle dynamics (Hoogerbrugge and Koelman 1992) and smoothed dissipative particle dynamics (Español and Revenga 2003).

1.9.1 DISSIPATIVE PARTICLE DYNAMICS

Dissipative particle dynamics (DPD) is a meshless, coarse-grained, particle-based method used to simulate systems at mesoscopic length and timescales (Coveney and Español 1997; Español and Warren 1995). In simple terms, DPD can be interpreted as coarse-grained MD. Atoms, molecules, or monomers are grouped together into mesoscopic clusters, or "beads," that are acted on by conservative, dissipative, and random forces. The interaction forces are pairwise additive in nature and act between bead centers. Connections between DPD and the macroscopic (hydrodynamic, Navier-Stokes) level of description (Espanol 1995; Groot and Warren 1997), as well as microscopic (atomistic MD) have been well established (Marsh and Coveney 1998). DPD has been used to model a wide variety of systems such as lipid bilayer membranes (Groot and Rabone 2001), vesicles (Yamamoto et al. 2002), polymersomes (Ortiz et al. 2005), binary immiscible fluids (Coveney and Novik 1996), colloidal suspensions (Boek et al. 1997), and nanotube polymer composites (Maiti et al. 2005).

1.10 MONTE CARLO METHODS

MC is a stochastic method that relies on probabilities. It samples from $3N$-dimensional space represented by the positions of the particles. Unlike MD simulations, it does not calculate the time-dependent properties because particle momentum is not involved. This method is useful for calculating the excess thermodynamic properties that result in deviation from ideal gas behavior. Fluids are simulated by generating different points in configuration space with a relative probability proportional to a corresponding Boltzmann factor by accepting or rejecting the trial configuration using a Metropolis-like scheme that uses the generation of random numbers. To determine the properties accurately in the finite time available, it is very important to sample those states which contribute most significantly. This can be achieved by generating a Markov chain. For a detailed description of the MC method and basic algorithms, the reader is requested to refer to some of the standard textbooks (Allen and Tildesley 1987; Frenkel and Smit 1996; Sadus 1999).

1.10.1 GRAND CANONICAL MONTE CARLO

Grand canonical Monte Carlo (GCMC) simulations (Allen and Tildesley 1987; Norman and Filinov 1969) are performed in the grand canonical ensemble of statistical mechanics, in which the chemical potential of the molecules (μ), temperature (T), and volume (V) are fixed. For mixtures, the chemical potentials of each species are also fixed. GCMC models a system that is in equilibrium with a bulk fluid at a specified chemical potential that is achieved by insertions and deletions of molecules. It is important to note that GCMC fails or becomes difficult to converge at a high density. Gelb et al. (1999) have reported on its application in adsorption. Some of the applications where GCMC can be applicable include:

- Solvent vapours in equilibrium with the bulk solvent
- Fluid adsorbed on a surface or in a pore in equilibrium with a bulk fluid phase
- A fluid phase in equilibrium with another fluid phase

1.10.2 GIBBS ENSEMBLE MONTE CARLO

Gibbs ensemble Monte Carlo (GEMC) method (Panagiotopoulos 1987a, 1987b; Panagiotopoulos et al. 1988) allows effective computing of phase equilibria for both pure fluids and mixtures. The technique involves the direct simulation of two fluid phases in equilibrium, though not in physical contact. In GEMC, the use of two boxes with direct molecule exchange yields the correct equilibrium point for two phases in coexistence. This method is efficient for mixtures and pure fluids. In fact, the computations for mixtures can be performed at constant pressure. However, they fail at high densities and near the critical point. Large fluctuations in the density may lead to instabilities. It has been applied to bulk fluid–phase equilibria and phase equilibria of confined fluids and can be combined with the Gibbs–Duhem method (Kofke 1993; Mehta and Kofke 1993) to map out full-phase diagrams. Panagiotopoulos et al. have reported extensively on GEMC (Orkoulas and Panagiotopoulos 1999; Panagiotopoulos 1992, 1996, 2000).

1.10.3 KINETIC MONTE CARLO

Kinetic MC models reaction kinetics within a molecular simulation framework by stochastic transitions using rate constants (Gillespie 1968, 1976). Several reviews of application areas in which kinetic MC has been used have been reported recently (Levi and Kotrla 1997; Martin 1998; Stoltze 2000).

1.10.4 CONTINUUM CONFIGURATIONAL BIAS MONTE CARLO

The continuum configurational bias MC (CCBMC) method is a technique that can be used to overcome the problem of random insertion of long-chain molecules in dense liquid phases (de Pablo et al. 1992, 1993, 1999; Mooij et al. 1992). The CCBMC method introduces a bias into the insertion procedure by placing the molecule into the liquid phase one monomer at a time, choosing the location of each new monomer in an optimal fashion (Frenkel and Smit 1996).

1.10.5 REACTIVE MONTE CARLO

Reactive MC is a technique that permits the incorporation of chemical equilibrium into MC simulations (Johnson et al. 1994; Smith and Triska 1994). It can be combined with GEMC to evaluate the impact of phase equilibria on chemical equilibria and to probe the impact of confinement to pores on chemical equilibrium.

1.11   QUANTUM MECHANICS

The postulates and theorems of quantum mechanics form the rigorous foundation for the prediction of observable chemical properties from first principles. In simple terms, *quantum mechanics* assumes that microscopic systems are described by wave functions that completely define all the physical properties of the system. In particular, there are quantum mechanical operators corresponding to each physical observable that, when applied to the wave function, allow one to predict the probability of finding the system to exhibit a particular value or range of values.

Electronic structure theory, which is based on quantum mechanics, is the most accurate method to calculate the structure, energy, and properties of a molecule or an assembly of molecules. It helps in computing many of the thermochemical, physical, spectroscopic, and kinetic (rate of reactions) properties of the molecules.

A brief description of the fundamental concepts underlying the *ab initio* molecular orbital computations is presented here. The reader is requested to refer to some of the textbooks providing comprehensive discussion of molecular orbital theory (Atkins 1991; Atkins and Friedman 1997; Cook 1998; Levine 1983; Simons and Nichols 1997). In electronic structure theory, given the position of atomic nuclei, R [under the Born-Oppenheimer approximation (Born and Oppenheimer 1927)], the Schrödinger equation for motion of electrons *(r)* is solved as:

$$H(r,R)\psi(r,R) = E(R)\psi(r,R) \tag{1.22}$$

where $E(R)$ is the potential energy surface (PES), which determines the equilibrium and transition state structures of a molecule, as well as vibrational normal coordinates and frequencies, and $\psi(r,R)$ is the wave function that determines electron density and various properties such as dipole moment and electrostatic potential. For systems of more than two interacting particles, the Schrödinger equation cannot be solved exactly. Therefore, all *ab initio* calculations for many-body systems (molecules) involve some level of approximation. Various approximate methods for solving the Schrödinger equation include semiempirical molecular orbital (MO) methods, *ab initio* MO methods, and density functional theory (DFT) methods.

1.11.1   SEMIEMPIRICAL METHODS

Semiempirical methods increase the speed of computation by using approximations of *ab initio* techniques—for example, by limiting choices of molecular orbitals or considering only valence electrons, which have been fitted to experimental data. Depending on the differential overlap of the two center integrals involved in the wave functions, they are categorized as complete neglect of differential overlap (CNDO) and neglect of diatomic differential overlap (NDDO) (Murrell and Harget 1972; Pople and Beveridge 1970; Pople et al. 1965; Pople and Segal 1965), intermediate neglect of differential overlap (INDO) (Clark 1985), and modified neglect of differential overlap (MNDO) (Clark 1985). These methods can reproduce geometric properties with reasonable accuracy. However, thermodynamic properties such as heat of formation or heat of reaction are not predicted very accurately. With the recent advancements in more sophisticated *ab initio* methods, these methods may be suitable to model only very large biological molecules (Dewar et al. 1985; Stewart 1989; Thiel 1996). Zerner's INDO (ZINDO) (Zerner 1995) method has been used extensively in industrial applications. Some recent semiempirical methods, such as the MOZYME method (Stewart and Stewart 1999), perform direct calculation of electron density and energy without diagonalizing the semiempirical self-consistent field (SCF) matrix and can be used for geometry optimization of large molecules like polypeptides. The divide-and-conquer method (Dixon and Merz 1996) divides a large molecule into smaller segments, performs semiempirical SCF calculations for segments, and obtains the electron density of the entire molecule by adding the density of segments, which is used to calculate the energy of the entire system. The self-consistent charge-density functional based tight binding (SCC-DFTB) (Frauenheim et al. 2000) method takes into account self-consistent redistribution of charges and accomplishes a substantial improvement in energies. Despite these developments, semiempirical methods are quick and inexpensive yet qualitative tools for industrial and biomolecular applications.

1.11.2   *AB INITIO* ELECTRONIC STRUCTURE METHODS

In the *ab initio* methods, MOs are usually expressed as the linear combinations of a finite number of basis functions:

$$\varphi_i = \sum_{v=1}^{N\,\text{basis}} C_{ir}\chi_r \tag{1.23}$$

where χ_r are the basis functions and the coefficients C_{ir} are adjustable parameters. For a molecular wave function, the electronic orbitals of the constituent atoms form a natural set of basis functions. These atomic orbitals can be represented by different types of mathematical functions. A highly accurate set of atomic orbitals (e.g., Slater-type orbitals) are based on hydrogenic wave functions having the form:

$$\chi_{\text{STO}}(r) \sim Ce^{-\alpha\gamma} \tag{1.24}$$

The total electronic wave function is expressed as a linear combination of Slater determinants (SDs):

$$\psi = \sum_I D_I\phi_I; \phi_I = (N!)^{-\frac{1}{2}} \sum_P P(1)^P \{\varphi_1 S_1(1)\varphi_2 S_2(2)\phi_3 S_3(3)\ldots\ldots\phi_N S_N(N)\} \tag{1.25}$$

Hartree and Fock (Fock 1930; Hartree 1928) formalism uses the single SD form of the total wave function, which is solved under self-consistent field (SCF) approximation. This involves an iterative process in which the orbitals are improved cycle to cycle until the electronic energy reaches a constant minimum and the orbitals no longer change. Upon convergence of the SCF method, the minimum-energy MOs produce an electric field that generates the same orbitals and hence the self-consistency.

1.11.2.1   Basis Sets

The basis set can be minimal (one basis function per valence shell) but can be improved systematically by going to double zeta (two per valence shell), to double zeta plus polarization, to triple zeta plus double polarization plus higher angular momentum plus diffuse functions, and finally to the complete set. Some of the improvements on minimal basis sets are described next.

1.11.2.1.1   Split-Valence Basis Sets

In split-valence basis sets, additional basis functions (one contracted Gaussian plus some primitive Gaussians) are allocated to each valence atomic orbital. Split-valence basis sets are characterized by the number of functions assigned to a valence orbital. Some examples of split-valence sets include 6-21G, 3-21G, 4-31G, and 6-311G.

1.11.2.1.2   Polarized Basis Sets

Polarization functions can be added to basis sets to allow for nonuniform displacement of a charge away from the atomic nuclei, thereby improving descriptions of chemical bonding. Polarization functions describe orbitals of higher angular momentum quantum number than those required for the isolated atom and are added to the valence electron shells. For example, the 6-31G(d) basis set is constructed by adding six d-type Gaussian primitives to the 6-31G description of each nonhydrogen atom. In 6-31G(d, p), in addition to six d-type Gaussian primitives for heavy atoms, a set of Gaussian p-type functions is also added to hydrogen and helium atoms.

1.11.2.1.3 Diffuse Basis Sets

Species with significant electron density far removed from the nuclear centers such as anions, lone pairs, and excited states require diffuse functions to account for the outermost weakly bound electrons. The addition of diffuse s- and p-type Gaussian functions to nonhydrogen atoms is denoted by a plus sign, as in 3-21 + G.

1.11.2.1.4 High Angular Momentum Basis Sets

High angular momentum basis sets augmented with diffuse functions represent the most sophisticated basis sets available. Basis sets with multiple polarization functions include:

- 6-31G(2d): Two d-functions are added to heavy atoms.
- 6-311G(2df, pd): Two d-functions and one f-function are added to heavy atoms, and p- and d-functions are added to hydrogen.
- 6-311G(3df, 2df, p): Three d-functions and one f-function are added to atoms with Z > 11, two d-functions and one f-function are added to first-row atoms (Li to Ne) and one p-function is added to hydrogen.

1.11.3 ELECTRON CORRELATION

The Hartree–Fock (HF) method, being a single-particle approximation method, neglects electron correlations, thereby leading to systematic errors such as underestimated bond lengths and overestimated vibrational frequencies. To account for correlations, more than one SD is required to represent the total wave function. Some of the electron correlation treatments are summarized next.

1.11.3.1 Configuration–Interaction Method

Configuration–interaction (CI) methods incorporate excited-state configurations into the wave function by constructing new determinants from the original HF determinant. New determinants are created by replacing one or more occupied orbitals with unoccupied (virtual) orbitals of higher energy. The number of replacements within the determinants designates the level of CI. For example, CI single substitution (CIS) switches one pair of occupied orbitals with one pair of virtual orbitals and is equivalent to a one-electron excitation. Accordingly, a full CI calculation forms the molecular wave function as a linear combination of the HF determinant and all possible substituted determinants. It provides the most complete nonrelativistic treatment possible for a molecular system, but due to its extremely computer-intensive quality, it may be applicable to only simple molecules described by the smallest possible basis sets.

1.11.3.2 Møller–Plesset Perturbation

Møller–Plesset (MP) perturbation theory assumes that the effects of electron correlation can be described by small corrections (perturbations) to the HF solution. MP methods assume the true molecular Hamiltonian to consist of two parts:

$$\widehat{H}_{\text{mol}} = \widehat{H}_{\text{one}-e} + \lambda \widehat{P}_{\text{many}-e} \tag{1.26}$$

where $\hat{H}_{one-e}$ denotes single-electron energy contributions that can be solved exactly by HF–SCF and $\hat{P}_{many-e}$ represents contributions due to electron correlation. The coefficient λ is used to write the power series expansion of energy and wave function:

$$E_{mol} = E^{(0)} + \lambda E^{(1)} + \lambda^2 E^{(2)} + \lambda^3 E^{(3)} + \ldots$$

$$\psi_{mol} = \psi^{(0)} + \lambda \psi^{(1)} + \lambda^2 \psi^{(2)} + \lambda^3 \psi^{(3)} + \ldots$$

(1.27)

The first two energy terms constitute the HF energy that is used to evaluate wave function $\psi^{(1)}$. The first-order correction to the wave function $\psi^{(1)}$ is then used to calculate the second-order correction to the total energy $E^{(2)}$ and so on until the desired order of correction is achieved. Accordingly, the MP method is designated as MP2, MP3, MP4, and so forth, because the second order $E^{(2)}$ is the first correction to HF energy.

1.11.4 EFFECTIVE CORE POTENTIAL

Effective core potential (ECP) or pseudo-potential approximation, has been proved to be very useful for modeling of heavy atoms in the *ab initio* methods (Hay and Wadt 1985). In this approximation, core electrons are replaced by an effective potential, thereby reducing the number of electrons to be considered and hence requiring fewer basis functions. The ECP method takes into account the relativistic effect on valence electrons, thus making it applicable to heavy atoms (e.g., second- and third-row transition metals, lanthanides and actinides). It is relatively cheap, works very well, and has very little loss in reliability.

1.11.5 DENSITY FUNCTIONAL THEORY

According to Hohenberg and Kohn (1964), the electron density $\rho(r)$ uniquely determines the wave function, $\rho(r) \rightarrow V(r) \rightarrow H \rightarrow \psi$, and the energy is a functional of $\rho(r)$:

$$E[\rho] = T[\rho] + \int \rho(r) V(r) dr + \iint \frac{\rho(r)\rho(r')}{r - r'} dr dr' + E_{ex}[\rho] + E_{corr}[\rho]$$

(1.28)

To determine density, Kohn–Sham equations for orbitals are solved (Joubert 1998; Parr and Yang 1989); however, the fundamental problem is that the correct form of the functional is unknown. Various approximate exchange $E_{ex}[\rho]$ and correlation $E_{corr}[\rho]$ functionals are used. The simplest is the local density functional approximation (LDA), in which functionals depend only on ρ. In the more accurate gradient correction method (Becke 1985), functionals also depend on the gradient of density. Some examples of the gradient-corrected functionals are Becke88-LYP and Becke88-Perdue86. Hybrid functionals such as B3LYP (Becke 1993) and B3PW9 (Perdew et al. 1996) mix the exact (HF) exchange with DFT in an empirical fraction, improving the energetics further. DFT, in its original form, is applicable for the ground-state calculations only. However, in the recent years, many new methods such as the time-dependent perturbation method (TD-DFT) (Jamorski et al. 1996) have been developed for excited-state calculations.

1.12 HYBRID METHODS

Hybrid methods learn from the pros and cons of classical and quantum methods and try to combine the best features of both. These methods are of two categories: (1) methods that are hybrid in a temporal sense, mixing electronic structure calculations with MD methods and (2) methods that are spatially hybrid, applying different methods in different physical regions of a molecule or computational domain. Car–Parrinello MD (CPMD) (Car and Parrinello 1985) and related methods such as

Vienna *ab initio* Simulation Package (VASP) (Kresse and Furethmüller 2000), and Car-Parrinello MD Program (CAMP-Atami) (Ohnishi 1994) combine molecular dynamics with DFT under a periodic boundary condition with the orbitals expanded in the plane wave. They have been routinely used in industrial applications such as heterogeneous catalysis.

The Quantum Mechanics/Molecular Mechanics (QM/MM) method (Warshel and Karplus 1972) belongs to the second category, which combines quantum mechanics and molecular mechanics by dividing the Hamiltonian into the form:

$$H = H_{QM} + H_{MM} + H_{QM-MM} \tag{1.29}$$

The QM/MM method has been applied mainly to biological problems such as enzyme reactions and reactions in solution. Hybrid methods are also useful in modeling solvent effects. Some examples are the self-consistent reaction field (SCRF) method (Miertus et al. 1981) and the Our own N-layered Integrated molecular Orbital & molecular Mechanics (ONIOM) method (Svensson et al. 1996).

1.13 OTHER CONCEPTS IN MATERIALS MODELING

1.13.1 CONFORMATIONAL SEARCH

Conformational search methods explore the conformational space (or energy hypersurface) for identifying low-energy conformations. A number of conformational search algorithms have been developed and periodically reviewed (Böhm 1996; Eisenhaber et al. 1995; Floudas et al. 1999; Leach 1991; Neumaier 1997; Scheraga et al. 1999; Vásquez et al. 1994). Because it is generally assumed that the native, or naturally occurring, conformation of a molecule corresponds to the point on the potential energy surface with the lowest energy value (global energy minimum), the main goal of a conformational search is to identify this point. Conformational search techniques can be classified broadly into two categories: stochastic and deterministic methods. Stochastic methods such as the MC method (Li and Scheraga 1987; Ripoll and Scheraga 1989), simulated annealing (Kirkpatrick et al. 1983; Morales et al. 1991; Wilson and Cui 1990), and genetic algorithm (Le Grand and Merz 1994; Schulze-Kremer 2000) rely on probabilistic descriptions to aid in locating the global minimum, and there is no natural endpoint to the procedure. In contrast, deterministic methods such as the systematic search method (Gibson and Scheraga 1987) provide a certain level of assurance in locating the global minimum, and there is a defined endpoint to the procedure. Conformational search is more applicable for the large biomolecules such as peptides and proteins.

1.13.2 QUANTITATIVE STRUCTURE–ACTIVITY RELATIONSHIPS

Quantitative structure–activity relationship/quantitative structure–property relationship (QSAR/QSPR) (Hansch and Leo 1995) attempts to correlate chemical structure with activity using statistical approaches. The QSAR models are useful for various purposes, including the prediction of activities of untested chemicals. QSARs and other related approaches have attracted broad scientific interest, particularly in the pharmaceutical industry for drug discovery and in toxicology and environmental science for risk assessment. The construction of a QSAR/QSPR model typically comprises two main steps: (1) description of molecular structure (independent variables) and (2) multivariate analysis for correlating molecular descriptors with observed activities and properties (dependent variables). Molecular descriptors transform the physicochemical properties of a molecule into several variables such as constitutional, electronic, geometric, hydrophobic, lipophilic, soluble, steric, quantum chemical, and topological. In-depth explanations of molecular descriptors can be found in the literature (Helguera et al. 2008; Karelson et al. 1996; Katritzky and Gordeeva 1993; Labute 2000; Randić 1990; Randić and Razinger 1997; Todeschini and Consonni 2000; Xue and Bajorath 2000).

The goal of cheminformatics and materials informatics (Rodgers and Cebon 2006) is to catalog, store, manipulate, and analyze the vast databases of materials and molecules. The tasks can be broadly divided into two main parts: data management and knowledge discovery. Some of the techniques of interest in materials informatics are standard, such as quantum methods for computing the stability of materials or information techniques such as data mining, statistical data analysis, and visualization techniques. However, combining these techniques to exploit materials informatics is not standard and offers a novel approach to materials design. Though techniques of this category have penetrated other scientific areas such as life sciences, they remain in the embryonic stage in materials science. The time is ripe to explore the diverse techniques of data mining with quantum methods for designing and discovering new materials.

1.14　HARDWARE AND SOFTWARE

Four elements enable scientific computing: (1) theoretical and mathematical formulation, (2) algorithmic design, (3) software development and implementation, and (4) compatibility with system architecture. Typically, these are not well integrated, and the consequence is that development of these enablers is rate limiting. Recent advances in both hardware and software have enabled significant applications of molecular modeling in industrial scenarios. The programs applied range from academic to open source to commercial to proprietary in-house codes. Some of the widespread commercial codes include those developed by Accelrys Inc., San Diego, USA (http://www.accelrys.com/), Gaussian Inc., Walling ford, CT, USA (Gaussian 09, http://www.gaussian.com/), Schroedinger Inc., Cambridge, MA, USA (http://www.schrodinger.com/), CPMD (CPMD 3.15, IBM Corp., USA, and Max Planck Institute, Stuttgart, http://cpmd.org/) and VASP (VASP 5.2, Materials Design Inc., USA, http://www.materialsdesign.com/medea/medea-vasp-52). There also exist some public cooperative efforts such as ESF-PsiK (http://psi-k.dl.ac.uk/), NISTCTCMS (http://www.ctcms.nist.gov/), Quantum ESPRESSO (http://www.quantum espresso.org/), and LAMMPS (http://lammps.sandia.gov/) with the goal of developing and disseminating community code. A complete list of various open source and freeware codes can be found at http://cmm.info.nih.gov/modeling/software.html. Even though x supercomputers have played an important role in the past owing to their high accuracy and efficiency in case of large systems, but since the late 1980s, workstations have been valuable for low-level calculations and visualization. Commodity processor–based workstations and clusters are now taking over even the largest tasks. Powerful single- and dual-processor PCs now achieve computational speeds faster than most previous workstations, and with disk and memory costs much lower, they are taking over many routine calculations. For large problems that benefit from parallel computation, multiple Intel Pentium, Intel Inc., Santa Clara, CA, USA or Compaq Alpha, Hewlett-Packard Development Company, CA, USA processors are coupled by a fast Ethernet switch and are often powered by the Linux operating system, giving impressive parallel computing performance for much less expense than a dedicated parallel supercomputer. However, most of the codes do not necessarily parallelize well, and not all parallel problems are suited to this architecture. In addition, considerable programing challenges exist at the hardware–software interface. The development of scalable software is a significant challenge. It is enormously difficult to write programs that run well on 256 processors or more. Indeed, there is a sizeable and growing gap between ideal peak performance and achievable performance with scientifically useable code.

Different approaches in MM place different demands on hardware and software. The mix of supercomputers, personal computers or workstations, and distributed computing (grid and cloud) will be the choice of the future. Accordingly, the integration of platforms, visualization, user interfaces, and cross-platform visual programing tools are the need of the hour.

1.15　SUMMARY

MM provides an important avenue for accelerating the development of advanced materials and chemicals and for turning them into new products. By validating theories with critical experiments,

MM can speed up the discovery process, develop scientific understanding at the atomic and molecular levels, and solve engineering problems in a wide range of disciplines in which conventional approaches have proved inadequate.

1.16 LOOKING AHEAD

With respect to MM, some of the pressing demands posed by most industries include, "bigger, better, faster," extensive validation and multiscale techniques. Bigger means being able to model larger systems with greater complexity. Better reflects the need for a greater accuracy. Faster enables the simulation of rare-event processes such as predicting the thermal and oxidative stability of plastic subjected to a wide variety of weathering. Multiscale modeling is at the heart of technological or engineering modeling. Modeling methods that are essential at one scale may be useless at another. For example, all chemical reactions happen at a molecular level and require information from quantum chemistry. However, a dynamic physical process such as conductive heat transfer can use molecular simulations effectively while convective heat transfer requires continuum-scale physics, and designing a bridge to carry a certain load in summer or winter requires a combination of engineering statics and materials properties that may only be measured. The holy grail of materials and chemicals design thus remains to seamlessly couple these domains with each other. In the future, these techniques will become an integral part of the materials and product design and development process as scientists and engineers continue to face increasingly more complex problems requiring fundamental solutions. Increasing economic and environmental pressures to optimally use the resources will be the key drivers in the realization of these goals.

REFERENCES

Alder, B. J., and Wainwright, T. E. J. 1957. Phase transition for a hard sphere system. *J. Chem. Phys.* 27: 1208.
Alder, B. J., and Wainwright, T. E. J. 1959. Studies in molecular dynamics. I. General method. *J. Chem. Phys.* 31: 459.
Allen, M. P., and Tildesley, D. J. 1987. *Computer Simulation of Liquids.* Oxford University Press, Oxford, UK.
Andersen, H. C. 1980. Molecular dynamics simulations at constant pressure and/or temperature. *J. Chem. Phys.* 72: 2384.
Andersen, H. C. 1983. Rattle: A "velocity" version of the Shake algorithm for molecular dynamics calculations. *J. Comput. Phys.* 52: 24.
Atkins, P. W. 1991. *Quanta,* 2nd ed. Oxford University Press, Oxford, UK.
Atkins, P. W., and Friedman, R. S. 1997. *Molecular Quantum Mechanics,* 3rd ed. Oxford University Press Inc. New York.
Bachrach, S. M. 1993. Population analysis and electron densities from quantum mechanics. *Rev. Comput. Chem.* 5: 171.
Bader, J., and Chandler, D. J. 1992. Computer simulation study of the mean forces between ferrous and ferric ions in water. *J. Phys. Chem.* 96: 6423.
Becke, A. 1985. Local exchange-correlation approximations and first-row molecular dissociation energies. *Int. J. Quant. Chem.* 27: 585.
Becke, A. 1993. Density functional thermochemistry. III. The role of exact exchange. *J. Chem. Phys.* 98: 5648.
Berendsen, H. J. C., Postma, J. P. M., van Gunsteren, W. F., DiNola, A., and Haak, J. R. 1984. Molecular dynamics with coupling to an external bath. *J. Chem. Phys.* 81: 3684.
Boek, E. S., Coveney, P. V., Lekkerkerker, H. N. W., and van der Schoot, P. 1997. Simulating the rheology of dense colloidal suspensions using dissipative particle dynamics. *Phys. Rev. E: Stat. Phys. Plasmas Fluids Relat. Interdisciplin. Top.* 55: 3124.
Böhm, G. 1996. New approaches in molecular structure prediction. *Biophys. Chem.* 59: 1.
Born, M., and Oppenheimer, J. R. 1927. Zur Quantentheorie der Molekeln. *Ann. Phys.* 84: 457.
Brady, J. F., and Bossis, G. 1988. Stokesian dynamics. *Annu. Rev. Fluid Mech.* 20: 111.
Car, R., and Parrinello, M. 1985. Unified approach for molecular dynamics and density-functional theory. *Phys. Rev. Lett.* 55: 2471.
Casewit, C. J., Colwell, K. S., and Rappe', A. K. 1992. Application of a universal force field to organic molecules. *J. Am. Chem. Soc.* 114: 10035.

Casewit, C. J., Colwell, K. S., and Rappe', A. K., 1992. Application of a universal force field to main group compounds. *J. Am. Chem. Soc.,* 114, 10046.

Chandler, D. 1987. *Introduction to Modern Statistical Mechanics.* Oxford University Press, New York.

Clark, T. 1985. *A Handbook of Computational Chemistry.* John Wiley & Sons, New York.

Cook, D. B. 1998. Handbook of Computational Quantum Chemistry, New York, Oxford University Press.

Coveney, P. V., and Español, P. 1997. Dissipative Particle Dynamics for Interacting Multicomponent Systems, *J. Phys. Chem. A* 30: 779.

Coveney, P. V., and Novik, K. E. 1996. Computer Simulations of Domain Growth and Phase Separation in Two-Dimensional Binary Immiscible Fluids Using Dissipative Particle Dynamics, *Phys. Rev.* E 54: 5134.

Cummings, P. T., and Evans, D. J. 1992. Molecular approaches to transport properties and non-Newtonian rheology. *Ind. Eng. Chem. Research.* 31: 1237.

Darden, T., York, D., and Pederson, L. 1993. Particle mesh Ewald: An $N.\log(N)$ method for Ewald sums in large systems. *J. Chem. Phys.* 98: 10089.

de Pablo, J. J., and Escobedo, F. A. 1999. Monte Carlo methods for polymeric systems. In *Monte Carlo Methods in Chemical Physics.* Eds. D. M. Ferguson, J. I Siepman, and D. G. Truhlar. Wiley, New York.

de Pablo, J. J., Laso, M., Siepmann, J. I., and Suter, U.W. 1993. Continuum-configurational-bias Monte-Carlo simulations of long-chain alkanes. *Mol. Phys.* 80: 55.

de Pablo, J. J., Laso, M., and Suter, U.W. 1992. Estimation of the chemical potential of chain molecules by simulation. *J. Chem. Phys.* 96: 6157.

Deem, M. W., Newsam, J. M., and Sinha, S. K. 1990. The $h = 0$ term in Coulomb sums by the Ewald transformation. *J. Phys. Chem.* 94: 8356.

Dewar M. J. S., Zoebisch, E. G., and Healy, E. F. 1985. Development and use of quantum mechanical molecular models. AM1: A new general purpose quantum mechanical molecular model. *J. Am. Chem. Soc.* 107: 3902.

Dixon, S. L., and Merz, K. M., 1996. Semiempirical molecular orbital calculations with linear system size scaling. *J. Chem. Phys.* 104: 6643.

Eisenhaber, F., Persson, B., and Argos, P. 1995. Protein structure prediction: Recognition of primary, secondary, and tertiary structural features from amino acid sequence. *Crit. Rev. Biochem. Mol. Biol.* 30: 1.

Ermak, D. L., and McCammon, J. A. 1978. Brownian dynamics with hydrodynamic interactions. *J. Chem. Phys.* 69: 1352.

Espanol, P. 1995. Hydrodynamics from dissipative particle dynamics. *Phys. Rev. E* 52: 1734.

Español, P., and Revenga, M. 2003. Smoothed dissipative particle dynamics. *Phys. Rev. E* 67: 026705.

Español, P., and Warren, P. 1995. Statistical mechanics of dissipative particle dynamics. *Europhys. Lett.* 30: 191.

Evans, D. J., and Morriss, G. P. 1990. *Statistical Mechanics of Nonequilibrium Liquids.* Academic Press, London.

Ewald, P. P. 1921. Evaluation of optical and electrostatic lattice potentials. *Ann. Phys.* 64: 253.

Floudas, C. A., Klepeis, J. L., and Pardalos, P. M. 1999. Global optimization approaches in protein folding and peptide docking. In *DIMACS Series in Discrete Mathematics and Theoretical Computer Science.* Eds. Farach-Colton M., Roberts, F. S., Vingron, M., and Waterman, M., American Mathematical Society, 47: 141–171.

Fock, V. 1930. Selfconsistent field mit Austausch für Natrium. *Z. Phys.* 62: 795.

Frauenheim, T., Seifert, G., Estner, M., Hajnal, Z., Jungnickel, G., Porezag, D., et al. 2000. *Phys. Stat. Sol. (b)* 217: 41.

Frenkel, D., and Smit, B. 1996. *Understanding Molecular Simulation: From Algorithms to Applications.* Academic Press, San Diego.

Gasteiger, J., and Marsili, M. 1980. Iterative partial equalization of orbital electronegativity—A rapid access to atomic charges. *Tetrahedron.* 36: 3219.

Gelb, L. D., Gubbins, K. E., Radhakrishnan, R., and Sliwinska-Bartkowiak, M. 1999. Phase separation in confined systems. *Rep. Prog. Phys.* 62: 1573.

Gibson, K. D., and Scheraga, H. A. 1987. Revised algorithms for the build-up procedure for predicting protein conformations by energy minimization. *J. Comput. Chem.* 8: 826.

Gillespie, D. T. 1968. Some aspects of resonance production and diffraction-dissociation in 5.44 GeV/c Ktp interactions, Ph.D. dissertation. Johns Hopkins University.

Gillespie, D. T. 1976. General method for numerically simulating stochastic time evolution of coupled chemical reactions. *J. Comput. Phys.* 22: 403.

Groot, R. D., and Rabone, K. L. 2001. Mesoscopic simulation of cell membrane damage, morphology change and rupture by nonionic surfactants. *Biophys. J.* 81: 725.

Groot, R. D., and Warren, P. B. 1997. Dissipative particle dynamics: Bridging the gap between atomistic and mesoscopic simulation. *J. Chem. Phys.* 107: 4423.

Hansch, C., and Leo A. 1995. *Exploring QSAR*. American Chemical Society: Washington, DC.

Hartree, D. 1928. The wave mechanics of an atom with a non-coulomb central field. Part III. Term values and intensities in series in optical spectra. *Proc. Cambridge Philos. Soc.* 24: 426.

Hay, P. J., and Wadt, W. R. 1985. *Ab initio* effective core potentials for molecular calculations. Potentials for K to Au including the outermost core orbitals. *J. Chem. Phys.* 82: 299.

Helguera, A. M., Combes, R. D., Gonzalez, M. P., and Cordeiro, M. N. 2008. Applications of 2D descriptors in drug design: A DRAGON tale. *Curr. Top. Med. Chem.* 8: 1628.

Hermansson, K., Lie, G. C., and Clementi, E. 1988. On velocity scaling in molecular dynamics simulations. *J. Comp. Chem.* 9: 200.

Hohenberg, P., and Kohn, W. 1964. Inhomogeneous electron gas. *Phys. Rev.* 136: B864.

Hoogerbrugge, P. J., and Koelman, J. M. V. A. 1992. Simulating microscopic hydrodynamic phenomena with dissipative particle dynamics. *Europhys. Lett.* 19: 155.

Hoover, W. 1985. Canonical dynamics: Equilibrium phase-space distributions. *Phys. Rev. A* 31: 1695.

Jamorski, C., Casida, M. E., and Salahub, D. R. 1996. Dynamic polarizabilities and excitation spectra from a molecular implementation of time dependent density functional response theory: N2 as a case study. *J. Chem. Phys.* 104: 5134.

Johnson, J. K., Panagiotopoulos, A. Z., and Gubbins, K. E. 1994. Reactive canonical Monte-Carlo: A new simulation technique for reacting or associating fluids. *Mol. Phys.* 81: 717.

Joubert, D. 1998. *Density Functionals: Theory and Applications*. Springer-Verlag, Heidelberg.

Karasawa, N., and Goddard, W. A. III 1989. Acceleration of convergence for lattice sums. *J. Phys. Chem.* 93: 7320.

Karelson, M., Lobanov, V. S., and Katritzky, A. R. 1996. Quantum-chemical descriptors in QSAR/QSPR studies. *Chem. Rev.* 96: 1027.

Katritzky, A. R., and Gordeeva, E. V. 1993. Traditional topological indices vs electronic, geometrical, and combined molecular descriptors in QSAR/QSPR research. *J Chem. Inf. Comput. Sci.* 33: 835.

Kirkpatrick, S., Gelatt, C. D., and Vecchi, M. P. 1983. Optimization by simulated annealing. *Science.* 220: 671.

Kofke, D. A. 1993. Gibbs-Duhem Integration: A new method for direct evaluation of phase coexistence by molecular simulation. *Mol. Phys.* 78: 1331.

Kresse, G., and Furethmüller, J. 2000. VASP: The Guide. http://cms.mpi.univie.ac.at/vasp/vasp.html; Technical. http://cms.mpi.univie.ac.at/vasp/

Labute, P. 2000. A widely applicable set of descriptors. *J. Mol. Graphics Modell.* 18: 464.

Le Grand, S. M., and Merz, K. M. 1994. The genetic algorithm and protein structure prediction. In *The Protein Folding Problem and Tertiary Structure Prediction.* Eds. K. M. Merz and S. M. Le Grand. Birkhäuser, Boston.

Leach, A. R. 1991. A survey of methods for searching the conformational space of small and medium-sized molecules. In *Reviews in Computational Chemistry.* Eds. K. B. Lipkowitz and D. B. Boyd. VCH Publishers, New York.

Levi, A. C., and Kotrla, M. 1997. Theory and simulation of crystal growth. *J. Phys. Condens. Matter.* 9: 299.

Levine, I. N. 1983. *Quantum Chemistry.* Allyn and Bacon Inc., Boston.

Li, Z., and Scheraga, H. A. 1987. Monte Carlo–minimization approach to the multiple-minima problem in protein folding. *Proc. Natl. Acad. Sci. USA.* 84: 6611.

Maiti, A., Wescott, J., and Kung, P. 2005. Nanotube-polymer composites: Insights from Flory–Huggins theory and mesoscale simulations. *Mol. Simul.* 31: 143.

Marsh, C. A., and Coveney, P. V. 1998. Detailed balance and H-theorems for dissipative particle dynamics. *J. Phys. A* 31: 6561.

Martin, G. 1998. Modelling materials driven far from equilibrium. *Curr. Opin. Solid State Mater. Sci.* 3: 552.

Mayo, S. L., Olafson, B. D., and Goddard, W. A. III. 1990. Dreiding: A generic forcefield for molecular simulations. *J Phys. Chem.* 94: 8897.

McCabe, C., Cummings, P. T., and Cui, S. T. 2001. Characterizing the viscosity-temperature dependence of lubricants by molecular simulation. *Fluid Phase Equilibr.* 183–184: 363.

McCammon, J. A., Gelin, B. R., and Karplus, M. 1977. Dynamics of folded proteins. *Nature (Lond.)* 267: 585.

McQuarrie, D. 1976. *Statistical Mechanics.* Harper & Row, New York.

Mehta, M., and Kofke, D. A. 1993. Implementation of the Gibbs ensemble using a thermodynamic model for one of the coexisting phases. *Mol. Phys.* 79: 39.

Miertus, S., Scroccco, E., and Tomasi, J. 1981. Electrostatic interaction of a solute with a continuum. A direct utilizaion of ab initio molecular potentials for the prevision of solvent effects. *Chem. Phys.* 55: 117.

Milik, M., and Skolnick, J. 1993. Insertion of peptide chains into lipid membranes: an off-lattice Monte Carlo dynamics model. *Proteins*. 15: 10.

Mooij, G. C. A. M., Frenkel, D., and Smit, B. 1992. Direct simulation of phase-equilibria of chain molecules. *J. Phys. Condens. Matter*. 4: L255.

Moore, J. D., Cui, S. T., Cummings, P. T., and Cochran, H. D. 1997. Lubricant characterization by molecular simulation. *A.I.Ch.E. Journal*. 43: 3260.

Moore, J. D., Cui, S. T., Cochran, H. D., and Cummings, P. T. 1999. The transient rheology of a polyethylene melt under shear. *Phys. Rev. E* 60: 6956.

Moore, J. D., Cui, S. T., Cochran, H. D., and Cummings, P. T. 2000a. Rheology of lubricant basestocks: A molecular dynamics study of C30 isomers. *J. Chem. Phys*. 113: 8833.

Moore, J. D., Cui, S., Cochran, H. D., and Cummings, P. T. 2000b. Molecular dynamics study of a short-chain polyethylene melt. I. Steady-state shear. *J. Non-Newtonian Fluid Mech*. 93: 83.

Moore, J. D., Cui, S., Cochran, H. D., and Cummings, P. T. 2000c. Molecular dynamics study of a short-chain polyethylene melt. II. Transient response upon onset of shear. *J. Non-Newtonian Fluid Mech*. 93: 101.

Morales, L. B., Garduno-Juárez, R., and Romero, D. 1991. Applications of simulated annealing to the multiple-minima problem in small peptides. *J. Biomol. Struct. Dyn*. 8: 721.

Mulliken, R. S. 1955. Electronic population analysis on LCAO-MO molecular wave functions. *J. Chem. Phys*. 23: 1833.

Murrell, J. N., and Harget, A. J. 1972. *Semi-Empirical Self-Consistent-Field Molecular Orbital Theory of Molecules*. Wiely-Interscience, London.

Neumaier, A. 1997. Molecular modeling of proteins and mathematical prediction of protein structure. *SIAM Rev*. 39: 407.

Norman, G. E., and Filinov, V. S. 1969. Investigations of phase transitions by a Monte Carlo method. *High Temp. (USSR)* 7: 216.

Nosé, S. 1984a. A molecular dynamics method for simulations in the canonical ensemble. *Mol. Phys*. 52: 255.

Nosé, S. 1984b. A unified formulation of the constant temperature molecular dynamics methods. *J. Chem. Phys*. 81: 511.

Nosé, S. 1991. Constant temperature molecular dynamics methods. *Prog. Theor. Phys. Suppl*. 103: 1.

Ohnishi, S. 1994. CAMP Project: What is CAMP-Atami. http://www.camp.or.jp/new/doc/camp-atami/

Ojeda, P., Garcia, M., Londono, A., and Chen, N. Y. 2009. Monte Carlo simulations of proteins in cages: Influence of confinement on the stability of intermediate states. *Biophys. J*. 96: 1076.

Orkoulas, G., and Panagiotopoulos, A. Z. 1999. Phase behaviour of the restricted primitive model and square--well fluids from Monte Carlo simulations in the grand canonical ensemble. *J. Chem. Phys*. 110: 1581.

Ortiz, V., Nielsen, S. O., Discher, D. E., Klein, M. L., Lipowsky, R., and Shillcock, J. 2005. Dissipative particle dynamics simulations of polymersomes. *J. Phys. Chem. B*. 109: 17708.

Panagiotopoulos, A. Z. 1987a. Direct determination of phase coexistence properties of fluids by Monte Carlo simulation in a new ensemble. *Mol. Phys*. 61: 813.

Panagiotopoulos, A. Z. 1987b. Adsorption and capillary condensation of fluids in cylindrical pores by Monte Carlo simulation in the Gibbs ensemble. *Mol. Phys*. 62: 701.

Panagiotopoulos, A. Z. 1992. Direct determination of fluid-phase equilibria by simulation in the Gibbs ensemble A review. *Mol. Simul*. 9: 1.

Panagiotopoulos, A. Z. 1996. Current advances in Monte Carlo methods. *Fluid Phase Equilib*. 116: 257.

Panagiotopoulos, A. Z. 2000. Monte Carlo methods for phase equilibria of fluids. *J. Phys. Condens. Matter*. 12: R25.

Panagiotopoulos, A. Z., Quirke, N., Stapleton, M., and Tildesley, D. J. 1988. Phase equilibria by simulation in the Gibbs ensemble. Alternative derivation, generation, and application to mixture and membrane equilibria. *Mol. Phys*. 63: 527.

Parr, R. G., and Yang, W. 1989. Density-functional theory of atoms and molecules. In *International Series of Monographs on Chemistry*. Eds. R. Breslow, J. B. Goodenough, J. Halpern, and J. Rolinson. Oxford University Press, New York.

Parrinello, M., and Rahman, A. 1982. Strain fluctuations and elastic constants. *J. Chem. Phys*. 76: 2662.

Pensak D. A. 1989. Molecular modelling: Scientific and technological boundaries. *Pure Appl. Chem*. 61: 601.

Perdew, J. P., Burke, K., and Wang, Y. 1996. Generalised gradient approximation for the exchange-correlation hole of a many-electron system. *Phys. Rev. B: Condens. Matter*. 54: 16533.

Pople, J. A., Santry, D. P., and Segal, G. A. 1965. Approximate self-consistent molecular orbital theory. I. Invariant Procedures. *J. Chem. Phys*. 43: S129.

Pople, J. A., and Beveridge, D. L. 1970. *Approximate Molecular Orbital Theory*. McGraw-Hill, New York.

Pople, J. A., and Segal, G. A. 1965. Approximate self-consistent molecular orbital theory. II. Calculations with complete neglect of differential overlap. *J. Chem. Phys*. 43: S136.

Press, W. H., Flannery, B. P., Teukolsky, S. A., and Vetterling, W. T. 1986. *Numerical Recipes: The Art of Scientific Computing*. Cambridge University Press, Cambridge.

Rahman, A. 1964. Correlations in the motion of atoms in liquid argon. *Phys. Rev.* 136: A405.

Randić, M. 1990. The nature of chemical structure. *J. Math. Chem.* 4: 157.

Randić, M., and Razinger, M. 1997. On characterization of 3D molecular structure. In *From Chemical Topology to Three Dimensional Geometry*. Ed. A. T. Balaban. Plenum Press, New York.

Rappe, A. K., and Goddard, W. A. III. 1991. Charge equilibration for molecular dynamics simulations. *J. Phys. Chem.* 95: 3358.

Rappe', A. K., Casewit, C. J., Colwell, K. S., Goddard, W. A. III, and Skiff, W. M. 1992. UFF: A full periodic table force field for molecular mechanics and molecular dynamics simulations. *J. Am. Chem. Soc.* 114: 10024.

Rappe', A. K., Colwell, K. S., and Casewit. C. J. 1993. Application of a universal force field to metal complexes. *J. Inorg. Chem.* 32: 3438.

Ray, J. R. 1988. Elastic constants and statistical ensembles in molecular dynamics. *Comput. Phys. Rep.* 8: 109.

Ripoll, D. R., and Scheraga, H. A. 1989. The multiple-minima problem in the conformational analysis of polypeptides. III. An electrostatically driven Monte Carlo method: Tests on enkephalin. *J. Protein Chem.* 8: 263.

Rivet, J. P., and Boon, J. P. 2001. *Lattice Gas Hydrodynamics*. Cambridge University Press, Cambridge.

Rodgers, J. R., and Cebon, D. 2006. Materials informatics. *MRS Bull.* 31: 975.

Ryckaert, J.-P., Ciccotti, G., and Berendsen, H. J. C. 1977. Numerical integration of the Cartesian equations of motion of a system with constraints: Molecular dynamics of n-alkanes. *J. Comput. Phys.* 23: 327.

Sadus, R. J. 1999. *Molecular Simulation of Fluids: Theory, Algorithm and Object-Orientation*. Elsevier, Amsterdam.

Sarman, S., Evans, D. J., and Cummings, P. T. 1998. Recent developments in non-equilibrium molecular dynamics. *Phys. Rep.* 305: 1.

Scheraga, H. A., Lee, J., Pillardy, J., Ye, Y. J., Liwo, A., and Ripoll, D. 1999. Surmounting the multiple-minima problem in protein folding. *J. Global Optim.* 15: 235.

Schreiber, H., and Steinhauser, O. 1992. Cutoff size does strongly influence molecular dynamics results on solvated polypeptides. *Biochem.* 31: 5856.

Schulze-Kremer, S. 2000. Genetic algorithm and protein folding. In *Protein Structure Prediction—Methods and Protocols*. Ed. D. M. Webster. Humana Press Inc., Totowa, NJ.

Simons, J., and Nichols, J. 1997. *Quantum Mechanics in Chemistry*. Oxford University Press, Oxford.

Smith, W. R., and Triska, B. 1994. The reaction ensemble method for the computer-simulation of chemical and phaseequilibria.1. Theory and basic examples. *J. Chem. Phys.* 100: 3019.

Stewart, J. J. P. 1989. Optimization of parameters for semi-empirical methods I—Method. *J. Comput. Chem.* 10: 209.

Stewart, J. J. P., and Stewart, A. C. 1999. Developing a semiempirical method. In *Crystal Engineering: The Design and Application of Functional Solids,* NATO ASI Ser., C. 539, Kluwer, 83.

Stillinger, F. H., and Rahman, A. J. 1974. Improved simulation of liquid water by molecular dynamics. *J. Chem. Phys.* 60: 1545.

Stoltze, P. 2000. Microkinetic simulation of catalytic reactions. *Prog. Surf. Sci.* 65: 65.

Succi, S. 2001. *The Lattice Boltzmann Equation: For Fluid Dynamics and Beyond*. Oxford, Clarendon.

Sun, H. 1998. COMPASS: An *ab initio* force-field optimized for condensed-phase applications-overview with details on alkane and benzene compounds. *J. Phys. Chem. B* 102: 7338.

Svensson, M., Humbel, S., Froese, R. D. J., Matsubara, T., Sieber, S., and Morokuma, K. 1996. ONIOM: A multilayered integrated MO + MM method for geometry optimizations and single point energy predictions. A test for Diels−Alder reactions and Pt(P(t-Bu)3)2 + H2 oxidative addition. *J. Phys. Chem.* 100: 19357.

Thiel, W. 1996. Perspectives on semiempirical molecular orbital theory. *Adv. Chem. Phys.* 93: 703.

Todeschini, R., and Consonni, V. 2000. *Handbook of Molecular Descriptors,* Vol. 11. Wiley-VCH, Weinheim.

Tosi, M. P. 1964. Cohesion of ionic solids in the born model. *J. Phys. C: Solid State Phys.* 16: 107.

Vásquez, M., Némethy, G., and Scheraga, H. A. 1994. Conformational energy calculations on polypeptides and protein. *Chem. Rev.* 94: 2183.

Verlet, L. 1967. Computer experiments on classical fluids. I. Thermodynamical properties of Lennard−Jones molecules. *Phys. Rev.* 159: 98.

Warshel, A., and Karplus, M. 1972. Calculation of ground and excited state potential surfaces of conjugated molecules. I. Formulation and parametrization. *J. Am. Chem. Soc.* 94: 5612.

Wilde, R. E., and Singh, S. 1998. *Statistical Mechanics, Fundamentals and Modern Applications*. John Wiley & Sons, New York.

26 Molecular Modeling for the Design of Novel Performance Chemicals and Materials

Willium, D. E. 1991. Atomic charge and multipole models for the ab initio molecular electric potential. *Rev. Comput. Chem.* 4: 219.

Wilson, S. R., and Cui, W. L. 1990. Applications of simulated annealing to peptides. *Biopolymers.* 29: 225.

Xue, L., and Bajorath, J. 2000. Molecular descriptors in chemoinformatics, computational combinatorial chemistry, and virtual screening. *Comb. Chem. High Throughput Screening.* 3: 363.

Yamamoto, S., Maruyama, Y., and Hyodo, S. 2002. Dissipative particle dynamics study of spontaneous vesicle formation of amphiphilic molecules. *J. Chem. Phys.* 116: 5842.

York, D., Wlodawer, A., Pedersen, L., and Durden, T. 1994. Atomic-level accuracy in simulations of large protein crystals. *Proc. Natl. Acad. Sci. USA.* 91: 8715.

Zerner, M. C. 1995. Neglect of differential overlap calculations of the electronic spectra of transition metal complexes. Metal-ligand interactions: Structure and reactivity. *NATO ASI Ser., Ser.* C. 474.

2 Rational Design of Selective Industrial Performance Chemicals Based on Molecular Modeling Computations

Beena Rai and Pradip

CONTENTS

2.1   INTRODUCTION

A variety of surface-active reagents, also known as *surfactants*, are employed in industry to achieve certain desirable changes in the interfacial properties of a system (Rosen 2004). The surfactants are used with as diverse a group of products as motor oils and lubricants, laundry detergents, personal care products and cosmetics, food additives, pharmaceutical formulations, additives for petroleum recovery, paints, leather processing, building materials, ceramic process-ing, and mineral processing materials. The success of all surfactant-based processes depends on the ability of the surfactant to possess the desired properties—sometimes the properties can even be conflicting (e.g., aqueous solubility and ability to impart maximum hydrophobicity)—required for that particular application. The increasing demands on enhancing the efficacy of the surfactants for a particular application and decreasing its cost at the same time has necessitated a search for a scientifically robust framework for designing tailor-made surfactants customized for each application, preferably based on first principles. The availability of such a powerful framework will eventually replace the current trial-and-error methodology of surfactants design (which is time-consuming and extremely costly in terms of effort) with a more robust rational design of surfactants. Such a design framework must incorporate the current understanding about the adsorption of surfactants at interfaces.

2.1.1   Adsorption of Surfactants at Interfaces

When present in relatively low concentrations in a system, a surfactant adsorbs onto the surfaces, or interfaces, and alters the surface, or interfacial, free energies of the surfaces. The *interface* usually refers to a boundary between any two immiscible phases while *surface* denotes an inter-face in which one phase is a gas (usually air). Surfactants have an amphipathic molecular struc-ture that consists of a group having very little attraction for the solvent, known as a *lyophobic group* (or *hydrophobic group* in case of aqueous systems) or *tail*, and a functional group that has a strong attraction for the solvent, called the *lyophilic group* (or *hydrophilic group* in case of aqueous systems) or *head group* (Figure 2.1a). The chemical structures of groups suitable as the lyophobic/hydrophobic and lyophilic/hydrophilic portions of the surfactant molecule depend on the nature of the solvent. For example, in a highly polar solvent like water, the hydrophobic group may be a hydrocarbon chain ($>C_8$), and ionic or highly polar groups may act as hydrophilic groups whereas in a nonpolar solvent like heptane the opposite may be true. Thus, for achieving the desired surface activity in a particular system, the surfactant molecule must have a chemical structure that is amphipathic in that solvent under the conditions of its use. This amphipathic structure of the surfactant leads to a specific orientation of the molecule at the surface with its

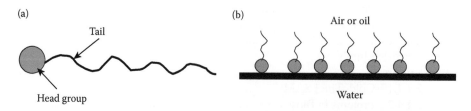

FIGURE 2.1   Schematic representation of (a) surfactant molecule and (b) adsorption of surfactant molecules.

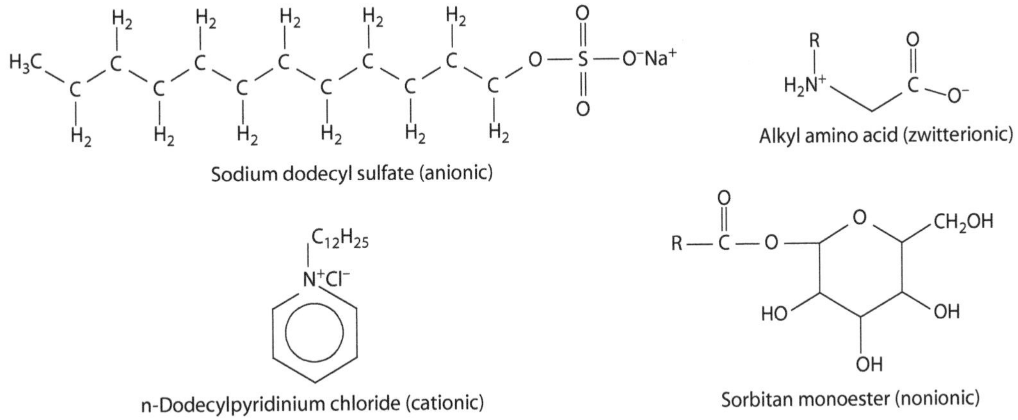

FIGURE 2.2 Structure of some common surfactant molecules.

lyophilic group in the solvent phase and lyophobic group oriented away, thereby increasing the concentration of the surfactant at the surface (known as the *surface activity*) and reducing the surface interfacial tension (Figure 2.1b).

For most of the commonly used surfactants, the hydrophobic group is usually a long-chain hydrocarbon (aliphatic/aromatic) and the hydrophilic group is an ionic or polar group. Depending on the nature of the hydrophilic group, surfactants are classified as anionic, cationic, zwitterionic, and nonionic (Figure 2.2).

Most commonly used surfactants are aqueous soluble, and unless otherwise needed, the terms *hydrophobic* and *hydrophilic* will be used in the following sections with an understanding that in general it could be any solvent. The adsorption of surfactants at interfaces is evidently a function of the interactions at the interface and in the bulk. The efficacy of the surfactant thus depends on the following three important interactions:

1. The chemical nature of the surfactant species, which could be in neutral or ionized form, being adsorbed: the character of its functional group (i.e., cationic, anionic, nonionic, zwitterionic), the nature of its nonpolar (hydrophobic) portion (i.e., the length and the degree of branching in case of hydrocarbon chains and also the solution conformation in case of high molecular weight polymers), and the associative complexes in bulk, if any.
2. The nature of the interface: the solid–water interface (or solid–liquid interface in general) in systems involving particles (e.g., minerals and ceramics) or the air–water interface or liquid–liquid interfaces in systems having bubbles or oil droplets, respectively: the surface charge, its hydrophobicity, and the nature of adsorption sites at the interface (e.g., exposed metal ions at the interface providing sites for chelation at interfaces). In the case of crystalline solids, the surface crystal structure of the interface plays an important role in surfactant adsorption.
3. The nature of the aqueous environment: the pH, the ionic strength, the reduction potential (E_h) (in certain cases), and the presence of other competing ions or other species that can complex surfactants in the bulk solution.

The interactions could be based on electrostatic interactions determined by the electric double-layer forces or specific interactions (e.g., chemical bonding, chelation, covalent-coordinate bonds, hydrophobic interactions, salvation forces, hydrogen bonding, and steric forces). The adsorption of the surfactant at the interface results in certain desirable changes in macroscopic properties such as wettability and colloidal stability. For example, the self-assembled monolayers at the interface

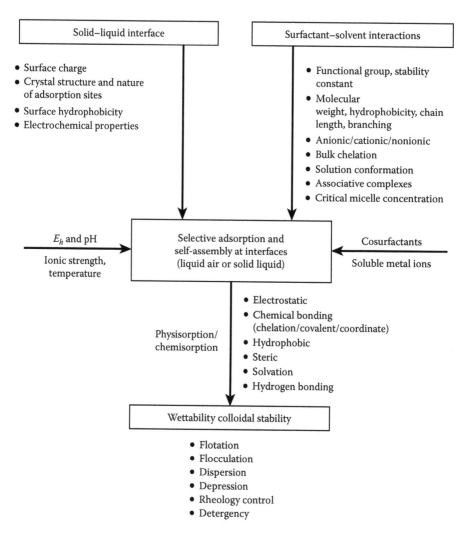

FIGURE 2.3   A schematic diagram showing the factors governing adsorption of surfactants at interfaces.

(or the conformation of the adsorbed surfactant species at the interface) govern the changes in the surface (interfacial) properties. In turn, the differences in the macroscopic properties created as a consequence of the intervention of the surfactant drive the particular end-application (e.g., flotation, dispersion, flocculation, changes in rheology, detergency, etc.).

As illustrated in Figure 2.3, the surfactant design framework thus must be able to capture this set of different molecular interactions to a sufficiently high degree of complexity and with adequate scientific rigor. We shall illustrate in the following sections with several examples taken from our own work that it is possible to come up with such a framework based on molecular-modeling (MM) computations that it is sufficiently robust and reliable so that one can use it for designing surfactants tailor-made for a variety of industrial applications. The proposed framework is capable of addressing both components of surfactant design, the functional group and the molecular architecture of the chain (or the polymeric chain as the case may be). Our framework is sufficiently flexible so that one can incorporate the continuous advances being made in the field of computational chemistry and thus be able to address even the finer nuances of surfactants design with the enhanced theoretical understanding of the different interactions occurring in the systems under investigation.

2.1.2   A Framework for Rational Design of Surfactants

The design, development, and selection of surfactants for different industrial applications are currently based on a trial-and-error methodology, rules of thumb, intuition, and past experience. The time and resources required to come up with an acceptable formulation is therefore sometimes prohibitively expensive. Additionally, because of the high cost of empirical search for novel surfactants, the search envelope is severely restricted to a few well-known groups and molecules. A quantitative methodology to screen out and identify the appropriate molecular architectures based on the quantitative assessment of the relative efficacy of different possible structures based on theoretical computations is evidently an economically attractive and elegant methodology as compared with the conventional trial-and-error approach. Selecting the most promising surfactants from a wide variety of possibilities based on computer-aided surfactant design tools—and thus screening out only a few designs, for subsequent time-consuming experimental test work (i.e., synthesis, characterization, lab and bench scale testing followed by pilot plant and plant trials)—will certainly lead to enormous savings in time and efforts needed to arrive at new formulations and novel surfactants.

With continuous advances being made in our understanding of molecular level phenomena that govern the adsorption of surfactants at interfaces, the accessibility of more user-friendly MM tools (that capture an ever-growing number of complex interactions underlying surfactant adsorption at interfaces) and the availability of increasingly more inexpensive computing power, it is possible to refine our framework further and to design surfactants customized for specific applications based on theoretical MM computations.

Various MM tools such as quantitative structure property relationship (Hu et al. 2010), quantum mechanics (Yuana et al. 2010), molecular dynamics (Arya and Panagiotopoulos 2005; Fodi and Hentschke 2000; Gupta et al. 2008; Iacovella et al. 2005; Larson 1994; Minami et al. 2011; Nielsen et al. 2005; Salaniwal et al. 2001; Siperstein and Gubbins 2003; Soddemann et al. 2001; Srinivas et al. 2006; von Gottberg et al. 1997, 1998), Brownian dynamics (Bedrov et al. 2002; Bourov and Bhattacharya 2003), Monte Carlo simulations (Larson 1996; Lisal et al. 2003; Reimer et al. 2005; Zehl et al. 2009) and coarse-grained mesoscale (Prinsen et al. 2002; Shinoda et al. 2008) methods have been applied successfully to model, predict, and validate properties, phenomena, and phases of surfactants at various length scales. Several research groups have also reported on the utility of molecular-modeling tools for understanding surfactant–surface (solid) interactions (Akim et al. 1998; Almora-barrios et al. 2009; Almora-barrios and de Leeuw 2010; Bromely et al. 1993, 1994; Claire et al. 1997; Coveney and Humphries 1996; de Leeuw and Rabone, 2007; Du and Miller 2007; Fa et al. 2006; Filgueiras et al. 2006; Filho et al. 2000; Heinz et al. 2007; Henao and Mazeau 2008; Hirva and Tikka 2002; Jacobs et al. 2006; Kundu et al. 2003; Liu et al. 2008; Lu et al. 2009; Natarajan and Nirdosh 2008; Shevchenko and Bailey 1998; Zeng et al. 2010). However, these studies have been mainly focused on the understanding of surfactant–surface interactions rather than being used as an engineering tool to design or develop novel surfactants. Since the early 1990s, we have demonstrated through our research work in several different industrial domains that molecular-modeling computations provide an elegant method to quantify the crystal structure specificity of different surfactant molecules for a given solid (i.e., crystal) surface (Pradip 1992, 1994; Pradip and Rai 2002a, 2002b, 2003, 2009; Pradip et al. 2002a, 2002b, 2002c, 2004a, 2004b, 2004c; Rai and Pradip 2003, 2008, 2009; Rai et al. 2004; Sathish et al. 2007; Singh et al. 2010). The two key features of our approach are (1) identification of the molecular recognition (crystal structure specificity) mechanisms underlying the adsorption of surfactants at the interface and (2) use of advanced molecular-modeling techniques for theoretical computations of the relative magnitude of interaction. The relative strength of surfactant interactions, as quantified through interaction energies, thus provides a quantitative search technique for screening and identifying the most promising molecular architectures from a large set of candidates available for a particular application, resulting in considerable saving in time and efforts needed for developing new formulations. The design framework incorporates tailoring the molecular architecture of the surfactant to take advantage of the molecular recognition mechanisms, if already known or identified, such as the surface-templating effects and the host–guest interactions at the interface. We have successfully applied this novel paradigm to design and develop surfactants for various industrial applications such as selective reagents for mineral separation, the design of effective dispersants for ceramics applications including dispersion of nanoparticles, and additives for paint formulations.

It is well known that the selective adsorption of surfactants at the solid–water interface imparts hydrophobicity to the surface of the solid. The relative hydrophobicity of the solid surface is responsible for various macroscopic properties observed experimentally. For example, in mineral separation, the hydrophobicity of the solid surface leads to selective bubble–particle attachment, which accounts for the selective flotation of minerals in large scale industrial plants. The relative measure of mineral surface hydrophobicity is usually quantified in terms of contact angle measurements and flotation experiments (Fuerstenau 1957, 1970, 2000; Fuerstenau and Herrera-Urbina 1989; Fuerstenau and Pradip 2005; Pradip 1988). Molecular-modeling tools can be successfully employed to compute the interaction energies and contact angle on both virgin and surfactant-covered mineral surfaces. The relative flotation efficacy of different surfactants can thus be related to their molecular structure and properties.

Dispersion of colloidal particles, including nanoparticles, in aqueous and nonaqueous suspensions is crucial for the success of the manufacturing processes for many applications such as advanced ceramics, coatings and paints, and pharmaceuticals. Colloidally stable suspensions produce higher average packing densities and finer pore size distributions in green or sintered ceramic compacts than do poorly dispersed or flocculated suspensions (Bergstrom et al. 1997; Kirby et al. 2004; Pradip et al. 1994; Subbanna et al. 2002). The selection of appropriate dispersants is thus important for achieving the advantages of lower sintering temperature and significant enhancement of performance in industrial applications as a consequence of employing colloidal and nanoparticles.

One of the important steps in the paint-making process is pigment dispersion, in which the pigment particles undergo wetting followed by colloidal stabilization with the help of surfactants in a suitable mechanical device. The extent of dispersion determines the final optical properties of the paint (Doroszkowski and Lambourne 1978). Thus, relative wettability and colloidal stability in a given multicomponent formulation, as quantified through interaction energies, is an important parameter that could be theoretically computed and related to the extent of dispersion and stability of these systems.

In this chapter, we present a brief overview of this rational approach for the design and development of novel performance chemicals and illustrate the utility and power of this methodology through some of the case studies drawn from our research work.

2.2 COMPUTATIONAL METHODOLOGY

Both atomistic and quantum simulations were employed to model the surfactant–surface interactions. Atomistic simulations were carried out using commercial software (Cerius^2/Material Studio, Accelrys Inc., San Diego, USA (http://accelrys.com). Density functional theory (DFT) computations were performed using the plane-wave self-consistent field (PWscf) code contained in the Quantum Espresso (http://www.quantum-espresso.org) distribution running on an EKA supercomputer at Computational Research Laboratory, Pune, India. Because we were interested in modeling surfactant interactions with a wide variety of surfaces and solids, we used generalized force fields like universal force field (UFF) (Rappe' et al. 1992), polymer-consistent force field (PCFF) (Sun 1995), and Dreiding (Mayo et al. 1990) for the computations reported in this study. However, in some cases, we have also derived specific force field parameters from more accurate quantum chemical calculations (Pradip et al. 2004a). The general methodology followed to model the surfactant–solid interactions is summarized in the following subsections.

2.2.1 SOLID (MINERAL/CERAMICS) SURFACE

A two-dimensional (2D) periodic surface cell was created from the crystal unit cell of the mineral/ceramics at a given Miller plane (usually the cleavage plane) and then optimized. The surface energy was then calculated for this optimized mineral surface:

$$[Us - Ub] \div A$$

where Us is the energy of the surface block of the crystal, Ub is the energy of an equal number of atoms of the bulk crystal, and A is the surface area. A detailed methodology of modeling three-dimensional (3D) mineral/ceramics crystals and the creation of 2D surfaces is reported elsewhere (Pradip and Rai 2002a; Pradip et al. 2002a; Rai and Pradip 2003).

2.2.2 SURFACTANT MOLECULE

The geometry of surfactant molecules was optimized using an appropriate force field. The partial charges on the constituent atoms were either computed using the charge equilibration method (Rappe' and Goddard 1991) or derived from quantum chemical methods. Experimental data on the bond lengths, bond angles, and other properties are used to validate the optimized structure (Pradip and Rai 2003).

2.2.3 SOLID–SURFACTANT COMPLEX

2.2.3.1 In Vacuum

The optimized surfactant molecule was docked on the mineral/ceramics surface. The initial geometry of the mineral/ceramics surface–surfactant complex was created physically on the screen with the help of molecular graphics tools, taking into consideration the possible interactions of surfactant functional groups with surface atoms. The surfactant molecule was then allowed to relax completely on the surface. Approximately 20 initial conformations were assessed to locate the minimum energy conformation of surfactant molecule at the mineral/ceramics surface.

2.2.3.2 In Solvent or Dispersing Medium

To realistically simulate the surfactant–mineral/ceramics interactions, we also included solvent molecules explicitly in our simulations (Pradip et al. 2004a, 2004b). A 3D periodic box of solvent molecules was constructed, optimized, and equilibrated with molecular dynamics (MD) so as to match the density in the box to that of the experimental density of the selected solvent. Bulk solvent molecules thus were introduced over the mineral/ceramics surface and optimized. Optimized surfactant molecules were then introduced in the box and the mineral/ceramic-surface–solvent–surfactant complexes thus created were subjected to energy minimization and MD simulation.

2.2.4 STATIC ENERGY MINIMIZATION

The intramolecular van der Waal interactions were calculated only between atoms that were located at distances greater than their fourth nearest neighbors. A modified Ewald summation method (Karasawa and Goddard 1989) was used for calculating the nonbonded coulomb interactions, and an atom-based direct cutoff method was used for van der Waal interactions. Smart minimizer, as implemented in Cerius^2/Material Studio, was used for geometry optimization. The optimization was considered to be converged when a gradient of 0.0001 kcal/mole was reached.

2.2.5 MOLECULAR DYNAMICS SIMULATIONS

Most often, the structure of the complex obtained through the static energy minimization method represents only a local minimum energy structure. This complex was used as the initial configuration to find a global minimum energy structure through MD simulations (Pradip et al. 2004a, 2004b). Both constant energy microcanonical ensemble [number of atoms, volume, and energy (NVE)] and constant temperature canonical ensemble [number of atoms, volume, and temperature (NVT)] methods were employed for MD runs at 300 K. For different case studies, time steps used were in the range of 0.5 to 2.5 fs and run lengths varied from 500 ps to 1 ns. Standard methods as reported by Hermansson et al. (1988) and Berendsen et al. (1984) were used for temperature control during the MD run.

2.2.6 COMPUTATION OF INTERACTION ENERGY

The relative affinity of interaction of surfactant molecule with different mineral/ceramics surfaces was quantified in terms of the interaction energy ΔE (Pradip and Rai 2002a; Pradip et al. 2002a, 2002b, 2004a, 2004b) in vacuum and in solvent/dispersing medium:

In Vacuum

$$\Delta E = E_{complex} - \left(E_{surface} + E_{surfactant} \right) \tag{2.1}$$

where $E_{complex}$ is the total energy of the optimized mineral/ceramics surface – surfactant complex, and $E_{surface}$ and $E_{surfactant}$ are the total energies of free surface and surfactant molecule, computed separately.

In Solvent/Dispersing Medium

$$IE_{(in\text{-}solvent)} = \Delta E_{combined} - \sum \left(\Delta E_{solvent\text{-}surfactant} + \Delta E_{surface\text{-}solvent} \right) \tag{2.2}$$

$$\Delta E_{combined} = E_{complex,solvent} - \sum \left(E_{surface} + E_{surfactant} + E_{solvent} \right) \tag{2.3}$$

where $\Delta E_{combined}$ is the total interaction energy calculated using Equation 2.3. $E_{complex,solvent}$ is the total energy of the optimized surface–surfactant complex in the presence of solvent, $E_{surface}$, $E_{surfactant}$, and $E_{solvent}$ are the total energies of free surface, surfactant, and solvent molecules, computed separately. $\Delta E_{solvent\text{-}surfactant}$ is the interaction energy computed for the interaction of solvent and surfactant molecule, and $\Delta E_{surface\text{-}solvent}$ is the contribution due to interaction of solvent molecules with the surface. These energies are subtracted from $\Delta E_{combined}$ to get the final interaction energy ($IE_{(in\text{-}solvent)}$) of surfactant molecule with the surface (Equation 2.2).

It is worth noting that a more negative magnitude of the interaction energy indicates more favorable interactions between surfactant molecule and the surface. The magnitude of this quantity is thus an excellent measure of the relative intensity or efficiency of the interaction of the surfactant with different surfaces.

2.2.7 SELF-ASSEMBLED MONOLAYERS AND INTERFACIAL PROPERTIES

To study the effect of packing of surfactant molecules at the surface on the interfacial properties (wettability or hydrophobicity), a monolayer of surfactant molecules was placed on the surface (Rai and Pradip 2003; Rai et al. 2004). The surfactant molecules were placed as per the most stable conformation obtained through optimized single molecule–surface complex (Rai and Pradip 2003). To find the equilibrium structures of the adsorbed monolayers, the clusters thus created were subjected to geometry optimization followed by MD simulations. The surface atoms were kept fixed during the entire simulation run and only adsorbed surfactant molecules were allowed to relax. Adsorption and self-assembly of surfactants at the air–water interfaces also modeled to study the interfacial properties (Tanwar et al. 2011).

2.2.8 CONTACT ANGLE AND WETTABILITY

Contact angle is an inverse measure of wettability (Sharfrin and Zisman 1960). MD simulations were employed to study the wettability of adsorbed self-assembled monolayers (SAMs) as characterized though computation of contact angle of a water droplet placed on them (Sathish et al. 2007; Singh et al. 2010). Water molecules were modeled using extended simple point charge (SPC/E) potential, a rigid three-point model with fixed charges assigned to hydrogen and oxygen (Berendsen et al. 1987). A 3D periodic cubic cell of water molecules was created, optimized, and equilibrated using MD simulation. A spherical droplet was cut from this equilibrated 3D periodic box of water molecules and placed on the SAMs and solid surfaces for wetting studies (Rai and Pradip 2008).

The methodology proposed by Fan and Cagin (1995) was used to extract the microscopic parameters—namely, the drop volume (V) and the interfacial area (S) of the water droplet from the MD simulations (Figure 2.4). The height (h) of the droplet from the terminal surface layer and the radius of the droplet (R) have been calculated using the following equations:

$$h^3 + 3S\frac{h}{\pi} - \frac{6V}{\pi} = 0 \tag{2.4}$$

$$R = \frac{h}{2} + \frac{S}{2}\pi h \tag{2.5}$$

The contact angle (θ) was further computed from the following relationship:

$$\cos\theta = 1 - \frac{h}{R} \tag{2.6}$$

It is important to note that a higher contact angle (θ) value of the water droplet indicates relatively more hydrophobic surface.

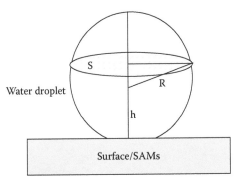

FIGURE 2.4 Schematic representation of the water droplet depicting various parameters computed through MD simulations for estimation of contact angle.

2.3   DESIGN OF PERFORMANCE CHEMICALS

The methodology outlined in Section 2.2 has been used to model surfactant–surface interactions and to design performance chemicals in many industrial applications. The examples are presented in the following sections to illustrate the utility of the proposed methodology.

2.3.1   ADSORPTION AT AIR–WATER INTERFACE

The dynamic behavior of surfactants at interfaces (air–water and solid–liquid) plays a critical role in determining their effectiveness for a given application. Quite often the experimental measurement of the surface tension of a given formulation might not be possible or economically viable. For example, while designing new drug formulations, the availability of the drug in a small quantity, the drug's cost, and subsequent testing of the formulations may not permit many trial-and-error experiments. Thus, a theoretical methodology to screen out various options *a priori* will lead to faster and cheaper formulations. As part of our larger research program of design and development of surfactant combinations for specific industrial applications, we also developed a simple, quick, direct, and robust methodology for the prediction of the surface tension of solvent–surfactant interfaces (Tanwar et al. 2011) This methodology offers a quick and reliable tool for computation of surface tension as a function of concentration, including the prediction of critical micelle concentration for any given surfactant–solvent system.

MD simulations were used to compute the surface tension of a sodium dodecyl sulfate (SDS)–water system. A 3D periodic simulation box ($L_x = L_y = 32\text{Å}$, $L_z = 96\text{Å}$) consisting of a rectangular parallelepiped cell with 800 water molecules located at the center of the cell was modeled (Figure 2.5). A layer of optimized SDS molecules was placed at the air–water interface. The numbers of SDS molecules adsorbed at the interface were gradually increased to study the effect of bulk surfactant concentration on interfacial properties. This was achieved by placing 2, 4, 8, 16, 24, and 30 SDS molecules at the air–water interface. A dissociated SDS molecule was modeled using an all-atom model. Partial atomic charges were assigned to each SDS molecule (Weng et al. 2000). An equivalent numbers of sodium ions were placed randomly in bulk water with a charge of +1 to maintain charge neutrality in the system (Figure 2.5). Water molecules were modeled using SPC/E potentials (Berendsen et al. 1987), and all other atoms were modeled using PCFF (Sun 1995) as implemented in Materials Studio 4.1, Accelrys Inc., San Diego, USA. MD simulations were run with a time step of 2.5 fs in NVT ensemble at 298 K for 500 ps. Stress tensor data were collected for

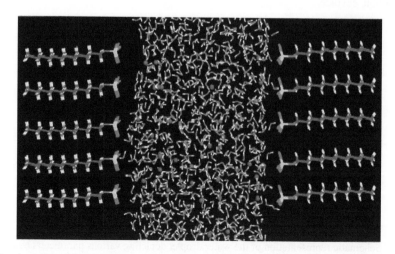

FIGURE 2.5   A model of the simulation cell consisting of 30 SDS and 800 water molecules.

each 1000 time steps and averaged over the entire production run. Surface tension was calculated using the following equation:

$$\gamma = \frac{1}{2}\left\langle P_{zz} - \frac{1}{2}(P_{xx} + P_{yy})\right\rangle \frac{V}{A} \tag{2.7}$$

where P represents the elements of the pressure tensor in x, y, and z direction, z axis is normal to the interface. $A = L_x L_y$ is the surface area and $V = AL_z$ is the volume of the box. The factor of 1/2 outside the bracket arises from the two water–vapor interfaces in the system. The pressure tensor components were computed using the virial expression (Allen and Tildesley 1987):

$$P_{\alpha\beta} = \frac{1}{V}\left[\sum_{i=1}^{N} m_i v_{i\,\alpha} v_{i\,\beta} + \sum_{i=1}^{N-1}\sum_{j>i}^{N} (r_{ij})_\alpha (f_{ij})_\beta\right] \tag{2.8}$$

where N is the number of molecules; V is the volume of the system; m_i and v_i are the mass and velocity of particle i, respectively; and f_{ij} is the force between particles i,j, and $r_{ij} = r_i - r_j$, with r_i being the position of particle.

From MD simulations the amount of surfactant adsorbed at the interface (Γ) and the corresponding surface tension (γ) values were obtained. These were used to compute the relationship between surface tension (γ) and bulk concentration (c) using the following equations:

$$\Gamma = -\frac{1}{RT}\frac{d\gamma}{d\ln c} \tag{2.9}$$

$$\gamma = \alpha_1 c + \alpha_2 \ln c + \alpha_3 \tag{2.10}$$

This relationship was further used to compute the critical micelle concentration (CMC) of SDS molecules. As presented in Table 2.1, the theoretically computed properties of the SDS–water system were found to be in close agreement with those measured experimentally (Nakahara et al. 2008; Rosen 2004; Zhang et al. 2004).

2.3.2 MINERAL/CERAMICS–WATER INTERFACE (WETTING CHARACTERISTICS)

The molecular structure and behavior of water near different substrates determines the hydrophobicity and hydrophilicity of the surface. Contact angle (θ) is an inverse measure of wettability. The relationship between contact angle and floatability of minerals is well established

TABLE 2.1

Theoretically Computed and Experimental Properties of Water–SDS System at 25°C

Property	Experimental			Our Simulations (Tanwar et al. 2011)
	Rosen 2004	Nakahara et al. 2008	Zhang et al. 2004	
CMC (mM)	8	8.2	10.5	10.9
Surface tension at CMC (m Nm^{-1})	33	39	38	33
Surface saturation (mol/Å^2)	2.6×10^{-10}	–	–	2.43×10^{-10}

CMC, critical micelle concentration.

(Fuerstenau 1957, 1970, 2000; Fuerstenau and Herrera-Urbina 1989). Based on the measured contact angle values, the surfaces have been classified as super hydrophobic ($\theta > 150$), hydrophobic ($65 < \theta < 150$), hydrophilic ($0 < \theta < 65$) and super hydrophilic ($\theta = 0$). Several experimental studies have been reported in the literature to measure the contact angle and to understand the structural and dynamic properties of water in contact with various hydrophobic and hydrophilic surfaces (Binggeli and Mate 1994; Mamontov et al. 2007). We have employed MD simulations to study the wetting properties of the four different mineral surfaces: α-alumina (corundum, Al_2O_3), fluorite (CaF_2), graphite, and kaolinite ($Al_2O_3 \cdot 2SiO_2 \cdot 2H_2O$) in presence of water (Singh et al. 2010). The atom positions in the unit cell of the minerals were taken from the experimental structural reports based on x-ray studies (Dana and Ford 1949). A periodic surface cell was created from the unit cell of the mineral crystal at its cleavage plane, optimized, and extended in x and y directions to create a larger surface slab. The surface slabs modeled for α-alumina {001}, fluorite {111} (Deer et al. 1992), graphite {100} and kaolinite {001} surfaces were of the dimensions 99.94×99.94, 75.24×75.24, 98.4×98.4, and 100×100 Å^2, respectively. A spherical droplet of water, cut from the equilibrated 3D periodic box of water, was placed on the mineral surfaces for wetting studies. A vacuum of 100 Å (in the z direction) above the top atomic layer of the surface cell was created to screen the interactions of water droplet from the bottom layers of the surface slab.

MD simulations were carried out using Materials Studio 4.1. UFF (Rappe' et al. 1992) as implemented in Materials Studio was used to model the mineral surface–water interactions. Nonbonded interactions (i.e., van der Waals and columbic) were modeled using a modified Ewald summation method (Karasawa and Goddard 1989). Partial charges on the atoms were calculated using QEq method (Rappe' and Goddard 1991). MD simulations were run in microcanonical (NVT) and canonical (NVE) ensemble at 298K. A time step of 1 fs was used. The surface atoms were kept fixed during interactions with water droplet. The system was equilibrated for 200 ps, followed by a production run of 500 ps.

It was observed that in the beginning of the MD run, the water droplet is almost spherical on the surface. As the simulation proceeded, the droplet started to wet the surface and the shape became anisotropic. The extent of wetting varied with the relative hydrophobicity of the surfaces. For example, on the graphite surface the droplet remained almost spherical while on the α-alumina surface a complete wetting of the surface was observed (Figure 2.6). Microscopic contact angles of a water droplet were computed using the approach outlined in Section 2.2.8 and compared with experimentally measured values reported in the literature (Table 2.2). The computed contact angles were in close agreement with those measured experimentally.

2.3.3 ADSORPTION AT MINERAL/CERAMICS SURFACE

2.3.3.1 Screening of Functional Groups

We have modeled interactions of different families of surfactants—namely, carboxylic/fatty acid (CA), hydroxamic acid (HXMA), iminobismethylene phosphonic acid (IMPA), and hydroxyalkylidene-1, 1-diphosphonic acid (Flotol)—with different calcium mineral surfaces. The molecular modeling method described in Section 2.2 was used to compute the interaction energies for the interaction of a single surfactant molecule with fluorite {111}, calcite {110}, fluorapatite {100} and dolomite {100} surfaces. In Table 2.3, the computed interaction energies are summarized and compared with the conventional fatty acid reagent used in the separation of these minerals. Based on the theoretically computed interaction energies, the IMPA reagent is likely to be the most selective for separation among these minerals. Indeed, it is heartening to note that the computed interaction energies correlated well with the experimental flotation response of the reagents, as observed in microflotation experiments (Pradip and Rai 2002a; Pradip et al. 2002a, 2002b).

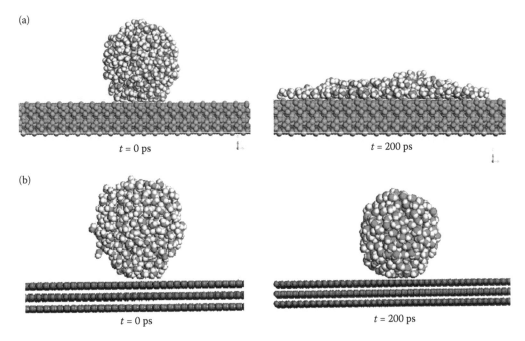

FIGURE 2.6 Spreading of water droplet on the (a) alumina {001} and (b) graphite {100} surfaces—snapshots of MD simulations.

TABLE 2.2

Comparison of Computed Contact Angle Values of a Water Droplet on Different Mineral Surfaces with Those Reported in Literature

	Contact Angle (degree)	
Mineral Surface	Our Simulations (Singh et al. 2010)	Experimental
Alumina {001}	38	25–37 (Somasundaran 2006)
Kaolinite {001}	42	46 (Wu 2001)
Fluorite {111}	48	10–55 (Zawala et al. 2007)
Graphite {100}	85	83 (Lundgren et al. 2002)

2.3.3.2 Conformation of Adsorbed Molecules

These computations were also used to screen the most favorable configuration of a molecule on a given surface. For example, the most stable conformation of a carboxylic acid (oleate) molecule on a fluorite {111} surface was found to be a bidentate conformation (Mielczarski et al. 2002). As shown in Figure 2.7, the theoretically simulated conformation, as indicated through adsorption angles, matched well with that measured experimentally, as obtained through a sophisticated *in situ* infrared external reflection spectroscopy technique (Mielczarski et al. 1998).

2.3.3.3 Optimization of Molecular Architecture

Because both functional group and molecular architecture of a surfactant molecule govern its selectivity toward a given surface, it is also important to design appropriate molecular architecture along with the most selective functional group so as to arrive at highly selective reagents. It is evident

TABLE 2.3

Computed Interaction Energies for Various Functional Groups

Group	Structure	Interaction Energy (kcal/mole)			
		Fluorite {111}	Calcite {110}	Apatite {100}	Dolomite {104}
Carboxylic acid (CA)	R—C(=O)—OH	−53	−40	−47	−39
Hydroxamic acid (HXMA)	R—C(=O)—NH OH	−67	−43	−39	−22
Iminobismethylene phosphonic acid (IMPA)	R—N(CH₂—P(=O)(OH)(OH))(CH₂—P(OH)(=O)OH)	−197	−103	−91	−167
Hydroxyalkylidene-1, 1-diphosphonic acid (Flotol)	R—C(OH)(P(=O)(OH)OH)—P(OH)(=O)OH	−200	−147	−141	−184

R, octyl chain.

from the results presented in Table 2.3 that the IMPA family of reagents was likely to be the most selective for calcium mineral separation. We further modeled the effect of molecular architecture on the interactions of the IMPA molecule with two calcium minerals, calcite ($CaCO_3$) and dolomite [$(Ca,Mg)CO_3$] (Rai et al. 2008b). For apatite and dolomite minerals, the surface planes modeled were {100} and {104}, respectively. The Ca–Ca distances and arrangement of anions (carbonates or phosphates) were found to be significantly different on both of these mineral surfaces (Figure 2.8). Because of these differences, on the apatite surface the PO_4 functional groups of IMPA molecules fit very well between triads of three calcium ions whereas on dolomite surface they interact with only one calcium atom (Figure 2.9). The computed interaction energies of some of the IMPA molecules with varying molecular architecture are summarized in Table 2.4. To validate our theoretical computations, some of these reagents were also tested for separation efficiency in the flotation of rock phosphate ore (Pradip et al. 2005a, 2005b).

2.3.3.4 SAMs

As illustrated in the Sections 2.3.3.1 to 2.3.3.3, interaction energies computed on the basis of a single surfactant molecule adsorbing on the mineral surface provides a very useful quantitative and theoretical measure for assessing the relative affinity of various mineral–surfactant combinations. However, these computations do not capture the associative interactions among the adsorbed molecules. To study the structure of the adsorbed monolayers so as to delineate the more subtle template effects of the substrate in determining the macroscopic behavior of the reagents in actual application, we also studied the packing of surfactants at the solid surface (Rai and Pradip 2003; Rai et al. 2004). MD simulations (Rai and Pradip 2003) were carried out to study the adsorbed oleate (9-octadecenoic acid) monolayer on the three calcium mineral surfaces—fluorite {111}, calcite {104}, and fluorapatite {100—using UFF potentials. The oleate molecules were placed in bidentate, unidentate, or bridged binding conformations near the surface calcium atoms (Figure 2.10).

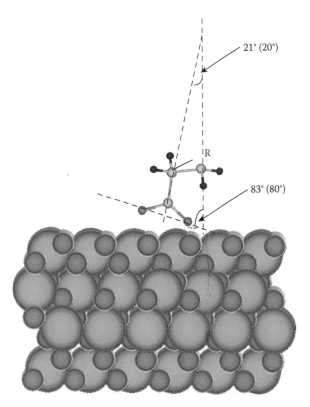

FIGURE 2.7 Conformation of adsorbed oleate molecule on fluorite {111} surface. The numbers in bracket denote experimental angles obtained through *insitu* infrared external reflection spectroscopy technique reported by (From Mielczarski, E., Mielczarski, J. A., and Cases, J. M., *Langmuir*, 14: 1739, 1998.)

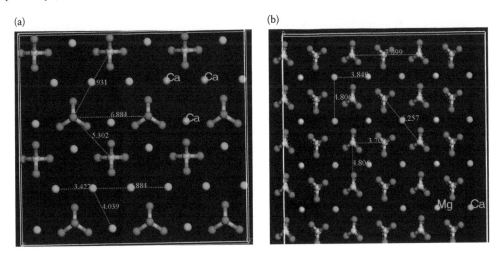

FIGURE 2.8 Top view of (a) apatite {100} and (b) dolomite {104} surfaces showing arrangements of calcium and phosphate/calcite anions.

The monolayers with different occupancy (25–100%) of all the calcium sites available on the surface (i.e., one oleate molecule adsorbed on each calcium site) were modeled. The structures of most stable oleate SAMs—that is, 67%, 100%, and 75% occupancy on fluorite {111}, calcite {104}, and fluorapatite {100} surfaces, respectively—are presented in Figure 2.11. Some of the characteristic properties of these SAMs are summarized in Table 2.5. It is evident form these results that the

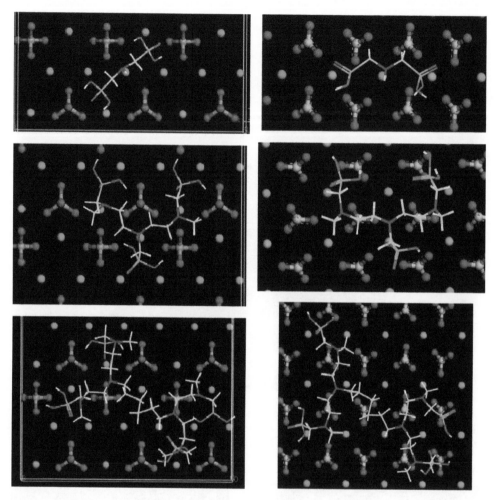

FIGURE 2.9   Top view of optimized complexes of various IMPA molecules on (a) apatite {100} and (b) dolomite {104} surfaces.

calcite {104} surface is the best template for an oleate monolayer because a 100% packing is possible on this surface. The computed area occupied per chain of oleate for this well-packed monolayer (20.2 Å²) matches well with the value calculated to be the parking area per oleate molecule (20 Å²) for a fully packed monolayer based on strictly geometric calculations.

2.3.3.5   Wettability of Adsorbed SAMs (Contact Angle)

The computational results presented in Section 2.3.3.4 suggest that an oleate molecule is likely to be more selective toward a calcite {104} surface as compared with fluorite {111} and fluorapatite {100} surfaces. However, higher packing of surfactant molecules may not necessarily render the surface most hydrophobic. Thus, to characterize the hydrophobicity or wettability of these adsorbed SAMs as quantified through contact-angle measurements, we also computed the contact angle of a water droplet placed on these SAMs (Rai and Pradip 2008). MD simulations were carried out with a water droplet placed near the terminal end of adsorbed oleate SAMs. To create a water droplet of experimental density, a 3D periodic box of water molecules was equilibrated at 300 K and a sphere of 20 Å radius was cut out of this 3D periodic box. UFF, as implemented in Materials Studio 4.1, was used to model the interactions of the water droplet with the adsorbed SAMs; SPC/E potentials (Berendsen et al. 1987) were used for water molecules. A time step of 1 fs was used. The temperature during MD simulations was controlled by scaling the velocities. A direct cutoff at 21.5 Å was

TABLE 2.4

Computed Interaction Energies of IMPA Molecules with Varying Architecture

Molecule	Interaction Energy (kcal/mole)		Energy Difference (kcal/mole)
	Apatite {100}	Dolomite {104}	
	−230	−100	−130
	−236	−103	−133
	−240	−104	−136
	−229	−98	−131
	−225	−91	−134
	−241	−104	−137
	−246	−131	−115
Where x = 2–12	−242	−116	−126
x = 2, C2P4	−246	−131	−115
x = 2, C3P4			
x = 4, C4P4			
x = 5, C5P4			
x = 6, C6P4			
x = 7, C7P4			
x = 8, C8P4			
x = 10, C10P4			
x = 12, C12P4			
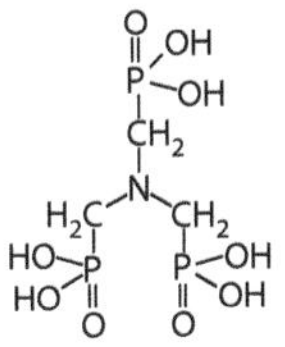 (PA1)	−92	−23	−69
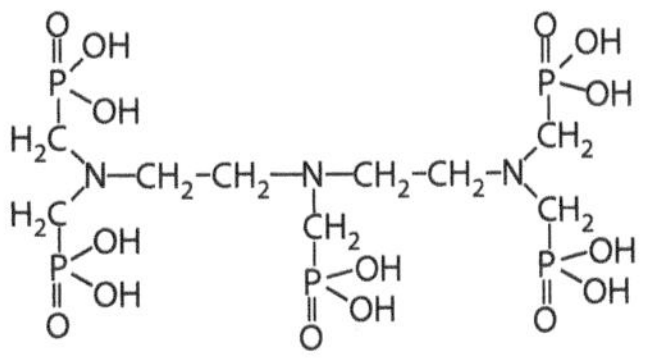 (PA2)	−160	−130	−30
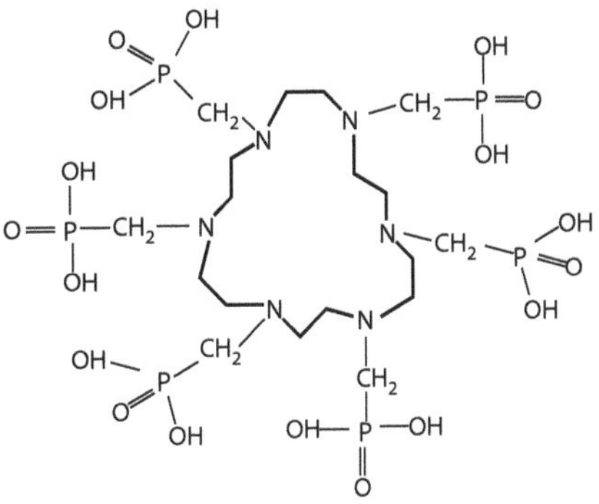	−211	−80	−131

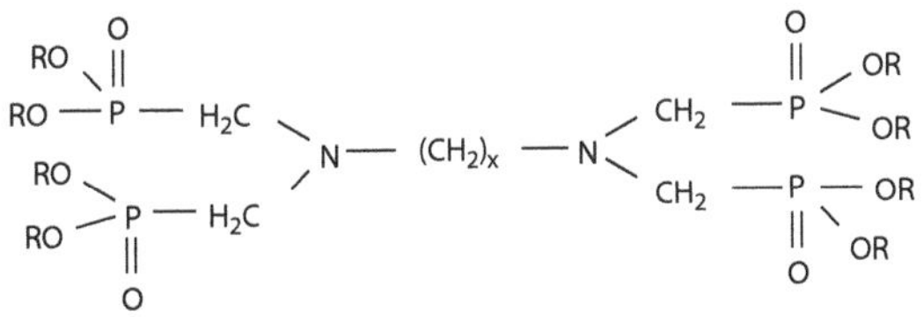

(Continued)

TABLE 2.4 (CONTINUED)

Computed Interaction Energies of IMPA Molecules with Varying Architecture

Molecule	Interaction Energy (kcal/mole)		Energy Difference (kcal/mole)
	Apatite {100}	Dolomite {104}	
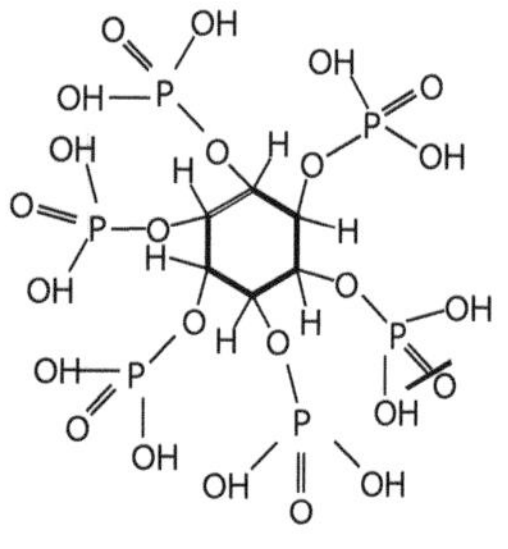	−183	−98	−85
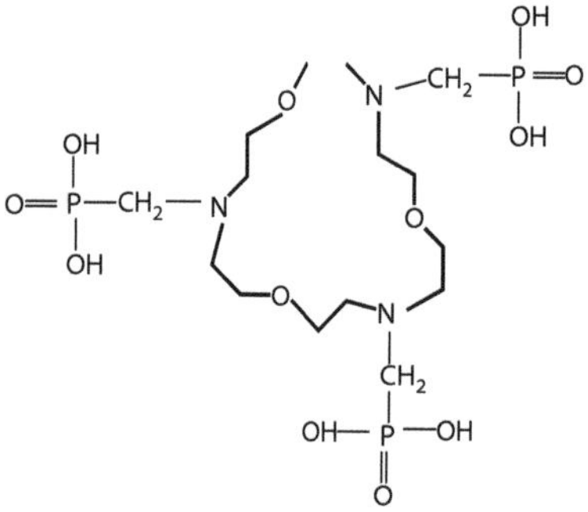	−207	−138	−69

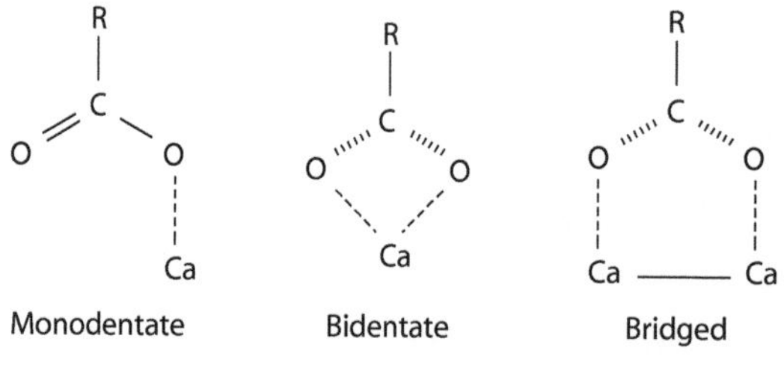

FIGURE 2.10   Schematic representation of different binding conformations of oleate with surface calcium atoms.

used to screen the nonbonded interactions. The mineral surface atoms were spatially constrained during the entire simulation. The system was allowed to equilibrate at 300 K for about 300 ps, followed by a production run of 300 ps.

In Figure 2.12, theoretically computed contact angles of water with the adsorbed oleate SAMs on fluorite {111}, calcite {104}, and fluorapatite {100} surfaces are compared with corresponding interaction energies (computed for the adsorption of a single oleate molecule) and experimental flotation results reported by Pugh and Stenius (1985). It is evident from the computed contact angle values that oleate adsorption on the fluorite {111} surface renders the surface more hydrophobic as compared with calcite {104} or fluorapatite {100} surfaces. Thus, even though calcite tends to be the best template for self-assembly, the wetting characteristics of adsorbed SAMs (as characterized through computed contact angles) predict better floatability for fluorite with oleate. Indeed, the

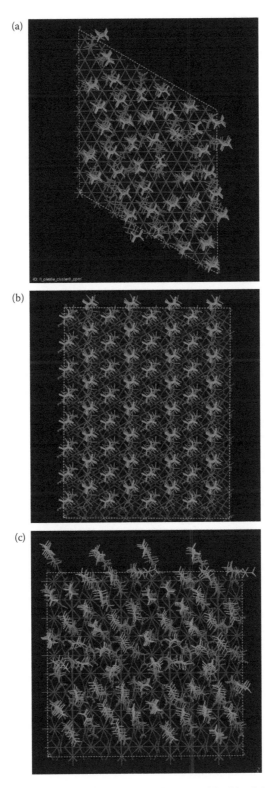

FIGURE 2.11   Top view of optimized oleate SAMs on (a) fluorite {111}, (b) calcite {104}, and (c) fluorapatite {100} surfaces.

TABLE 2.5

Computed Properties of Oleate SAMs Adsorbed on Different Calcium Mineral Surfaces

Mineral Surface	% of Calcium Sites Occupied	Parking Area per Oleate Chain (Å²)	Binding Conformation	Total Energy (kcal/mol)
Fluorite {111}	67	22.2	Bidentate	−2415.6
Calcite {104}	100	20.2	Bridged	−5107.8
Fluorapatite {100}	75	21.5	Monodentate and bridged	−2931.4

experimental flotation response as reported by Pugh and Stenius (1985) follows the same order as the predicted wettability (contact angle) by our MD simulations (Figure 2.12).

2.3.4 BULK COMPLEXATION

Complexation of surfactant molecules in bulk is important for applications like solvent extraction. Salicylaldoxime (SALO) and its derivatives (Figure 2.13) are known solvent extraction reagents for purification and concentration of leach liquor containing copper, lead, and zinc ions (Baird 1991; Ritcey 1980). These reagents are also reported to be excellent flotation collectors (Nagaraj and Somasundaran 1981; Ritcey and Ashbrook 1984). Flotation experiments carried out at our laboratory have shown that it is possible to achieve selective separation among complex sulphide ores containing copper, lead, and zinc minerals using SALO derivatives with appropriate alkyl group substitution in the main chain or the side chain (Aliaga and Somasundaran 1987; Das et al. 1995). We have employed DFT simulations to model the interaction of SALO and its derivatives with copper and zinc ions to understand the role of molecular architecture in the observed selectivity of these reagents (Rai et al. 2008a).

DFT simulations were performed using BP86 functional (Becke 1988; Perdew 1986) and DeMon codes (http://www.demon-software.com/public_html/index.html). Transition metal ions were described using Watchers basis set, and for the rest of the atoms standard 6-31G** basis sets were used. In addition to SALO, two types of SALO derivatives—alkyl group substitution (C_2H_5, C_5H_{10} and $C_{10}H_{20}$) in the side chain (CS-SALO) or main chain (CM-SALO)—were considered (Figure 2.14). The molecules were allowed to bind with Cu^{2+} and Zn^{2+} metal ions. Because these molecules dissociate in aqueous solution and form metal complexes with 1:2 and 1:1 stoichiometry, we considered three possible modes of metal–SALO complexation (Figure 2.14).

The computed interaction energies for various complexes of copper and zinc ions with SALO and its derivatives are summarized in Table 2.6. It is evident from these results that the interaction energy of a copper ion with SALO and its derivatives is always more negative than that of a zinc ion. This implies higher affinity of SALO and its derivatives toward copper as compared with zinc. As far as the differences between main chain (CM) and side chain (CS) derivatives are concerned, while a copper ion behaves very similar with both types of derivatives, a zinc ion shows higher efficiency for CM derivatives. These results are in good agreement with experimental observations (Ramesh et al. 1998).

We also computed the Infra Red (IR) frequencies for optimized molecules and their complexes. The theoretically simulated IR frequencies are compared with experimentally measured values in Tables 2.7 and 2.8. Theoretically computed IR frequencies compare well with the experimental assignments, thus validating our results.

Our DFT computations of SALO and its derivatives with copper and zinc metal ions provided us insights into the basis of selectivity in the system. Currently, we are extending these computations to model the interactions of SALO-based derivatives with sulphide mineral surfaces.

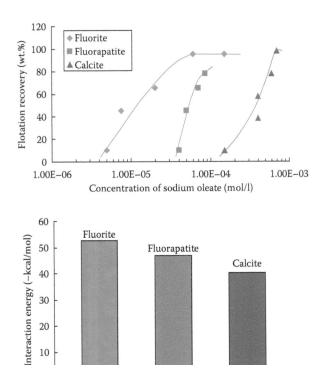

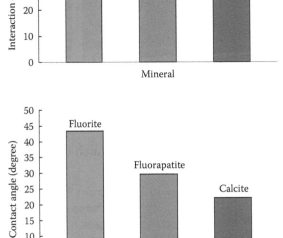

FIGURE 2.12 A comparison of computed contact angles with theoretically computed interaction energies and experimental flotation response of oleate with different calcium minerals. (Data from Pugh, R. and Stenius, P., *Int. J. Min. Process.* 15: 193, 1985.)

2.4 APPLICATIONS

2.4.1 MINERAL PROCESSING REAGENTS

For the beneficiation of multicomponent, highly disseminated, and difficult-to-treat ore deposits, novel reagents are needed. Considering the demands both with respect to the cost and their efficiency and selectivity, and coupling that with environmental constraints, the scientific challenge facing the mineral engineers today lies in developing highly selective and customized reagents for specific problems (Pradip 1994). Moreover, there is a perceptible shift now in the mining chemicals industry to design and market tailor-made performance chemicals customized for specific

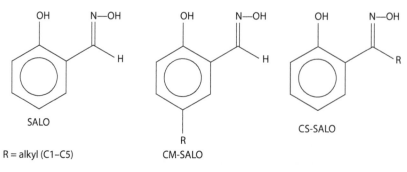

SALO

R = alkyl (C1–C5)

CM-SALO

CS-SALO

FIGURE 2.13   Molecular structure of SALO and its derivatives.

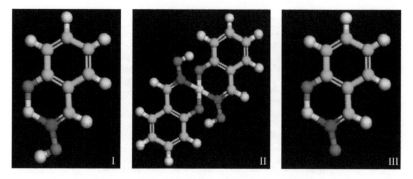

FIGURE 2.14   Types of surfactant–metal complexes modeled. Color codes: C, cyan; Cu, yellow; H, white; N, blue; O, red.

TABLE 2.6

Computed Interaction Energies (kcal/mol) of Metal–SALO Complexes

SALO–metal	Complex I	Complex II	Complex III
SALO–Cu^{2+}	−510.25	−700.92	−729.63
CM-SALO–Cu^{2+}	−517.62	−698.42	−727.99
CS-SALO–Cu^{2+}	−518.30	−699.78	−728.18
SALO–Zn^{2+}	−455.04	−644.86	−673.95
CM-SALO–Zn^{2+}	−459.89	−664.66	−694.76
CS-SALO–Zn^{2+}	−460.84	−600.92	−669.51

applications rather than offer conventional, generic, multipurpose commodity chemicals as flotation reagents (Cappuccitti 1994; Malhotra 1994; Nagaraj et al. 1999). The MM approach highlighted in Section 2.2 offers a sound, robust, and quantitative methodology to achieve these goals. We illustrate the utility of this methodology in the design and development of selective reagents for several mineral systems by taking examples from our work.

2.4.1.1   Phosphate Ores

Processing dolomitic phosphate ores (Cook and Shergold 2005) remains a challenging problem. For example, the beneficiation plant at Jhamarkotra in Rajasthan, India, is currently using a conventional double-stage flotation process (Prasad et al. 2010). In the first stage, a fatty acid collector is used to produce a combined dolomite–apatite bulk concentrate separated from associated silicate gangue. Dolomite is then separated from this bulk concentrate by flotation with fatty acids collectors and depressing apatite with phosphoric acid and sulfuric acid. The process requires the

TABLE 2.7

A Comparison of Theoretically Computed IR Frequencies from Our Simulations with Experimental Data

	SALO		CM–SALO		CS–SALO	
Assignments	Expt.	Our Simulation	Expt.	Our Simulation	Expt.	Our Simulation
O-H Stretch	3380	3376	3353	3350	3318	3383
C-H Stretch	–	–	2929	2955	2950	2956
C=N Stretch	1621	1616	1626	1613	1630	1625
C=C Stretch	1576	1562	1594	1597	1612	1600
O-H Bend I	1290	1290	1290	1275	1290	1289
C-O Aromatic	1254	1259	1259	1268	1254	1256
O-H Bend II	989	959	1006	–	1054	1033
N-O Stretch	896	885	883	863	860	–
Chelate Ring	–	–	563	524	563	535

Source: From Rai et al. 2008a; data from Ramesh, V., Umasundari, P., and Das, K. K., *Spectrochim. Acta, Part A,* 54: 285, 1998.

TABLE 2.8

Simulated and Experimental IR Frequencies for Metal–SALO Complexes

	Cu^{+2}				Zn^{+2}			
		Our Simulation				Our Simulation		
Assignment	Expt.	I	II	III	Expt.	I	II	III
C=N Stretch	1650	1534	1477	1635	1610	1499	1502	1628
C–O Stretch	1310	1326	1265	1291	1300	1253	1240	1290
N–O Stretch	920	999	–	1016	920	1022	–	1013
Cu/Zn–O Stretch	485	487	496	490	450	650	657	448
Cu/Zn–N Stretch	310	311	327	301	300	333	335	312

Source: Simulation data from Rai et al. 2008a; experimental data from Ramesh, V., Umasundari, P., and Das, K. K., *Spectrochim. Acta,* Part A, 54: 285, 1998.

addition of very large dosages of reagents to depress apatite and to maintain highly acidic pH during flotation. Furthermore, several cleaning and scavenging stages are needed to obtain the product of desired quality.

Our molecular-modeling computations presented in Section 2.3.3 revealed that IMPA-type reagents are likely to be more selective for this system. Accordingly, we conducted flotation separation experiments on a typical rock phosphate ore sample (Jhamarkotra, Rajasthan, India) that contained apatite, dolomite, and quartz as the primary constituents, with some of the promising IMPA molecules. The details of experimental test work are reported elsewhere (Pradip et al. 2005a, 2005b; Rai et al. 2008b). In the Figure 2.15, the flotation results (recovery-grade plots) obtained for IMPA reagents are compared with the conventional phosphoric acid (H_3PO_4) reagent currently being used in industrial practice. IMPA reagents exhibit better separation efficiency as compared with existing plant practice. Usage of IMPA regents leads to considerable reduction in MgO contents (from 13.3% to 1.5%) in a single step of rougher flotation, suggesting that IMPA reagents are highly selective. In addition, as compared with conventional

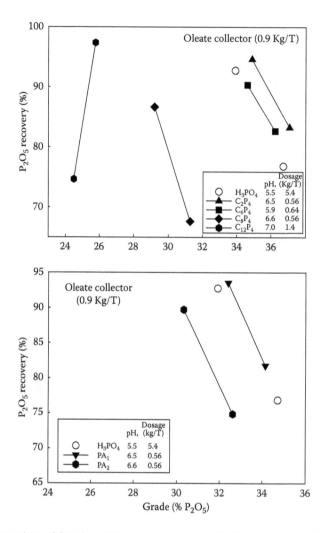

FIGURE 2.15   Comparison of flotation efficiency obtained with IMPA depressants with conventional phosphoric acid depressant in the flotation of dolomitic rock phosphate.

H_3PO_4 depressant (pH 5.5 and 6Kg/T), these reagents work at a higher pH (7.0) and lower dosage (0.6 Kg/T), leading to considerable savings in the cost of reagents.

2.4.1.2   Iron Ores

India is endowed with rich resources of iron ores (approximately 24 billion tons) (Indian Minerals Year Book 2005) in the form of hematite and magnetite ore deposits. Indian hematite ores are typically rich in iron (>60% iron) but contain unusually high alumina (4–7%). The blast furnace productivity is significantly affected due to the presence of high alumina in the feed (Kumar and Mukherjee 1994). The current practice of iron ore processing in India leads to the generation of 6 to 10% of slimes, which are currently being discarded as waste. It amounts to an approximate loss of 15 to 20 million tons per year of iron values. In addition, these slimes, stored in massive water ponds, pose an enormous environmental hazard. Thus, finding suitable means for using and safely disposing of slimes is urgent.

    The key to solving the problem of iron ore slimes lies in developing selective reagents (e.g., flocculants, dispersants, and flotation collectors) that will separate iron oxide (hematite) from other impurities like kaolinite, gibbsite, and silica. The MM-based rational design paradigm appears to be promising for the design of highly selective, tailor-made reagents for this purpose.

We have employed first-principles DFT commutations to model the interactions of starch reagent with two main constituent minerals of iron ore, hematite (α-Fe_2O_3) and kaolinite [$Al_2Si_2O_5(OH)_4$] (Jain et al. 2011). DFT simulations were performed using PWscf code contained in the Quantum Espresso distribution (http://www.quantum-espresso.org/) running on an EKA supercomputer at Computational Research Laboratories, Pune, India. The generalized gradient approximation (GGA) of Perdew, Burke, and Ernzerhof (Perdew et al. 1996) was used for the exchange-correlation functional. Vanderbilt ultrasoft pseudo-potentials (Vandebilt 1990) were used for describing the ionic cores. The Kohn–Sham wave functions were expanded using a plane wave basis set upto a kinetic energy cutoff of 30 Ry and charge density with a cutoff of 180 Ry. Structural relaxations were performed until the total force on each atom was less than 0.01 eV/bohr. The bulk hematite and kaolinite structures were fully optimized with Brillouin zone integrations sampled on Monkhorst pack grids of $3 \times 3 \times 2$ and $5 \times 3 \times 4$ k-points, respectively and surface slabs were created at hematite {0001} and kaolinite {001} planes. The adsorption studies were performed by docking the starch molecule at various sites on optimized hematite {0001} and kaolinite {001} surfaces. Several different conformations were modeled. Interaction energies were computed using the procedure outlined in Section 2.2.6.

The computed interaction energies are summarized and compared with experimental flocculation results (Pradip et al. 1993) in Figure 2.16. Our calculations successfully captured the selectivity of starch toward hematite, arising due to similar Fe–Fe and O–O (starch) spatial distances

Mineral surface	Complex	Interaction energy (kcal/mol)
Hematite {0001}	I	−43
	II	−39
	III	−60
	IV	−74
Kaolinite {001}	$Al_{(O)}$ terminated	−3
	Si terminated	−12

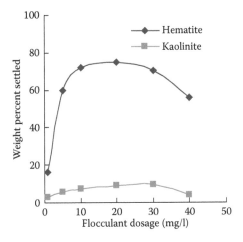

FIGURE 2.16 A comparison of computed interaction energies of starch on different mineral surfaces with experimental flocculation results on separation of hematite from its synthetic mixture with kaolinite using a natural maize starch flocculants (pH 10.5) in presence of sodium silicate dispersant. (Data from Pradip, Ravishankar, S. A., Sankar, T. A. P., and Khosla, N. K., *Proceedings of XVIII International Mineral Processing Congress*, Australasian Institute of Mining and Metallurgy, Parkville, 1289, Carlton South, Australia, 1993.)

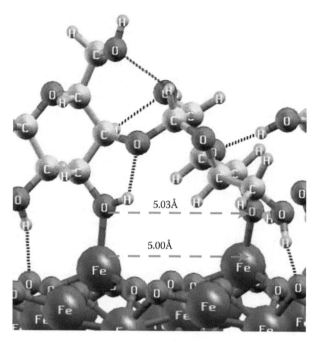

FIGURE 2.17   Optimized complex of starch on hematite {0001} surface showing underlying templating effect.

(Figure 2.17). We have applied this methodology further for the design and selection of reagents for separation of hematite from other aluminium-bearing minerals (e.g., gibbsite and alumina). Tests with some of the promising reagents for the separation of iron minerals from aluminium minerals is currently underway at our laboratory.

2.4.2   DISPERSANTS

Paint is a colloidal system consisting of pigments (inorganic or organic) and fillers dispersed in an aqueous or nonaqueous continuous phase. Titanium dioxide (rutile or anatase) is the most important pigment used in the paint industry. A critical step in the paint-making process is pigment dispersion, in which the pigment particles undergo wetting, followed by colloidal stabilization with the help of surfactants and additives in a suitable mechanical device. The extent of dispersion and rheology of paints determine the final optical properties of the paint. We have employed MM-based methodology for the selection and design of additives for nonaqueous titanium dioxide–based paints (Pradip and Rai 2002a). The experimentally observed dispersion, rheological properties, and stability of the selected paint system were correlated with the calculated interaction energy obtained from MM methodology. For example, as presented in Figure 2.18, the effect of molecular architecture of long chain fatty acid dispersants on the dispersion properties of titania in nonaqueous media as characterized by flocculation factor (Doroszkowski 1978) correlates very well with the computed interactions energies. A number of dispersants with varying molecular structure and architecture were screened for their efficacy as a dispersant. The methodology was thus applied to arrive at an optimized dispersant for the paint system under study for subsequent plant trials (Pradip et al. 1997).

2.4.3   CORROSION INHIBITORS

The development of a methodology to design efficient corrosion inhibitors and predict their performance has been an area of active research (Duda et al. 2005; Elliot and Cook 2008). Black et al.

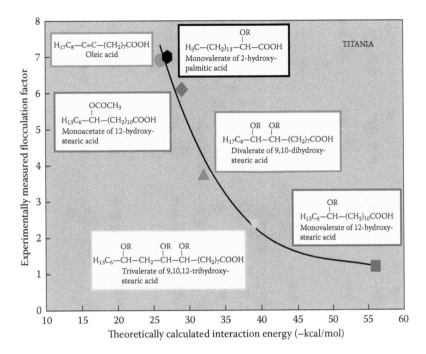

FIGURE 2.18 Correlation of computed interaction energies for fatty acid dispersants with their experimental flocculation factor. Flocculation factor is the slope of the plot of log (viscosity) against the reciprocal square root of shear rate. (Data from Doroszkowski, A. and Lambourne, R., *Faraday Discuss. Chem. Soc.,* 65: 252, 1978.)

(1991) and Davey et al. (1991) have reported extensively on the possibilities of designing highly efficient corrosion inhibitors belonging to diphosphonate family of surfactants based on the molecular recognition between the structure of the molecules and the distribution of surface binding sites on the surface of barite ($BaSO_4$) crystals (Figure 2.19). We modeled the interaction of phosphonic acids with varying N–N link length (C_0 to C_{12}) on barite {011} surface (Pradip and Rai 2002a). The computed interaction energies are compared with the experimentally observed values of the degree of corrosion inhibition as reported by Davey et al. (1991) in Figure 2.20. The theoretical predictions based on MM computations are consistent with the observed trends.

2.4.4 Sensors

2.4.4.1 Gold–Thiol SAMs

SAMs of alkanethiols on gold are key elements for building many systems and devices with applications in sensors, biomolecular chips, molecular electronics, electrochromic devices, and resists for soft lithography. Despite the considerable progress made in the knowledge of these fascinating 2D molecular systems, there are still gaps of knowledge in several areas, such as the type of self-assembly on planar and irregular surfaces (Vericat et al. 2008) and the chemical reactivity and thermal stability in ambient and aqueous solutions, which need to be addressed so as to understand and control their physical and chemical properties at the molecular level. Vericat et al. (2010) have articulated some of these topics in a recently published critical review. We have employed MD simulations to study the influence of the Au–S–C bond on the structural properties and stability of alkylthiol SAMs (Rai et al. 2004). We considered two situations: one in which the Au–S–C bond is fixed (FF I) and the other in which the bond is flexible (FF II). FF I resulted in greater tilt angles and smaller film thickness compared with FF II (Figure 2.21). Both force fields predicted that the tilt

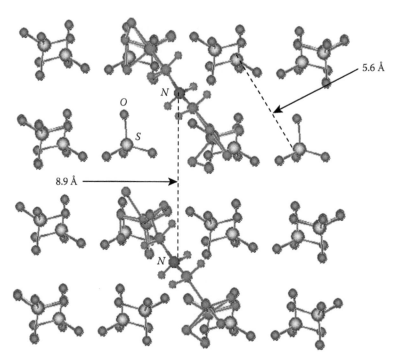

FIGURE 2.19 Diphosphonic acid molecule on barite {011} surface showing the structural compatibility. (From Pradip, Rai, B., Rao, T. K., Krishnamurthy, S., Vetrivel, R., Mielczarski, J. A., et al., *Langmuir,* 18, 932, 2002a. With permission.)

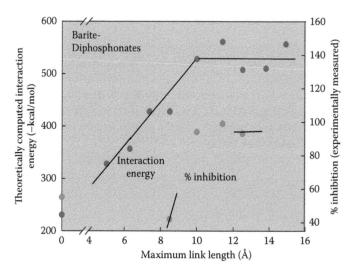

FIGURE 2.20 Comparison of computed interaction energy of diphosphonates on barite {011} surface with their experimental inhibitory activity as reported in literature (Data from Davey, R. J., et al. *Nature* 353: 549, 1991; Pradip, et al., *Langmuir,* 18, 932, 2002a. With permission.)

angles do not follow a monotonic decrease with temperature but show minima around 200 K. MD simulations carried out for different chain length (C_8 to C_{15}) SAMs at 300 K revealed that FF II led to higher film thickness than FF I (Figure 2.22).

To study the stability of SAMs in the presence of water, we also applied MD simulations to investigate their wetting characteristics (Sathish et al. 2007). The computed contact angles, which

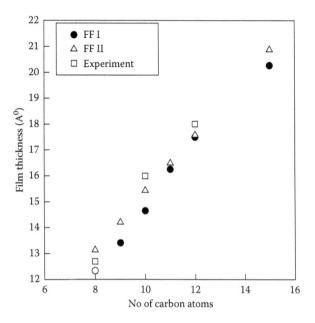

FIGURE 2.21 Comparison of the SMAs thickness as a function of the chain length with the experimental values. All simulations are reported at 300 K. (Data from Porter, M. D., et al. *J. Am. Chem. Soc.*, 109: 3559, 1987. from Rai, B., et al. *Langmuir*, 20: 3138, 2004. With permission.)

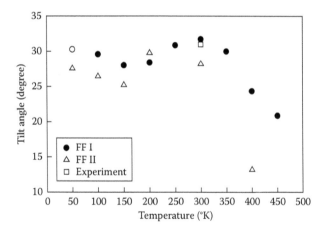

FIGURE 2.22 Variation of the tilt angle for C_{15}-SAMs with temperature. (Data from Camillone, N., III, Chidsey, C. E. D., Fenter, P., Eisenberger, Li, J., Liang, K. S., and Liu, G. Y., Scoles, G., *J. Chem. Phys.*, 99, 744, 1993; Fenter, P., Eisenberger, P., and Liang, K. S., *Phys. Rev. Lett.*, 70, 2447, 1993; from Rai, B., Sathish, P., Malhotra, C. P., Pradip, and Ayappa, K. G., *Langmuir*, 20, 3138, 2004. With permission.)

were obtained by using the methodology described in Section 2.2.8, for the C15-SAM and C8-SAM were 121° and 63°, respectively indicating more wetting for shorter chain SAM than long chain SAM (Figure 2.23). The computed contact angle values were in line with those reported in the literature. For example, experimentally measured contact angles for C_{12}-SAM, 115° for advancing drops (Pertsin and Grunze 2000) and 104° for receding drops (Pertsin et al. 2002)—compared well with the theoretically computed value of 135°.

(a)   (b)

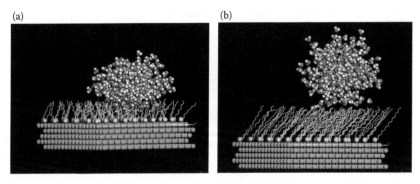

FIGURE 2.23 Snapshots of wetting of alkylthiol SAMs adsorbed on Au (111) surface with water droplet (a) C_8-SAM and (b) C_{15}-SAM.

TABLE 2.9

Computed Interaction Energies of BSA Molecule with Different Surfaces

Surface	Interaction Energy (kcal/mol)
Hydrophilic	−5286.0
Hydrophobic	−360.5

2.4.4.2 Proteins in Biosensors

Most of the immunochemical biosensor techniques involve the immobilization of proteins on a surface (Keiki-Pua et al. 1999; Lu et al. 1996; Tan et al. 2007). However, the nonspecific binding of other proteins or other biomolecules to unoccupied sites during subsequent processing steps is detrimental to the specificity and sensitivity of the system. This nonspecific binding can be minimized by saturating the unoccupied sites with a blocking reagent that does not participate in the immunochemical reactions of the assay (S&S 1997). Bovine serum albumin (BSA) is used as a traditional blocking agent to prevent nonspecific protein–surface binding on both hydrophobic and hydrophilic surfaces (Brorson 1997; Gibbs 2001).

Jeyachandran et al. (2009) studied the adsorption of BSA on hydrophilic and hydrophobic surfaces through attenuated total reflection–Fourier transform infrared spectroscopy (ATR-FTIR) and MM. Experimental Infra Red (IR) studies revealed that, even though the BSA molecules adsorbed to both surfaces with an R-helical structure, their molecular conformation and orientation differed on these surfaces. The conformation of adsorbed BSA molecules depended on the concentration of the BSA solution and the incubation time. It was also observed that the BSA surface coverage was only 53% of a complete monolayer on the hydrophobic surface, whereas, on the hydrophilic surface, almost 100% coverage was achieved. The adsorption and interaction of BSA molecules were found to be much stronger on the hydrophilic surface than on the hydrophobic surface. MM computations further substantiated these observations. The computed interaction energies (Table 2.9) indicated strong interactions on hydrophilic surface as compared with hydrophobic surface. MM studies also showed that BSA molecules interacted with hydrophobic (PS) surfaces through CH_3 groups, whereas on hydrophilic surfaces, polar -COOH groups participated in the interaction. Our results thus highlighted the utility of a combined experimental and MM approach in the design and development of immunochemical biosensors.

2.4.5 Drug Formulations

Hydrogels are cross-linked polymeric networks that have the ability to hold water within the cross-linked polymeric chains. They have been used extensively in biomedical applications such as drug

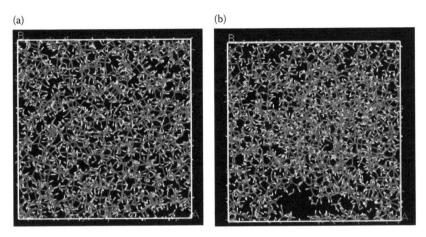

FIGURE 2.24 Molecular models of crosslinked PVA gel with corsslinking ratio (a) 0.01 and (b) 0.1.

TABLE 2.10

A Comparison of Theoretically Computed Properties of PVA Hydrogels with Those Reported in Literature

Cross Linking Ratio	Average Mesh Size (Å)	Diffusion Coefficient (Water) (cm^2/sec x 10^{-7})	
		Our Simulations	Experimental*
0.01	112	4.93	3.84
0.1	10	0.35	0.20

* Data from Brazel, C. S. and Peppas, N. A., *Polymer,* 40: 3383, 1999.

delivery, wound management, and tissue engineering (Pal et al. 2009; Park et al. 1993). The functionality of the hydrogels depends on both its structural and dynamic properties. The molecular structure of the hydrogels is usually characterized by cross-linking density. Cross-linking density of hydrogels is a function of several parameters such as type and amount of cross-linking agent, temperature, and method of cross-linking. The cross-linking density affects gel properties such as swelling, diffusion and release of molecules, stability, and density. Experimental estimation of cross-linking density is expensive and time consuming. To design gels for a given application such as drug-release, estimation of cross-linking density is important. Molecular simulations offer an alternative to the experiments. Khandelwal et al. (2009) have employed MD simulations to model cross-linking of polyvinyl alcohol (PVA) chains in the presence of glutardehyde as the cross-linking agent. The PVA gels with varying cross-linking density were formed using reactive MD (formation of bonds between polymer chains) simulations (Figure 2.24). The swelling (water uptake) and diffusion properties of gels were computed and compared with those reported in literature (Table 2.10). The methodology thus developed is currently being used by us for the design and development of hyrdorgel-based drug delivery systems.

2.5 CONCLUDING REMARKS

The power and utility of currently available MM tools in the design, screening of different molecular architectures, and development of tailor-made performance chemicals is demonstrated through our work. The proposed methodology is very generic in nature and provides a scientifically robust framework for the design and selection of appropriate reagent, solvent, and formulation for a given industrial application. We believe that with steady advancements in the theory and practice of MM, a rational design framework, as presented here, will facilitate addressing more complex design problems of industrial relevance.

2.6 ACKNOWLEDGMENTS

We thank all our past and present colleagues and collaborators for their direct or indirect contributions to the work presented here. We also thank Mr. K. Ananth Krishnan, corporate technology officer, Tata Consultancy Services Ltd., for his encouragement and support during the compilation of this chapter. Financial support from Indo-French Centre for Promotion of Advanced Research (IFCPAR) is gratefully acknowledged for some of the work incorporated here.

REFERENCES

Akim, L. G., Bailey, G. W., and Shevchenko, S. M. 1998. In *A computational chemistry approach to study the interactions of humic substances with mineral surfaces. Humic substances: Structures, properties and uses.* Eds. G. Davies and E. A. Ghabbour. The Royal Society of Chemistry, Cambridge.

Aliaga, W., and Somasundaran, P. 1987. Molecular-orbital modeling and UV spectroscopic investigation of adsorption of oxime surfactants. *Langmuir.* 3: 1103.

Allen, M. P., and Tildesley, D. J. 1987. *Computer Simulations of Liquids.* Oxford, Clarendon.

Almora-barrios, N., and De Leeuw, N. H. 2010. A density functional theory study of the interaction of collagen peptides with hydroxyapatite surfaces. *Langmuir.* 26: 14535.

Almora-barrios, N., Austen, K. F., and De Leeuw, N. H. 2009. Density functional theory study of the binding of glycine, proline, and hydroxyproline to the hydroxyapatite (0001) and (011⁻0) surfaces. *Langmuir.* 25: 5018.

Arya, G., and Panagiotopoulos, A. Z. 2005. Molecular modeling of shear-induced alignment of cylindrical micelles. *Comput. Phys. Commun.* 169: 262.

Baird, M. H. I. 1991. Solvent extraction—The challenges of a "mature" technology. *Can. J. Chem. Eng.* 69: 1287.

Becke, A. D. 1988. Density-functional exchange-energy approximation with correct asymptotic behaviour. *Phys. Rev. A* 38: 3098.

Bedrov, D., Smith, G. D., Freed, K. F., and Dudowicz, J. 2002. A comparison of self-assembly in lattice and off-lattice model amphiphile solutions. *J. Chem. Phys.* 116: 4765.

Berendsen, H. J. C., Postma, J. P. M., van Gunsteren, W. F., DiNola, A., and Haak, J. R. 1984. Molecular dynamics with coupling to an external bath. *J. Chem. Phys.* 81: 3684.

Berendsen, H. J. C., Grigera, J. R., and Straatsma, T. P. 1987. The missing term in effective pair potentials. *J. Phys. Chem.* 91: 6269.

Bergstrom, L., Shinozaki, K., Tomiyama, H., and Mizutani, N. 1997. Colloidal processing of a very fine $BaTiO_3$ powder—Effect of particle interactions on the suspension properties, consolidation, and sintering behavior. *J. Am. Ceram. Soc.* 80: 291.

Binggeli, M., and Mate, C. M. 1994. Influence of capillary condensation of water on nanotribology studied by force microscopy. *Appl. Phys. Lett.* 65: 415.

Black, S. N., Bromley, L. A., Cottier, D., Davey, R. J., Dobbs, B., and Rout, J. E. 1991. Interactions at the organic/inorganic interface: binding motifs for phosphonates at the surface of barite crystals. *J. Chem. Soc. Faraday Trans.* 87: 3409.

Bourov, G. K., and Bhattacharya, A. 2003. The role of geometric constraints in amphiphilic self-assembly: A Brownian dynamics study. *J. Chem. Phys.* 119: 9219.

Brazel, C. S., and Peppas, N. A. 1999. Mechanisms of solute and drug transport in relaxing, swellable, hydrophilic glassy polymers. *Polymer.* 40: 3383.

Bromely, L. A., Cottier, D., Davey, R. J., Dobbs, B., Smith, S., and Heywood, B. R. 1993. Interactions at the organic/inorganic interface: Molecular design of crystallization inhibitors for barite. *Langmuir.* 9: 3594.

Bromely, L. A., Buckley, A. M., Chlad, M., Davey, R. J., Drewe, S., and Filan, G. T. 1994. Interactions at the inorganic/organic interface: Discriminatory binding of hydroxybenzenes by lepidocrocite surfaces. *J. Colloid Interface Sci.* 164: 498.

Brorson, S.-H. 1997. Bovine serum albumin (BSA) as a reagent against non-specific immunogold labeling on LR-White and epoxy resin. *Micron*. 28: 189.

Camillone, N., III, Chidsey, C. E. D., Fenter, P., Eisenberger, Li, J., Liang, K. S., and Liu, G. Y., Scoles, G. et al. 1993. Structural defects in self-assembled organic monolayers via combined atomic beam and x-ray diffraction. *J. Chem. Phys.* 99: 744.

Cappuccitti, F. R. 1994. Current trends in the marketing of sulphide mineral collectors. In *Reagents for Better Metallurgy*. Ed. P. S. Mulukutla, SME-AIME, Littleton, CO [Chap. 8, 67].

Chechik, V., and Stirling, C. J. M. 1999. Gold-thiol self-assembled monolayers. In *The Chemistry of Organic Derivatives of Gold And Silver*. Eds. S. Patai, and Z. Rappoport. Wiley, Hoboken, NJ, 561.

Claire, P. de Sainte, Hass, K. C., Schneider, W. F., and Hase, W. L. 1997. Simulation of hydrocarbon adsorption and subsequent water penetration on an aluminum oxide surface. *J. Chem. Phys.* 106: 7331.

Cook, P. J., and Shergold, J. H. 2005. *Phosphate Deposits of the World: Proterozoic and Cambrian Phosphorites*, Vol. 1. Cambridge University Press, Cambridge.

Coveney, P. V., and Humphries, W. 1996. Molecular modelling of the mechanism of phosphonate retarders on hydrating cements. *J. Chem. Soc. Faraday Trans.* 92: 831.

Dana, E. S., and Ford, W. F. 1949. *Textbook of Mineralogy, with an Extended Treatise on Crystallography and Physical Mineralogy*, 4th ed. Wiley Eastern, New Delhi.

Das, K. K., and Suresh, B. P. 1995. Role of molecular architecture and chain length in the flotation separation of oxidized copper–lead–zinc minerals using salicylaldoxime derivatives. In *Proceedings of XIX International Mineral Processing Congress*, San Francisco, SME-AIME, 245.

Davey, R. J., Black, S. N., Bromley, L. A., Cottier, D., Dobbs, B., and Rout, J. E. 1991. Molecular design based on recognition at inorganic surfaces. *Nature*. 353: 549.

de Leeuw, N. H., and Rabone, J. A. L. 2007. Molecular dynamics simulations of the interaction of citric acid with the hydroxyapatite (0001) and (011$\cdot$0) surfaces in an aqueous environment. *CrystEngComm*. 9: 1178.

Deer, W. A., Howie, R. A., and Zussman, J. 1992. *An Introduction to the Rock Forming Minerals*. Longman. Harlow, UK.

Doroszkowski, A., and Lambourne, R. 1978. Effect of molecular architecture of long chain fatty acids on the dispersion properties of titanium dioxide in non-aqueous liquids. *Faraday Discuss. Chem. Soc.* 65: 252.

Du, H., and Miller, J. D. 2007. A molecular dynamics simulation study of water structure and adsorption states at talc surfaces. *Int. J. Miner. Process.* 84: 172.

Duda, Y., Govea-Rueda, R., Galicia, M., Beltrán, H. I., and Zamudio-Rivera, L. S. 2005. Corrosion inhibitors: Design, performance, and computer simulations. *J. Phys. Chem. B.* 109: 22674.

Elliott, J., and Cook, R. 2008. Design of novel corrosion inhibiting additives. *Corrosion*, NACE International, New Orleans.

Fa, K., Nguyen, A. V., and Miller, J. D. 2006. Interaction of calcium dioleate collector colloids with calcite and fluorite surfaces as revealed by AFM force measurements and molecular dynamics simulation. *Int. J. Miner. Process.* 81: 166.

Fan, C. F., and Cagin, T. 1995. Wetting of crystalline polymer surfaces—A molecular dynamics simulation. *Chem. Phys.* 103: 9053.

Fenter, P., Eisenberger, P., and Liang, K. S. 1993. Chain-length dependence of the structures and phases of $CH_3(CH_2)n$-1 SH self-assembled on Au(111). *Phys. Rev. Lett.* 70: 2447.

Filgueiras, M. R. T., Mkhonto, D., and de Leeuw, N. H. 2006. Computer simulations of the adsorption of citric acid at hydroxyapatite surfaces. *J. Cryst. Growth* 294: 60.

Filho, L. S., Seidl, P. R., Correia, J. C. G., and Cerqueira, L. C. K. 2000. Molecular modelling of reagents for flotation processes. *Miner. Eng.* 13: 1495.

Fodi, B., and Hentschke, R. 2000. Simulated phase behavior of model surfactant solutions, *Langmuir*. 16: 1626.

Frasconi, M., Mazzei, F., and Ferri, T., 2010. Protein immobilization at gold–thiol surfaces and potential for biosensing. *Anal. Bioanal. Chem.* 398: 1545.

Fuerstenau, D. W. 1957. Correlation of contact angles, adsorption density, zeta potentials, and flotation rate. *Trans. Am. Inst. Min. Metall. Pet. Eng. Soc. Min. Eng. AIME*. 208: 1365.

Fuerstenau, D. W. 1970. Interfacial processes in mineral/water systems. *Pure Appl. Chem.* 24: 135.

Fuerstenau, D. W. 2000. *A Century of Developments in the Chemistry of Flotation Process. Froth Flotation—A Century of Innovation*. Eds. M. C. Fuerstenau, G. J. Jameson, and R. H. Yoon. SME, Littleton, CO, 3.

Fuerstenau, D. W., and Herrera-Urbina, R. 1989. In *Mineral Separation by Froth Flotation. Surfactant Based Separation Processes*, Eds. J. F. Scamehorn and J. H. Harwell. Marcel Dekker, New York, 259.

Fuerstenau, D. W., and Pradip, 2005. Zeta potentials in the flotation of oxide and silicate minerals. *Adv. Colloid Interface Sci.* 114: 9.

Gibbs, J. 2001. Effective Blocking Procedures: ELISA. *Technical Bulletin—No. 3,* Corning Incorporation Life Sciences, Lowell, Massachusetts.

Gupta, A., Chauhan, A., and Kopelevich, D. I. 2008. Molecular modelling of surfactant covered oil–water interfaces: Dynamics, microstructure, and barrier for mass transport. *J. Chem. Phys.* 128: 234709.

Heinz, H., Vaia, R. A., Krishnamoorti, R., and Farmer, B. L. 2007. Self-assembly of alkylammonium chains on montmorillonite: Effect of chain length, head group structure, and cation exchange capacity. *Chem. Mater.* 19: 59.

Henao, L., and Mazeau, K. 2008. The molecular basis of the adsorption of bacterial exopolysaccharides on montmorillonite mineral surface. *Mol. Simul.* 34: 1029.

Hermansson, K., Lie, G. C., and Clementi, E. 1988. On velocity scaling in molecular dynamics simulations. *J. Comp. Chem.* 9: 200.

Hirva, P., and Tikka, H. 2002. *ab initio* study on the interaction of anionic collectors with calcite and dolomite surfaces. *Langmuir* 18: 5002.

Hu, J., Zhang, X., and Wang, Z. 2010. A review on progress in QSPR studies for surfactants. *Int. J. Mol. Sci.* 11: 1020.

Iacovella, C. R., Horsch, M. A., Zhang, Z., and Glotzer, S. C. 2005. Phase diagrams of self-assembled mono-tethered nanospheres from molecular simulation and comparison to surfactants. *Langmuir.* 21: 9488.

Indian Minerals Year Book 2005. 2006 Indian Bureau of Mines Publication.

Jacobs, J. D., Koerner, H., Heinz, H., Farmer, B. L., Mirau, P. A., Garrett, P. H., et al. 2006. Dynamics of alkyl ammonium intercalants within organically modified montmorillonite: dielectric relaxation and ionic conductivity. *J. Phys. Chem. B.* 110: 20143.

Jain, V., Singh, S. S., Pradip, and Waghmare, U. V. 2011. A first-principles study of the adsorption of starch on hematite and kaolinite mineral surfaces. Unpublished results.

Jeyachandran, Y. L., Mielczarski, E., Rai, B., and Mielczarski, J. A. 2009. Quantitative and qualitative evaluation of adsorption/desorption of bovine serum albumin on hydrophilic and hydrophobic surfaces. *Langmuir.* 25: 11614.

Karasawa, N., and Goddard, W. A. III. 1989. Acceleration of convergence for lattice sums. *J. Phys. Chem.* 93: 7320.

Keiki-Pua, S. D., Greiner, D. P., and Sailor, M. J. 1999. A porous silicon optical biosensor: detection of reversible binding of IgG to a protein A-modified surface. *J. Am. Chem. Soc.* 121: 7925.

Khandelwal, R., Singh, S. S., and Rai, B. 2009. Molecular dynamics simulations of polyvinyl alcohol hydrogels—Effect of cross-linking density on diffusion properties. In *Proceedings of Asian Particle Technology Symposium (APT 2009),* New Delhi, 171.

Kirby, G. H., Harris, D. J., Li, Q., and Lewis, J. A. 2004. Poly(acrylic acid)–Poly(ethylene oxide) comb polymer effects on BaTiO$_3$ nanoparticle suspension stability. *J. Am. Ceram. Soc.* 87: 181.

Koster, A. M., Calaminici, P., Casida, M. F., Flores-oreno, R., Geudtner, G., Goursot, et al. 2006. *deMon2k, demon Developers,* http://www.demon-software.com/public_html/index.html

Kumar, A., and Mukherjee, T. 1994. Role of raw materials and technology in the performance of blast furnaces, *Tata Search*: 1.

Kundu, T. K., Hanumantha Rao, K., and Parker, S. C. 2003. Atomistic simulation of the surface structure of wollastonite and adsorption phenomena relevant to flotation. *Int. J. Miner. Process.* 72: 111.

Larson, R. G. 1994. Molecular simulation of ordered amphiphilic phases. *Chem. Eng. Sci.* 49: 2833.

Larson, R. G., 1996. Monte Carlo simulations of the phase behavior of surfactant solutions. *J. Phys. II* 6: 1441.

Lisal M., Hall, C. K., Gubbins, K. E., and Panagiotopoulos, A. Z. 2003. Formation of Spherical Micelles in a supercritical solvent: Lattice Monte Carlo simulation and multicomponent solution model. *Mol. Simul.* 29: 139.

Liu, G., Zhong, H., Dai, T., and Xia, L. 2008. Investigation of the effect of N-substituent's on performance of thionocarbamates as selective collectors for copper sulfides by *ab initio* calculations. *Miner. Eng.* 21: 1050.

Lu, B., Smyth, M. R., and O'Kennedy, R. 1996. Oriented immobilization of antibodies and its applications in immunoassays and immunosensors. *Analyst.* 121: 29R.

Lu, G., Zhang, X., Shao, C., and Yang, H. 2009. Molecular dynamics simulation of adsorption of an oil–water–surfactant mixture on calcite surface. *Pet. Sci.* 6: 76.

Lundgren, M., Allan, N. A., Cosgrove, T., and George, N. 2002. Wetting of water and water/ethanol droplets on a non-polar surface: a molecular dynamics study. *Langmuir.* 18: 10462.

Malhotra, D. 1994. Reagents in the mining industry: Commodities or specialty chemicals. In *Reagents for Better Metallurgy.* Ed. P. S. Mulukutla. SME-AIME, Littleton, CO, [Chap. 9, 75].

Mamontov, E., Vicek, L., Wesolowski, D. J., Cummings, P. T., Wang, W., Anovitz, L. M., et al. 2007. Dynamics and structure of hydration water on rutile and cassiterite nanopowders studied by quasielastic neutron scattering and molecular dynamics simulations. *J. Phys. Chem. C.* 111: 4328.

Mayo, S. L., Olafson, B. D., and Goddard, W. A. III. 1990. Dreiding: A generic forcefield for molecular simulations. *J. Phys. Chem.* 94: 8897.

Mielczarski, E., Mielczarski, J. A., and Cases, J. M. 1998. Molecular recognition effect in monolayer formation of oleate on fluorite. *Langmuir.* 14: 1739.

Mielczarski, E., Mielczarski, J. A., Cases, J. M., Rai. B., and Pradip. 2002. Influence of solution conditions and mineral surface structure on the formation of oleate adsorption layers on fluorite. *Colloids Surf., A* 205: 73.

Minami, D., Horikoshi, S., Sakai, K., Sakai, H., and Abe, M. 2011. Simulation of dynamic behaviour of surfactants on a hydrophobic surface using periodic-shell boundary molecular dynamics. *J. Oleo Sci.* 60: 171.

Nagaraj, D. R., and Somasundaran, P. 1981. Chelating agents as collectors in flotation: Oximes-copper minerals systems. *Min. Eng.* 33: 1351.

Nagaraj, D. R., Day, A., and Gorken, A. 1999. Nonsulphide minerals flotation: An overview. In *Advances in Flotation Technology*. Eds. D. R., Parekh, and J. D., Miller, SME. Littleton, Colo, 245.

Nakahara, H., Shibata, O., Rusdi, M., and Moroi, Y. 2008. Examination of surface adsorption of soluble surfactants by surface potential measurement at the air/solution interface. *J. Phys. Chem. C.* 112: 6398.

Natarajan, R., and Nirdosh, I. 2008. Quantitative structure–activity relationship (QSAR) approach for the selection of chelating mineral collectors. *Min. Eng.* 21: 1038.

Nielsen, S. O., Srinivas, G., Lopez, C. F., and Klein, M. L. 2005. Modelling surfactant adsorption on hydrophobic surfaces. *Phys. Rev. Lett.* 94: 228301.

Pal, K., Banthia, A. K., and Majumdar, D. K. 2009. Polymeric hydrogels: Characterization and biomedical applications—A mini review. *Des. Monomers Polym.* 12: 197.

Park, K., Shalaby, W. S. W., and Park, H. 1993. In *Biodegradable Hydrogels for Drug Delivery*, Technomic, Lancaster, PA.

Perdew, J. P. 1986. Density-functional approximation for the correlation energy of the inhomogeneous electron gas. *Phys. Rev. B* 33: 8822.

Perdew, J. P., Burke, K., and Ernzerhof, M. 1996. Generalized gradient approximation made simple. *Phys. Rev. Lett.* 77: 3865.

Pertsin, A. J., and Grunze, M. 2000. Computer simulation of water near the surface of oligo(ethylene glycol)-terminated alkanethiol self-assembled monolayers. *Langmuir.* 16: 8829.

Pertsin, A. J., Hayashi, T., and Grunze, M. 2002. Grand canonical Monte Carlo simulations of the hydration interaction between oligo(ethylene glycol)-terminated alkanethiol self-assembled monolayers *J. Phys. Chem. B.* 106: 12274.

Porter, M. D., Bright, T. B., Allara, D. L., and Chidsey, C. E. D. 1987. Spontaneously organized molecular assemblies. 4. Structural characterization of n-alkyl thiol monolayers on gold by optical ellipsometry, infrared spectroscopy, and electrochemistry. *J. Am. Chem. Soc.* 109: 3559.

Pradip. 1988. Applications of chelating agents in mineral processing. *Miner. Metall. Process.* 5: 80.

Pradip. 1992. Design of crystal structure-specific surfactants based on molecular recognition at mineral surfaces. *Curr. Sci.* 63: 180.

Pradip. 1994. Reagents design and molecular recognition at mineral surfaces. In *Reagents for Better Metallurgy*, Ed. P. S. Mulukutla, SME-AIME, Denver, USA.

Pradip, and Rai, B. 2002a. Design of tailor-made surfactants for industrial applications using a molecular modelling approach. *Colloids Surf. A.* 205: 139.

Pradip, and Rai, B. 2002b. Molecular modelling and rational design of flotation reagents, In *Proceedings of International Seminar on Mineral Processing Technology (MPT 2002)*, Eds. S. Subramanian, K. A. Natarajan, B. S. Rao and T. R. R. Rao, IIME, Bangalore, India, 126.

Pradip, and Rai, B. 2003. Molecular modelling and rational design of flotation reagents. *Int. J. Miner. Process.* 72: 95.

Pradip, and Rai, B. 2009. Selective flocculation of alumina rich Indian iron ore slimes. In *Proceedings of ICBeneFiT-2007*, Eds. A. K., Mukherjee, and D. Bhattacharjee, Tata McGraw-Hill Education, New Delhi, 123.

Pradip, Ravishankar, S. A., Sankar, T. A. P., and Khosla, N. K. 1993. In *Proceedings of XVIII International Mineral Processing Congress*, Australasian Institute of Mining and Metallurgy, Parkville, Carlton South, Australia, 1289.

Pradip, Premachandran, R. S., and Malghan, S. G. 1994. Electrokinetic behaviour and dispersion characteristics of ceramic powders with cationic and anionic polyelectrolytes. *Bull. Mat. Sci.* 17: 911.

Pradip, Subanna, M., Rai, B., and Ganvir, V. 1997. Design of dispersant for Titania based points, Internal Research Report, Tata Consultancy Services Ltd., India.

Pradip, Rai, B., Rao, T. K., Krishnamurthy, S., Vetrivel, R., Mielczarski, J. A., et al. 2002a. Molecular modelling of interactions of diphosphonic acid based surfactants with calcium minerals. *Langmuir.* 18: 932.

Pradip, Rai, B., Rao, T. K., Krishnamurthy, S., Vetrivel, R., Mielczarski, J. A., et al. 2002b. Molecular modelling of interactions of alkyl hydroxamates with calcium minerals. *J. Colloid Interface Sci.* 256: 106.

Pradip, Rai, B., and Rao, T. K. 2002c. Design and development of novel reagents for the minerals Industry, In Proceedings of *Russian-Indian Symposium on Metallurgy of Nonferrous and Rare Metals.* Eds. L. I. Leontiev, A. I. Kholkin, and V. V. Belova. Russian Academy of Sciences, Moscow, 25.

Pradip, Rai, B., Sathish, P., and Krishnamurty, S. 2004a. Design of dispersants for colloidal processing of barium titanate suspensions by molecular modelling. *Ferroelectr.* 306: 195.

Pradip, Rai, B., and Sathish, P. 2004b. Rational design of dispersants by molecular modelling for advanced ceramic processing applications. *KONA.* 22: 151.

Pradip, Rai, B., and Sathish, P. 2004c. A molecular modelling based rational design of dispersants. In *Proceedings of International Conference on Nano-Materials: Synthesis, Characterisation and Application (Nano-2004)*, Eds. S. Bandyopadhyay, N. R. Bandopadhyay, H. S. Maiti, and S. P. Sengupta. Tata McGraw Hill, New Delhi, 198.

Pradip, Rai, B., Rao, T. K., Sathish, P., and Sandhya, K. 2005a. Indian Patent Grant 227499. Selective separation of phosphate minerals from other minerals, using aminotris (methylenephosphonic acid), and diethylenetriaminepentakis (methylenephosphonic acid) as depressants.

Pradip, Rai, B., Rao, T. K., Sathish, P., and Sandhya, K. 2005b. Indian Patent Grant 238197. Selective separation of phosphate minerals from other minerals using alkylimino-bis-methylenephosphonic acid based depressant.

Prasad, A., Krishna, B., Ravindran, I., Ravi, B. P., Haran, N. P., and Majumdar, A. 2010. Optimization flotation studies on a phosphorite sample from Jhamarkotra, Udaipur district, Rajasthan, India. In *Proceedings of the XI International Seminar on Mineral Processing Technology (MPT-2010)*, IIME, Jamshedpur, India, Vol. 2, 720.

Prinsen, P., Warren, P. B., and Michels, M. A. J. 2002. Mesoscale simulations of surfactant dissolution and mesophase formation, *Phys. Rev. Lett.* 89: 148302.

Pugh, R., and Stenius, P. 1985. Solution chemistry studies and flotation behaviour of apatite, calcite and fluorite minerals with sodium oleate collector. *Int. J. Min. Process.* 15: 193.

Rai, B., Krishnamurty, S., Sathish, P., and Pradip. 2008a. Salicylaldoxime derivatives as flotation collectors: Quantum mechanics calculations (density functional theory) for the design of selective reagents. In *Proceedings of Mineral Processing Technology—2008*, IIME, Trivandrum, India.

Rai, B., and Pradip 2003. A molecular modelling study of oleate monolayers adsorbed at calcium mineral surfaces. In Proceedings of *XXII International Mineral Processing Congress (IMPC) Cape Town*, Eds. L. Lorenzen, and D. J. Bradshaw. Cape Town, South African Institute of Mining & Metallurgy, Vol. 2, 1085.

Rai, B., and Pradip. 2008. Design of highly selective industrial performance chemicals: A molecular modelling approach. *Mol. Simul.* 34: 1209.

Rai, B., and Pradip. 2009. Development of innovative technologies for the treatment, recycling and utilisation of bauxite residue (Red Mud) in India. *Indian Chem. Eng.* 51: 28.

Rai, B., Sathish, P., and Pradip. 2008b. Molecular modeling based design of selective depressants for beneficiation of dolomitic phosphate ores. In *Proceedings of XXIV International Mineral Processing Congress*, Eds. W. D. Zuo, S. C. Yao, W. F. Liang, Z. L. Cheng, and H. Long. Science Press, Beijing, Vol. 1, 1518.

Rai, B., Sathish, P., Malhotra, C. P., Pradip, and Ayappa, K. G. 2004. Molecular dynamics simulations of self-assembled alkylthiolate monolayers on an Au (111) surface. *Langmuir.* 20: 3138.

Ramesh, V., Umasundari, P., and Das, K. K. 1998. Study of bonding characteristics of some new metal complexes of salicylaldoxime (SALO) and its derivatives by far infrared and UV spectroscopy. *Spectrochim. Acta, Part A* 54: 285.

Rappe', A. K., and Goddard, W. A. 1991. Charge equilibration for molecular dynamics simulations. *J. Phys. Chem.* 95: 3358.

Rappe', A. K., Casewit, C. J., Colwell, K. S., Goddard, W. A., and Skiff, W. M. 1992. UFF, a full periodic table force field for molecular mechanics and molecular dynamics simulations. *J. Am. Chem. Soc.* 114: 10024.

Reimer, U., Wahab, M., Schiller, P., and Mögel, H.-J. 2005. Monte Carlo study of surfactant adsorption on heterogeneous solid surfaces. *Langmuir.* 21: 1640.

Ritcey, G. M. 1980. Crud in solvent extraction processing—A review of causes and treatment. *Hydrometallurgy* 5: 97.

Ritcey, G. M., and Ashbrook, A. W. 1984. *Solvent Extraction: Principles & Applications to Process Metallurgy,* Part I. Elsevier, New York.

Rosen, M. J. 2004. Surfactants and interfacial phenomena, 3rd ed. Wiley Interscience, John Wiley & Sons, Amsterdam.

Salaniwal, S., Cui, S. T., Cochran, H. D., and Cummings, P. T. 2001. Molecular simulation of a dichain surfactant water carbon dioxide system. 1. Structural properties of aggregates, *Langmuir.* 17: 1773.

Sathish, P., Rai, B., and Pradip 2007. Molecular dynamics (MD) simulations to study the wettability of self-assembled alkylthiolate monolayers (SAMs) adsorbed on Au(111) Surface. In *Proceedings of International Conference on Advanced Materials Design and Development (ICAMDD-2005),* Eds. M. Chakraborty, D. L. McDowell, S. Ghosh, F. Mistree, and D. Bhattacharya. Elsevier, Goa, India, 107.

Sawyer, E. S., and Sawyer, P. J. 1997. Fish serum as blocking reagent. United States Patent No. 5602041.

Sharfrin, E., and Zisman, W. 1960. Constitutive relations in the wetting of low energy surfaces and the theory of the retraction method of preparing monolayers. *J. Phys. Chem.* 64: 519.

Shevchenko, S. M., and Bailey, G. W. 1998. Non-bonded organo-mineral interactions and sorption of organic compounds on soil surfaces: A model approach. *J. Mol. Struct. (THEOCHEM)* 422: 259.

Shinoda, W., DeVane, R., and Klein, M. L. 2008. Coarse-grained molecular modelling of non-ionic surfactant self-assembly. *Soft Matter.* 4: 2454.

Singh, S. S., Pradip, and Rai, B. 2010. Wetting characteristics of mineral surfaces: contact angle measurements through molecular dynamics simulations. In *Proceeding of XXV International Mineral Processing Congress (IMPC 2010),* AUSIMM, Brisbane, Australia, 2357.

Siperstein, F. R., and Gubbins, K. E. 2003. Phase separation and liquid crystal self-assembly in surfactant-inorganic-solvent systems. *Langmuir* 19: 2049.

Soddemann, T., Dunweg, B., and Kremer, K. 2001. A generic computer model for amphiphilic systems. *Eur. Phys. J. E* 6: 409.

Somasundaran, P. 2006. *Encyclopaedia of surface and colloid science,* 2nd ed. Ed. A. T. Hubbard. Taylor and Francis, New York, 1540.

Srinivas G., Nielsen S. O., Moore P. B., and Klein M. L. 2006. Molecular dynamics simulations of surfactant self-organization at a solid-liquid interface. *J. Am. Chem. Soc.* 128: 848.

Subbanna, M., Kapur, P. C., and Pradip. 2002. Role of powder size, packing, solid loading and dispersion in colloidal processing of ceramics. *Ceram. Int.* 28: 401.

Sun, H. 1995. *ab initio* calculations and force field development for computer simulation of polysilanes. *Macromol.* 28: 701.

Tan, L., Xie, Q., and Yao, S. 2007. Electrochemical piezoelectric quartz crystal impedance study on the interaction between concanavalin A and glycogen at Au electrodes. *Bioelectrochem.* 70: 348.

Tanwar, J., Anubhav, Pradip, and Rai, B. 2011. A molecular dynamics simulation study for prediction of surface tension of sodium dodecyl sulfate–water systems. *Ind. Eng. Chem.:* submitted.

Vandebilt, D. 1990. Soft self-consistent pseudopotentials in a generalized eigenvalue formalism. *Phys. Rev. B.* 41: 7892.

Vericat, C., Benitez, G. A., Grumelli, D. E., Vela, M. E., and Salvarezza, R. C. 2008. Thiol-capped gold: From planar to irregular surfaces. *J. Phys.: Condens. Matter* 20: 184004.

Vericat, C., Vela, M. E., and Benitez, G., 2010. Self-assembled monolayers of thiols and dithiols on gold: New challenges for a well-known system. *Chem. Soc. Rev.* 39: 1805.

von Gottberg, F. K., Smith, K. A., and Hatton, T. A. 1997. Stochastic dynamics simulation of surfactant self-assembly. *J. Chem. Phys.* 106: 9850.

von Gottberg, F. K., Smith, K. A., and Hatton, T. A. 1998. Dynamics of self-assembled surfactant systems. *J. Chem. Phys.* 108: 2232.

Weng, J. G., Park, S. H., Lukes, J. R., and Tien, C. L. 2000. Molecular dynamics investigation of thickness effect on liquid films. *J. Chem. Phys.* 113: 5917.

Wu, W. 2001. Baseline studies of the clay minerals society source clays: Colloid and surface phenomena. *Clays Clay Miner.* 49: 446.

Yuana, S.-M., Yana, H., Lva, K., Liua, Cheng-B., and Yuana, S.-L. 2010. Surface behaviour of a model surfactant: A theoretical simulation study. *J. Colloid Interface Sci.* 348: 159.

Zawala, J., Drzymala, J., and Malysa, K. 2007. Natural hydrophobicity and flotation of fluoride. *Physicochemical problems of mineral processing.* 41: 5.

Zehl, T., Wahab, M., Schiller, P., and Mögel, H.-J. 2009. Monte Carlo simulation of surfactant adsorption on hydrophilic surfaces. *Langmuir.* 25: 2090.

Zeng, J., Zhang, S., Gong, X., and Wang, F. 2010. Molecular dynamics simulation of interaction between calcite crystal and phosphonic acid molecules. *Chin. J. Chem.* 28: 337.

Zhang, R., Zhang, L., and Somasundaran, P. 2004. Study of mixtures of n-dodecyl-β--maltoside with anionic, cationic, and nonionic surfactant in aqueous solutions using surface tension and fluorescence techniques. *J. Coll. Interfacial Sci.* 278: 453.

3 Molecular Modeling of Mineral Surface Reactions in Flotation

K. Hanumantha Rao, T. K. Kundu, and S. C. Parker

CONTENTS

3.1 INTRODUCTION

Flotation is the single most important method of mineral processing and is widely used for the concentration of metal ores, industrial minerals, and coals. Several investigations were carried out during the last century, and the flotation literature was immense, particularly on selective collectors and their interaction mechanisms on sulfide, oxide, and silicate minerals. However, it is often difficult to determine experimentally the fundamental processes affecting the separation procedures. The summary presented by Kitchener (1984) on the innovations and discoveries of various collector molecules in flotation shows that the reagents' selectivity toward certain mineral surfaces is clueless. It is invariably a process of trial-and-error experimentation to reach an optimized reagent scheme with different types of organic molecules and their modified forms in order to separate valuable minerals from ores by flotation process.

Principally, the success of the flotation process depends on selectively making the desired mineral hydrophobic by adding surface-active reagents (collectors) while keeping all the other minerals hydrophilic. The hydrophobic solid particles thus obtained attach to bubbles dispersed through the liquid, which can be generated in a variety of methods, and rise to the surface where they are skimmed off with the froth. For the thermodynamic criterion (Shergold 1984) to be satisfied via flotation, the work of adhesion of water molecule (W_{AD}) onto the solid surface must be less than the work of cohesion between two water molecules ($W_{CO} = 149$ mJ m^{-2}). van der Waals dispersion forces (W_d), hydrogen bonding of water with polar sites (W_h), and ionic (electrostatic–columbic) interactions (W_i) contribute mostly toward W_{AD}. Because W_d is less than W_{CO} for all solids, adhesion to an air bubble can only be obtained if the ($W_h + W_i$) contribution diminishes appreciably. Collector adsorption at the solid–water interface shields polar sites that do not ionize or take part in hydrogen bonding and thereby reduce the ($W_h + W_i$) contribution and make flotation feasible.

The term *floatability* is positively connected to a contact angle subtended by minerals with bubbles against water. A finite contact angle indicates that W_{AD} is less than W_{CO} and principally relates to the nature of the intermolecular forces that bind water molecules together or to solid surfaces. From the combined form of Dupre and Young's equations, the expression of free energy changes (ΔG) with contact angle (θ), [$\Delta G = \gamma_{LG} (\cos \theta - 1)$] implies that for any finite value of the contact angle there will be a free energy decrease upon attachment of a mineral particle to an air bubble. The criteria used in flotation practice provide conditions for obtaining a large difference of hydrophobicity between the valuable and gangue minerals after reagents conditioning. The criteria that is being applied to date is to exploit the differences in the physicochemical properties of individual mineral constituents, their response to all possible reagents (i.e., collectors and modifiers), the effect of pH on surface charge, possible surface complexation, and the action of chelating reagents.

Thousands of reagents have been proposed for sulfide and nonsulfide mineral flotation, but the number of reagents used in practice is quite small (Nagaraj 1994). In sulfides flotation, xanthates have been used as collectors since the 1920s. The usage of dithiophosphates and thionocarbamates has steadily increased since their introduction in the 1960s. Xanthates are active toward the entire class of sulfides without any major selectivity. Several new chelating agents—especially in numerous combinations with the three important donor atoms (i.e., sulfur, nitrogen, and oxygen)—in the basic bonding group have been tested, but the status of the flotation agents in industry has not changed much.

Fatty acids and soaps dominate the nonsulfide flotation. These collectors also tend to be nonselective because almost all minerals can be floated. The use of appropriate activators or depressants alleviates the problem to some extent. The lack of selectivity is often related to the tendency of fatty acids to form a precipitated phase with dissolved multivalent ions. The more tolerant ether carboxylic acids have been tested in some cases. Amines are used invariably for silicate minerals. Other collectors such as sulfates, sulfonates, phosphonic acids, hydroxamates, and some amphoteric surfactants have gained little importance.

The scientific explanation for reagents' selectivity in mineral separations still leaves many questions unanswered and is based on knowledge available from analytical separation literature. The increasing need to treat today's lean and complex ores demands new selective and tailor-made reagents. No theory is currently available in the literature to answer the question, what would be the functional group and molecular structure of a collector that could achieve the desired separation? An alternative concept for selective reagents adsorption on mineral surfaces is proposed here based on the electrostatic, geometric, and stereochemical matching of functional groups of reagents to the lattice sites of the exposed surfaces of mineral particles similar to the underlying molecular recognition mechanisms at inorganic–organic interfaces occurring in biomineralization.

A review of the biomineralization literature on the influence of organic additives on crystal growth processes suggests that molecules that have two functional groups can either influence the morphology of surfaces or inhibit the growth, whereas no marked changes have been observed with molecules containing a single functional group (Black et al. 1991; Davey et al. 1991; Heywood and Mann 1992; Mann 1993). The spacing between the functional groups is found to be an important factor in the efficacy of crystal growth inhibition. The concept of molecules consisting of two groups having appropriate spacing between them to achieve structural compatibility with the surface is of direct relevance to the reagents selectivity in flotation processes.

Traditionally, reagents have been designed and improved by a trial-and-error approach, and this approach is still being employed. A modern version of this approach is combinatorial chemistry, in which the reagents with specific molecular structure can be modeled and their interaction with inorganic surfaces is checked within a short period. The development of new techniques and the increased power of computers have allowed for a rapid advance in the use of atomistic simulation techniques for modeling the structural and mechanical properties of materials. Experimental and computational techniques have converged to give a detailed understanding of interfacial chemistry on an atom-by-atom basis. As a three-dimensional atom probe is capable of restructuring the chemistry of solids atom by atom with subnanometer resolution, atomistic simulation techniques can calculate similar quantities for ensembles of hundreds of thousands of particles. Atomistic simulations represent a valuable methodology and have been successful in accurately predicting the structures and properties of a wide range of minerals including crystal morphology, surface structure, and adsorption behavior (de Leeuw et al. 1995, 1996, 1998a, 1998c; Lawrence and Parker 1989; Parker et al. 1993; Pradip et al. 2002; Watson et al. 1997). Many areas of application have benefited from atomic simulation, including the structure and dynamics of molecular crystals and liquids, complex crystal structures, micelles and colloids, aqueous solutions and electrolytes, adsorption in porous media, and properties of surfaces and effects of impurities (Schleyer et al. 1998). Computer modeling and simulation have become increasingly recognized as important tools in the study and development of materials. Computational techniques are well placed to investigate at the atomic level not only the surface structure and the influence of additives but also the interactions at the mineral–water interface.

Our recent atomistic modeling calculations on the interaction of carboxylic and amine molecules on pure and hydroxylated mineral surfaces exemplify the utility of computer simulations in flotation practice (de Leeuw et al. 1998c; Hanumantha Rao and Kundu 2006; Kundu et al. 2003a, 2003b, 2004, 2005). However, a lot more work is needed to develop a theoretically robust framework that can provide a rational basis for the design and selection of highly selective flotation reagents. This chapter consists of a summary of our molecular modeling computations on quartz and wollastonite minerals and their interaction with water, methanoic acid, and methylamine molecules.

3.2 METHODOLOGY

3.2.1 ATOMISTIC SIMULATION TECHNIQUES

Atomistic simulation techniques based on Born model of solids (Born and Mayer 1932) are employed for both static lattice simulations and molecular dynamics calculations. The basis of this approach is to calculate the total interaction energy, often called the *lattice energy*. The simple parameterized analytical functions that describe the interactions between all species in the crystal are used to calculate the total interaction energy of the system, which is defined as the energy released when the component ions from infinity are brought together on their lattice sites. The interactions in a crystal can conveniently be divided into two parts: (1) short-range attraction and repulsions, which contain adjustable parameters, and (2) long-range electrostatic interactions, which contribute approximately 80% of the total interaction energy.

3.2.2   THE BORN MODEL OF SOLIDS AND THE POTENTIAL MODEL

The *Born model of solids* assumes that the sum of all pairwise interactions between atoms i and j produce the lattice energy of a crystal. The lattice energy is given by:

$$U(r_{ij}) = \sum_{ij}^{\prime} \frac{q_i q_j}{r_{ij}} + \sum_{ij}^{\prime} \phi_{ij}(r_{ij}) + \sum_{ijk} \phi_{ijk}(r_{ijk}) + \ldots \tag{3.1}$$

The first component defines the long-range electrostatic interactions, and the second and third components define short-range two-body and many-body interactions, respectively. The dash above the summation shows that the interaction where $i = j$ is not included. It is considered adequate to calculate only the two-body interactions for systems in which the interactions are nondirectional, such as ionic solids. However, when studying systems that contain a degree of covalent bonding, the evaluation of higher-body terms such as *bond bending* and *bond stretching* is necessary to include their directionality (Catlow 1997).

The *potential model* describes the variation of energy of the system as a function of the atomic coordinate. This energy is derived from the long-range electrostatic forces and short-range attractive and repulsive forces, or the coulombic contribution, whereas the short-range interactions are described using simple parameterized functions. A potential model that accurately describes the lattice properties is essential if quantitative results are to be obtained. This is particularly important for surfaces for which it is necessary to describe the interaction at distances possibly far removed from those found in the bulk lattice (Allan et al. 1993).

The functional forms of the potential used are:

$$\text{Coulombic } E_{lr} = \sum_{ij} \frac{q_i q_j}{4\pi\varepsilon_0 (r_{ij} + I)} \tag{3.2}$$

where q_i and q_j are the charges on the ions, ε_0 is the permittivity of free space, r_{ij} is the separation distance between the ions, and I is the set of lattice vectors representing the periodicity of the crystal lattice.

$$\text{Buckingham } \phi_{ij}(r_{ij}) = C_{ij} \exp\left(-\frac{r_{ij}}{\rho_{ij}}\right) - \frac{D_{ij}}{r_{ij}^6} \tag{3.3}$$

where C, ρ, and D are parameters that differ for each pair of ions interacting.

$$\text{Harmonic angle } \Phi_{ijk} = \frac{1}{2} k_{ijk} (\theta - \theta_0)^2 \tag{3.4}$$

where k_{ijk} is the harmonic force constant, θ_0 is the equilibrium bond angle, and θ is the bond angle.

$$\text{Lennard-Jones } \phi_{ij}(r_{ij}) = \frac{A_{ij}}{r_{ij}^{12}} - \frac{B_{ij}}{r_{ij}^6} \tag{3.5}$$

where A_{ij} and B_{ij} are variable parameters for each pair of ions interaction. The first term describes the repulsive forces, dependent on r_{ij}^{-12} and therefore dominant at very short range and the second term describes the attractive interactions, which dominate the interaction at longer separations.

$$\text{Morse } V(r_{ij}) = E_{ij}(1 - \exp[-F_{ij}(r_{ij} - r_{0,ij})])^2 - E_{ij} \tag{3.6}$$

where E_{ij} is the bond dissociation energy, $r_{0,ij}$ is the equilibrium bond distance, and F_{ij} is related to the curvature of the slope of the potential energy well.

$$\text{Electronic polarizability } \alpha = \frac{Y^2}{K} \tag{3.7}$$

where Y and K are the shell charge and spring constant, respectively.

3.2.3 Types of Surfaces

The major consideration while generating a surface is that the repeat unit, which is duplicated to create the crystal, must not have a dipole moment perpendicular to the surface or else the surface energies will diverge. As an aid, Tasker (1979) identified three different types of surfaces shown in Figure 3.1.

A *type I surface* (Figure 3.1a) has a stoichiometric ratio of anionic and cationic planes. Therefore each plane has an overall zero charge, and there is no dipole moment perpendicular to the surface. A *type II surface* (Figure 3.1b) has a stacking sequence of charged planes, but the repeat unit consists of several planes that, when considered together have, no dipole moment perpendicular to the surface. In both types of surfaces, the electrostatic energy converges with increased region size.

However, in the case of a *type III surface* (Figure 3.1c), alternately charged planes are stacked and produce a dipole moment perpendicular to the surface if cut between any plane of atoms. In nature, these surfaces are stabilized by defects and adsorbed species. We can generate a stoichiometric surface by removing half of the ions in the surface layer at the top of the repeat unit and transferring them to the bottom, thereby producing a highly defective surface structure. Once the dipole has been removed, the surface energy can be calculated.

3.2.4 Calculation of Surface Energy

Static lattice energy minimizations have been used throughout this work to calculate surface and bulk block energies. This is effectively carried out at 0 K, because it ignores the vibrational

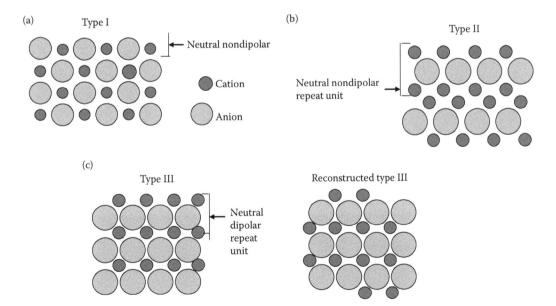

FIGURE 3.1   (a) Type I, (b) type II, (c) type III, and reconstructed type III stacking sequences.

properties of the crystal, including the zero-point energy contribution. However, these energies are small, and simulations have been shown to be in good agreement with experiments (Parker 1983).

The METADISE code (Minimum Energy Techniques Applied to Dislocation, Interface, and Surface Energies) (Watson et al. 1996) was used when studying surfaces, which takes the two-dimensional periodicity into account when calculating their energies. In this program, the crystal is regarded as a series of charged planes parallel to the surface and periodic in two dimensions. The crystal block is divided into two regions, Region I and Region II (Figure 3.2). Region I contains all the ions near the surface, or interface, and Region II contains all the bulk ions. During the minimization process, the ions in Region I are explicitly relaxed whereas the ions in Region II are held fixed. The block energy E, can be split into two parts:

$$E = E_I + E_{II} \tag{3.8}$$

where E_I is the total energy of the ions in Region I, including coulombic, short-range, polarization, and three body bond energy. E_{II} is the energy of the ions in Region II.

To calculate the surface energy of the crystal face, the energy of the surface block, E_S, and bulk block E_B must be calculated. These energies can be broken down into interaction energies as shown in Equations 3.9 and 3.10:

$$E_S = E'_{I-II} + E'_{I-I} + E'_{II-I} + E'_{II-II} \tag{3.9}$$

where E'_{I-I} is the interaction energy of ions in Region I with others in Region I, E'_{I-II} is the interaction energy of all the ions in Region I with all the ions in Region II, and so on. Similarly, the bulk energy is:

$$E_B = E''_{I-II} + E''_{I-I} + E''_{II-I} + E''_{II-II} \tag{3.10}$$

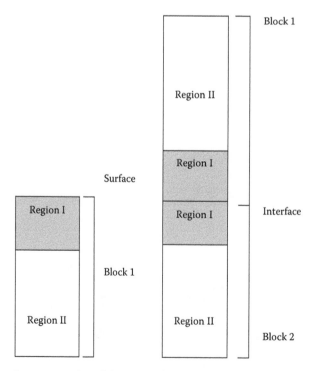

FIGURE 3.2   Schematic representation of the two-regions approach used for (a) surfaces and (b) interfaces in METADISE.

The surface energy of the crystal faces takes the form of:

$$\gamma = \frac{(E_S - E_B)}{AREA} \tag{3.11}$$

Because the ions in Region II do not relax, the total interaction energy of all the ions in Region II with all the other ions in Region II does not change, hence these cancel in the surface energy calculation.

3.2.5 CALCULATION OF ADSORPTION ENERGY

The adsorption energies (U_{ads}) were calculated by comparing the energy of the pure surface (U_s) and that of an isolated adsorbed molecule (U_{mol}) with the energy of the covered surface (U_{def}) as given here:

$$U_{ads} = U_{def} - (U_s + U_{mol}) \tag{3.12}$$

The most common, low-energy, low Miller index surfaces of wollastonite were considered, and these were close packed planes with large interplanar spacing. The final structure and energy was achieved by allowing the ions in the surface region to relax to the point where they experienced zero force. Defects and impurities were accommodated in the surface as we assumed periodic boundary conditions while maintaining charge neutrality because otherwise coulombic energy would be divergent if the repeat cell was charged.

Wollastonite crystal does not possess any center of inversion symmetry element, and thus no inverse asymmetric surface was found. A symmetric repeat unit has the same surface characteristics at the top and bottom atomic layers. Figure 3.3 shows wollastonite symmetric stacking for {100}, {001} surfaces and asymmetric stacking of {102}, {10$\bar{1}$}, {01$\bar{1}$}, {1$\bar{1}\bar{1}$}, and {1$\bar{1}$0} surfaces.

The simulation of the hydroxylated surfaces was achieved by adsorbing dissociated water molecules on to surface cation–oxygen pairs in such a way that the OH group was bonded to a surface cation and the hydrogen atom to a surface oxygen atom. This can be considered as a replacement of surface oxygen by two hydroxyl groups in Kröger–Vink notation (Kröger 1973, 1974):

$$H_2O + O_O^x \rightarrow (OH)_O^{\bullet} + (OH)_i^{/} \tag{3.13}$$

where O_O^x is an oxygen at a lattice oxygen site with zero charge with respect to the lattice oxygen, $(OH)_O^{\bullet}$ is a hydroxyl group at an oxygen lattice site with charge +1, and $(OH)_i^{/}$ is a hydroxyl group at an interstitial site with charge −1. For the hydration energy calculation a value for the energy of dissociation of a water molecule is required:

$$2H_2O + O_{(g)}^{2-} \rightarrow 2OH_{(g)}^{-} \tag{3.14}$$

However, this requires the material dependent on second electron affinity of oxygen (Harding and Pyper 1995), which is overcome by using experimental heats of formation of metal hydroxide from wollastonite and water and computing the lattice energy of wollastonite, $Ca(OH)_2$ and $Si(OH)_4$, by using the Hess law of thermodynamics. The dissociation energy of water on wollastonite and quartz was found to be −8.10337 and 8.285 eV, respectively.

For wollastonite dissolution, we modeled the nonstoichimetric release of Ca^{2+} from the surface, keeping the Si^{4+} cation content the same, and the charge neutrality was maintained by adding

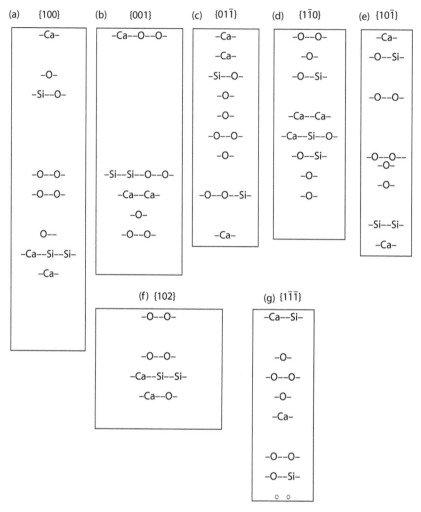

FIGURE 3.3   Schematic representation of stacking sequence in unreconstructed neutral nondipolar cuts of symmetric (a) {100} and (b) {001} and the lowest energy (c) asymmetric {01$\bar{1}$}, (d) asymmetric {1$\bar{1}$0}, (e) asymmetric {10$\bar{1}$}, (f) asymmetric {102}, and (g) asymmetric {1$\bar{1}\bar{1}$} surfaces. The vertical distance between two closest layers shown is 0.2 Å.

2H$^+$ to the surface Si–O ligand by making it Si–OH. In essence, we can write this phenomenon in neutral molecular form as follows:

$$(CaO)_X[s] + H_2O[l] \rightarrow (CaO)_{X-1}(H_2O)[s] + (CaO)[aq] \tag{3.15}$$

Under acidic conditions, the initial dissolution of calcium from the wollastonite surface may be described by the following surface reaction:

$$>(SiO)_2Ca[s] + 2H^+[aq] \rightarrow 2 > SiOH[s] + Ca^{2+}[aq] \tag{3.16}$$

where $>(SiO)_2Ca$ is a bidentate surface complex on the wollastonite surface occupied by a calcium attached to two oxygens that are each bonded to a silicon atom, and $>SiOH$ is the neutrally charged site formed by adsorption of H$^+$ to the $>SiO^-$ site that can be shown schematically as a simplified manner of the bonding structure between the calcium atom and surface oxygen atoms, as in

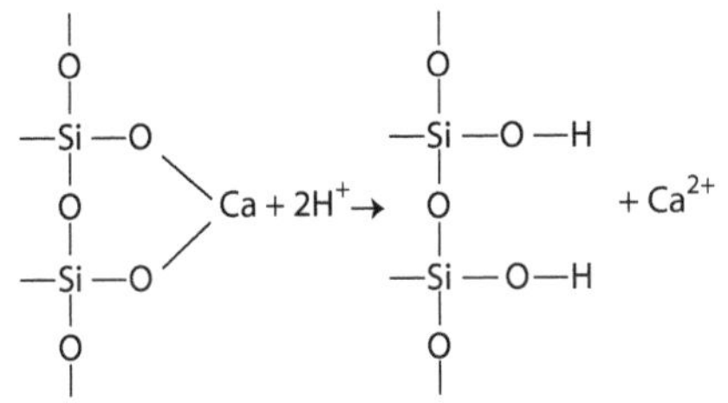

FIGURE 3.4 Schematic diagram of the $Ca^{2+} - 2H^+$ exchange reaction on the wollastonite surface.

Figure 3.4, though this is in contrast with the findings of Rimstidt and Dove (1986) of stoichiometric wollastonite dissolution.

We can also exchange Ca^{2+} and $2H^+$ under alkaline conditions by following this type of surface reaction:

$$>(SiO)_2Ca[s] + 2H_2O[l] \rightarrow 2 >SiOH[s] + Ca^{2+}[aq] + 2OH^-[aq] \tag{3.17}$$

Equations 3.16 and 3.17 depict the same type of replacement as simplified by Equation 3.15, which the simulation program follows, but differ in the excess energy requirement for the dissociation of water on wollastonite surface that have an acidic and alkaline environment. Using these two different energy contributions to U_{def}, we have calculated the replacement energy for $Ca^{2+} - 2H^+$ exchange under acidic and alkaline conditions by following Equation 3.12.

3.2.6 INTERATOMIC POTENTIAL PARAMETERS

The method for obtaining the parameters is to empirically fit the parameters to reproduce experimental data. The parameters k_{ij} (bond-stretch force constant), k_{ijk} (angle-bend force constant), D_e (bond dissociation energy), $r_{ij,0}$ (equilibrium bond length), θ_0 (equilibrium angle), and D (dielectric constant) are obtained from spectroscopic and thermodynamic literatures. Torsional energy is used as a correction factor to match the experimental energy value. Parameter A (control depth of the potential energy well) can be obtained quantum mechanically or from polarisability measurement. Parameters C and ρ, which determine interatomic distance of the potential energy well, are obtained by matching the crystallographic data. The values of α, Y, and K are obtained by fitting experimental dielectric data. In mineral systems, there is often insufficient experimental data to be used for fitting potential parameters. Transferability of potential parameters adopted in an atomistic simulation is a useful concept, and it has proved highly successful for modeling silicate minerals (Parker and Price 1989).

Solids and their surfaces were modeled by the Born model potential as discussed in the following paragraph. Adsorbed molecules were modeled by the molecular mechanics approach in which energy for a covalently bonded molecule is dependant mainly on bond stretching term (E_r) along other contributing terms.

The potential parameters for wollastonite were derived by reproducing crystal structure parameters. The potential parameters used for the simulation of quartz crystal were derived empirically by Sanders et al. (1984), which have been used successfully for many previous studies, including α-quartz surface stability studies by de Leeuw et al. (1999b) and a study on the effect of lattice relaxation on cation exchange in zeolite A by Higgins et al. (1997). The potential models for the intramolecular and intermolecular water interactions used were those derived and used by de Leeuw et al. (1998b, 1999a). Potential parameters describing the interactions between a water molecule and quartz surfaces are the same as those used by de Leeuw et al. (1999b) and were derived by following the approach by Schröder et al. (1992).

TABLE 3.1

Interacting Ions Used and Their Mass and Charges

		Charge (au)		
Ions	Mass (amu)	Core	Shell*	K (eV/Å²)
Si	28.09	+4.0		
Ca	40.08	+2.0		
H	1.008	+0.4		
O	15.99	+0.848	−2.848	74.92038
Oh	15.99	+0.90	−2.30	74.92038
Ow	15.99	+1.25	−2.05	209.4496
C_f	12.01	+0.31		
O_{fC}	15.99	−0.38		
O_{fH}	15.99	−0.38		
H_{fC}	1.008	+0.1		
H_{fO}	1.008	+0.35		
N	14.007	−0.5		
C_a	12.01	−0.08		
H_{aC}	1.008	+0.1		
H_{aN}	1.008	+0.14		

* Shell is the interacting part for oxygen. Ca, calcium of wollastonite; H, hydrogen of hydroxyl and water; H_{fC}, hydrogen of formic acid connected to C_f; H_{aC}, hydrogen of methylamine connected to C_a; HaN, hydrogen of methylamine connected to N; H_{fO}, hydrogen of formic acid connected to O_{fH}; O, oxygen of wollastonite/quartz, O_{fC}, oxygen of formic acid double-bonded with C_f; O_{fH}, oxygen of formic acid single-bonded with C_f; Oh, oxygen of hydroxyl; Ow, oxygen of water; Si, silicon of quartz/wollastonite.

The potential parameters of the hydroxide ions are the same as those used successfully by de Leeuw (1995), which were modified by Baram and Parker (1996) to include a polarizable oxygen ion and were applied in their work on hydroxide formation at quartz and zeolite surfaces. Methanoic acid and methylamine molecules were modeled using the Consistent Valence Force Field (CVFF) from the Insight II package (Molecular Simulations Inc, San Diego, CA). The interaction between methanoic acid and methylamine with a quartz surface was adopted according to the partial charges on the atoms.

The interacting ions with their mass and charges is presented in Table 3.1, and the potential energy function parameters for adsorbed molecules is presented in Table 3.2. Table 3.3 shows the respective potential parameters used for quartz and wollastonite mineral systems.

3.3   RESULTS AND DISCUSSION

3.3.1   QUARTZ–WOLLASTONITE MINERAL SYSTEM

Quartz is one of the predominant polymorphs of silicon dioxide (SiO_2) and one of the most abundant minerals. It occurs as an essential constituent of many igneous, sedimentary, and metamorphic rocks (Deer et al. 1992) and is used in many electronic devices, in the ceramic industry, and in the synthesis of zeolites. α-Quartz has a hexagonal crystal structure with space group P3$_1$21 (left side)

TABLE 3.2

Potential Energy Function Parameters for Adsorbed Molecules

Water			

Buckingham Potential

Ion Pairs	C (eV)	ρ (Å)	D (eV/Å^6)
H^{+0}–$O^{-0.8}$	396.27	0.25	0

Lennard-Jones Potential

	A (eVÅ^{12})	B (eVÅ^6)	r_0 (Å)
$O^{-0.8}$–$O^{-0.8}$	39344.98	42.15	

Morse Potential

	E (eV)	F (Å^{-1})	r_0 (Å)
$H^{+0.4}$–$H^{+0.4}$	0.0	2.8405	1.5
$H^{+0.4}$–$O^{-0.8}$	6.203713	2.22003	0.92

Three Body Potential

	k (eV/rad)	θ_0 (deg.)
H–Ow–H	4.19978	108.69

Hydroxide Buckingham Potential

Ion Pairs	C (eV)	ρ (Å)	D (eV/Å^6)
$H^{+0.4}$–$O^{-1.4}$	311.97	0.25	0.0
$O^{-1.4}$–$O^{-1.4}$	22764.3	0.149	6.97

Morse Potential

	E (eV)	F (Å^{-1})	r_0 (Å)
$H^{+0.4}$–$O^{-1.4}$	7.0525	3.1749	0.93

Formic Acid Lennard-Jones Potential

	A (eVÅ^{12})	B (eVÅ^6)	r_0 (Å)
$O^{-0.38}$–$C^{+0.31}$	38994.306	35.232	1.6
$O^{-0.38}$–$H^{+0.1}$	1908.103	5.55	2.4
$O^{-0.38}$–$H^{+0.35}$	1908.103	5.55	1.4
$C^{+0.31}$–$C^{+0.31}$	128614.7211	57.43274	
$H^{+0.1}/H^{+0.35}$–$H^{+0.1}/H^{+0.35}$	307.9583033	1.425184	

Morse Potential

	E (eV)	F (Å^{-1})	r_0 (Å)
$C^{+0.31}$–$H^{+0.1}$	4.66	1.77	1.10
$O^{-0.38}$–$H^{+0.35}$	4.08	2.28	0.96
$C^{+0.31}$–$O^{-0.38*}$	4.29	2.00	1.37
$C^{+0.31}$–$O^{-0.38}$	6.22	2.06	1.23

Three Body Potential

	k (eV/rad)	θ_0 (deg.)
$O^{-0.38}$–$H^{+0.35}$–$C^{+0.31}$	4.29	112.0
$C^{+0.31}$–$H^{+0.1}$–$O^{-0.38}$	4.72	110.0
$C^{+0.31}$–$O^{-0.38*}$–$H^{+0.1}$	4.72	120.0
$C^{+0.31}$–$O^{-0.38*}$–$O^{-0.38}$	12.45	123.0

(Continued)

TABLE 3.2 (CONTINUED)

Potential Energy Function Parameters for Adsorbed Molecules

Methyl Amine

Morse Potential

	E (eV)	F (Å^{-1})	r_0 (Å)
N–C$_a$	2.95	2.29	1.47
C$_a$–H$_{aC}$	4.71	1.77	1.11
N–H$_{aN}$	3.82	2.28	1.03

Three Body Potential

	k (eV/rad)	θ_0 (deg.)
H$_{aN}$–N–C$_a$	1.71	106.4

TABLE 3.3

Potential Parameters for Wollastonite and Quartz

	Buckingham Potential		
Ion Pairs	C (eV)	ρ (Å)	D (eV/Å^6)
Si–Oh	983.907	0.321	10.662
Si–O/Ow	1283.91	0.321	10.662
O–O	22764.0	0.149	27.88
Ca–O	1227.70	0.337	0.00
Ca–Oh	859.39	0.337	0.00
Ca–Ow	490.8	0.337	0.00
O–Oh	22764.0	0.149	13.94
O–Ow	22764.0	0.149	28.92
O–H	311.97	0.25	0.00
Ca–O$_{fC}$/O$_{fH}$	233.26	0.337	0.00
Ca–N	663.26	0.337	0.00

	Lennard-Jones Potential	
	A (eVÅ^{12})	B (eVÅ^6)
O–O$_{fC}$/O$_{fH}$	23430.0	32.12
O–H$_{fC}$/H$_{fO}$	5600.0	12.00
O–C$_f$/C$_a$	87327.5	56.32
O–H$_{aC}$/H$_{aN}$	5600.0	12.0
O–N	67528.4	50.45

	Three Body Potential	
	k (eV/rad)	θ_0 (deg.)
Oh/O–Si–O/Oh	2.09724	109.47

and P3$_2$21 (right side). It is a tectosilicate mineral in which four oxygen atoms tetrahedrally coordinate silicon atoms in the bulk. Corner-sharing SiO$_4$ form a three-dimensional network of Si–O–Si bonds with essentially covalent character. All the Si–O bond lengths in α-quartz are not equal; each silicon is bonded to two oxygen atoms at a distance of 1.60 Å and with two oxygen atoms at a distance of 1.61 Å.

Wollastonite is a natural calcium metasilicate ($CaSiO_3$) formed by the metamorphism of siliceous limestones at approximately 450°C and higher. Since the early 2000s, market demand for various wollastonite products has increased dramatically. Wollastonite is used in resins and plastics as filler material and in ceramics, metallurgy, paint, frictional products, biomaterial, and other industrial application owing to its chemical purity, low loss of ignition, high aspect ratio, and bright whiteness coupled with its low thermal coefficient of expansion and fluxing capability (Power 1986). In our simulation, we modeled low-temperature polymorph α-wollastonite, which has triclinic crystal structure with space group $P\bar{1}$. Wollastonite is a single-chain silicate of the pyroxenoid-type in which three tetrahedrons per unit cell are arranged parallel to the *b*-axis and the repeated unit consists of a pair of tetrahedrons joined apex to apex as in the $[Si_2O_7]$ group, alternating with a single tetrahedron with one edge parallel to the chain direction (Deer et al. 1992). The structure constitutes two chains pair repetition. Each pair is composed of inversion-related chains and represents one basic slab of the structure (Mazzucato and Gualtieri 2000). This structure is also pseudo-symmetrical in which the atoms are near but not at the same positions, which keeps it from having a center of symmetry. Wollastonite in this interpretation is a rare example of a crystal with no symmetry elements (Bragg et al. 1965).

3.3.2 Quartz Surface Structures and Adsorption

We modeled the $P3_121$ space group α-quartz crystal with the starting unit cell parameters of $a = 4.913$ Å, $b = 4.913$ Å, $c = 5.404$ Å, $\alpha = 90.0°$, $\beta = 90.0°$, and $\gamma = 120.0°$. The quartz cell parameters obtained after minimization were $a = 4.88$ Å, $b = 4.88$ Å, $c = 5.354$ Å, $\alpha = 90.0°$, $\beta = 90.0°$, and $\gamma = 120.0°$.

α-Quartz morphology is very complex, with more than 500 noted forms; it does not have any excellent cleavages but does have a number of preferred cleavage planes. The best cleavage planes are $\{10\bar{1}1\}$, $\{01\bar{1}1\}$, and $\{11\bar{2}2\}$, followed by less-pronounced cleavages on $\{0001\}$, $\{11\bar{2}1\}$, $\{51\bar{6}1\}$, and $\{10\bar{1}0\}$ and then perhaps by cleavages on $\{30\bar{3}2\}$ and $\{11\bar{2}0\}$. The common growth planes of α-quartz are $\{10\bar{1}0\}$, $\{01\bar{1}1\}$, $\{10\bar{2}1\}$, and $\{11\bar{2}1\}$. We modeled $\{0001\}$ because it has been found to be the most predominant surface by de Leeuw (1999b). The other surfaces studied were $\{10\bar{1}0\}$, $\{10\bar{1}1\}$, $\{10\bar{1}\bar{1}\}$, $\{01\bar{1}1\}$, $\{11\bar{1}0\}$, and $\{11\bar{2}1\}$.

Clean surfaces (i.e., the surfaces created in a vacuum) that are formed by breaking Si−O bonds generate highly reactive Si* and Si−O* free radical species. During energy minimization, bonds are formed between radicals and with other atoms, and the first few atomic layers show reorientation in order to obtain minimum energy position. In most cases, these layers pull inward, toward the bulk of the structure, and thereby reduce or extend (in the case of new bond formation) the spacing among first few layers so that it differs from ideal spacing. Lateral reconstruction is found to occur near surface regions, changing the coordination of surface atoms as compared with that of bulk coordination.

3.3.3 Pure Surfaces of Quartz

In pure $\{0001\}$ surface simulation, we started with a knife cut structure of the bulk crystal. The top silicon atom before minimization was coordinated with three oxygen atoms with bond distances of 1.60 Å, 1.61 Å, and 1.60 Å and O−Si−O angles of 108.9°, 108.7°, and 110.5°, respectively. The two Si−O−Si angles in which the top silicon was connected to two lower-layer silicons were 141.5° each. This configuration, as shown in Figure 3.5a, is the same as the bulk material, which is very close to the ideal tetrahedral angles (Levein et al. 1980), as opposed to having one three-coordinated silicon and one coordinated nonbridging oxygen. The six bridging oxygen atoms between the six silicon atoms form one complete ring, and the bulk structure is formed by their periodic repetition in three dimensions.

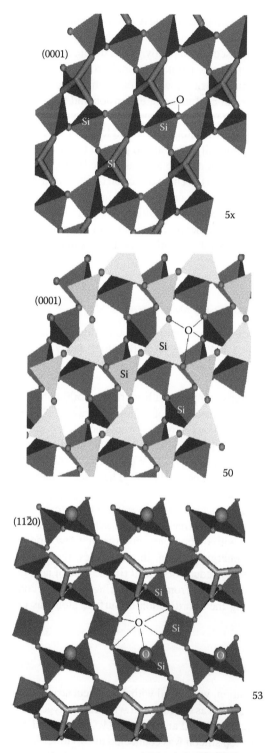

FIGURE 3.5 Top view of a quartz {0001} surface. (a) Before energy minimization where a three-coordinated silicon is shown as a stick and a four-coordinated silicon as a tetrahedron. (b) After energy minimisation where all surface silicon is four coordinated. (c) The {11$\bar{2}$0} surface showing a three-coordinated silicon as a stick, a four-coordinated silicon as a tetrahedron, and a single-bonded oxygen as a ball. Colour key: Si, gray; O, black ball.

TABLE 3.4

Surface Energies (J/m²) of Pure Quartz and in the Presence of Water, Methanoic Acid, and Methylamine

Surfaces	Pure Surface	Water	Dissociated Water	Formic Acid	Methylamine
{0001}	2.28	1.94	0.08	0.99	–
{10$\bar{1}$0}	2.93	2.35	0.12	2.04	–
{10$\bar{1}$1}	2.61	2.02	0.11	2.18	1.64
{10$\bar{1}\bar{1}$}	2.46	1.86	0.14	1.65	1.47
{01$\bar{1}$1}	2.46	1.55	0.14	1.49	–
{11$\bar{2}$0}	2.80	1.97	0.13	1.54	1.51
{11$\bar{2}$1}	3.10	2.63	–	2.84	–

After energy minimization, bridging oxygen between the second and third layers of silicon broke the bond with the second layer and bonds with the top three-coordinated silicon. However, the nonbridging oxygen from the top layer connected with the second layer silicon. Thus on the top surface, Si–O–Si bridges were formed and all silicon became fully coordinated, as shown in Figure 3.5b. Because the surface atoms coordination was fully satisfied, this surface corresponded to the minimum surface energy cut, as evident in Table 3.4. The lower the surface energy, the higher the stability, and pure quartz morphology should be more or less predominant by the {0001} surface. On the surface region, three silicon atoms with three oxygen atoms between them formed complete rings, but below that the structure was same as that of the bulk. This finding mirrors that of Ringnanese et al. (2000), who found an unexpected densification of the two uppermost layers of SiO_2 tetrahedron by three-member and six-member rings. At the same time, they found through *ab initio* molecular dynamics simulations that the dense surface {0001} is the most stable structure obtained upon cleaving. After relaxation, the Si–O bond distances from the top silicon in first layer were 1.54 Å, 1.58 Å, 1.59 Å, and 1.72 Å and those of the top layer silicon were 1.58 Å, 1.61 Å, 1.63 Å, and 1.65 Å. The changed minimum energy configuration O–Si–O angle values, in which silicon was the top layer, were 81.8°, 107.8°, 107.9°, and 116.1° and Si–O–Si angles were 117.8°, 129.1°, 129.2°, and 142.4°.

For other Miller index surface cuts, the surface layer atoms rattled to some extent, but after energy minimization none of them satisfied the complete coordination of the top silicon. This was evident from the surface energy values summarized in Table 3.4, which are all higher than the {0001} surface. For example, on the fresh cut {11$\bar{2}$0} surface there were one two-coordinated silicon and two single-coordinated oxygen atoms per unit cell, but after energy minimization one of the dangling oxygen was bonded with a two-coordinated silicon as shown in Figure 3.5c. Because of the presence of unsaturated silicon sites, these surfaces were highly reactive and stabilized by bonding with other species (e.g., –OH, H_2O, and CO).

3.3.4 ADSORPTION OF WATER ON PURE QUARTZ SURFACES

The {0001} surface of quartz was not reactive, and hence the surface configuration did not change by the molecular adsorption of water. All the surface silicon atoms were still coordinated to four lattice oxygen atoms forming Si–O–Si bridges as shown in Figure 3.6a. The oxygen atom of the water molecule was at a distance of 2.54 Å, while one hydrogen atom coordinated at a hydrogen-bond distance of 1.86 Å to a surface oxygen atom. The other hydrogen points were slightly away from the surface, with the lattice oxygen and hydrogen bond distance of 2.08 Å. The hydrated {0001} plane was more stable than the pure surface (Table 3.4). The low adsorption energy of molecular liquid water (−41.87 kJmol⁻¹; Table 3.5) indicated that the interaction between molecular water and this surface was in the region of physisorption rather than chemisorption.

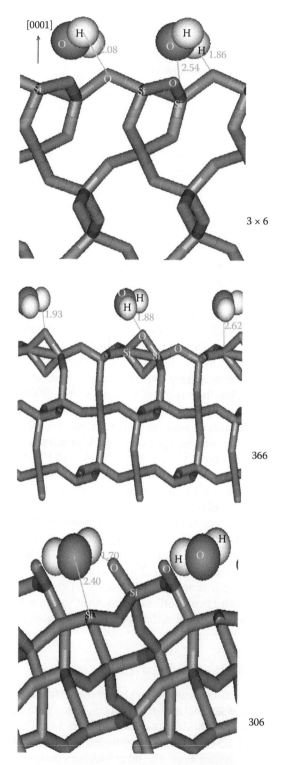

FIGURE 3.6 (a) Energy minimized {0001} after water adsorption showing Si–O–Si bridge on the surface. (b) Side view of water adsorption on {01$\bar{1}$1} showing a five-coordinated silicon on the surface. (c) Adsorption of water in the crevice of {11$\bar{2}$1} showing a three-coordinated silicon and a dangling oxygen on the surface. Crystal shown as a stick and the water as space filled. Color key: Si, gray; O, black; H, light gray.

TABLE 3.5

Adsorption Energies (kJ/mole) for Water, Methanoic Acid, and Methylamine on Quartz Surfaces

Surfaces	Water	Dissociated Water	Methanoic Acid	Methylamine
$\{0001\}$	−41.87	−136.52	−159.74	−
$\{10\bar{1}0\}$	−92.41	−221.25	−141.02	−
$\{10\bar{1}1\}$	−119.03	−250.66	−125.06	−195.30
$\{10\bar{1}\bar{1}\}$	−119.86	−233.36	−163.82	−198.92
$\{01\bar{1}1\}$	−182.19	−233.33	−195.63	−
$\{11\bar{2}0\}$	−224.99	−171.04	−343.52	−351.53
$\{11\bar{2}1\}$	−141.86	−	−78.06	−

In presence of the water molecules, the $\{01\bar{1}1\}$ surface reconstructed itself whereby the top dangling oxygen atoms of the lattice connected to a nearby three-coordinated silicon. The top two silicon atoms came very close (a distance 2.5 Å) and became five coordinated (Figure 3.6b). These reorientations stabilized the hydrated $\{01\bar{1}1\}$ surface as compared with the pure surface, which was evident from the decrease in surface energy (Table 3.4). The oxygen atom of the water molecule was kept at a distance of 2.62 Å from the five-coordinated silicon, and the two hydrogen atoms were connected to the same bridging oxygen by a hydrogen-bonding distance of 1.88 Å and 1.93 Å. The higher adsorption energy (−182.19 kJ mol⁻¹; Table 3.5) corresponds to chemisorption.

On the $\{10\bar{1}0\}$ surface, one of the hydrogen atoms of water was oriented toward the dangling oxygen at a hydrogen bond distance of 1.64 Å; the other hydrogen was kept away from the surface, having a distance of 2.61 Å from a bridging surface oxygen. The oxygen atom was kept 2.44 Å away from the three-coordinated surface silicon.

The $\{10\bar{1}1\}$ surface did not show much reorientation in surface structure, except for little bond length and angle variation in presence of water. One of the hydrogen atoms of water was at a hydrogen bond distance of 1.54 Å from the dangling surface oxygen while the other hydrogen was kept away from the surface.

The surface relaxation on the $\{10\bar{1}\bar{1}\}$ surface was similar to the $\{01\bar{1}1\}$ surface. The dangling oxygen bonded to three-coordinated silicon, two surface silicon atoms came closer (2.5 Å), and both silicons became five coordinated. One of the hydrogen atoms of the water molecule was at the hydrogen bond distance 1.72 Å from one bridging oxygen, while the other hydrogen was away from the surface. The adsorption energy (−194.89 kJ mole⁻¹) indicated chemisorption of water on this surface.

The $\{11\bar{2}0\}$ surface shows a different kind of reorganization in the presence of water where the dangling surface oxygen bonds with nearby three-coordinated silicon atoms and all the surface atoms become fully coordinated like a bulk. One of the hydrogen of the water molecule is connected to one bridging oxygen at a hydrogen bond distance of 1.83 Å, the oxygen of the water is 2.61 Å from the surface silicon, and the other hydrogen is away from the surface. This surface shows the highest adsorption energy (−224.99 kJ mole⁻¹) of water among the quartz surfaces. Water adsorption also stabilizes this surface, as evident from the decrease in surface energy (Table 3.5).

The adsorption of water on the $\{11\bar{2}1\}$ surface was very different from all other surfaces. Both the hydrogen atoms of the water molecule were at a hydrogen bond distance 1.70 Å and 1.81 Å from the two dangling quartz surface oxygens, and the oxygen of the water was 2.40 Å from the three-coordinated surface silicon. The surface configuration remained similar to the pure surface. The water molecule fit very well inside the surface valley, as seen from the side-view picture (Figure 3.6c).

3.3.5   HYDROXYLATION OF QUARTZ SURFACES

The dissociated water as H⁺ and OH⁻ was placed on the quartz surface for adsorption, and the assembly was energy minimized. We found that OH⁻ bonded to unsaturated silicon (hydroxylation) and that H⁺ bonded to dangling oxygen (protonation) on fresh cut unrelaxed surfaces, thus forming isolated (>SiOH) or vicinal (two silanol with a common oxygen atom of the silica network) and geminal (>Si(OH)₂) hydroxyl groups. This configuration was then energy minimized. This situation depicted hydroxylation of quartz under neutral conditions, and the surface energy values (Table 3.4) indicate that this made the quartz surface the most stable. The surface energy values, along with reaction energy for dissociative (Table 3.5) water adsorption, are indicative of why hydroxyl groups were invariably present on crushed α-quartz powders when measured in air using reflectance infrared spectroscopy (Koretsky et al. 1998). Hydroxylated surface energy (Table 3.4) was almost the same for all the Miller index surfaces. These very small and close ranges of surface energies obtained for hydroxylated surfaces, along with the tendency of silanol group formation, indicate that there is no preferred morphological plane for naturally occurring quartz. One possible cause of highly stabilized hydroxylated quartz surfaces, as argued from these simulations, could be that the atomic arrangement on or near the surface region does not change much as compared with pure bulk quartz crystal (Figure 3.7a and b).

The hydroxylated {0001} surface (Figure 3.7a) contained geminal hydroxyl groups as reported by de Leeuw et al. (1999b). The distance between the top silicon and the hydroxide oxygen atom was 1.59 Å, whereas the distances between the lower layer silicon atoms (with same x- and y-coordinates) and oxide (bulk) oxygen atoms were 1.61 Å and 1.60 Å, respectively. Two geminal hydroxyl groups were separated by 2.9 Å. Though the hydroxylated {0001} surface was the most stable, its hydroxylation energy was the least negative among all the surfaces, indicating less reactivity of this surface in presence of water.

On the {10$\bar{1}$0} surface, vicinal hydroxyl groups interacted with the nearby O–H group with a separation of 1.95 Å. The hydrogen atoms also interacted with surface oxygen atoms (having an O–H distance of 2.65 to 2.83 Å). The surface Si–O bond distance differed by 0.01 Å from that of the bulk. The hydroxylated {10$\bar{1}$1} surface ended up with two isolated hydroxyl groups after minimization. These hydroxyl groups did not interact with each other because they were more than 2.7 Å apart, but one hydrogen atom of one of the hydroxyl groups interacted with surface oxygen because the hydroxyl groups were closer (2.46 Å). This hydroxylated surface was very stable, as it did not require any significant reconstruction. The {10$\bar{1}$1} and {01$\bar{1}$1} surfaces ended up with same type of surface configuration as did the pure surface. The hydroxylated surface energy and reaction energy

(a)   (b)

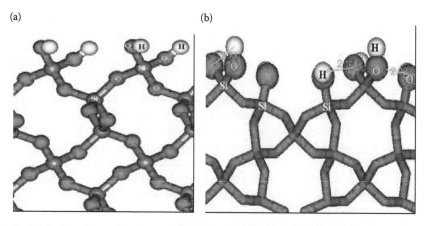

FIGURE 3.7   (a) Hydroxylated {0001} quartz surface showing geminal hydroxyl group, shown as ball and stick. (b) Hydroxylated {11$\bar{2}$0} showing both vicinal and geminal hydroxyl group on the surface, crystal shown asa cage, and the hydroxyl group as space filled. Color key: Si, gray; O, black; H, light gray.

for hydroxylation were same and indicative of surface stability and high tendency to surface hydroxylation (Tables 3.4 and 3.5). There was no interaction among the isolated surface hydroxyl groups and within hydroxyl hydrogen and crystal oxygen atoms because they were more than 2.7 Å apart.

The hydroxylated {11$\bar{2}$0} surface had an altogether different surface configuration (Figure 3.7b). It contained four hydroxyl ions per unit cell in which the top two-coordinated surface silicon atoms were involved in hydroxylation and two dangling oxygen atoms in protonation. This surface possessed both geminal and vicinal hydroxyl groups. As with the hydroxylated {0001} surface, there was no interaction among geminal hydroxyl groups because the hydrogen and oxygen atoms were separated by more than 2.71 Å. Two vicinal hydroxyl groups also did not show any interaction because they were more than 3.3 Å apart. There were interactions between vicinal and geminal hydroxyl groups because the hydrogen and oxygen atoms were separated by 2.08 Å.

One interesting finding from our computations is that the principal surface site–types expected for different quartz surfaces match very well with the findings from Koretsky et al. (1998). As reported by them in our simulations, the {0001} surface ended up with the geminal hydroxyl group only and both isolated and geminal hydroxyl groups were found on the {11$\bar{2}$0} surface. The number of coordinately unsaturated atoms obtained by them also corresponds very well with our finding of the number of hydroxyl groups per nm^2 formed through protonation or hydroxylation of unsaturated surface oxygen and silicon, respectively. For example, their finding of the number of unsaturated atoms on {0001}, {10$\bar{1}$0}, {11$\bar{2}$0}, and {01$\bar{1}$1} surfaces (and 9.6/nm^2, 7.5/nm^2, 8.7/nm^2, 5.9/nm^2, respectively) corresponds very well with the values obtained through our calculations (9.7/nm^2, 7.6/nm^2, 8.8/nm^2, and 6.0/nm^2, respectively).

3.3.6 Adsorption of Water on Hydroxylated Quartz Surfaces

A layer of silanol groups ($\equiv$ SiOH) is always present on quartz surface, as evident from our calculations and from a vast number of experimental findings (Gregg and Sing 1982). Complete dehydration is difficult; a temperature as high as 1100°C is reported to be insufficient to completely drive the water off from the quartz surface (Lowen and Broge 1961). In previous sections, we presented the maximum number of hydroxyl ions possible on different Miller index quartz surfaces as obtained though the molecular simulations. Ernstsson and Larsson (2000) reported on the presence of 4 to 5 OH groups/nm^2 on a vacuum-dried quartz surface by three different methods: solution depletion adsorption isotherms, adsorption microcalorimetry, and desorption by solvent extraction followed by surface analysis using electron spectroscopy for chemical analysis (ESCA). Xiao and Lasaga (1996) used disilicic acid [$(OH)_3Si-O-Si(OH)_3$] to represent an appropriate model for a neutral site of the quartz surface in quantum mechanical calculation. Their calculations show that hydroxyl ions are prevalent on quartz surface, especially in aqueous media. Accordingly, we attempted to model adsorption of water and surfactants on hydroxylated quartz surfaces, because it is believed that in flotation cells most of the quartz surface will remain hydroxylated.

Water on a hydroxylated {0001} surface bound more strongly than it did on a pure surface, as our calculations showed more negative adsorption energy (Table 3.6). The surface silanol group interacted with water by van der Waals interaction and forms hydrogen bonds. One of the hydrogen atoms of water came closer (within the hydrogen bond distance of 1.88 Å) to the oxygen of surface hydroxyl group while the oxygen of water was positioned 2.22 Å from the hydrogen of surface hydroxyl group.

On the hydroxylated {10$\bar{1}$0} quartz surface, hydrogen atoms of the water molecule were 2.19 Å and 2.22 Å from the two oxygen atoms of the surface hydroxyl groups. Oxygen of the water molecule remained 2.15 Å from the surface hydroxyl's hydrogen atom. Hydrogen of the water molecule also interacted with the bridging oxygen atom of the quartz surface at a distance of 2.13 Å (Figure 3.8a). On the hydroxylated {10$\bar{1}$1} surface, the oxygen atom of the water molecule oriented 2.07 Å and 2.4 Å away from the two nearby hydrogen atoms of the hydroxyl group, while hydrogen atoms bent toward oxygen atoms of hydroxyl (1.73 Å) and bulk crystal (1.89 Å), respectively.

TABLE 3.6

Adsorption Energies (kJ/mole) of Water, Methanoic Acid, and Methylamine on Hydroxylated Quartz Surfaces

Surfaces	Water	Methanoic Acid	Methylamine
{0001}	−77.9	−41.31	−1479.59
{10$\bar{1}$0}	−74.88	14.61	−1019.30
{10$\bar{1}$1}	−123.53	−41.44	−149.40
{10$\bar{1}\bar{1}$}	−137.79	−42.07	−528.25
{01$\bar{1}$1}	−131.82	−44.35	−118.82

(a)   (b)

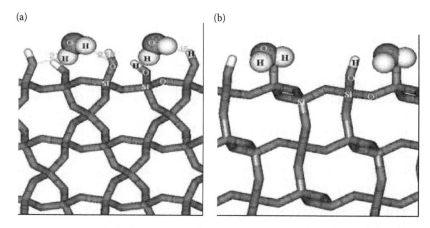

FIGURE 3.8 (a) Side view of water adsorption on hydroxylated {10$\bar{1}$0} and (b) {10$\bar{1}\bar{1}$} quartz surfaces. Crystal is shown as cage and water as space filled. Color key: Si, gray; O, black; H, light gray.

During adsorption on the hydroxylated {10$\bar{1}\bar{1}$} surface, hydrogen atoms of water bent toward the surface and the oxygen atom stayed away from it. One of the hydrogen atoms of the water molecule was 1.99 Å and 2.03 Å from two nearby hydroxyl oxygen atoms, and the other hydrogen was 2.15 Å from one of the hydoxyl oxygens and 2.35 Å from the bulk oxygen atom (Figure 3.8b). Oxygen of the water molecule oriented itself 1.99 Å and 2.07 Å from two nearby hydroxyl hydrogen atoms. The computed adsorption energy indicates that the water adsorbed strongly on this hydroxylated surface. The interaction of water on the hydroxylated {01$\bar{1}$1} surface was also quite strong, as evident from the adsorption energy values (Table 3.6). The oxygen atom of the water molecule was 2.06 Å and 2.22 Å from the two nearby hydroxyl group's hydrogen. One of the hydrogen atoms was 2.04 Å from one of the hydroxyl's oxygen. The other hydrogen atom was positioned 2.23 Å from the bulk oxygen and 2.35 Å from the hydroxyl's oxygen.

3.3.7 Adsorption of Methanoic Acid on Pure Quartz Surfaces

There was no significant change observed in the surface configuration after methanoic acid adsorption on the {0001} quartz surface. The adsorption energy was more negative than the water adsorption (Table 3.5), indicating that methanoic acid would be able to replace the water molecule from the surface. The oxygen atom of methanoic acid (O_{fC}) bound to the fully coordinated silicon at 2.93 Å, and the hydrogen (H_{fO}) atoms were situated at almost equal distances (2.65 Å, 2.66 Å, and 2.83 Å) from three bridging oxygens. The hydrogen atom connected to carbon atom stayed away from the surface (Figure 3.9a).

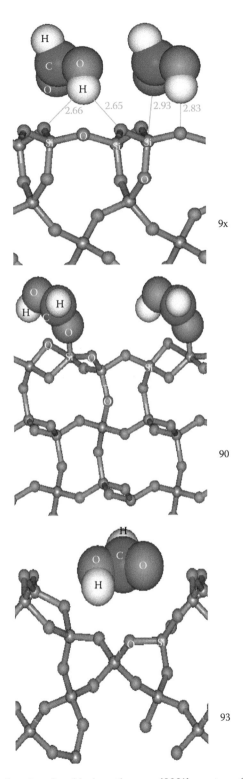

FIGURE 3.9 (a) Side view of methanoic acid adsorption on a {0001} quartz surface showing Si–O–Si bridge. (b) Methanoic acid bond formation on {01$\bar{1}$1} formation leading to a five-coordinated surface silicon. (c) Side view of methanoic acid adsorption on a crevice of the {11$\bar{2}$0} surface. The crystal is shown as ball-stick cage and the acid molecule as space filled. Color key: Si, gray; O, deep gray; H, light gray; C, black.

Methanoic acid adsorption on the $\{01\bar{1}1\}$ surface stabilized the surface where dangling oxygen bonded to the three-coordinated silicon and the oxygen of methanoic acid (O_{fC}) bonded to the silicon with a bond distance of 1.72 Å (Figure 3.9b). The silicon atoms in the top layer were five coordinated. In the near surface region, two silicon atoms with two oxygen atoms between them formed one ring while the rest of the structure consisted of six silicon rings. Methanoic hydrogen (H_{fO}) was located at 2.61 Å and 2.79 Å from two bridging oxygen atoms, while the other hydrogen (H_{fC}) atoms were turned away from the surface.

On the $\{10\bar{1}0\}$ surface, the double-bonded oxygen (O_{fC}) of methanoic acid bound (bond distance 1.97 Å) with the three-coordinated silicon atoms, and the hydrogen atom was away from the surface. The presence of methanoic acid did not change the surface structure, except for making the top silicon atoms four coordinated. One of the methanoic hydrogens (H_{fO}) oriented itself 2.55 Å away from the dangling oxygen while the other hydrogen atom (H_{fC}) was kept away from the surface. On the $\{10\bar{1}1\}$ surface, no change in surface configuration was observed; the surface energy was lower than it was in the presence of water (Table 3.4), but the computed adsorption energy for adsorption of methanoic acid was slightly more negative (Table 3.5) than that for water. The dangling oxygen and three-coordinated silicon remained present in the original surface, and the oxygen (O_{fC}) atom of methanoic acid was 2.55 Å from the three-coordinated silicon, while the hydrogen (H_{fO}) atom was 2.63 Å from the dangling oxygen.

The oxygen atom (O_{fC}) of methanoic acid bound (bond distance 1.78 Å) with three-coordinated silicon on the $\{10\bar{1}1\}$ quartz surface, making the surface silicon atoms fully coordinated. The hydrogen atom of methanoic acid (H_{fO}) leaned toward the dangling surface oxygen at a distance of 2.90 Å. The dissociative water adsorption energy was more negative than that of methanoic acid, so water hydroxylation was stronger than methanoic acid in this case. The $\{11\bar{2}0\}$ surface minimized to its lowest energy conformation in the presence of methanoic acid, as was the case with water adsorption. The dangling oxygen bonded to the three-coordinated silicon and the methanoic acid molecule lay in the valley of the crystal (Figure 3.9c).

3.3.8   ADSORPTION OF METHYLAMINE ON PURE QUARTZ SURFACES

The presence of methylamine on quartz $\{10\bar{1}1\}$ stabilized the surface more as compared with those in the presence of water and methanoic acid (Table 3.4). The adsorption energy for an amine molecule was more negative than for water and methanoic acid (Table 3.5), indicating that in an equilibrium condition methylamine will adsorb preferentially on this surface by displacing water and methanoic acid molecules. Methylamine adsorption did not induce any noticeable surface reconstruction. Nitrogen of amine oriented itself at the bump distance (2.52 Å with 70% of the sum of covalent radii) from the three-coordinated surface silicon, and the methyl group of the amine molecule was located away from the surface (Figure 3.10a).

The methylamine interaction with the quartz $\{10\bar{1}1\}$ surface also did not induce any significant structural changes. This surface exhibited less surface energy and more negative adsorption energy in the presence of methylamine in comparison to that of water and methanoic acid adsorption. Both nitrogen and carbon atoms of methylamine were within the bump distance (2.53 Å and 2.96 Å, respectively) from the three-coordinated surface silicon. The quartz $\{11\bar{2}0\}$ surface minimized in the presence of methylamine, just as it did in the presence of water and methanoic acid. The carbon and nitrogen atoms of methylamine molecule were not within the bump distance from any quartz surface atom. The methylamine molecule oriented itself into the crevice of the surface after minimization (Figure 3.10b).

3.3.9   ADSORPTION OF METHANOIC ACID ON HYDROXYLATED QUARTZ SURFACES

The adsorption energies calculated for the adsorption of methanoic acid on the minimized hydroxylated quartz surfaces are summarized in Table 3.6. The methanoic acid adsorption energy was

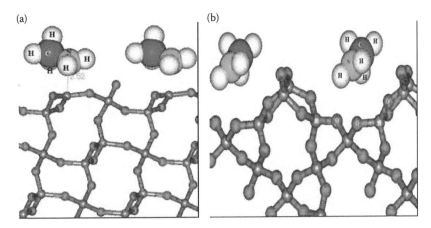

FIGURE 3.10 (A, left) Side view of methylamine adsorption on the quartz $\{10\bar{1}1\}$ and (B, right) $\{11\bar{2}0\}$ surfaces. The crystal is shown as ball-and-stick cage and amine as space filled. Color key: Si, dark gray; O, deep gray; H, light gray; N, gray; C, black.

less negative than that of water adsorption on all hydroxylated surfaces, indicating that the methanoic acid interactions were less strong than water and could not replace water from these surfaces. Methanoic acid on the hydroxylated $\{0001\}$ quartz surface showed no strong bonding with surface atoms. Oxygen of methanoic acid (O_{fC}) was 2.69 Å and 2.72 Å away from the two nearby geminal hydroxyl's hydrogen atoms. The hydrogen atom connected to the carbon atom of methanoic acid was 2.48 Å and 2.95 Å away from the surface oxygen of hydroxyl groups. The hydroxylated $\{10\bar{1}0\}$ quartz showed repulsion from methanoic acid, as evident from the positive adsorption energy value. The methanoic acid molecule was located 2.93 Å from surface hydroxyl groups (Figure 3.11a).

On the hydroxylated $\{10\bar{1}1\}$ quartz surface, the methanoic acid molecule oriented itself in such a manner so that the −COOH group lay parallel to the surface. Once again, the adsorption was not very strong and adsorption energy was less than water adsorption. The double-bonded oxygen atom of methanoic acid oriented itself 2.58 Å away from the surface hydroxyl's hydrogen. However, the hydrogen connected to oxygen of methanoic acid remained 2.60 Å away from the oxygen of the surface hydroxyl group. On the hydroxylated $\{10\bar{1}\bar{1}\}$ quartz surface, methanoic acid was located away from the surface, though it showed little adsorption tendency as evident from the negative adsorption energy. The minimum distance between adsorbent and adsorbate was 2.48 Å. The methanoic acid on the hydroxylated $\{01\bar{1}1\}$ surface adsorbed in the same fashion as it did on other hydroxylated surfaces.

3.3.10 ADSORPTION OF METHYLAMINE ON HYDROXYLATED QUARTZ SURFACES

The adsorption energies of methylamine with hydroxylated quartz surface are summarized in Table 3.6. The energy was most negative on the $\{0001\}$ surface. On the hydroxylated $\{0001\}$ quartz surface, the hydrogen atom connected to the nitrogen atom of methylamine bound with the surface hydroxyl group. In the presence of methylamine, one of the hydrogen atoms of the isolated hydroxyl ion (per unit cell) dissociated from the surface. The nitrogen atom of the amine remained at the hydrogen bond distance of 1.93 Å from another hydrogen atom of the hydroxyl group. The amine hydrogen atom connected itself to the oxygen of the dissociated hydroxyl group at a distance of 2.47 Å. The surface atomic arrangement also changed. The Si−O bond length in the near surface region increased to 1.63 Å (with a corresponding bulk bond length of 1.61 Å), and the O−Si−O bond angle changed to 106.8° and 111.5°, respectively (with corresponding bulk angles of 109.5° and 108.9°). The dissociated hydrogen atom remained within the bump distance of the crystal oxygen atom and the oxygen atom of the undissociated hydroxyl group (Figure 3.11b).

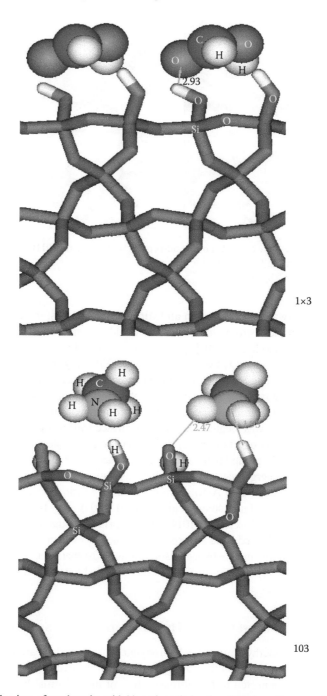

FIGURE 3.11 Side view of methanoic acid (a) and methylamine (b) adsorption on a hydroxylated {10$\bar{1}$0} quartz surface, showing hydroxylated crystal as a cage and the adsorbed molecule as space filled. Color key: Si, gray; O, deep gray; H, very light gray; C, black; N, light gray; H-dissociated, medium gray.

The hydroxylated {10$\bar{1}$1} surface in the presence of methyamine did not exhibit significant surface reconstruction. The amine hydrogen and hydroxyl oxygen atoms remained more than the hydrogen-bond distance away (2.55 Å). The nitrogen atom of methylamine was connected to the surface hydroxyl's hydrogen atom at 1.93 Å. The carbon atom tilted toward the surface at a distance of 2.38 Å. Methylamine adsorbed on the hydroxylated {01$\bar{1}$1} quartz surface by hydrogen bonding between the hydrogen atoms connected to the nitrogen of amine molecule and the two surface hydroxyl oxygen atoms at distances of 1.88 Å and 1.93 Å, respectively. The nitrogen atoms were also connected to the surface hydroxyl hydrogen atoms at 2.01 Å.

3.3.11 WOLLASTONITE SURFACE STRUCTURES AND ADSORPTION

We used a unit cell of $a = 7.926$ Å, $b = 7.320$ Å, and $c = 7.0675$ Å and $\alpha = 90.05°$, $\beta = 95.22°$, and $\gamma = 103.43°$ for wollastonite (Dana and Dana 1997). Using the potential model described in previous sections, the bulk crystal was allowed to relax to a minimum energy configuration (a = 8.00 Å, b = 7.43 Å, and c = 7.13 Å with angles as $\alpha = 90.39°$, $\beta = 93.57°$, and $\gamma = 103.74°$).

3.3.12 PURE SURFACES OF WOLLASTONITE

We modeled seven low indexed planes reported to be the most stable surfaces: {100}, {001}, {102}, {10$\bar{1}$}, {01$\bar{1}$}, {1$\bar{1}\bar{1}$}, and {1$\bar{1}$0}. Table 3.7 summarizes the surface characteristics of the minimized surfaces obtained as charge and dipole free cuts by the layer peeling and stacking process in METADISE without any surface reconstruction. In most of the Miller index planes before minimization, the surface consisted of low coordinated silicon atoms (coordination number 2,3) mixed with usual four-coordinated silicon atoms, low coordinated (3,4,5) calcium atoms, and nonbonded dangling oxygen atoms. After minimization a lot of surface reconstruction took place whereby some bonds broke and more new bonds were formed to minimize the energy. The surface atoms tended toward near or same coordination number as in the bulk coordination.

The cut {100} surface looked like a surface having a CaO and SiO_2 cluster consisting of one silicon atom connected to the two oxygen atoms and one nonbonded oxygen per unit cell surface area. After minimization, two-coordinated surface silicon atoms bound with the lone oxygen atom and with the lower layer silicon atom through its oxygen. However, another oxygen connected to the lower layer silicon atoms formed a bond with the top silicon atom. Finally, the surface consisted of two edge-sharing surface silicons having five-coordinated silicon in the lower layer (Figure 3.12a). Surface calcium atoms remained four coordinated, and their distance with nearby oxygen varies between 2.22 Å and 2.46 Å, as compared with 2.33 Å and 2.56 Å in the bulk crystal. Si–O bond distances in the surface region were between 1.58 Å and 1.91 Å while they were between 1.57 Å and 1.69 Å in the bulk.

On the {001} surface, nonbonded oxygen atoms bound with the silicon atom, making it fully coordinated. {10$\bar{1}$} was the only surface where surface silicon remained three coordinated. On

TABLE 3.7
Surface Characteristics and Energies for Wollastonite of Top Layer Silicon and Calcium Atom After Minimization

{hkl}	Area/Å^2	E_{unrel}/J m^{-2}	E_{rel} /J m^{-2}	Symmetry	Type*	Si$_{CN}$	Ca$_{CN}$
{100}	52.93	5.97	1.43	as	II	4,5	4
{001}	57.76	5.97	1.36	as	II	4	3
{102}	130.20	5.21	2.04	as	II	4,5	2, 4
{10$\bar{1}$}	75.73	3.44	1.70	as	II	3,4	3,4,5
{01$\bar{1}$}	80.17	5.02	1.54	as	II	4	3,4
{1$\bar{1}\bar{1}$}	87.66	3.91	1.39	as	II	4	3,3
{1$\bar{1}$0}	67.86	2.59	1.18	as	II	4	5,5,5

*Tasker classification. as, asymmetric surface; Ca$_{CN}$, surface calcium coordination number; rel, relaxed surface energy; Si$_{CN}$, surface silicon coordination number; unrel, unrelaxed surface energy.

(a)

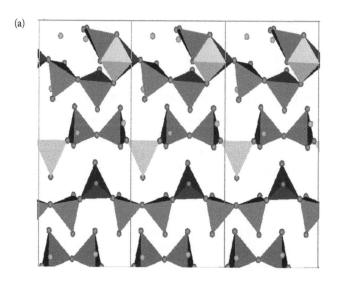

(b)

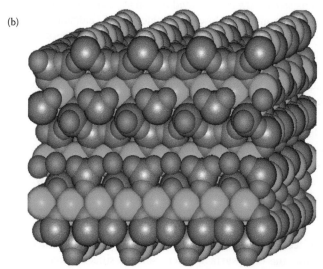

FIGURE 3.12   (a) The side view of relaxed surface structure of wollastonite {100} showing silicon as a gray tetrahedron parallel to the *b*-axis. A five-coordinated silicon is shown in gray, calcium atom as light gray, and oxygen as black. (b) Perspective view of relaxed {100} after rearrangement of lone surface oxygen and calcium. Top layer oxygen is shown as light gray and other oxygen as black, calcium as deep gray, and silicon as gray.

the $\{1\bar{1}\bar{1}\}$ and $\{1\bar{1}0\}$ surfaces, two Si–O tetrahedron chains running side by side met at a common oxygen atom near the surface region, thereby condensing the surface. As evident from Table 3.8, $\{1\bar{1}0\}$ was the lowest energy surface obtained. On this surface, all three calcium atoms exhibit near-bulk coordination with the surrounding oxygen atoms. The {100} and {102} surfaces possess edge-sharing five-coordinated silicon atoms near the surface region.

Some of the fresh cut surfaces ({100}, {001}, $\{10\bar{1}\}$, and {102}) where silicon was found to be four coordinated were subjected to the rearrangement that is carried out before minimization. Each non-bonded oxygen atom, if present, is transferred along with one top surface calcium atom (to maintain charge neutrality) to the bottom of Region I of Block 1. For the $\{10\bar{1}\}$ plane, we obtained two such cuts, with one and two lone oxygen atoms. However, after transformation and minimization,

TABLE 3.8

Surafce Energy of Reconstructed Wollastonite*

{hkl}	$E_{unrel}{}^a$/J m^{-2}	$E_{rel}{}^b$ /J m^{-2}	$Ca_{CN}{}^e$
{100}	1.10	0.72	4,5
{001}	2.37	1.42	3,5
{10$\bar{1}$}	1.51	0.86	4,5
{102}	7.42	1.50	4,4,4

*Before minimization, the surface silicon is tetrahedrally coordinated and the surface is devoid of lone oxygen.

these exhibited similar surface orientations and energies. On the {102} surface, one lone silicon, one calcium, and three lone oxygen atoms were transferred to the bottom layer so as to have four-coordinated silicon atoms on the surface. The surface energies of these reconstructed surfaces are summarized in Table 3.8. The surface energies of {100} and {10$\bar{1}$} decreased to a great extent due to the reconstruction. As expected, the {100} surface exhibited the lowest energy, which also is known to be the perfect cleavage plane for wollastonite.

A perspective view of the {100} surface is shown in Figure 3.12b, in which the Si–O tetrahedral chains run parallel to the *b*-axis, and ridges and valleys are formed parallel to the *c*-axis as one tetrahedron in every three protrudes upward along with two single-coordinated surface oxygen atoms, which are shown in yellow. The top calcium atoms remained between the two chains running side by side, having connection with four oxygen atoms with a distance of 2.25 Å to 2.41 Å. The distance between the two calcium atoms was 7.43 Å. The oxygen and calcium atoms were exposed and were mainly accessible to the interactions with adsorbing molecules, whereas silicon atoms were shielded by oxygen atoms (Figure 3.13a). The distance between silicon and oxygen atoms in the tetrahedron was between 1.60 Å and 1.67 Å, and the minimum oxygen–oxygen distance between the two chains running parallel to the *b*-axis was 3.18 Å.

For the {001} plane, we created a surface with four-coordinated silicon atoms and no lone oxygen atoms. The energy-minimized surface configuration is shown in Figure 3.13b, and the surface energy values are tabulated in Table 3.8. This surface was similar to the {100} surface where chains were parallel to the *b*-axis but two of the three repeating Si–O tetrahedra protruded upward, the calcium atom between them constitutd the ridge, and the third tetrahedron was oriented downward and kept its base parallel to the surface that constituted the valley region. The exposed surface consisted of calcium and oxygen atoms, and silicon atoms were shielded by oxygen atoms. The surface energy for the minimized reconstructed surface was lower than that for the as-cut surface. The reason for such behavior is that in the as-cut surface case we started with two lone oxygen atoms and two three-coordinated silicon atoms, which bonded together after minimization. Because the Si–O bond formation was exothermic, it decreased the block energy sufficiently, thereby decreasing the surface energy to a great extent.

The reconstructed {102} surface (after energy minimization) is shown in Figure 3.13c. This surface shows that silicon atoms, along with calcium and oxygen atoms, were exposed on the surface. Two of the top calcium atoms in the surface layer were connected with three and four number oxygen atoms, respectively at 2.09 Å to 2.75 Å. The chain axis was not parallel to the top surface, and no well-defined ridge and valleys were observed.

3.3.13 Adsorption of Water on Pure Wollastonite Surfaces

The {100}, {001}, and {102} surfaces were further selected for the adsorption studies in which before minimization the surface consisted of four-coordinated silicon atoms with no lone oxygen

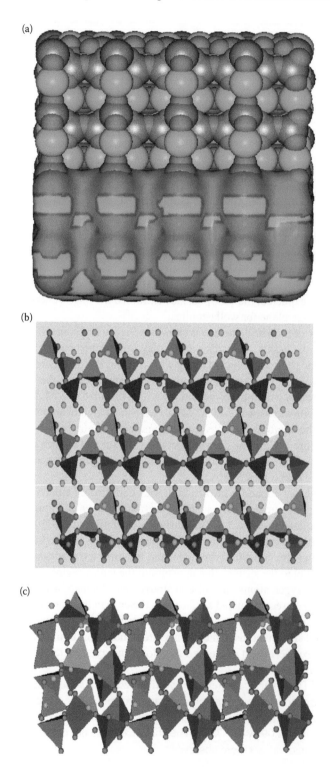

FIGURE 3.13 (a) Topview of a wollastonite {100} surface shown as space filled where the lower half shows, surface construction to show exposed atoms prevail on surface. (b) Side view of wollastonite {001} showing three repeating tetrahedral units, parallel to the b-axis, two protruding upward and one downward. (c) Sideview of wollastonite {102} surface. Color key: Ca, light gray; O, black; Si, gray tetrahedral.

TABLE 3.9

Surface Energies and Adsorption Energies for Water, Hydroxyl, Methanoic Acid, and Methylamine on Low Energy Pure Wollastonite Surfaces

Surface	Surface Energy (J m^{-2})					Adsorption Energy (kJ mole^{-1})			
	SE$_P$	SE$_W$	SE$_H$	SE$_M$	SE$_A$	AE$_W$	AE$_H$	AE$_M$	AE$_A$
{100}	0.72	0.44	0.69	0.43	0.40	−87.4	−462.23	−92.43	−100.88
{001}	1.42	1.18	0.46	1.02	−0.28	−85.34	−232.41	−139.92	−593.34
{102}	1.33	1.22	1.76	1.20	1.19	−93.23	−418.27	−102.18	−109.78

A, for methylamine on wollastonite surfaces; AEv, adsorption energy; H, for hydroxyl; M, for methanoic acid; SE, surface energy; W, for water.

atoms. Adsorption of water stabilized all three surfaces, as indicated by the decrease in surface energies (Table 3.9). On the {100} surface, one water molecule per unit cell area was placed so as to keep the distance between the top surface calcium atom and the oxygen atom of water at 1.85 Å and the distance between the top oxygen atom of wollastonite surface and the hydrogen atom of water at 1.74 Å. After minimization the distance between these atoms became 2.34 and 1.55 Å, respectively (Figure 3.14a). Water bound through one of the hydrogen atoms to the surface oxygen by hydrogen bonding.

Water on the {001} surface was placed so as to keep the distance between the water hydrogen atom and the crystal oxygen atom at 1.63 Å. After minimization the water was adsorbed in a flat orientation, having distance of 1.94 Å and 2.19 Å between two hydrogen atoms of water and two protruding crystal oxygen atoms, respectively.

On the {102} surface, water was placed by keeping the oxygen atom of water at 1.38 Å away from the surface calcium atom. Water did not stabilize this surface too much because the decrease in surface energy was much less than other adsorbates (Table 3.9), but negative hydration energy indicated that water adsorption was spontaneous. Here one of the hydrogen atoms of water is at a hydrogen-bond distance from the surface oxygen atom while another hydrogen was located away from the surface. The perpendicular type of water adsorption is shown in Figure 3.14c; the oxygen atom of water bends toward the surface calcium atom at 2.38 Å.

3.3.14 HYDROXYLATION OF WOLLASTONITE SURFACES

While grinding wollastonite, the possible type of exposed broken bonds were −O−Si$^+$ and −O−Ca$^+$, which are electron pair acceptors (Lewis acids), and −Si−O$^-$ and −Ca−O$^-$, which are electron pair donors, or bases (Pugh et al. 1995). It is plausible that Si−O bonds along the chain direction had stronger linkage energy, as of partial covalent nature, than Ca−O bonds between the chains, which are weak ionic bonds. Thus, Ca−O bonds broke off earlier than Si−O bonds. As a result, the crystal preferentially cleaved along the planes of high silicon content so that the exposed surface was probably closer to −O−Si$^+$ and −Si−O$^-$. In aqueous media, H$^+$ ions were attracted toward −Si−O$^-$ and OH$^-$ and made the surface charge neutral. Accordingly, we followed the procedure presented in the Equations 3.13 and 3.14 for surface hydroxylation, in which OH$^-$ is added to a low coordinated silicon atom and H$^+$ is attached to a dangling oxygen atom on a fresh cut surface. The lone oxygen, if present, is transferred, along with an equivalently charged calcium atom, to the bottom of the surface to make the surface more realistic.

Table 3.9 summarizes the surface energy and reaction energy for hydroxylation on three predominant wollastonite surfaces. Hydroxylation stabilized all the surfaces, as evident from their lower surface energy in the presence of the hydroxyl group than in the presence of pure surfaces.

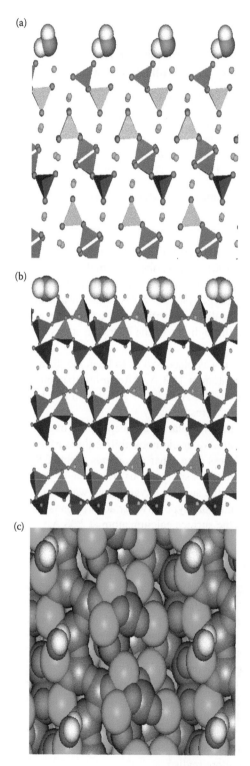

FIGURE 3.14 (a and b) Side view of water adsorption on wollastonite {100} and {001} surfaces, respectively where water is shown as space filled and the crystal as Si tetrahedral unit and oxygen and calcium as balls. (c) Top view of water adsorbed on wollastonite (shown as space filled) {102} surface. Color key: Ca, light gray; O, black; Si, gray; H, whitish gray.

Highly negative reaction energy is indicative of a greater tendency toward hydroxylation. For {100}, the surface energy decrease was very small, but the reaction energy was highly negative because its surface conformation resembled the bulk and the coordinate saturation tendency was highest in the presence of dissociated water (Figure 3.15a). The hydroxyl ions were isolated as they are more than 2.98 Å away from each other. Hydrogen bent toward the bridging oxygen and kept a minimum distance of 2.75 Å in between.

The surface characteristics of hydroxylated {001} were similar to that of hydroxylated {100}, but the decrease in surface energy was almost three-fold. On {102}, hydroxyl ions were isolated, with a minimum interaction distance of 2.67 Å. The chains were broken, and two Si–O tetrahedra containing one hydroxyl ion each remained as an isolated group while the three silicon with oxygen between them made one full ring. One of the silicons in this ring was in a five-coordination state and remained as an edge shared with the nearby tetrahedron. The hydroxylated surface assembly is shown in Figure 3.15b.

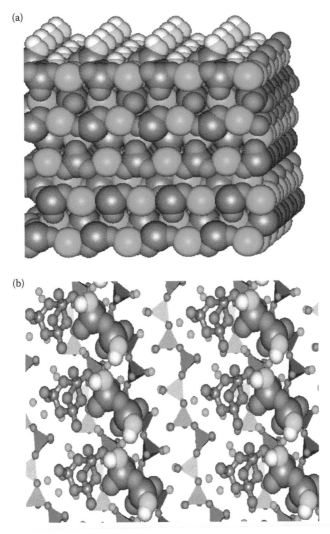

FIGURE 3.15 (a) Perspective view of dissociative water adsorption on wollastonite {100} surface all shown as space filled. (b) Top view of hydroxylated {102} wollastonite surface showing isolated Si–O–OH group as space filled, surface Si–O rings as ball and sticks, and underlying chains as silicon tetrahedral unit with calcium and crystal as balls. Color key: Ca, light gray; O, black; Si, gray; Ohydroxyl, light gray; H, whitish gray.

3.3.15   Adsorption of Methanoic Acid on Pure Wollastonite Surfaces

The methanoic acid was placed on reconstructed low-energy {100} surface with its double-bonded oxygen (O_{fC}) at 1.42 Å from the surface calcium atom. Methanoic acid stabilized the surface, and its adsorption energy was more negative than pure water adsorption but less negative than dissociative water adsorption. Thus, methanoic acid was able to replace pure water but not acidic or alkaline water. The energy-minimized assembly is shown in Figure 3.16a, in which double-bonded oxygen (O_{fC}) of methanoic acid is adsorbed at 2.18 Å from surface calcium and its hydrogen (H_{fO}) is 2.44 Å from the surface oxygen. The methanoic acid is adsorbed on the ridge portion and covers all exposed top calcium atoms.

On the {001} surface, methanoic acid was placed so as to keep its double-bonded oxygen (O_{fC}) 1.54 Å from the surface calcium, and the assembly was energy minimized. The surface was stabilized and adsorption energy was more negative than pure water adsorption as reported in Table 3.9. As shown in Figure 3.16b, methanoic acid molecule lay flat on the surface with its two oxygen atoms (O_{fC} and O_{fH}) interacting with two surface calcium atoms at 2.17 Å and 2.94 Å, respectively. Methanoic acid's hydrogen (H_{fO}) bent toward the surface oxygen, and the distance between those was 2.71 Å.

On the {102} surface, methanoic acid was placed in such a way that double-bonded oxygen (O_{fC}) and surface calcium atom were 1.46 Å apart. The minimization stabilized the surface and its adsorption energy was more negative than pure water adsorption but less negative than dissociated water adsorption. Energy minimized adsorption assembly is shown in Figure 3.16c, in which double-bonded oxygen (O_{fC}) is connected to surface calcium atoms at 2.30 Å. Hydrogen (H_{fO}) of methanoic acid interacted with two surface oxygen atoms and oriented itself at 2.63 Å and 3.01 Å from them. Only 20% of the exposed calcium atom was covered by one methanoic acid per unit cell surface area.

3.3.16   Adsorption of Methylamine on Pure Wollastonite Surfaces

Methylamine adsorption stabilized all the wollastonite surfaces as evident from Table 3.9. For the {001} surface, the surface and adsorption energies were highly negative as compared with other surfaces. On the {100} surface, methylamine was placed in such a way that hydrogen connected to nitrogen (H_{aN}) was 1.82 Å from the surface oxygen and the $-CH_3$ group was perpendicular to the surface. The relaxed surface structure is shown in Figure 3.17a, in which methylamine was adsorbed by keeping its C$-$N bond parallel to the ridge region of the wollastonite surface. The distance between the surface calcium and the nitrogen atom of methylamine was 2.75 Å. The two hydrogen atoms of methylamine attached to carbon (H_{aC}) were 2.75 Å and 2.69 Å from the two surface oxygen atoms, respectively. The two hydrogen atoms connected to nitrogen (H_{aN}) remained far from the surface without interaction with the surface atoms.

Methylamine on the {001} surface was placed by its hydrogen connected to nitrogen (H_{aN}) at 1.35 Å from the surface oxygen. The minimized configuration of the assembly is shown in Figure 3.17b; the very large negative adsorption energy indicates the higher tendency of adsorption. The C$-$N bond remained parallel to the surface, and the molecule occupied one valley region of the surface. Methylamine did not form any chemical bond but was attracted by the disperse force experienced by the two surface calcium situated at 2.84 Å and 3.06 Å. One hydrogen (H_{aN}) atom attached to the nitrogen remained in the same plane of C$-$N, and the interacting distance between this and the surface oxygen was 3.07 Å. The other H_{aN} remained away from the surface. One hydrogen (H_{aC}) connected to carbon atom of the methylamine molecule was situated between the two surface oxygens at distances of 2.53 Å and 2.80 Å.

On the {102} surface, the nitrogen atom of methylamine was placed at 2.10 Å from the surface calcium atom and the assembly was subjected to energy minimization. The surface exhibited stabilization, and the adsorption energy was near that of methanoic acid adsorption. As shown in Figure 3.17c, methylamine adsorbed on the valley region of the surface with its C$-$N axis between the

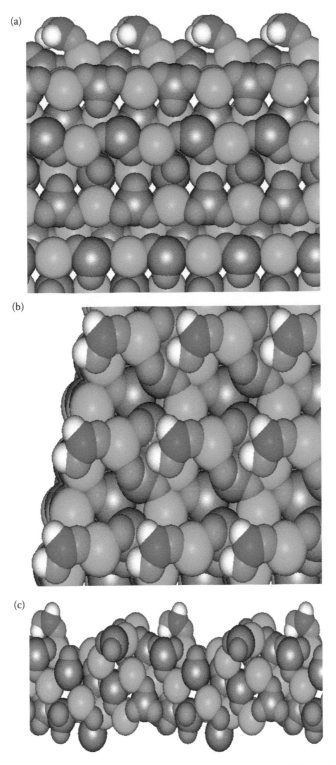

FIGURE 3.16 (a) Side view of methanoic acid adsorption on wollastonite {100} surface. (b) Top view of methanoic acid adsorption on wollastonite {001} surface. (c) Side view of methanoic acid adsorption on wollastonite {102} surface, shown as space filled. Color key: Ca, light gray; O, deep gray; Si, gray; C, black; O (methanoic acid), deep gray; H, white.

(a)

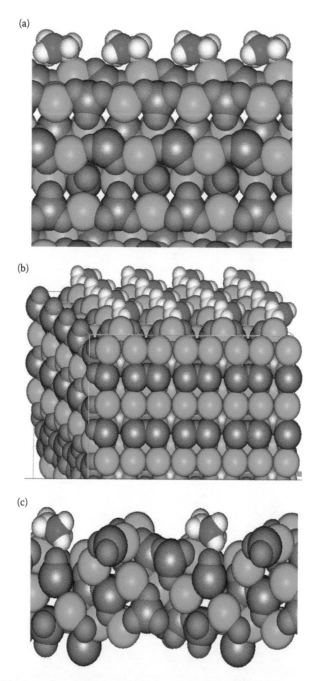

(b)

(c)

FIGURE 3.17   (a) Side view of methylamine adsorption on wollastonite {100} surface. (b) Perspective view of wollastonite {001} surface in presence of methylamine. (c) Side view of methylamine adsorption on wollastonite {102} surface shown as space filled. Color key: Ca, light gray; O, deep gray; Si, gray; C, black; N, very light gray; H, white.

grooves of two sets of exposed surface calcium running parallel to each other. The nitrogen atoms were 3.01 Å and 3.75 Å from the two surface calcium atoms. Two of the three hydrogen atoms connected to carbon (H_{aC}) were 2.52 Å and 2.61 Å from two surface oxygen atoms, and one hydrogen connected to nitrogen (H_{aN}) was 2.76 Å away from one surface calcium while the other hydrogen (H_{aN}) was far from the surface.

3.3.17 Dissolution of Calcium from Wollastonite Surfaces

It is well known that when wollastonite is brought into contact with pure water, Ca^{2+} ions release rapidly from the wollastonite surface in exchange of solution H^+ ions and that the solution pH increases to about 10 within a few minutes (Rimstidt and Dove 1986; Weissbart and Rimstidt 2000; Xie and Walther 1994). We chose those surfaces that have fully coordinated silicon without any lone oxygens on the surface for our dissolution studies. The hydroxylated surface that used for those surfaces were terminated with a three-coordinated silicon. The top calcium atom was removed from the surfaces, and two hydrogen atoms were added in place of each calcium atom to simulate the non-stoichiometric dissolution of wollastonite. The resultant structure was energy minimized, and the dissolution energy was obtained according to Equation 3.12, in which the self-energy of adsorbed species is replaced by the correction factor term for the removal of Ca^{2+} ions and the subsequent addition of two H^+ ions in place of each Ca^{2+} ion. According to Equation 3.16, the addition of H^+ can now be achieved for replacement of Ca^{2+} ions under acidic conditions or, as shown in Equation 3.17, under alkaline aqueous conditions, and the correction factors corresponding to acidic and alkaline condition are -46.65 eV and -45.498 eV, respectively. The dissolution energies for replacing each Ca^{2+} ion with two H^+ ions on the wollastonite surface under acidic and alkaline conditions are summarized in Table 3.10. We first considered six calcium ions for replacement, because these constituted the first layer of the wollastonite surface in our cell. Dissolution energies under acidic conditions are more negative than under alkaline conditions, and for almost all the Miller index planes, its negative value indicates that the wollastonite mineral is very prone to such a dissolution. The $\{01\bar{1}\}$ and $\{1\bar{1}0\}$ surfaces did not show any tendency toward dissolution under alkaline conditions. The wollastonite $\{01\bar{1}\}$ and $\{1\bar{1}\bar{1}\}$ surfaces showed dissolution tendencies only for the first Ca^{2+} from the surface under acidic and alkaline conditions, respectively. For most of the surfaces, dissolution energies remained negative after depletion of all Ca^{2+} from the first layer, and further leaching was also expected. Dissolution energy did not tend toward any fixed value, and from this

TABLE 3.10

Dissolution Energy of Ca^{2+} from Wollastonite Surfaces Substituted by the Addition of $2H^+$ Under Acidic and Alkaline Conditions

{hkl}	Dissolution Energy (kJ/mole)	Number of Ca²⁺/2H⁺ Replacement per Unit Cell Area					
		1	2	3	4	5	6
$\{100\}$	E_{acidic}	-53.23	-34.84	-67.57	-41.56	-57.37	-52.80
	$E_{alkaline}$	2.34	20.73	-11.99	14.02	-1.80	2.78
$\{001\}$	E_{acidic}	-118.14	-43.34	-80.23	-61.34	-66.54	-72.01
	$E_{alkaline}$	-62.57	12.24	-24.65	-5.77	-10.97	-16.43
$\{102\}$	E_{acidic}	-214.36	-145.53	-134.04	-129.26	-123.86	-117.43
	$E_{alkaline}$	-158.79	-89.96	-78.46	-73.69	-68.28	-61.86
$\{10\bar{1}\}$	E_{acidic}	-142.54	-332.23	-125.37	-120.73	-88.45	-90.31
	$E_{alkaline}$	-31.39	-221.08	-69.80	-65.15	-32.88	-34.73
$\{01\bar{1}\}$	E_{acidic}	-72.67	250.92	197.61	101.95	76.44	64.89
	$E_{alkaline}$	38.48	306.49	142.04	157.53	132.02	120.46
$\{1\bar{1}\bar{1}\}$	E_{acidic}	-161.39	51.84	18.24	-17.26	-21.16	-21.18
	$E_{alkaline}$	-50.24	107.42	73.82	38.32	34.42	34.40
$\{1\bar{1}0\}$	E_{acidic}	-10.38	-39.88	-42.94	-41.40	-46.86	-41.82
	$E_{alkaline}$	45.20	15.70	12.63	14.18	8.71	13.76

E_{acidic}, dissolution energy under acidic conditions as given by Equation 3.19; $E_{alkaline}$, issolution energy under alkaline conditions as given by Equation 3.20.

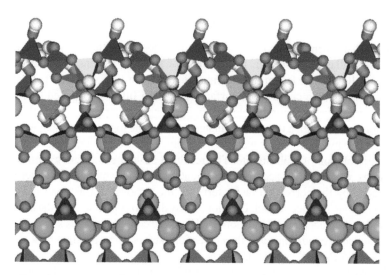

FIGURE 3.18 Side view of 100% Ca^{2+} depleted surface of wollastonite {100} showing the cluster formation and densification of leached surface due to connection between two side by side run chains. A five-coordinated silicon is shown as pentagon, Si(4) as gray tetrahedra, calcium as gray ball, hydroxides aas ball and stick, and oxygen is red hydrogen as white.

we concluded that leaching of Ca^{2+} is not a steady state; this corresponds well with experimental observation (Weissbart and Rimstidt 2000). The fluctuation in energy values, which is also evident from Table 3.10, was attributed to surface modification and densification due to replacement of Ca^{2+} with H^+ ions and the formation of a leach layer. As shown in Figure 3.18, after leaching out the entire first layer of Ca^{2+} from the {100} surface and the subsequent protonation, two Si–O tetrahedra running side by side met each other through a common bridging oxygen, and clusters were formed that constituted rings of three silicons with oxygen between them. This type of network, which has four silicon rings in a surface configuration, has been found by Casey et al. (1993) by Raman spectroscopic studies on leached wollastonite. This reconstruction also leads to crazing and poros- ity, which can contribute to further surface protonation and Ca^{2+} leaching. At the same time, dense clusters near the surface region formed due to the lack of attractive interactions between the network as leaching process continued. It is quite likely that whole chunk of Si–O–OH polymer-type cluster came out of the surface and broke into monomer in solution, increasing the surface Si^{4+} concentra- tion just after the depletion of the first few layers of Ca^{2+}. This kind of surface dissolution model was proposed by Weissbart and Rimstidt (2000). Our atomistic simulations indicate that Si^{4+} dissolution from the wollastonite surfaces was the most likely mechanism.

3.4 GENERAL DISCUSSION OF QUARTZ AND WOLLASTONITE FLOTATION SYSTEM

Using the static energy minimization technique, we have modeled most of the predominant surfaces and studied their active sites and coordination states and the stereochemistry of α-quartz. We have investigated adsorption behavior of molecular water, methanoic acid, and methylamine on pure α-quartz. Dissociated water adsorption depicting the situation of protonation and hydroxylation onto unsaturated surface has also been modeled. Using our knowledge of the persistent presence of surface hydroxyl group in aqueous media, we have studied adsorption of molecular water, metha- noic acid, and methylamine on hydroxylated surface.

Silicon and oxygen atoms were in a full-coordination state on the {0001} surface and formed Si–O–Si bridges, and the corresponding surface energy was at its lowest point. The plane near surface region was compact due to the formation of rings consisting of three silicon and three oxygen atoms as opposed to those found in the bulk, which consisted of six silicon and six oxygen atoms. This kind of structure prevailed for the associative adsorption of water, while for the dissociative adsorption the surface structure was more like bulk. This observation was true for all other surfaces, and the low surface energy observed could be attributed to no reconstruction because all surface reaction sites were fully coordinated due to hydroxylation and protonation. Associative water, methanoic acid, and methylamine adsorption stabilized all Miller index surfaces (Table 3.4).

The α-quartz showed all three kinds of surface hydroxyl groups as reported experimentally (de Leeuw et al. 1999b). Geminal hydroxyls were on the $\{0001\}$ surface. On the $\{10\bar{1}1\}$, $\{10\bar{1}\bar{1}\}$, and $\{01\bar{1}1\}$ surfaces, isolated hydroxyls were observed. The $\{10\bar{1}0\}$ surface exhibited vicinal hydroxyls, whereas on the $\{11\bar{2}0\}$ surface both geminal and vicinal hydroxyl groups were found. The number of hydroxyl ions on the surface indicates surface reaction site density, and our findings match very well with reported values (Koretsky et al. 1998).

Methanoic acid adsorption on a pure surface did not change surface configuration significantly. In most cases, it was bonded with its double-bonded oxygen with unsaturated surface silicon while its alkyl part remained away from the surface. Methylamine also did not bring any significant change in surface configuration after adsorption, and in most cases it oriented the amine group of the molecule toward the surface, while the alkyl chain was kept away from the surface.

It is evident from the adsorption energies presented in Table 3.5, that except for the $\{0001\}$ and $\{11\bar{2}0\}$ surfaces, dissociative water adsorption was more energetically favorable than other adsorbate molecules. The $\{10\bar{2}0\}$ surface was the most reactive for associative adsorption, and the preference was for methylamine, followed by methanoic acid, followed by water. The $\{10\bar{1}1\}$ and $\{10\bar{1}\bar{1}\}$ surfaces did better when hydroxylated, but for associative adsorption the preference was more toward methylamine, followed by methanoic acid, followed by water.

Water on hydroxylated quartz was adsorbed by orienting its hydrogen toward hydroxyl's oxygen. Methanoic acid on hydroxylated quartz was kept away from the surface with the double-bonded oxygen slightly attracted toward hydroxyl's hydrogen. Methylamine induced dissociation of the hydroxyl group from the $\{10\bar{1}0\}$ surface, and the amine group bends toward the surface while the alkyl chain remained away from the surface. A comparison of adsorption energy values (Table 3.5) indicates that, on a hydroxylated quartz surface, methanoic acid was least reactive, which follows the practical trends of the unfloatability of quartz without metal ion activation by organic acid group collectors. Except on the $\{01\bar{1}1\}$ surface, methylamine adsorption was energetically more preferable than water. This finding also corroborates industrial practice of quartz flotation with an amine group collector in aqueous medium.

We modeled morphologically dominant low indexed planes of an α-wollastonite crystal. An elaborate stereochemistry description of the surfaces and their inherent asymmetric atomic stacking in bulk and near the surface region was attained. Surface energy values obtained from minimization of freshly cut surfaces did not correspond to the morphologically most dominant surfaces in terms of lowest energies. We consider that this was due to irregularities present near the surface region of the neutral and nondipolar fresh cut surfaces. So, we attempted to reconstruct surfaces by transforming tramp oxygen and calcium and keeping the surface neutral. By doing this, we came up with highly stabilized surfaces. The $\{100\}$ surface became the lowest energy surface in a pure state, which we presume will be the perfect cleavage plane; this corresponds well with experimental findings. We also observed a parallel-running ridge-and-valley structure; these two chains running side by side met at a common oxygen and edge and shared a five-coordinated silicon in low energy surfaces. These findings have yet to be substantiated by experimentation.

We used three morphologically predominant wollastonite surfaces for adsorption studies. For molecular adsorption studies, we chose those neutral apolar surface cuts on which surface silicon was fully coordinated. The stabilization of wollastonite surfaces containing a three-coordinated silicon by adsorption of water in dissociated form was attempted, and a great extent of surface stabilization with lowering in surface energy was observed. Reaction energy for hydroxylation was highly negative for all the cases, indicating a higher hydroxylation tendency in order to obtain bulk coordination state. Hydroxylation on the {100} surface was the most reactive, and it made the surface configuration bulklike. Hydroxylation on the {102} surface formed a detached cluster that contained two silicons and isolated a three silicon–ring structure.

A comparison of surface energies in the presence of adsorbed molecules showed that in all the cases the wollastonite surface was stabilized and methylamine was the best candidate for adsorption. Among the added molecules, dissociated water adsorption showed the lowest adsorption energy on the {100} and {102} surfaces, whereas methylamine showed the least adsorption energy on the {001} surface. Thus, if all the molecules were present in a mixture, the {100} and {102} wollastonite surfaces would prefer to be hydroxylated. This finding is indicative of why wollastonite is not floatable by an amine or fatty acid collector molecule. If we consider only the adsorption in molecular form, the adsorption energy for methylamine was the most negative except on the {001} surface and the difference in the adsorption energies for methanoic acid, water, and methylamine were very close to each other. The preferred adsorption tendency in molecular form would be methylamine, followed by methanoic acid, followed by water. Because the {100} and {102} were the two most predominant surfaces, we can say from the consideration of the earlier observations that methylamine would replace either methanoic acid or water from the wollastonite surface. In actual practice where long-chain alkyl collector molecules are used to give the desired hydrophobicity, we can speculate with confidence that the difference in adsorption energies would be too insignificant to impart the desired selectivity if the surface activation procedure is not applied. This corresponds well with the reported observations in wollastonite flotation literature. Thus we can conclude that our simulation work predicts a well-established trend by considering the head group of the molecule, which is primarily responsible for adsorption.

The incongruent dissolution behavior of Ca^{2+} ions from the surfaces and the subsequent H^+ addition on the leached Si–O tetrahedron mirrored experimental finding of the high dissolution tendency of wollastonite in acidic and from some surfaces in an alkaline atmosphere. Here we observed that surface reconstruction and densification lead to crazing and porosity in the leached layer and that the dissolution energy of Ca^{2+} ion from the first layer did not correspond to the wollastonite stead steady-state situation. We found three-silicon ring formations in leached hydrogen-rich surfaces and a chunk of Si–O–OH polymeric chains formed that were detached with the bulk and had less interaction with the network modifying Ca^{2+} ion. We propose that with time this may detach from the surface and increase the Si^{4+} concentration in the solution.

After comparing the surface and adsorption energies for the most predominant naturally occurring surfaces for wollastonite {100} and hydroxylated quartz {0001} surfaces (Tables 3.6 and 3.9), we suggest that methylamine adsorbs more strongly on quartz than on wollastonite and should be able to float quartz preferentially, and the reverse is true with the carboxylic acid group molecule with selective wollastonite flotation from quartz. These observations are in agreement with the flotation literature.

3.5 SUMMARY

The idealistic cases of surfaces considered in this chapter had no resultant charges and dipole moment that were perpendicular to the surface. Though transforming a lone oxygen and a cation from the surface performs some reconstruction, natural surfaces invariably contain impurities and extended defects such as dislocation, stacking fault, and etch pits. Molecular dynamics (MD) simulations that take into account all these defects will more closely mimic natural situations. It is also plausible that some condensation reaction takes place on the leached wollastonite surface whereby water molecules are detached from the hydroxylated silica-like structure. We intend to consider and pursue these aspects in our future work.

In addition to investigating different wollastonite surfaces and their hydroxylation and dissolution behavior by static simulation, it would be interesting to study this further using MD simulations to add activation energies to these processes. Initial wollastonite dissolution is incongruent in terms of preference to calcium, and our modeled calculation was able to predict very well of the dissolution process. In the later stages of dissolution, silicon and calcium release, and both have their contribution in the overall dissolution process. The major steps in dissolution kinetics on wollastonite comprise the diffusion of calcium, silicon, and hydrogen atoms; the breaking of lattice bonds; and the formation of bonds with hydrogen ions. The first step is also dependent on the surface structure of hydrated silica in the calcium-depleted area, and it is very likely that the leached layer undergoes some reconstruction reactions whereby the connectivity of the network-forming species increases, which leads to greater resistance to the release of both silicon and calcium and the accommodation of hydrogen. However, the release of a network modifier (e.g., Ca) reduces the surface volume, and the reconstruction reaction increases local density and produces porosity and crazing, which exposes additional surface area to the solution. These two opposing processes play dominant roles in dissolution kinetics aiding our present knowledge, which to date is still incomplete. Employing MD simulation to all these cases would bring more light on wollastonite dissolution.

Further we have considered adsorbates that represent only the polar group of actual collector molecules, and our observations match the flotation practice. In future work, we look forward to include long-chain alkyl groups together with functional group for adsorption modelling. In actual case collectors are not in a molecular state but are in either cationic or anionic state, depending on the solubility product and the pH of the solution. We also intend to incorporate dissociated collector groups in our future studies. Metal ion activation plays a vital role in oxide and silicate mineral flotation by fatty acid collectors, and in subsequent work we shall address these processes. Furthermore, to simulate actual flotation practice, we will use MD simulations in which both water and collector molecules will be incorporated for their competition for surface sites and selectivity in the adsorption process.

We have shown that the atomistic simulation techniques are well placed to provide insight at the atomic level into the interactions between the substrate and adsorbate molecules. The calculations, albeit on the model systems, showed actual mineral flotation tendencies and provided scientific explanations for the experimental findings. Thus, in the long run, the studies aimed at discovering the basic mechanisms of surfactant interactions on mineral surfaces by computational techniques will be beneficial to the overall flotation process. The fundamental principles that govern the geometric and electronic structure of mineral surfaces and the processes occurring on these surfaces in flotation will be affected by computer simulations. It is our aim to use atomistic simulations as a successful predictive tool in the design of mineral-specific collector molecules for the efficient separation of valuable minerals in flotation. After addressing the mineral–surface reactions at the atomic level, we plan to design the molecular architecture of reagents specific to a mineral surface while addressing the criteria of electrostatic, stereochemical, and geometric match requirements of organic molecules at the mineral–water interface.

3.6   ACKNOWLEDGMENT

The financial support from the Centre for Advanced Mining and Metallurgy (CAMM) under Swedish Strategic Research Initiative program is gratefully acknowledged.

REFERENCES

Allan, N. L., Rohl, A. L., Gay, D. H., Davey, R. J., and Mackrodt, W. C. 1993. Calculated bulk and surface properties of sulfates. *Faraday Discuss.* 95: 273.

Baram, P. S., and Parker, S. C. 1996. Atomistic simulation of hydroxide ions in inorganic solid. *Philos. Mag. B.* 73: 49.

Black, S. N., Bromley, L. A., Cottier, D., Davey, R. J., Dobbs, B., and Rout, J. E. 1991. Interactions at the organic/inorganic interface: Binding motifs for phosphonates at the surface of barite crystals. *J. Chem. Soc. Faraday Trans.* 87: 3409.

Born, M., and Mayer, J. E. 1932. The lattice theory of ion crystals, *Zeits f. Physik* 75: 1.

Bragg, L., Claringbull, G. F., and Taylao, W. H. 1965. *Crystal Structure of Minerals.* G Bell and Sons, London.

Casey, W. H., Westrich, H. R., Banfield, J. F., Ferruzzi, G., and Arnold, G. W. 1993. Leaching and reconstruction at the surfaces of dissolving chain-silicate minerals. *Nature* 366: 253.

Catlow, C. R. A. 1997. *Computer Modelling in Inorganic Crystallography*. Academic Press, London.

Dana, J. D., and Dana, E. S., 1997. *Dana's new mineralogy: The system of mineralogy*. Wiley, New York.

Davey, R. J., Black, S. N., Bromley, L. A., Cottier, D., Dobbs, B., and Rout, J. E. 1991. Molecular design based on recognition at inorganic surfaces. *Nature* 353: 549.

Deer, W. A., Howie, R. A., and Zussman, J. 1992. *An Introduction to the Rock-Forming Minerals*. Longman Scientific & Technical, Harlow, Essex, UK.

de Leeuw, N. H., Watson, G. W., and Parker, S. C. 1995. Atomistic simulation of the effect of dissociative adsorption of water on the surface structure and stability of calcium and magnesium oxide. *J. Phys. Chem.* 99: 17219.

de Leeuw, N. H., Watson, G. W., and Parker, S. C. 1996. Atomistic simulation of adsorption of water on three-, four- and five-coordinated surface sites of magnesium oxide, *J. Chem. Soc. Faraday Trans.* 92: 2081.

de Leeuw, N. H., and Parker, S. C. 1998a. Surface structure and morphology of calcium carbonate polymorphs calcite, aragonite, and vaterite: An atomistic approach. *J. Phys. Chem. B* 102: 2914.

de Leeuw, N. H., and Parker, S. C. 1998b. Molecular dynamics simulation of MgO surfaces in liquid water using a shell model potential for water. *Phys. Rev. B*, 58: 13901.

de Leeuw, N. H., Parker, S. C., and Hanumantha Rao, K. 1998c. Modelling the competitive adsorption of water and methanoic acid on calcite and fluorite surfaces. *Langmuir* 14: 5900.

de Leeuw, N. H., and Parker, S. C. 1999a. Computer simulation of dissoiative adsorption of water on CaO and MgO surfaces and the relation to dissolution. *Res. Chem. Intermed.* 25: 195.

de Leeuw, N. H., Higgins, F. M., and Parker, S. C. 1999b. Modelling the surface structure and stability of alpha-quartz. *J. Phys. Chem. B.* 103: 1270.

Ernstsson, M., and Larsson, A. A. 2000. A multianalytical approach to characterize acidic adsorption sites on a quartz powder. *Colloids Surf. A.* 168: 215.

Gregg, S. J., and Sing, K. S. W. 1982. *Adsorption, Surface Area and Porosity*. Academic Press, London.

Hanumantha Rao, K., and Kundu, T. K. 2006. Atomistic simulation studies: Competitive adsorption of water, methanoic acid and methylamine on pure and hydroxylated quartz. In *Proceeding of XXIII Int Miner Process Cong*. Eds. G. Önal, N. Acarkan, M. S. Celik, F. Arslan, G. Atesok, A. Guney, et al. Promed Advertizing Limited, Isthanbul, Vol 3, 1729.

Harding, J. H., and Pyper, N. C. 1995. The meaning of the oxygen second electron affinity and oxide potential models, *Philos. Mag. Lett.* 71: 113.

Heywood, B. R., and Mann, S. 1992. Crystal recognition at inorganic-organic interface: Nucleation and growth of oriented $BaSO_4$ under compressed Langmuir monolayers. *Adv. Mater.* 4: 278.

Higgins, F. M., Watson, G. W., and Parker, S. C. 1997. Effect of relaxation on cation exchange in zeolite A using computer simulation. *J. Phys. Chem. B.* 101: 9964.

Kitchener, J. A. 1984. The froth flotation process: past, present and future-in brief. In *The Scientific Basis of Flotation*. Ed. K. J. Ives. Martinus Nijhoff Publishers, The Hague, 3.

Koretsky, C. M., Sverjensky, D. A., and Sahai, N. 1998. Calculating site densities of oxides and silicates from crystal structures. *Am. J. Sci.* 298: 349.

Kröger, F. A. 1973/74. *The Chemistry of Imperfect Crystals*. North Holland Press, Amsterdam.

Kundu, T. K., Hanumantha Rao, K., and Parker, S. C. 2003a. Atomistic simulation of the surface structure of wollastonite. *Chem. Phys. Lett.* 377: 81.

Kundu, T. K., Hanumantha Rao, K., and Parker, S. C. 2003b. Atomistic simulation of the surface structure of wollastonite and adsorption phenomena relevant to flotation. *Int. J. Miner. Process.* 72: 111.

Kundu, T. K. 2004. Atomistic simulation techniques for modelling inorganic/organic interface and flotation collector design, Doctoral thesis. University Press, Luleå, Sweden.

Kundu, T. K., Hanumantha Rao, K., and Parker, S. C. 2005. Competitive adsorption on wollastonite: An atomistic simulation approach. *J. Phys. Chem. B.* 109: 11286.

Lawrence P. J., and Parker, S. C. 1989. Computer modelling of oxide surfaces and interfaces. In *Computer Modelling of Fluids, Polymers and Solids*. Eds. C.R.A. Catlow, S. C. Parker, and M. P. Allen, NATO ASI Series, 219, Kluwer Academic Publishers, Dordrecht, The Netherlands.

Levein, L., Prewitt, C. T., and Weidner, D. J. 1980. Structure and elastic properties of quartz at pressure. *Am. Mineral.* 65: 920.

Lowen, W. K., and Broge, E. C. 1961. Effects of dehydration and chemisorbed materials on the surface properties of amorphous silica. *J. Phys. Chem.* 65: 16.

Mann, S. 1993. Molecular tectonics in biomineralization and biomimetic materials chemistry. *Nature* 365: 499.

Mazzucato, E., and Gualtieri, A. F. 2000. Wollastonite polytypes in the CaO-SiO2 system, Part I: crystallization kinetics. *Phys. Chem. Miner.* 27: 565.

Nagaraj, D. R. 1994. A critical assessment of flotation agents, in *Reagents for Better Metallurgy.* Ed. P. S. Mulukutla. Society for Mining, Metallurgy, and Exploration Inc, Littleton, Colo., 81.

Parker, S. C. 1983. Prediction of mineral crystal structures. *Solid State Ionics* 8: 179.

Parker, S. C., and Price, G. D. 1989. Computer modelling of phase transition in minerals. *Adv. Solid State Chem.* 1: 295.

Parker, S. C., Kelsey, E. T., Oliver, P. M., and Titiloye, J. O. 1993. Computer modelling of inorganic solids and surfaces. *Faraday Discuss.* 95: 75.

Pradip, Rai, B., Rao, T. K., Krishnamurthy, S., Vetrivel, R., Mielczarski, J. A., Cases, J. M. 2002. Molecular modelling of interactions of diphosphonic acid based surfactants with calcium minerals. *Langmuir* 18: 932.

Power, T. 1986. Wollastonite. *Ind. Mineral.* 220: 19.

Pugh, R., Kizling, M., Palm, C. O. 1995. Grinding of wollastonite under gaseous environments: influence on acidic/basic surface sites. *Miner. Eng.* 8: 1239.

Rimstidt, J. D., and Dove, P. M. 1986. Mineral/solution reaction rates in a mixed flow reactor: Wollastonite hydrolysis. *Geochim. Cosmochim. Acta* 50: 2509.

Ringnanese, G. M., Alessandro, D. V., Charlier, J. C., Gonze, X., and Car, R. 2000. First principles molecular-dynamics study of the (0001) α–quartz surface. *Phys. Rev. B* 6: 13250.

Sanders, M. J., Leslie, M., and Catlow, C. R. A. 1984. Interatomic potentials for SiO_2. *J. Chem. Soc. Chem. Commun.* 1271.

Schleyer, P. V. R. 1998. *Encyclopedia of Computational Chemistry,* John Wiley & Sons, Chichester, UK.

Schröder, K. P., Sauer, J., Leslie, M., and Catlow, C. R. A. 1992. Bridging hydroxyl groups in zeolitic catalysts: A computer simulation of their structure, vibrational properties and acidity in protonated faujasites (H-Y zeolites). *Chem. Phys. Lett.* 188: 320.

Shergold, H. L. 1984. Flotation in mineral processing. In *The Scientific Basis of Flotation,* Ed, K. J. Ives. Martinus Nijhoff, The Hague, 229.

Tasker, P. W. 1979. The stability of ionic crystal surfaces, *J. Phys. C. Solid State Phys.* 12: 4977.

Watson, G. W., Kelsey, E. T., de Leeuw, N. H., Harris, D. J., and Parker, S. C. 1996. Atomistic simulation of dislocations, surfaces and interfaces in MgO., *J. Chem. Soc. Faraday Trans.* 92: 433.

Watson, G. W., Tschaufeser, P., Wall, A., Jackson, R. A., and Parker, S. C. 1997. Lattice energy and free energy minimization techniques. In *Computer Modelling in Inorganic Chemistry,* Ed. C. R. A., Catlow. Academic Press, London, 55.

Weissbart, E. J., and Rimstidt, J. D. 2000. Wollastonite: Incongruent dissolution and leached layer formation. *Geochim. Cosmochim. Acta* 64: 4007.

Xiao, Y., and Lasaga, A. C. 1996. Ab-initio quantum mechanical studies of the kinetics and mechanisms of quartz dissolution: OH-catalysis. *Geochem. Cosmochem. Acta* 60: 2283.

Xie, Z., and Walther, J. V. 1994. Dissolution stoichiometry and adsorption of alkali and alkaline earth elements to the acid-reacted wollastonite surface at 25°C. *Geochim. Cosmochim. Acta* 58: 2587.

4 Molecular Dynamics Simulation Analysis of Solutions and Surfaces in Nonsulfide Flotation Systems

*Hao Du, Xihui Yin, Orhan Ozdemir, Jin Liu,
Xuming Wang, Shili Zheng, and Jan D. Miller*

CONTENTS

4.1   INTRODUCTION

Traditionally, mineral processing is an important area of the metallurgical engineering discipline and is, itself, a broad field that deals with the processing of mineral resources, including metallic, nonmetallic, and energy resources. Separation and concentration of mineral values generally are done from aqueous particulate suspensions; consequently, the study of the physicochemical properties of solutions and surfaces is critical for the development of improved processing technology (Fuerstenau et al. 2007). Similar to other scientific domains, advanced understanding is heavily dependent on the development of analytical methodologies and instrumentation. For example, froth flotation is one of the key technologies in mineral processing, and since its discovery in 1911, significant improvements have been made due to advanced analytical methodologies such as surface tension measurement, surface wettability determination, surface charge evaluation, and surface-state analysis by spectroscopic techniques (Bryant 1996; Dang 1997; Dillon and Dougherty 2002; Jiang and Sandler 2003; Kaminsky 1957; Luck 1973; Luck and Schioeberg 1979; Max and Chapados 2000, 2001; Nickolov and Miller 2005; Smith and Dang 1994a; Terpstra et al. 1990). Beginning in the 1990s, the rapid development of computer science and information technology has triggered the use of a large number of computer-aided instruments in scientific research, and some of them, such as atomic force microscopy (Bakshi et al. 2004; Miller and Paruchuri 2004; Nalaskowski et al. 2003) and sum frequency vibrational spectroscopy (Du et al. 2008; Nickolov et al. 2004; Wang et al. 2009), have been used in mineral-processing research due to their advantages in providing molecular level information regarding the physicochemical properties of solutions and surfaces.

Due to the inherent complexities of flotation separations in mineral-processing systems as well as past computational limitations, the use of computational chemistry in understanding the solution and surface physicochemical properties of these flotation systems has been limited. However, thanks to the rapid development of computer science and technology, especially the maturity of parallel-computing hardware and software since the early 2000s, this situation has changed significantly, as evidenced by the ever-increasing number of publications regarding the computational study of solutions and surfaces in the area of mineral processing (Brossard et al. 2008; Cao et al. 2010; Du and Miller 2007a, 2007b, 2007c; Du et al. 2007, 2008; Nalaskowski et al. 2007; Wang et al. 2009).

Computational chemistry uses the results of theoretical chemistry, with the aid of efficient computer programs, to simulate the structures and properties for systems of interest at a molecular level. While computational chemistry, normally functions as a complement to experiments, in some cases the simulations predict unobserved physicochemical phenomena and thus lead the experimental research. Computational chemistry ranges from the highly accurate *ab initio* quantum simulations, which are based entirely on theory from first principles, to the empirical or semiempirical simulations, which employ experimental results from acceptable models for atoms or related molecules, to the quite approximate simulations such as molecular dynamics (MD) simulations.

Ab initio methods solve the molecular Schrödinger equation associated with the molecular Hamiltonian based on different quantum-chemical methodologies that are derived directly from theoretical principles without inclusion of any empirical or semiempirical parameters in the equations. Though rigorously defined on first principles (quantum theory), the solutions from *ab initio* methods are obtained within an error margin that is qualitatively known beforehand; thus all the solutions are approximate to some extent. Due to the expensive computational cost, *ab initio* methods are rarely used directly to study the physicochemical properties of flotation systems in mineral processing, but their application in developing force fields for molecular mechanics (MM) and MD simulation has been extensively documented. (Cacelli et al. 2004; Cho et al. 2002; Kamiya et al.

2005; Larentzos et al. 2007; Leng et al. 2003; Liu et al. 2007; McNamara et al. 2004; Ponder et al. 2003; Zhou et al. 2007; Zhu et al. 2008).

The rigorous Hartree–Fock method without approximations is too expensive to treat large systems such as large organic molecules. Thus semiempirical quantum chemistry methods, which are based on approximated Hartree–Fock formalism by inclusion of some parameters from empirical data, have been introduced to study systems that do not necessarily require the exact quantum solutions to understand the physicochemical properties and are, therefore, very important in simulating large molecular systems.

The quantum simulation is capable of providing precise information on systems such as water structure, ion hydration state, and physical–chemical surface interactions. Due to the computational demand, this method is usually limited to small systems involving hundreds of atoms, which may not be sufficient to describe flotation systems of interest. Further, a successful quantum simulation needs a carefully designed initial state with energy achieving or approaching equilibrium, a condition that may not feasible due to the limited understanding of such systems.

In many cases, large molecular systems can be modeled successfully while avoiding quantum mechanical calculations entirely using molecular simulations. For example, MM simulations use Newtonian mechanics to model molecular systems and can be used to study small molecules as well as large biological systems or material assemblies that have many thousands to millions of atoms. The prototypical MM application is energy minimization, using a force field–appropriate algorithm to find the lowest energy conformation of a molecule or identifying a set of low-energy conformers that are in equilibrium with each other.

MD is closely related with MM, but instead of focusing on the study of potential energy surfaces for different molecular systems by implementing more "static" energy minimization methods as in MM simulations, the main purpose of MD simulation is the modeling of molecular motions, although it is also applied for optimization (e.g., using simulated annealing).

In MD simulation, atoms and molecules are allowed to interact for a period of time by approximations of known physics in order to explore the physicochemical properties of solutions and structures such as interfacial phenomena and the dynamics of water molecules and ions, thus providing detailed information and fundamental understanding on relationships between molecular structure, movement, and function (Brossard et al. 2008; Du and Miller 2007a, 2007b; Du et al. 2007a, 2007b; Lazarevic et al. 2007; Miller et al. 2007; Nalaskowski, et al. 2007). With MD simulation, scientists are able to examine the motion of individual atoms and molecules in a way not possible in laboratory experiments.

Monte Carlo (MC) molecular simulation is the application of MC methods to molecular problems. MC methods are a class of computational algorithms that rely on repeated random sampling to compute results. MC simulation methods are especially useful in studying systems with a large number of coupled degrees of freedom, such as fluids, disordered materials, strongly coupled solids, and cellular structures. The systems treated by MC simulations can also be modeled by MD methods. The difference is that the MC approach relies on statistical mechanics rather than MD. Instead of trying to reproduce the dynamics of a system, MC simulation generates the states of a system according to appropriate Boltzmann probabilities. Thus, it is the application of the MC algorithm to molecular systems.

Quantum simulation studies of surface and interfacial phenomena provide further insights regarding the relationships among structure, function, and dynamics at the atomic level. However, because surface chemistry problems in flotation systems and other mineral-processing systems are generally very complicated with many atoms involved, it is difficult to treat these systems using quantum mechanics (Stote et al. 1999). By using molecular simulation techniques including MM, MC, and MD simulations with proper selection of empirical potential energy functions (or force fields), which are much less computationally demanding than quantum mechanics, the problems become much more tractable. However, the saving of computational time using molecular simulation comes at a cost due to the numerous approximations introduced, leading to certain limitations. One of the most important problems is that no drastic changes in electronic structure are allowed; in other words, events like bond making and breaking (or chemical interactions) cannot be modeled (Stote et al. 1999). Discussion of different computational chemistry methods and their applications go beyond the scope of this chapter, and therefore, will not be detailed in following sections. For more detailed information, the reader is requested to refer the standard textbooks (Allen and Tildesley 2007; Cramer 2002; Leach 2001).

The accuracy of molecular simulation is heavily dependent on the force field used to describe the atomic and molecular interactions. The force field concept, originating from chemical physics and developed to define the interactions in a molecular system, is a set of functions that is crucial to the success of molecular simulation. Force fields have a wide variety of analytical forms and are parameterized to define the appropriate energy and forces (Stote et al. 1999). An optimized force field requires a reasonably good compromise between accuracy and computational efficiency. The force field is often optimized using experimental results and quantum mechanical calculations of small, model compounds. The ability of a force field to reproduce physical properties measured by experiment is tested. These properties include structural data obtained from x-ray crystallography and NMR, dynamic data obtained from spectroscopy and inelastic neutron scattering and thermodynamic data. Among the most commonly used potential energy functions are the AMBER (Pearlman et al. 1995), CHARMM (Brooks et al. 1983), GROMOS (van Gunsteren et al. 1996), CLAYFF (Cygan et al. 2004), and Dreiding (Mayo et al. 1990) force fields. MD and MM are usually based on the same classical force fields, but MD may also be based on quantum chemical methods like density functional theory. Furthermore, current potential functions are, in many cases, not sufficiently accurate to reproduce the dynamics of molecular systems, thus the much more computationally demanding *ab initio* MD method must be used.

From this discussion, it is clear that molecular simulation can provide essentially exact results for some statistical mechanics problems that would otherwise be insolvable or only tractable by approximation (Stote et al. 1999). By comparison of the simulation results with laboratory experimental results, the underlying model used in simulation can be validated. Eventually, if the model is valid, further information and insights are expected from the simulation of new systems. Thus, computer simulation works as a bridge connecting theoretical model prediction with experimental results. Some macroscopic properties of experimental interest such as equilibrium states and structural and dynamic properties can be rationalized using microscopic details of a system such as the masses of the atoms, the atomic interactions, and molecular geometries as provided from molecular simulation (Allen and Tildesley 2007).

In this chapter, the use of MD simulations in examining the solution chemistry and interfacial phenomena of selected flotation systems common to mineral processing are examined, including soluble salt minerals, phyllosilicate minerals, oxide minerals, and natural hydrophobic minerals. These initial MD simulation results are discussed in terms of their significance in the understanding of flotation technology for the separation and recovery of mineral resources.

4.2   SOLUBLE SALT MINERALS

As one of the most important soluble salt mineral systems, potash production is mainly achieved by selective flotation of sylvite (KCl) from halite (NaCl) and other gangue in concentrated brines with solution concentrations of about 5 M. Despite the successful industrial application of this flotation technology, the collector adsorption mechanisms at soluble salt mineral surfaces remain vague, mainly due to experimental limitations in the determination of the physicochemical properties of concentrated brines and to interfacial phenomena, including interfacial water structure and collector adsorption states. In this regard, MD simulation has been shown to be able to provide complementary information regarding soluble salt flotation systems, particularly sylvite, as evidenced by the large number of MD simulation publications dealing with the dynamic and thermodynamic characteristics of electrolyte solutions, interfacial phenomena, and collector adsorption states (Koneshan et al. 1998; Uchida and Matsuoka 2004). Also efforts have been extended to establish accurate force fields of various ions (Chowdhuri and Chandra 2001, 2003; Dang 1992, 1994, 1995; Dang and Smith 1993; Lee and Rasaiah 1994, 1996; Lynden-Bell and Rasaiah 1996; Smith and Dang 1994a, 1994b) for a more realistic description of brine systems.

In the following section, some of the MD simulation results on the physicochemical properties of concentrated alkali halide solutions, interfacial phenomena at salt crystal surfaces, and collector adsorption states are reported and discussed in detail. The DL_POLY_2, Daresbury laboratory, United Kingdom (Forester and Smith 1995) simulation package has been used, and the simulation details have been reported in previous publications (Cao et al. 2010; Du and Miller 2007; Du et al. 2007).

4.2.1 Concentrated Alkali Halide Solutions

Soluble salt flotation was carried out in saturated brine; therefore, an understanding of the physicochemical properties of concentrated alkali halide solutions is of particular significance. In the following section, an MD simulation analysis of concentrated LiCl, RbCl, and CsI systems is discussed.

4.2.1.1 Structural Properties

Because MD simulation analysis is based on particle interactions, extensive information can be obtained regarding the microscopic structures of aqueous alkali halide solutions, including ion hydration states, hydrogen-bonding network structure, and ion pairing structure, which can be obtained from the analysis of the radial distribution function (RDF), $g(r)$ (or pair correlation function), which describes how the atomic density varies as a function of distance from one particular atom.

4.2.1.1.1 Water–Ion Interactions

For aqueous alkali halide solutions, both cations and anions carry a mono charge regardless of the ion species and the sign of the ion charge. Therefore, the difference in the solution physicochemical properties is due to the difference in ion size, which can be described using the average radius of the primary hydration shell of selected ions. The sizes of hydrated ions determined from the distance to the first minimum in the corresponding pair–correlation functions in dilute solutions are listed in Table 4.1. It is obvious that the size of the hydrated Li^+ ion represented by the primary hydration shell radius of 2.775 Å is much smaller than that of water in its coordination shell, which is 3.325 Å. For larger cations such as Rb^+ and Cs^+, the size of the hydration shell increases accordingly and is larger than that of a cluster containing water molecules only.

Similarly, owing to their large size, the Cl^- and I^- hydration shells are much larger than that of a water cluster. The size of the hydrated ions significantly influences the structure of the system, as well as the dynamic properties of the system such as particle diffusion coefficients and system viscosity, which will be discussed in more detail in later sections.

Further, the hydration number N_h of ions (i) in the primary shell was calculated from the solute–oxygen distribution function (RDF) $g_{io}(r)$ using:

$$N_h = 4\pi\rho * \int_0^{R_1} g_{io}(r)r^2\,dr \tag{4.1}$$

where the upper limit corresponds to the first minimum in $g_{io}(r)$, which represents the radius for the first hydration sphere, and p^* is the density of particles in a given volume. The results are summarized in Table 4.2. As expected, the hydration number was dependent on ion size, and for the same salt concentration, the hydration number increases as the size increases. As the solution becomes more concentrated, the hydration number of cations decreases monotonically, with Cs^+ showing the

TABLE 4.1

Average Radii of the First Hydration Shell of Selected Particles

	Distance (Å)		
Particle Type	This Study (0.22 M)	Literature	Reference
O	3.33	3.30–3.45	Lee and Rasaiah 1994, 1996; Koneshan et al. 1998
Li^+	2.78	2.65, 3.10	Lee and Rasaiah 1994, 1996; Koneshan et al. 1998
Rb^+	3.75	3.75, 3.90	Lee and Rasaiah 1994, 1996; Koneshan et al. 1998
Cs^+	4.04	3.85, 4.20	Lee and Rasaiah 1996; Koneshan et al. 1998; Dang and Smith 1993
Cl^-	3.98	3.80, 3.85	Lee and Rasaiah 1996; Koneshan et al. 1998
I^-	4.22	4.30	Koneshan et al. 1998

Source: From Du, H., Rasaiah, J. C., and Miller, J. D., *J. Phys. Chem. B,* 111: 209, 2007. With permission.

TABLE 4.2

Hydration Numbers for Selected Cations and Anions as a Function of Salt Concentration

	Hydration Number of Ions					
Concentration (M)	Li^+/LiCl	Rb^+/RbCl	Cs^+/CsI	Cl^-/LiCl	Cl^-/RbCl	I^-/CsI
Dilute Solution	4.10	7.90	8.30	7.20	7.20	7.90
0.22	4.16	7.43	9.16	7.36	7.18	7.61
0.44	4.14	7.43	8.73	7.38	7.01	7.42
0.90	4.14	7.40	8.15	7.40	7.07	7.08
1.37	4.05	7.16	7.78	7.47	6.72	6.84
1.85	3.95	6.94	7.16	7.55	6.55	6.45
2.36	3.90	6.73	6.39	7.60	6.35	5.77
2.88	3.50	6.38	6.06	7.84	6.22	5.50
3.97	3.33	5.99	5.77	7.33	5.72	5.29

Source: From Du, H., Rasaiah, J. C., and Miller, J. D., *J. Phys. Chem. B,* 111: 209, 2007. With permission.

most significant drop and Li^+ the smallest decrease. For small ions like Li^+, the large local electric field binds water molecules in a tetrahedrally coordinated and tightly held hydrophilic hydration shell. As the solution becomes more concentrated, some of the water molecules participating in the Li^+ ion hydration shell are replaced by negatively charged Cl^- ions. Hence, the hydration number of Li^+ ions decreases as the solution concentration increases.

For very large cations, such as a Cs^+ ion, the local electric fields at the ion surface decrease significantly, and they behave more like uncharged particles (Koneshan et al. 1998). Hence, as the size of the cation increases, the dominating electrostatic hydrophilic hydration of small ions is gradually replaced by hydrophobic hydration of large ions in which hydrogen-bonded water molecules formed a disordered cage surrounding the ions.

When hydrophobic hydration dominates, water molecules are loosely bonded to the ion and the radius of the primary hydration shell is large, allowing more water molecules to be accommodated. As the number of ions increases in the solution, there are fewer water molecules available to complete the cages around ions, and the cations and anions pair up to include less water molecules in their hydration shells. Therefore, the hydration number around these ions also decreases with concentration and does so more dramatically than for the Li^+ ions.

4.2.1.1.2 Water–Water Interactions

The variation of water–water coordination as summarized in Figure 4.1 is observed to be strongly dependent on the hydration structures of ions in the solution. In pure water, tetrahedral structures of water molecules form dynamic networks, and the average hydration number is 4.5. When Li^+ ions are present, some of the water molecules in the water–hydrogen-bonding network will be replaced by tetrahedral hydrated Li^+ clusters, which are small in size and tight in structure, resulting in an increase in the water hydration number. As more Li^+ ions are present, the number of water molecules participating in the hydration of Li^+ ions increases accordingly, and one direct result is the water–water coordination number increases with salt concentration.

In contrast, when large Cs^+ ions are in the solution, water molecules are loosely caged around these ions. For the water molecules comprising the cages, one position has been taken by these large ions, and so only three other water molecules can be hydrogen-bonded to them and counted as primary waters of coordination; thus the water–water coordination number of these cation-bonded water molecules is less than that of a pure water cluster. As more ions are in the solution, more water molecules will be around ions, and consequently, for large alkali cations the water–water hydration number decreases monotonically with an increase in salt concentration.

4.2.1.1.3 Cation–Anion Interactions

The cation–anion coordination number as a function of salt concentration is summarized in Figure 4.2, in which both the primary and secondary coordination numbers are presented. Generally speaking, independent of ion species, both the primary and secondary coordination numbers between cation and anion increases when the salt concentration increases. Specifically, for the Li^+–Cl^- combination, an increase of the secondary coordination number, which is the interaction of hydrated or solvent separated ions, is much more significant than an increase of the primary coordination number, which is the naked ion–ion interaction. In contrast, for the Cs^+–I^- combination, the primary coordination increases much more significantly with concentration. And for the Rb^+–Cl^- combination, the change in the coordination number for both the primary and the secondary shells is not significant.

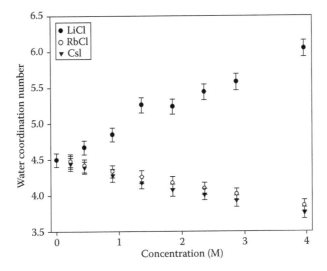

FIGURE 4.1 The first shell water–water coordination numbers in different salt solutions for various solution concentrations. (From Du, H., Rasaiah, J. C., and Miller, J. D., *J. Phys. Chem. B,* 111: 209, 2007. With permission.)

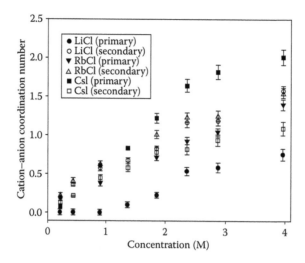

FIGURE 4.2 The coordination number of naked cation–anion pairs (direct cation–anion contact) and hydrated cation–anion pairs (a layer of water between cation–anion pair) in different salt solutions for various solution concentrations. (From Du, H., Rasaiah, J. C., and Miller, J. D., *J. Phys. Chem. B*, 111: 209, 2007. With permission.)

When Li$^+$ ions are in the solution, water molecules are tightly bonded by Li$^+$ ions—excluding the direct contact with Cl$^-$ ions—due to the strong local electric fields. When a large number of ions are in the solution, there is competition between ion–water and ion–ion interactions. The hydrophilic hydration of Li$^+$ ions dominates; therefore, driven by minimization of system energy, the coordination number between hydrated ions increases more significantly relative to the direct ion–ion contact.

For larger ions such as Cs$^+$ ions, loose hydrophobic hydration shells are formed around these ions. However, water molecules try to keep their integrity as pure water clusters. Because the direct cation–anion interaction, which will lead to fewer water molecules participating in the hydration shells, is energetically more favorable, the primary coordination number increases faster as a function of salt concentration. For the intermediate ion combination (e.g., Rb$^+$ and Cl$^-$), the ion–water and the ion–ion interactions are in close competition.

4.2.1.2 Dynamic Properties

The influence of ion size on solution properties is further revealed from the solution dynamics, which can be studied by analyzing particle residence time and diffusion coefficients as discussed in the following sections.

4.2.1.2.1 Residence Times

The residence time of water in the primary hydration shells of ions, τ is defined by

$$\tau = \int_0^\infty \langle R(t)\rangle dt \tag{4.2}$$

where $\langle R(t)\rangle$ is derived from time correlation functions (Berendsen et al. 1987; Chowdhuri and Chandra 2001; Koneshan et al. 1998) and is defined by

$$R(t) = \frac{1}{N_h}\sum_{i=1}^{N_h}[\theta_i(0)\theta_i(t)] \tag{4.3}$$

where $\theta_i(t)$ is the Heaviside unit step function, which has the value 1 if a water molecule i is in the hydration shell of the ion at time t and zero, otherwise N_h is the hydration number. The residence time of water in the primary hydration shells of ions and water clusters in different solutions is summarized in Table 4.3. First, the residence time for water molecules in the primary water shell increases inversely with ion size for both cations and anions. Small ions such as Li$^+$ ions strongly interact with water molecules in the hydration shell and therefore, significantly confine the movement of water molecules in the primary water shell, which results in a substantially longer residence time.

For large ions, such as Cs$^+$, Rb$^+$, and I$^-$ ions, the local electric fields are weak (as discussed earlier) and water molecules in the primary hydration shell are more loosely bonded to them. Thus, they are more mobile and the residence time of water is shorter. A similar observation has also been documented in the literature (Chowdhuri and Chandra 2001; Koneshan et al. 1998).

The residence times for the primary water shells around Li$^+$ ions increase significantly with solution concentration, while only a moderate increase was observed for Rb$^+$ ions and almost no noticeable change for Cs$^+$ ions. This can also be explained in terms of different hydration shell structures around these ions. Strong hydrophilic hydration is dominating for small ions.

At very high solution concentration, as much as one-fourth of the total water molecules are tetrahedrally bonded by Li$^+$ ions. Hence, when one water molecule diffuses away from an Li$^+$ ion, there is a very good chance for this water molecule to be tetrahedrally bonded by other Li$^+$ ions. Therefore, the diffusivity of water molecules around Li$^+$ ions decreases significantly with solution concentration. This is also the reason for the significantly increased residence times for water molecules around Cl$^-$ ions and within pure water clusters in an LiCl solution as a function of solution concentration.

When the ion size increases, hydrophobic hydration plays a more significant role; water molecules form cages and bond loosely around ions. Under these conditions, the diffusivity of water molecules in pure water clusters and around ions is of the same magnitude. In very concentrated solutions, despite the fact that almost all water molecules participate in the hydration of ions, the mobility of water molecules does not change significantly. Therefore, in CsI solutions the residence

---

TABLE 4.3

Residence Times of Water Molecules Around Various Particles as a Function of Solution Concentration

Concentration (M)	Residence Time (ps)								
	LiCl Solutions			RbCl Solutions			CsI Solutions		
	Li$^+$	Cl$^-$	H$_2$O	Rb$^+$	Cl$^-$	H$_2$O	Cs$^+$	I$^-$	H$_2$O
0.22	50.4	11	5.7	7.7	10.4	5.6	6.8	7.2	5.6
0.44	54.1	11.3	5.7	7.4	9.9	5.6	7.4	7.3	5.5
0.9	64.1	11.5	6.3	8.2	10.3	5.5	7.1	7.3	5.5
1.37	68.5	12.1	6.4	8.1	9.8	5.4	6.9	7.4	5.6
1.85	90.1	13.6	6.6	7.9	10.2	5.5	7	7.1	5.4
2.36	101	14.3	6.9	8.7	10.6	5.8	7.5	7.4	5.8
2.88	108.7	15.8	7.9	8.3	10.2	5.6	7.1	7.5	5.4
3.97	109.9	16.4	7.9	9.1	10.6	5.6	7.5	7.5	5.8

Source: From Du, H., Rasaiah, J. C., and Miller, J. D., *J. Phys. Chem. B*, 111: 209, 2007. With permission.

time of water does not have a noticeable change with a change in solution concentration. Only a marginal increase was noticed for water molecules in RbCl solutions.

4.2.1.2.2 Self-Diffusion Coefficients

The tracer diffusion coefficient D_i of an ion can be calculated from its mean square displacement (MSD) (Allen and Tildesley 1987; Chowdhuri and Chandra 2001; Koneshan et al. 1998) using the relation

$$D_i = \lim_{t \to \infty} \frac{1}{6}\langle [r_i(t) - r_i(0)]^2 \rangle \tag{4.4}$$

where $r_i(t)$ is the position of a particle i at time t.

The tracer diffusion coefficients for cations, anions, and water are shown as a function of salt concentration in Table 4.4. From the table, it can be seen that the diffusion coefficients for the ions and water molecules decrease with increasing salt concentration, and close analysis of the data in Table 4.4 indicates that, for the CsI solution, the decrease in diffusion coefficients with concentration is the smallest and largest for the LiCl solution. In both cases, as the concentration increases, the diffusion coefficients of the cation and anion become more nearly alike. A similar observation has been made in previous studies of aqueous solutions of NaCl and KCl at high concentrations (Allen and Tildesley 1987; Chowdhuri and Chandra 2001; Koneshan and Rasaiah 2000). As for the systems containing large ions (e.g., Cs^+, Rb^+, and I^- ions) with weaker electric fields, the observed decrease of ion diffusivity also can be explained in terms of increased ion pairing, which slows down the movement of ions and water molecules bound to them.

The self-diffusion of water molecules in LiCl solution changes dramatically in magnitude as the solution becomes denser. The self-diffusion coefficient of extended simple point charge (SPC/E) water at 25°C is 2.5×10^{-9} m^{2}s^{-1} and decreases by 42% as the concentration of the LiCl solution increases from 0.22 M to 3.97 M (see Table 4.4). Over the same concentration range, the diffusion coefficient of water in RbCl and CsI shows a slight increase (more apparent in CsI solutions) and finally an overall drop with factors of 21% and 6%, respectively, at 3.97 M from their

TABLE 4.4

Diffusion Coefficients for Cations, Anions, and Water Shown as a Function of Salt Concentration

Concentration, M	Diffusion Coefficient (10⁻⁹m²/sec)								
	LiCl Solutions			RbCl Solutions			CsI Solutions		
	Li⁺	Cl⁻	H₂O	Rb⁺	Cl⁻	H₂O	Cs⁺	I⁻	H₂O
Dilute solution	1.22	1.77	2.5	1.98	1.77	2.5	1.88	1.6	2.5
0.22	1.06	1.61	2.49	1.86	1.52	2.54	1.77	1.59	2.48
0.44	0.99	1.47	2.49	1.74	1.5	2.52	1.76	1.56	2.49
0.9	0.91	1.28	2.27	1.62	1.46	2.42	1.64	1.48	2.57
1.37	0.8	1.08	2.2	1.52	1.39	2.35	1.62	1.41	2.53
1.85	0.75	0.97	2.07	1.44	1.32	2.31	1.52	1.37	2.51
2.36	0.64	0.87	1.8	1.35	1.21	2.18	1.42	1.28	2.5
2.88	0.58	0.8	1.65	1.25	1.1	2.12	1.27	1.18	2.43
3.97	0.43	0.62	1.44	1.05	1.02	1.97	1.23	1.12	2.35

Source: Data from Dang, L. X., *J. Am. Chem. Soc.,* 117: 6954, 1995; Du, H., Rasaiah, J. C., and Miller, J. D., *J. Phys. Chem. B,* 111: 209, 2007; Koneshan, S., Rasaiah, J. C., Lynden-Bell, R. M., and Lee, S. H. *J. Phys. Chem. B,* 102: 4193, 1998. With permission.

value of 0.22 M. The decrease in the diffusion coefficient of water with a concentration of LiCl can be attributed to fewer unbound water molecules outside the tightly bound hydration sheaths of Li^+ ions and LiCl ion pairs.

The initial increase in the diffusion coefficient of water as the salt concentration of RbCl and CsI increases is interesting and may be attributed to the increase in number of water molecules in the more loosely bound hydration cages of these ions. As the concentration increases further, increased ion pairing slows down the movement of the ions and water molecules bound to them as discussed earlier. In CsI solutions, water molecules exist either in ion hydration shells or in pure water clusters. Nevertheless, the mobility of water molecules in these two situations does not have a significant difference, as discussed previously. Therefore, in these cases, the diffusivity of water molecules is relatively independent of the solution concentration.

4.2.1.3 Viscosity

In addition to these microscopic properties of the solutions, MD simulation analysis can be further extended to macroscopic properties. For example, the shear viscosity η, which describes resistance to flow, can be calculated using the equilibrium fluctuation of the off-diagonal components of the stress tensor (Daivis et al. 1996). Averaging the three off-diagonal components can improve the statistical convergence of the calculation. It was shown by Daivis and Evans (Daivis et al. 1996; Haile 1997) that, for an isotropic system, the statistical convergence can be further improved by including equilibrium fluctuations of the diagonal components of the stress tensor. Therefore, the generalized Green–Kubo formula can be expressed as:

$$\eta = \frac{V}{10k_BTt} \int_0^\infty \sum_\alpha \sum_\beta q_{\alpha\beta} \langle P_{\alpha\beta}(t)P_{\alpha\beta}(0)\rangle dt \tag{4.5}$$

where V and T are volume and temperature of the simulation cell, k_B is the Boltzmann constant, $q_{\alpha\beta}$ is a weight factor ($q_{\alpha\beta} = 1$ if $\alpha = \beta$, $q_{\alpha\beta} = 4/3$ if $\alpha \neq \beta$), and $P_{\alpha\beta}$ is defined as:

$$P_{\alpha\beta} = (\sigma_{\alpha\beta} + \sigma_{\beta\alpha})/2 - \frac{\delta_{\alpha\beta}}{3}\left(\sum_\gamma \sigma_{\gamma\gamma}\right) \tag{4.6}$$

where $\delta_{\alpha\beta}$ is the Kronecker delta ($\delta_{\alpha\beta} = 1$ if $\alpha = \beta$, $\delta_{\alpha\beta} = 0$ if $\alpha \neq \beta$).

By analogy with the Einstein relation for self-diffusion, the shear viscosity can be calculated using the mean-square "displacement" of the time integral of the shear components of the stress-tensor (Mondello and Grest 1997).

$$\eta = \lim_{t\to\infty} \frac{V}{20k_BTt} \left\langle \sum_\alpha \sum_\beta q_{\alpha\beta}[L_{\alpha\beta}(t) - L_{\alpha\beta}(0)]^2 \right\rangle$$

$$= \lim_{t\to\infty} \frac{V}{20k_BTt} \left\langle \sum_\alpha \sum_\beta q_{\alpha\beta}[\Delta L_{\alpha\beta}(t)]^2 \right\rangle \tag{4.7}$$

where

$$\Delta L_{\alpha\beta}(t) = \int_0^t P_{\alpha\beta}(t')dt' \tag{4.8}$$

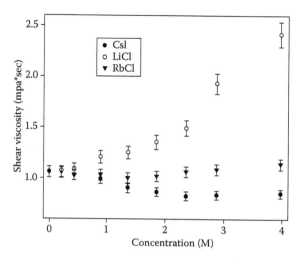

FIGURE 4.3 System shear viscosity changes as a function of solution concentration. (From Du, H., Rasaiah, J. C., and Miller, J. D., *J. Phys. Chem. B*, 111: 209, 2007. With permission.)

It has been shown by Mondello and Grest (1997) that Equations 4.7 and 4.8 give the same results as the Green–Kubo formulation for MD simulations of short-chain alkanes. The viscosities calculated using Equations 4.7 and 4.8 from our MD simulations of brine solutions as a function of solution concentration are shown in Figure 4.3. As expected, the size of the ions has a significant influence on the viscosity of the solution. When small ions are present (LiCl solution), the system shear viscosity increases monotonically with solution concentration. As the ion size increases (RbCl solution), the system viscosity shows very weak dependence on the solution concentration. Further increase in ion sizes (CsI solution) leads to an obvious decrease of viscosity as a function of solution concentration. This variation of solution viscosity as a function of ion size and solution concentration has also been observed experimentally by several research groups (Hancer et al. 2001; Jiang and Sandler 2003). The results from this simulation (shear viscosity) are clearly in good qualitative agreement with theoretical and experimental results reported in the literature (Jiang and Sandler 2003).

In an LiCl solution, Li^+ ions interact strongly with either water molecules or corresponding anions and form stable tetrahedral structures, which contribute substantially to a "stickier" system. Though Cl^- ions, due to their large size, do not form strong bonds with water molecules, the influence of cations is dominating. Consequently, the mobility of the solution decreases, and the system's shear viscosity, which describes mobility macroscopically, increases. The higher the solution concentration, the more significant the role that the Li^+ ions play, and the higher the viscosity. In contrast, when large ions (e.g., Cs^+ and I^-) are present in the solution, loose hydrophobic shells form around those ions, and the ion–water interaction is not as strong as water–water interaction, which accounts for the decrease in the system's shear viscosity.

4.2.2 Interfacial Phenomena of Alkali Chloride Salt Crystals

Soluble salt flotation occurs in saturated solution in which the salt crystal surface is in dynamic balance between crystallization and dissolution, making it difficult to examine the interfacial properties using traditional experimental measures. In this regard, the water structure at selected alkali halide salt surfaces has been studied using MD simulation. Equilibrium surface charge signs for these salt minerals in saturated solution have been calculated by considering the ion hydration and water dipole distribution at salt–saturated brine interfaces, and the results are compared with analytical modeling and experimental results (Hancer et al. 2001). Water structures at salt–brine interfaces have been examined in detail with respect to the ion structure making-and-breaking ability. Explanations for collector adsorption behavior at different alkali halide salt surfaces have been provided based on ion characteristics (e.g., size and charge) and ion–water interactions. The simulation details are discussed elsewhere (Du and Miller 2007).

4.2.2.1 Structural Properties

In this section, MD simulation results regarding water structures at selected alkali chloride salt surfaces are reported and discussed. The sign of the salt surface charge in saturated brine solutions is also discussed and compared with experimental results.

4.2.2.1.1 *Water Density Distribution at Salt Surfaces*

The relative number of density distributions along the surface normally obtained from MD simulation are shown in Figure 4.4 for selected alkali halide systems. As expected, the effect of cation size on the orientation of water molecules at the salt surface is studied by analyzing the position of the oxygen atom relative to the position of hydrogen atoms of the water molecules. From Figure 4.4a, it is noticed that at the LiCl salt surface, there are two layers of closely packed water molecules with distinguishable orientation suggested by the position of the two major oxygen atom density peaks relative to the corresponding hydrogen peak, which is sandwiched between the two water–oxygen layers.

Increasing the cation size to the Na^+ ion, the second major oxygen peak observed at the LiCl surface starts diminishing and the magnitude of the primary oxygen peak increases as seen in Figure 4.4b for the NaCl solution particle number distribution. This observation suggests the different orientation of water molecules at the LiCl and NaCl surfaces. The different orientation of water molecules at the interfacial region is directly related to the size and also to the electronegativity of the cations in the salt crystals. The hydrogen peak, which is located ahead of the primary oxygen peak, is obtained by further increase in the cation size to the K^+ ion and Rb^+ ion (Figure 4.4c for KCl and Figure 4.4d for RbCl), and the magnitude of this peak increases with cation size.

The obvious change in water molecule orientation at different alkali chloride crystal surfaces is due to the variation of cation size and consequently the columbic interaction between water molecules and lattice ions at the salt surface. When small Li^+ ions are present, the interaction between surface Li^+ ions and the oxygen atoms of water molecules dominates when compared with the interaction of water hydrogen and surface Cl^- ions due to the larger local electric field around the small Li^+ ions (Koneshan et al. 1998). Therefore, at the LiCl crystal surface, a layer of water molecules with oxygen atoms in close contact at the crystal surface, and hydrogen atoms stretched out to the water phase (represented schematically in Figure 4.5a) is energetically favorable. The presence of the second water layer with an opposite orientation neutralizes the dipole moment of water molecules in the first layer. As the cation size increases, the magnitude of the local electric field decreases accordingly. Thus, at the NaCl surface, water molecules tilt to such an orientation that water oxygen atoms are further away from the crystal surface while the water hydrogen atoms are closer to the surface, as seen in Figure 4.5b. Similar water structures at the NaCl crystal surface have also been obtained from *ab initio* calculation (Pramanik et al. 2005), density function calculation (Park et al. 2004), and MC simulations (Engkvist and Stone 2000). With further increase of cation size to K^+ and Rb^+ ions, an H–O–H configuration of interfacial water molecules, as seen in Figure 4.5c, is favored.

4.2.2.1.2 *Water Dipole Distribution at Salt Surfaces*

The interfacial water structure at different alkali halide salt surfaces can be further studied from the water dipole moment distributions at the crystal surface. Figure 4.6 describes the water dipole

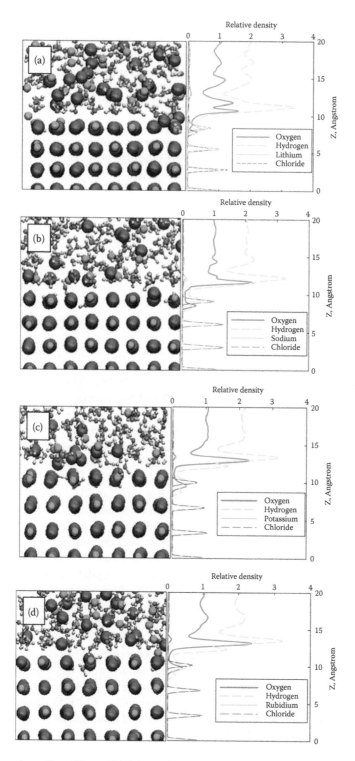

FIGURE 4.4 Snapshots of equilibrated LiCl (a), NaCl (b), KCl (c), and RbCl (d) systems and particle number density distributions along the surface normal. Color key: Purple, chloride ion; green, cation; red, oxygen; yellow, hydrogen. (From Du, H. and Miller, J. D., *J. Phys. Chem. C*, 111: 10013, 2007c. With permission.)

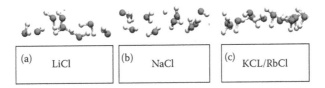

| (a) LiCl | (b) NaCl | (c) KCL/RbCl |

FIGURE 4.5 Schematic representations (from simulation snapshots) describing different orientations of interfacial water molecules at different alkali chloride surfaces. (From Du, H. and Miller, J. D., *J. Phys. Chem. C*, 111: 10013, 2007c. With permission.)

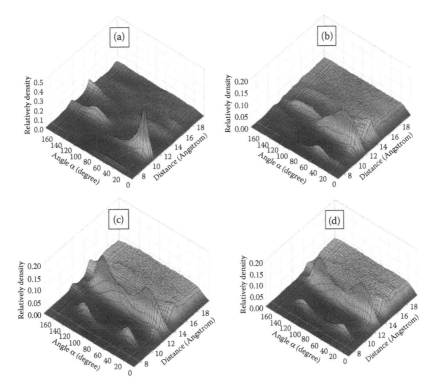

FIGURE 4.6 Water dipole moment density distribution along LiCl (a), NaCl (b), KCl (c), and RbCl (d) salt surface normal. (From Du, H. and Miller, J. D. *J. Phys. Chem. C*, 111: 10013, 2007c. With permission.)

moment density distribution profiles along the crystal surface normal for LiCl, NaCl, KCl, and RbCl. In Figure 4.6a, a sharp dipole moment distribution peak at the LiCl surface is found at around 20° followed by another major peak at around 120°, complementing the water molecule orientation proposed previously in Figure 4.5a. At the NaCl surface, seen in Figure 4.6b, the water dipole moment density distribution peaks become broader and flatter. Then, the primary dipole moment density peak shifts to larger angles. As previously discussed, the Na$^+$ ion is larger than the Li$^+$ ion. Consequently, the local electric field is weaker. On the other hand, attracted by surface Cl$^-$ ions, hydrogen atoms in water are closer to the NaCl salt surface, leading to the orientation proposed in Figure 4.5b. With further increase in the cation size, some dramatic changes are observed for the KCl system, as illustrated in Figure 4.6c. The two interfacial water layers at the LiCl and NaCl crystal surfaces have been replaced by one layer with two density peaks located at around 130° and 40°, respectively, and the magnitude of the peak at 40° is slightly larger.

The orientation of interfacial water molecules at the KCl crystal surface is a result of the balance between the K^+ ion–oxygen and Cl^- ion–hydrogen interactions and suggests that these two interactions are of comparable magnitude. At the RbCl crystal surface (see Figure 4.6d), the results are similar to those observed at a KCl crystal surface. The water molecules show significant disorder.

4.2.2.1.3 Equilibrium Surface Charge

The surface charge of alkali halide crystals in a saturated solution has been calculated as the summation of cation and anion number density at a designated distance from the surface. The crystal surface is positively charged if the accumulated ion density is greater than zero and is negatively charged if less than zero. Following this analogy, the sign of the surface charge for salt surfaces is calculated and summarized in Table 4.5. For comparison, theoretical and experimental results reported in the literature are also listed.

The sign of the surface charge for selected alkali chloride salt surfaces in saturated solutions obtained from this MD simulation study are in excellent agreement with the results from extended lattice hydration theory (analytical model), which take into consideration the lattice ion hydration free energy and ion size effects (Veeramasuneni et al. 1997). It has been suggested that surface lattice ions are partially hydrated; therefore, instead of considering ion hydration energy alone, which accounts for the hydration of the gaseous ions, an ion–water dipole interaction has to be considered (Veeramasuneni et al. 1997).

This ion–water dipole interaction accounts for the transaction of a partially hydrated surface lattice ion to a free (vacuum) ion (Veeramasuneni et al. 1997), of which the hydration energy is measured experimentally (Hunt 1963). In the LiCl system, due to the significant difference in hydration energy (-470.7 kJ/mole for Li^+ and -347.27 kJ/mole for Cl^-) (Hunt 1963; Miller et al. 1992) Li^+ ions in the LiCl crystal lattice have a greater tendency than Cl^- ions to be dissolved into water, therefore leaving the salt surface negatively charged (Miller et al. 1992; Veeramasuneni et al. 1997; Yalamanchili et al. 1993). In contrast, consider the NaCl crystal lattice, which, due to a much larger ion–water dipole interaction (-35.37 kJ/mole for Cl^- compared with 4.3 kJ/mole for Na^+) (Veeramasuneni et al. 1997), accounts for the more favorable dissolution of the partially hydrated surface Cl^- ion than the dissolution of the surface Na^+ ion despite the fact that Na^+ has a larger hydration energy (-371.54 kJ/mole) (Hunt 1963; Veeramasuneni et al. 1997). For KCl and RbCl salts, the hydration of the Cl^- ion is dominating. Consequently, the salt surfaces are positively

---

TABLE 4.5

Surface Charge of Selected Alkali Halide Crystals in Saturated Solution Compared with Analytical Model and Experimental Results

Salt	Cutoff Distance (Angstrom)	Surface Charge		
		This Study (MDS)	Analytical Model (Veeramasuneni et al. 1997)	Experimental Results (Miller et al. 1992, Yalamanchili et al. 1993)
LiCl	12.5	−	−	−
NaCl	13.0	+	+	+
KCl	14.0	+	+	−
RbCl	14.0	+	+	+

Source: Data from Du, H. and Miller, J. D., *The J. Phys. Chem. C*, 111: 10013, 2007c; Hancer, M., Celik, M. S., and Miller, J. D., *J. Colloid Interface Sci.*, 235: 150, 2001. With permission.

charged (Veeramasuneni et al. 1997). The obvious difference with respect to the sign of the surface charge for the KCl experimental results has been attributed to the presence of oxygen defects in the KCl crystal lattice (Veeramasuneni et al. 1997; Yalamanchili et al. 1993). Also, from the extended lattice ion hydration theory, the difference between the hydration energies of K^+ and Cl^- ions is only 3.46 kJ/mole (Veeramasuneni et al. 1997; Yalamanchili et al. 1993), and this might account for the anomalous behavior of KCl during the sensitive nonequilibrium electrophoresis measurement (Veeramasuneni et al. 1997; Yalamanchili et al. 1993).

4.2.2.2 Dynamic Properties

In this section, dynamic properties of water molecules are analyzed as a function of distance from the salt surface, and the results are discussed in terms of water–ion interactions.

4.2.2.2.1 Self-Diffusion Coefficients

The tracer diffusion coefficient of water along the salt surface normally can be calculated from its MSD (Allen and Tildesley 1987; Chowdhuri and Chandra 2001; Koneshan et al. 1998), and Figure 4.7 summarizes the self-diffusion coefficients of water molecules as a function of distance for different salts. First of all, in general, water molecules in KCl- and RbCl-saturated solutions diffuse significantly faster than they do in NaCl- and LiCl-saturated solutions. There are a few reasons for this. First, because in NaCl and LiCl solutions the cation–water interaction is stronger, and thus water molecules are immobilized to a greater extent. Similar observations have also been reported for a variety of alkali halide solutions (Allen and Tildesley 1987; Brossard et al. 2008; Chowdhuri and Chandra 2001; Koneshan et al. 1998; Koneshan and Rasaiah 2000). Second, the self-diffusion coefficient for water molecules in the salt–water interfacial region is significantly lower than in bulk water. In the interfacial region, where ion dissolution–deposition processes dominate, there is a substantial accumulation of ions. As a result, the movement of water molecules in this region is confined due to the ion–water interactions. Third, the diffusion of water molecules in bulk shows a strong concentration dependence. In an LiCl-saturated solution (~14 M), the self-diffusion coefficient of water molecules in the bulk solution (~0.8×10^{-9} m^2/sec) is significantly lower than the

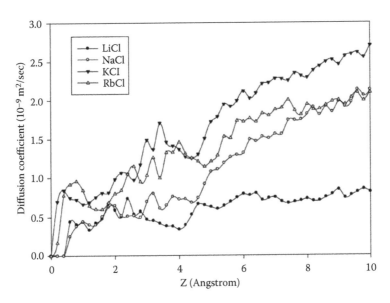

FIGURE 4.7 Comparison of water diffusion coefficients along surface normals for different alkali chlorides. (Zero distance corresponds to the position where the number density of water molecules is zero.) (From Du, H. and Miller, J. D., *J. Phys. Chem. C*, 111: 10013, 2007c. With permission.)

self-diffusion coefficient in pure water ($\sim 2.5 \times 10^{-9}$ m²/sec) (Berendsen et al. 1987; Du and Miller 2007) due to the large concentration of ions (especially Li^+ ions), which interact strongly with water molecules and immobilize them. In contrast, in a KCl-saturated solution (~ 4.8 M), the water self-diffusion coefficient in bulk solution ($\sim 2.5 \times 10^{-9}$ m²/sec) is similar to the value for pure water. A possible explanation is that a large number of ions are accumulated in the crystal–water interfacial region. Thus, ion concentration in bulk solution is relatively low when compared with the ion concentration in the interfacial region. Also, the ion–water interactions are weaker when compared with the LiCl system, and consequently water molecules are more mobile.

4.2.2.2.2 Residence Time

To understand how tightly water molecules are bonded to the salt surface, and how long a water molecule will stay in each water layer, the residence time of water molecules along the salt surface normally has been calculated using the residence-time correlation function, and residence times of water molecules at various salt surfaces as a function of distance using the exponential fit method were calculated and are presented in Table 4.6. Water molecules are classified into three regions:

1. Crystal lattice water: accounts for those water molecules accommodated in the crystal lattice vacancies
2. Interfacial water: accounts for those water molecules within the primary water layer at the salt surface
3. Bulk water: accounts for water molecules in bulk solution

For water molecules taken to be present at the surface lattice vacancy (produced to facilitate the dynamic dissolution–deposition process), the water residence time is very long, suggesting that water molecules are tightly bonded to lattice ions. Correspondingly, the self-diffusion coefficient of water molecules is low. The residence time of water molecules in this region does not seem to be dependent on cation properties, as suggested by the fact that $\tau_{NaCl} > \tau_{LiCl} > \tau_{RbCl} > \tau_{KCl}$. This is explained by the randomness of lattice defects at different salt surfaces, which produces different position barriers when water molecules take the lattice vacancy positions. For water molecules in the interfacial region, their residence time shows a clear cation dependence as $\tau_{LiCl} > \tau_{NaCl} > \tau_{KCl} \approx \tau_{RbCl}$. This is due to the fact that small cations such as Li^+ and Na^+ interact more strongly with water molecules when compared with large cations such as K^+ and Rb^+. Consequently, water molecules are immobilized more significantly around small cations. The inconsistency involving KCl and

---

TABLE 4.6

Comparison of Water Surface Residence Time and Self-Diffusion Coefficients in Saturated Solutions at Different Distances from the Surface for Selected Alkali Chloride Crystals

Salt	Residence Time (ps)			Diffusion Coefficient (10^{-9} m²/sec)		
	Water in Crystal Lattice	Primary Water Layer	Bulk Water	Water in Crystal Lattice	Primary Water Layer	Bulk Water
LiCl	196	102.0	44.0	0.49	0.61	0.78
NaCl	263	32.5	7.7	0.53	1.08	1.90
KCl	62	14.8	5.9	0.93	1.65	2.42
RbCl	126	16.5	6.8	0.83	1.39	1.93

Source: Data from Du, H. and Miller, J. D., *The J. Phys. Chem. C*, 111: 10013, 2007c. With permission.

RbCl interfacial water is due to a higher RbCl saturation solution concentration. The variation of interfacial water residence time is in excellent agreement with water self-diffusion coefficients. In bulk solutions, water molecules move faster when compared with water molecules at crystal lattice positions and interfacial water molecules as indicated by the short residence times and large diffusion coefficients. For water molecules in bulk KCl solution, the residence time (5.9 ps) and self-diffusion coefficient ($\sim 2.5 \times 10^{-9}$ m²/sec) are very close to the values for pure water (~ 5 ps and $\sim 2.5 \times 10^{-9}$ m²/sec, respectively) (Du and Miller 2007c; Koneshan et al. 1998; Koneshan and Rasaiah 2000).

4.2.3 ADSORPTION STATES OF AMINE

In soluble salt flotation, the attachment of bubbles at salt particle surfaces is achieved with the help of collector molecules, which provide the salt crystal hydrophobic properties that were originally hydrophilic. In this regard, the adsorption of a primary amine cationic surfactant octylammonium (ODA) hydrochloride at the NaCl and KCl crystal surfaces from their saturated brine solutions was simulated to investigate the organization of the surfactant at these salt surfaces. The adsorption of ODA and its organization at these salt crystal surfaces was examined by observation of their interaction at such surfaces. The significance of interfacial water structure, which is directly related with salt ion size and further ion–water structure making-and-breaking ability, is addressed with respect to the surfactant organization and adsorption state. The simulation details are discussed elsewhere (Cao et al. 2010).

Figure 4.8 describes the adsorption behavior of ODA collector molecules at a NaCl and a KCl crystal surface in their saturated brine, a hypothetical situation because ODA actually precipitates under these conditions. As seen in Figure 4.8a, at equilibrium the ODA collector molecules self-assemble to a spherical micellar structure with the hydrophobic tails associated together, forming the micelle core and the hydrophilic head groups that stretch out in contact with the aqueous phase. This kind of micellar structure minimizes the contact between the nonpolar ODA tails and the water molecules and thus keeps the ODA organization stabilized as a micelle structure in the brine solution. During the entire simulation, no significant ODA organization or attachment at the NaCl crystal surface was observed, suggesting that the adsorption of the ODA molecule at the NaCl surface is not an energetically stable state. This is consistent with previous literature and our experimental

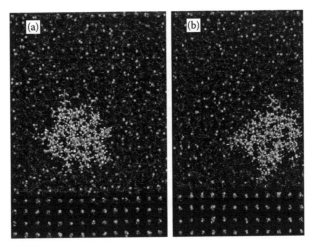

FIGURE 4.8 MDS snapshots of nine ODA molecules near NaCl (a) and KCl (b) surfaces, respectively. For clarity, only the skeletons of the water molecules are shown. Color key: red, oxygen atoms; white, hydrogen atoms; blue, nitrogen atoms; green, cations; purple, chloride ions; light blue, carbon atoms. (From Cao, Q., Du, H., Miller, J. D., Wang, X., and Cheng, F. *Miner. Eng.,* 23: 365, 2010. With permission.)

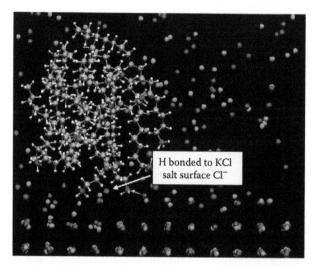

FIGURE 4.9   Close-up MDS snapshot of nine ODA molecules at a KCl surface. This snap shot describes the closest contact between the ODA colloid structure and the KCl crystal surface. For clarity, water molecules are not shown. Color key: red, oxygen atom; white, hydrogen atoms; blue, nitrogen atoms; green, sodium ions; purple, chloride ions; light blue, carbon atoms. (From Cao, Q., Du, H., Miller, J. D., Wang, X., and Cheng, F. *Miner. Eng.,* 23: 365, 2010. With permission.)

observation regarding the adsorption states and flotation behavior of the NaCl salt, which suggests that the strong interaction between the surface lattice ion, especially Na$^+$, and water molecules stabilizes the interfacial water molecules, and consequently, that replacement of the interfacial water molecules with organized collector molecules cannot be achieved.

At a KCl crystal surface, which is shown in Figure 4.8b, the ODA collector ions behave in a different manner. In general, the micellar organization similar to that observed for the NaCl system is formed in the KCl brine during most of the simulation time, and at equilibrium the ODA organization is attached to the KCl surface through a couple of polar head groups, suggesting that adsorption of ODA collectors at a KCl surface is an energetically favorable process. This adsorption is due to the K$^+$ cations in the crystal lattice, which have a weaker interaction with interfacial water molecules. Thus, the interfacial water molecules are not very stable, and the penetration of the surfactant molecules through the interfacial water layer is possible. Further, a close examination of Figure 4.9, which is a close-up of the ODA organization at the KCl surface, reveals that the ODA organization is achieved by the attachment of the polar head group of the ODA ion to Cl$^-$ ions, which are stabilized at the surface. This interesting observation suggests that the adsorption and organization of collector molecules at the KCl surface results from the instability of the interfacial water structure; thus water molecules can be replaced by collector organizations that have some stability in the interfacial region because of the presence of negatively charged surface defects where the positively charged ODA head group can be accommodated.

4.2.4   FLOTATION OF ALKALI HALIDE SALTS IN SATURATED SOLUTIONS

Previous research regarding soluble salt flotation has suggested that interfacial water structure may play a significant role in surfactant adsorption (Hancer et al. 2001; Veeramasuneni et al. 1997). For example, even in the absence of collectors, structure-breaking salts such as KCl and KI have been shown to have a hydrophobic surface character, whereas structure-making salts are completely wetted by their saturated brine. Table 4.7 is the summary of contact angles for saturated brines at the surfaces of selected alkali halide salt surfaces in the absence of a surfactant (Hancer et al. 2001).

TABLE 4.7

Advancing Contact Angles of Saturated Brines at Selected Alkali Halide Salt Surfaces

Salt	Contact Angle (Degree)
KI	25 ± 2
KCl	$7.9 \pm 0.5 \, (12.0 \pm 1.4)^*$
NaCl	0
NaF	0

Source: Data from Hancer, M., Celik, M. S., and Miller, J. D., *J. Colloid Interface Sci.*, 235: 150, 2001.

*Measured on the natural cleavage plane of a single crystal.

The contact angle systematically decreases as the ion size in the solution increases. NaCl and NaF salt surfaces are totally hydrated, as suggested by the zero contact angles, and in contrast, KI and KCl appear to be less hydrated, as indicated by the finite contact angles (Hancer et al. 2001).

The MD simulation results complement the contact angle measurement results very well. According to our MD simulation study, due to different cation sizes—and consequently different magnitudes of the local electric field—water structures at different alkali chloride crystal surfaces are very different. At LiCl and NaCl surfaces, water molecules are highly ordered, as suggested by the domination of particular orientations of interfacial water, which was discussed previously. Due to the strong interaction between water molecules and surface lattice ions (especially cations), the structure of water molecules at structure-making salt surfaces suggests a very stable thermodynamic state, which is confirmed by the small self-diffusion coefficients and long residence times of interfacial water molecules. When surfactant molecules are present, the replacement of surface water should be energetically unfavorable. Therefore, adsorption is not expected despite the sign of the salt crystal surface charge.

However, the organization of water molecules at KCl and RbCl crystal surfaces shows obvious randomness because of the relatively weak water–surface ion interactions, and these weak interactions are confirmed by the finite contact angles (Hancer et al. 2001), as well as by the large interfacial water diffusion coefficient and the short residence times from MD simulation analysis. The less-ordered interfacial water structure being an energetically less stable state suggests that surfactant adsorption would be possible with sufficient hydrophobicity created for bubble attachment and subsequent flotation.

The different adsorption behavior of the ODA surfactant molecules at the NaCl and KCl, surfaces demonstrates that the salt water structure making and breaking characteristics have a significant effect on the adsorption state. As the ion size increases, the interfacial water molecules experience a transition from hydrophilic hydration (in the case of the NaCl system) to hydrophobic hydration (in the case of the KCl system). When interfacial water molecules are stable due to the presence of a structure-making salt such as NaCl, replacement of these water molecules with surfactant molecules is difficult. In contrast, at the structure-breaking KCl crystal surface, which naturally shows a weak hydrophobic character, the penetration of the ODA surfactant structure through the interfacial water layer and the further replacement of interfacial water molecules is possible, and thus instantaneous adsorption is achieved.

4.2.5 Summary

MD simulation has been validated as a useful tool for providing more detailed information regarding the solution chemistry and interfacial phenomena of soluble salt flotation systems.

The study of structural and dynamic properties of concentrated LiCl, RbCl, and CsI aqueous solutions has found that a small ion such as Li$^+$, owing to strong electronegativity, significantly immobilizes the water molecules by tetrahedrally bonding with them, which leads to a more compact structure. The higher the salt concentration, the denser the water molecules are packed, as indicated by a monotonically increase of water–water coordination numbers as a function of salt concentration. However, large ions such as Rb$^+$, Cs$^+$, and I$^-$ form weak bonds with water molecules and a loose hydrophobic hydration shell. Thus the coordination between water molecules decreases with salt concentration. Also, as the ion size increases, the ion–water electrostatic interaction becomes less significant when compared with the hydrogen-bonding interaction of water molecules. This agrees satisfactorily with the observed dominating increase of naked Cs$^+$–I$^-$ ion pair contacts, and hydrated Li$^+$–Cl$^-$ ion pair contacts with respect to an increase in salt concentration.

This leads to the conclusion that ion size influences the mobility of water molecules in the solution. When small ions such as Li$^+$ ions are present, the mobility of water molecules in both the ion hydration shell and the water-only clusters decreases as a function of salt concentration. In contrast, for solutions containing large ions (CsI and RbCl), the ion–water electrostatic interaction does not contribute significantly to immobilizing the water molecules. Consequently the residence time of water molecules does not show a substantial change with solution concentration. The changes in residence times support the observations of self-diffusion coefficients of particles as a function of salt concentration. These observations show that when Li$^+$ ions are present the diffusion coefficients of water molecules in the solution decrease significantly with salt concentration and that when Cs$^+$ and I$^-$ ions are present there is no significant change in the water diffusion coefficients with salt concentration.

The change of system viscosity with solution concentration as determined from MD simulation successfully complements the experimental results reported in the literature (Jiang and Sandler 2003) qualitatively and provides an in-depth understanding regarding the variation of viscosity with respect to ion size and salt concentration from a molecular perspective. For LiCl solutions, the system viscosity increases monotonically with salt concentration due to the strong ion–water interactions in the solution. As the ion size increases, hydrophobic hydration becomes dominating, and in the case of RbCl leads to negligible variation of system viscosity with salt concentration. Further increase in the ion size to Cs$^+$ and I$^-$ ions revealed a noticeable decrease of system viscosity as a function of salt concentration. The excellent qualitative agreement between the simulation results and the experimental results for the variation of viscosity with ion size and concentration provides future information to phenomenologically describe behavior of particles in alkali halide solutions.

The surface charge signs obtained for selected alkali chloride salts using MD simulation complements both the extended lattice hydration theory based on ion hydration energy calculations and laser–Doppler electrophoretic measurements (with the exception of KCl due to the presence of oxygen defects). Due to their large hydration energy, Li$^+$ ions are preferentially released into bulk solution, leaving a negative charge at the LiCl crystal surface. However, NaCl, KCl, and RbCl salt surfaces are positively charged because Cl$^-$ ions have a greater tendency to be hydrated and accommodated in the bulk solutions.

MD simulations of selected alkali chloride salt surfaces in saturated solutions reveal that Li$^+$ and Na$^+$ lattice cations, due to their small sizes and large local electric fields, strongly interact with interfacial water molecules. Consequently interfacial water molecules are stabilized at these salt crystal surfaces with distinct orientations. When surfactant molecules are present in the solution, it is energetically unfavorable for these molecules to be adsorbed at the salt crystal surface by replacement of the interfacial waters. But at the same time the interaction between interfacial water and surface cations for KCl and RbCl salt is weak, which favors the surfactant molecule adsorption despite the sign of the salt surface charge.

4.3   PHYLLOSILICATE MINERALS

As one of the major categories of nonsulfide minerals, phyllosilicates, the sheet silicates, are important minerals and include aluminum silicate minerals such as kaolinite, pyrophyllite, and muscovite, and the corresponding magnesium silicate minerals, talc, sepiolite, and phlogopite. Due to their wide application in the paper, polymer, paint, lubricant, plastics, cosmetics, and pharmaceutical industries, understanding interfacial phenomena of phyllosilicate minerals is of particular interest to mineral processing engineers and has drawn much research attention for many decades. With the development of silicate and oxide crystal force fields (Cygan et al. 2004; Heinz et al. 2005), probing interfacial phenomena using MD simulation has become realistic, and a significant number of publications have been produced in this area in the past decade (Brossard et al. 2008; Heinz et al. 2003, 2006; Jacobs et al. 2006; Kalinichev and Kirkpatrick 2002; McNamara et al. 2004; Wang et al. 2005). In the following sections, MD simulation analysis of some of the typical industrial silicate minerals including talc, kaolinite, and sepiolite is reported and discussed.

4.3.1   Water Structure at Phyllosilicate Mineral Surfaces

Interfacial water structures at selected phyllosilicate minerals, including talc, kaolinite, and sepiolite, are analyzed in terms of water distribution and water dipole moment orientation. The behavior of water molecules and wetting characteristics at these surfaces are explained in terms of the mineral surface structure. Simulation details have been reported in previous publications (Du and Miller 2007a, 2007b; Miller et al. 2007; Nalaskowski et al. 2007).

4.3.1.1   Interfacial Water Structure at Talc Surfaces

Talc, which has the chemical formula $Mg_3(Si_4O_{10})(OH)_2$, is composed of three layers. Its middle layer is a brucite layer, a magnesium–oxygen/hydroxyl octahedral layer, while the two outer layers are silicon–oxygen tetrahedral layers. The brucite layer has the positive charge necessary to neutralize the two hexagonal networks of silica tetrahedra, resulting in a sandwich structure and providing a neutral charge for the three-layer talc mineral. These three-layer sheets are bonded to each other only by van der Waals forces; thus the layers are capable of slipping easily over one another, which accounts for the soft character of talc and its smoothness. The basal surfaces of the three-layer elementary structure do not contain hydroxyl groups or active sites, which provide the basal plane of talc with a natural hydrophobicity and accounts for its floatability (Fuerstenau and Huang 2003).

An MD simulation snapshot of water molecules near the talc basal plane surface is presented in Figure 4.10. From Figure 4.10, an obvious gap is observed between the aqueous phase and the talc basal plane surface, reflecting the dominance of the so-called excluded volume or hard wall effect for the system (Abraham 1978; Yu et al. 1999).

This phenomenon is due to the absence of specific hydrogen-bonding donor and acceptor sites on the basal plane of talc; the interaction between water molecules and the basal plane is weak. The relatively weak water–talc interaction on the molecular scale is the origin of the macroscopic hydrophobic character of the basal plane surface, and the incompatibility between the water and the hydrophobic surface accounts for the fluctuating voids at the surface as suggested in the literature (Ruckenstein and Churaev 1991).

The water density distribution along the talc basal plane surface normal further reveals that, at the talc basal plane surface, the water number density is zero, which is due to the natural hydrophobicity of the surface, and thus water molecules are excluded from this surface.

In addition, similar to previous reports in the literature (Wang et al. 2004a, 2004b) the primary water density peak is located approximately 3.1 Å from the surface, a distance larger than the distance between hydrogen bonded water–water molecules, which is approximately 2.8 Å (Dang and

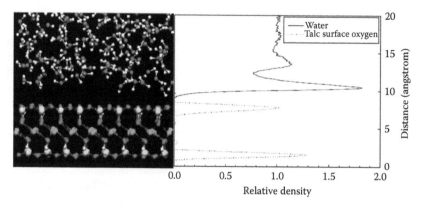

FIGURE 4.10 MD simulation snapshot of equilibrated water–talc basal plane surface (left) and interfacial water density distribution function at the talc basal plane surface (right). Color key: red, oxygen atom; white, hydrogen atoms; yellow, silicon atoms; green, magnesium atoms. (From Du, H. and Miller, J. D., *Int. J. Miner. Process.*, 84: 172, 2007a; Du, H. and Miller, J. D., *Langmuir*, 23: 11587, 2007b. With permission.)

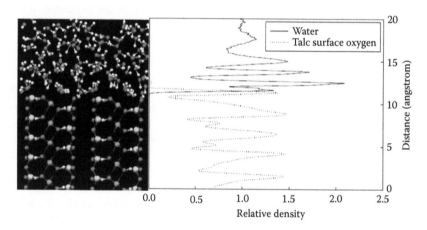

FIGURE 4.11 MD simulation snapshot of equilibrated water–talc edge surface (left) and interfacial water density distribution function at the talc edge surface (right). Color key: red, oxygen atom; white, hydrogen atoms; yellow, silicon atoms; green, magnesium atoms. (From Du, H. and Miller, J. D., *Int. J. Miner. Process.*, 84: 172, 2007a; Du, H. and Miller, J. D., *Langmuir*, 23: 11587, 2007b. With permission.)

Pettitt 1990; Dang and Smith 1993; Lee and Rasaiah 1996; Lynden-Bell and Rasaiah 1997; Rasaiah 1988). This result further suggests the weak interaction between water molecules and talc basal plane surface.

In contrast, water molecules are tightly bonded with atoms on the edge surface as indicated by the close contact between the water molecules and the edge atoms seen in Figure 4.11. Hydration of the edge surface is due to the breakage of the Si–O or Mg–O bonds, and consequently the exposure of polar surface atoms such as magnesium, oxygen, and silicon, which provide plenty of electron donor and acceptor sites to facilitate the formation of strong hydrogen bonds with interfacial water molecules. Thus, water is expected to wet the hydrophilic talc edge surface.

Also, due to the complex structures of the talc edge surface, some water molecules can even be accommodated into the top surface layer of the crystal lattice, as can be seen from the snapshot in Figure 4.11 and as indicated by the noticeable water density at the talc edge surface (the talc surface

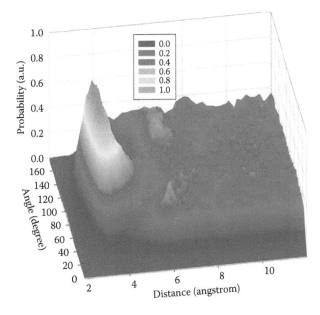

FIGURE 4.12 Water dipole moment distribution at the talc basal plane surface. (The angle is between the water dipole moment and the talc basal plane surface normal) (From Du, H. and Miller, J. D., *Langmuir,* 23: 11587, 2007b. With permission.)

is taken as the average position of the surface oxygen atoms). The complexity of edge surface structures explains the more complicated characteristic peaks of the water density distribution function, as well. The first two peaks are attributed to water molecules attracted by positively charged surface silicon and magnesium atoms, respectively, and the third peak originates from interaction between surface oxygen atoms and water molecules.

The water dipole moment distribution provides more detailed information regarding the interfacial water orientations at mineral surfaces. As seen in Figure 4.12, there is a broad water dipole angle distribution peak from 50° to 130° at the talc basal plane surface, suggesting the random orientation of interfacial water molecules due to the absence of hydrogen-bonding donor and acceptor sites at the talc basal plane surface. This observation is in good agreement with other MD simulations (Arab et al. 2003; Wang et al. 2005), quantum chemical calculations (Bridgeman et al. 1996), and MC simulations (Bridgeman and Skipper 1997) regarding water structures at a variety of hydrophobic surfaces.

Sum frequency vibrational spectroscopy (SFVS) studies also reveal the presence of a large number of free OH groups at hydrophobic surfaces due to the lack of bonding between the substrate and the water molecules (Miranda and Shen 1999), which is schematically illustrated in Figure 4.13. In contrast, as shown in Figure 4.14, when near the talc edge, water molecules clearly show preferred orientations. When the talc surface oxygen atoms participate in hydrogen bonding as electron donors, interfacial water dipoles point toward the surface, and the angle between the water dipole and the talc edge surface normal is larger than 90°. However, when interacting with magnesium or silicon atoms, the interfacial water dipoles point away from the surface and the angle is smaller than 90°. The large interfacial water density and the characteristic orientation of water dipoles reflect the strong interactions between the water and the talc edge surface, leading to a distinct hydrophilic character as observed experimentally (Fuerstenau and Huang 2003). Further, the SFVS studies suggest that the free OH groups observed at hydrophobic surfaces disappear at hydrophilic surfaces due to the strong interaction between water molecules and the substrate (Miranda and Shen 1999).

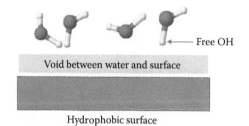

Hydrophobic surface

FIGURE 4.13   Schematic drawing illustrating the water structures at an ideal hydrophobic surface.

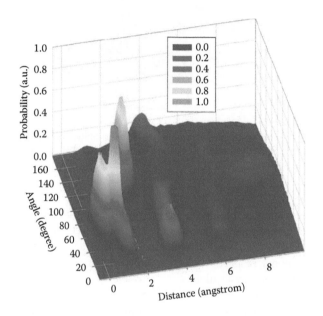

FIGURE 4.14   Water dipole moment distribution at the talc edge surface. (The angle is between the water dipole moment and the talc edge surface normal) (From Du, H. and Miller, J. D., *Langmuir,* 23: 11587, 2007b. With permission.)

4.3.1.2   Interfacial Water Structure at Kaolinite Surfaces

Kaolinite is a two-layer silicate mineral consisting of alternating sheets of silica tetrahedra and aluminum hydroxide octahedra. The hydrogen bonding between the hydroxyl ions of the octahedral sheet of one bilayer and the tetrahedral oxygen of the adjacent silica sheet of the next bilayer holds the two-layer structure together. Ideally the hexagonal surface structure of the silica tetrahedral sheet is expected to be of low polarity and unable to support hydrogen bonding with water molecules, as in the case with the three-layer silicates, such as talc, discussed previously. However, the aluminum hydroxyl octahedral surface is expected to be hydrophilic due to the terminal hydroxyl groups. Unlike three-layer silicates such as talc, the two-layer kaolinite has different face surfaces, the silica face and the alumina face.

An MD simulation snapshot of water molecules at kaolinite face surfaces is presented in Figure 4.15. Similar to talc basal plane surfaces, the gap due to the hydrophobic characteristics of the kaolinite silica face surface is observed. In contrast, water molecules are tightly bonded with the alumina face surface, indicated by the close contact between the water molecules and this surface, as seen in Figure 4.15. The explanation for such water organization is that the presence of

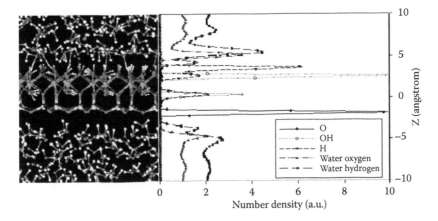

FIGURE 4.15 MD simulation snapshots of equilibrated water–kaolinite surfaces. Top, water at the alumina face. Bottom, water at the silica face. Color key: red, oxygen atom; white, hydrogen atoms; yellow, silicon atoms; green, aluminum atoms. (From Miller, J. D., Nalaskowski, J., Abdul, B., and Du, H. *Can. J. Chem. Eng.*, 85: 617, 2007. With permission.)

hydroxyl groups at the alumina face surface provides plenty of hydrogen-bonding sites and facilitates the formation of strong hydrogen bonds. Thus, it is expected that water will wet this surface very well.

Moreover, from the particle density distribution plot, it is evident that due to the natural hydrophobicity of the kaolinite silica face surface, water molecules are excluded from this surface, as indicated by the discontinuity observed in the density distribution curve, again revealing the absence of water molecules at the kaolinite silica face surface. Similar to previous reports in the literature for water structure at talc surfaces (Wang et al. 2004a, 2004b), the primary water oxygen density peak is located approximately 3 Å from the silica face surface oxygen peak, and this distance is larger than the oxygen–oxygen distance for hydrogen-bonded water molecules, demonstrating the weak interaction between water molecules and the silica tetrahedral face. However, water molecules interact strongly with the aluminum octahedral surface, as indicated by overlap of water hydrogens and kaolinite surface hydrogens from the density distribution curves.

The distance between surface hydroxyl oxygen atoms and oxygen atoms in the primary water layer is about 2.5 Å, which is less than the hydrogen bonded water–water distance, suggesting a strong affinity of water molecules to the aluminum octahedral surface. As previously discussed, because of the presence of surface hydroxyl groups, there are plenty of electron donor and acceptor sites, hydrogen bonding with the surface will stabilize the surface water structure and the aluminum octahedral surface will be wetted.

4.3.1.3 Interfacial Water Structure at Sepiolite Surfaces

Sepiolite is a complex magnesium phyllosilicate clay mineral. A typical formula for sepiolite is $Mg_4Si_6O_{15}(OH)_2 \cdot 6H_2O$. It can be present in fibrous, fine-particulate, and massive forms. It has many industrial applications due to its unusual modulated structure and lathlike morphology; for example, it is used in oil drilling and for cat litter and is also used in massive form for carving, where it is known as meerschaum. Recently, sepiolite has been used to adsorb various metal ions such as Cd, Cu, and Zn for the purpose of remediating polluted soils, and thus has received extensive research attention (Kara et al. 2003; Lazarevic et al. 2007).

Preliminary MD simulation of sepiolite interfacial water structure, as seen in Figure 4.16, clearly shows that sepiolite exhibits interesting interfacial properties due to its unique crystal structure, which is composed of a discontinuous magnesium–oxygen/hydroxyl octahedral sheet with alternating 2:1 magnesium–silicate modules (ribbons) and hydrated channels. Each ribbon exhibits a

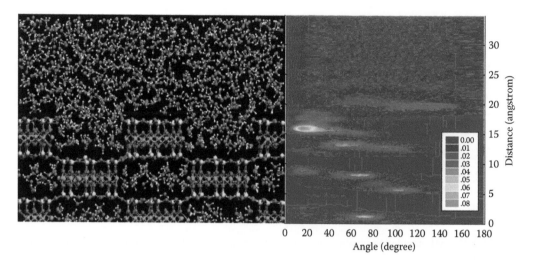

FIGURE 4.16 MD simulation snapshots of equilibrated water–sepiolite surface and water dipole moment distribution. Color key: red, oxygen atom; white, hydrogen atoms; yellow, silicon atoms; green, magnesium atoms.

talclike structure, has low polarity, and is hydrophobic in nature, as clearly suggested by the random interfacial water dipole moment distribution (~20 Å). Within each channel, due to the presence of naked oxygen, magnesium atoms, and hydroxyl groups, water molecules are strongly hydrogen bonded. Consequently, for these channels, specific dipole moment orientation is revealed by the sharp dipole moment peaks within the crystal.

4.3.2 Interaction Between Amphipathic Surfactants and Talc

The MD simulation interfacial water structure analysis provides further information regarding the wetting characteristics of phyllosilicate minerals, which is critical for understanding the adsorption states of surfactant molecules. For instance, the adsorption states at talc surfaces have been extensively studied, and Healy (1974), and later Pugh (1989), have suggested that chemical, electrostatic, hydrogen-bonding, and hydrophobic interactions are the major mechanisms that govern the polymer adsorption at talc surfaces. Other researchers also confirmed that the hydrophobic interactions play a critical role in polysaccharide adsorption at talc surfaces (Fuerstenau and Huang 2003; Jenkins and Ralston 1998; Morris et al. 2002). Some researchers suggest that hydrogen bonding, which happens between the hydroxyl groups in the organic molecule and the hydrogen-bonding sites at mineral surfaces, is the major reason for adsorption (Balajee and Iwasaki 1969; Rath et al. 1997). Chemical adsorption related to metallic impurities and consequent complexion effects have also been recognized as another mechanism that accounts for macromolecule adsorption (Liu and Laskowski 1989). In addition, surfactant adsorption is of interest and will be discussed first.

4.3.2.1 Adsorption of Cationic Surfactant at Talc Surfaces

Preliminary MD simulation analysis of decyltrimethylammonium bromide (DTAB) at a talc surface suggests that the adsorption states of this cationic surfactant are quite different at the basal plane surface and the edge surface. In Figure 4.17, snapshots of DTABs at a talc basal plane surface are presented for different simulation times. The DTABs were originally configured in such a way that the polar head groups were facing the basal plane surface (Figure 4.17a). As expected, after a short period of simulation, due to the hydrophobic nature of the talc basal plane surface, the surfactant molecules spontaneously reorient themselves so that their hydrophobic tails move toward the

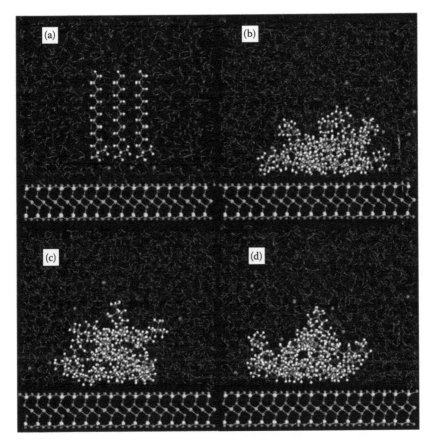

FIGURE 4.17   MDS snapshots of adsorption states for nine DTAB molecules at the talc basal plane surface for different simulation times. The snapshots are taken at 0 ps (a), 200ps (b), 500ps (c), and 1ns (d). Color key: red, oxygen atom; white, hydrogen atoms; blue, nitrogen atoms; purple, bromide atoms; yellow, silicon atoms; green, magnesium atoms; light blue, carbon atoms. For clarity, only the skeletons of the water molecules are shown. (From Du, H. and Miller, J. D., *Langmuir,* 23: 11587, 2007b. With permission.)

surface and displace the water molecules (Figure 4.17b). As simulation continues, the hydrophobic interaction between the substrate and the tail of the surfactant is so dominating that almost all water molecules are excluded between them (Figure 4.17c and d). Further study of the surfactants in Figure 4.17c and d reveals that at equilibrium the DTABs form a hemispherical like micelle structure, with the hydrophobic tails sticking to the substrate and the polar heads in contact with surrounding water molecules. Due to the orientation of adsorbed DTAB ions, a hydrophilic surface is produced. This observation from MD simulation complements the literature, in which contact angle measurements using dodecyl ammonium acetate as a surfactant show that surfactant adsorption at the basal plane decreases the contact angle from higher than 60° to approximately 30° (Fuerstenau and Huang 2003; Miller et al. 2007).

At talc edge surfaces, the DTAB molecules behave differently, as can be seen in Figure 4.18. First, due to the incompatibility of the DTAB hydrophobic tail and polar water molecules, the surfactants have a tendency to aggregate together and form micelle-like spherical structures (Figure 4.18b and c), which minimized contact between the surfactant tail and surrounding water molecules.

Second, a couple of layers of interfacial water molecules are sandwiched between the talc edge surface and the surfactant molecules. The presence of the stable interfacial water layers is due to the strong water–substrate interaction as discussed previously. Thus, no replacement of interfacial water

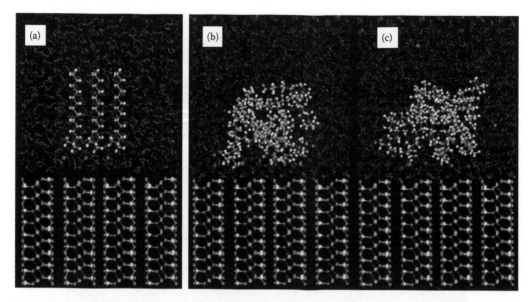

FIGURE 4.18   MDS snapshots of adsorption states of nine DTAB molecules at the talc edge surface for different simulation times. The snapshots are taken at 0 ps (a), 500ps (b), and 1ns (c). Color key: red, oxygen atom; white, hydrogen atoms; blue, nitrogen atoms; purple, bromide atoms; yellow, silicon atoms; green, magnesium atoms; light blue, carbon atoms. For clarity, only the skeletons of the water molecules are shown. (From Du, H. and Miller, J. D., *Langmuir,* 23: 11587, 2007b. With permission.)

molecules by DTAB surfactants is observed during the simulation. This phenomenon has also been reported during the study of adsorption of lipid bilayers at hydrophilic surfaces, which confirms the presence of a water layer (0.5 nm to 1.5 nm thick) between physically adsorbed lipid bilayers and the substrate (Bayerl and Bloom 1990; Kim et al. 2001; Koenig et al. 1996).

Third, the micellelike DTAB aggregate is stabilized at the talc edge surface during the entire simulation, suggesting a strong affinity to the surface. It has been proposed that the adsorption of cationic surfactants at talc edge surfaces is due to the electrostatic attraction between the negatively charged edges and the positively charged surfactant head groups, with the adsorption being strongly influenced by solution pH (Fuerstenau and Huang 2003).

The simulation results appear to be in good agreement with this adsorption mechanism, despite the fact that the talc edge surface is kept neutral in the simulation system due to the negligence of surface lattice ion dissolution and surface hydrolysis reactions.

In this regard, edge adsorption of the cationic surfactant, DTAB, is achieved through the interaction between exposed surface oxygen atoms, which are negatively charged, and the positively charged surfactant head group. Similarly, the presence of bromide anions in the interfacial region is caused by the electrostatic attraction associated with cations (magnesium and silicon) at the talc edge surface.

Regardless of the adsorption mechanism, the surfactants are stabilized at the vicinity of the talc edge during the entire simulation, suggesting the strong affinity of DTAB at the edge surface. At certain concentrations with their long hydrophobic chains extended into the solution, the DTABs may create a hydrophobic state at the edge surface, which has also been reported in the literature (Fuerstenau and Huang 2003).

4.3.2.2   Adsorption of Dextrin at Talc Surfaces

The MD simulation of a dextrin molecule at a talc basal plane surface is presented in Figure 4.19. Initially, the dextrin molecules are positioned in the middle of the solution (Figure 4.19a). As the

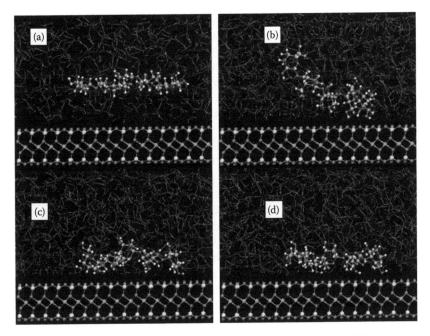

FIGURE 4.19   MD simulation snapshots of dextrin at the talc basal plane surface for different simulation times. The snapshots are taken at 0 ps (a), 200ps (b), 500ps (c), and 1ns (d). Color key: red, oxygen atom; white, hydrogen atoms; yellow, silicon atoms; green, magnesium atoms; light blue, carbon atoms. For clarity, only the skeletons of the water molecules are shown. (From Du, H. and Miller, J. D., *Langmuir*, 23: 11587, 2007b. With permission.)

simulation evolves, the dextrin organization migrates toward the talc basal plane surface (Figure 4.19a and b), and remains in full contact with the talc surface once it has achieved contact (Figure 4.19c and d), indicating its strong affinity for the basal plane surface.

In this simulation, the influence of metallic sites, which has been proposed to be complex and able to stabilize dextrin molecules at such a surface (Liu and Laskowski 1989a, 1989b, 1989c), is excluded due to the lack of such sites in the simulation system. Thus, possible explanations for the dextrin adsorption are the hydrophobic interaction between the substrate and the dextrin molecule (Haung et al. 1978; Miller et al. 1983; Wie and Fuerstenau 1974) and hydrogen bonding between the talc basal plane substrate and hydroxyl groups in the dextrin molecule (Balajee and Iwasaki 1969).

The dextrin molecule was originally positioned in the middle of the aqueous phase in favor of the hydrogen bonding between water molecules and hydroxyl groups in the dextrin molecule. The migration of the dextrin molecule to the talc basal plane surface, where hydrogen bonding is unfavorable (see Figure 4.19), indicates that hydrogen bonding is not the only influential mechanism that governs the dextrin adsorption. Thus, hydrophobic interaction appears to play a significant role regarding the adsorption behavior of the dextrin molecule. The driving force for the dextrin molecule to move to the talc basal plane surface and stabilize at the surface is the incompatibility between the polar water molecules and the nonpolar C-H–methylene groups in the dextrin molecule (Du and Miller 2007a; Fuerstenau and Huang 2003; Jenkins and Ralston 1998; Morris et al. 2002). However, despite the fact that the talc basal plane surface oxygen atoms are not ideal hydrogen-bonding sites because of the negative charge these oxygen atoms carry, they can weakly hydrogen-bond with water molecules; thus these tetrahedral oxygen atoms at the talc surface are able to accommodate hydroxyl groups in the dextrin molecule to some extent. Consequently, both the hydrophobic moieties and the hydroxyl groups are stabilized at the talc basal plane surface, as suggested by Figure 4.19c and d which show that both C-H–methylene and hydroxyl groups are very close to the substrate. In summary, hydrophobic interactions, as well as attraction between the talc basal plane surface oxygen atoms and hydrogen atoms in the hydroxyl groups, contributed to dextrin adsorption at the talc basal plane surface.

Typical hydrogen bonding is the interaction between polarized water oxygen atoms and polarized water hydrogen atoms. In contrast, the talc basal plane surface oxygen atoms, which participate in the silicon–oxygen tetrahedral structure, are nonpolar. Thus, the talc basal plane surface oxygen atoms are not ideal hydrogen-bonding sites. However, because of the negative charge these oxygen atoms carry, they can form weak hydrogen bonding when compared with typical hydrogen bonding (Du and Miller 2007a; Miller et al. 2007; Wang et al. 2005) with hydroxyl groups in the dextrin molecule, contributing to the accommodation of the dextrin molecule. Further, the incapability of the talc basal plane surface oxygen atoms to form traditional hydrogen bonds has been partially compensated for by the low mobility of the dextrin molecule at the talc basal plane surface (due to the hydrophobic interactions), consequently facilitating the electrostatic interaction between the hydrogen atoms of the dextrin hydroxyl group and the talc basal plane surface oxygen atoms. In summary, in conjunction with hydrophobic interactions, the weak hydrogen bonding between the talc basal plane surface oxygen atoms and the hydroxyl groups of the dextrin due to the electrostatic interaction accounts for the dextrin adsorption at the talc basal plane surface.

The dextrin organization has unique structural and polarization properties. On the one hand, the polar hydroxyl groups in the dextrin organization enable it to form hydrogen bonds with surrounding water molecules, which accounts for its stabilization in water. On the other hand, the hydrophobic moieties and the hydroxyl groups in the molecule can interact with the nonpolar talc basal plane surface and replace surface water molecules, which are at a high energy level due to the breakdown of hydrogen bonding. Therefore, the dextrin molecule behaves as a bridge connecting the polar water and nonpolar talc surface, making the surface hydrophilic and thus depressing the talc particle during flotation.

Snapshots showing the interaction between dextrin and the talc edge surface are presented in Figure 4.20. It is interesting to notice that during the entire simulation the dextrin organization is

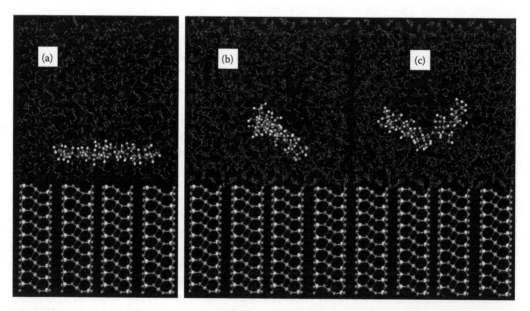

FIGURE 4.20 MD simulation snapshots of dextrin at the talc edge surface for different simulation times. The snapshots are taken at 0 ps (a), 500 ps (b), and 1 ns (c). For clarity, only the skeletons of the water molecules are shown. Color key: red, oxygen atom; white, hydrogen atoms; yellow, silicon atoms; green, magnesium atoms; light blue, carbon atoms. (From Du, H. and Miller, J. D., *Langmuir*, 23: 11587, 2007b. With permission.)

stabilized in the water phase and does not reach the talc edge surface. A close examination of Figure 4.20 shows that water molecules are tightly bonded at the talc edge surface and form a distinct local order attributed to the abundant surface electron donor and acceptor sites (see Section 4.7.1.1). For the dextrin to be adsorbed at the talc edge surface, the surface-bonded water molecules have to be displaced, which is energetically unfavorable. In addition, the hydroxyl groups in the dextrin organization hydrogen bonded with surrounding water molecules and stabilized the molecule in solution. Consequently, no obvious adsorption of dextrin at the talc edge surface is observed, which is in good agreement with an investigation into the adsorption of dextrin at hydrophilic mineral surfaces as revealed by atomic force microscopy (AFM) using the soft contact technique. In this AFM study, no obvious dextrin adsorption at an oxidized silicon surface were observed. Also, adsorption density measurements further indicated that dextrin molecules were not absorbed at the surfaces of hydrophilic minerals such as quartz and pyrite (Miller et al. 1983).

4.3.3 Summary

MD simulation of water structures at selected phyllosilicate mineral surfaces suggests that the wetting characteristics of these minerals is dependent on their crystal structure and that the substrate crystal structure plays an important role in defining the distribution of water normal to selected surfaces. When a silicon–oxygen tetrahedral face is exposed to the water phase, such as the talc basal plane surface, kaolinite silicon–oxygen tetrahedral surface, and sepiolite ribbon surface, due to the absence of electron donor and acceptor sites at the talc basal plane surface water molecules interact weakly with the surface atoms and arrange themselves randomly some distance from the surface. On the other hand, when polar groups such as naked oxygen, aluminum, and magnesium atoms and hydroxyl groups are exposed, abundant hydrogen bonding sites are provided. Therefore, interfacial water dipoles bond directly with edge surfaces and are orientated preferentially either away from or toward the surface. This strong hydrogen-bonding ability of edge surface atoms explains the hydrophilic characteristic of such surfaces.

It is further concluded from MD simulation that the hydrophobic characteristic of the talc basal plane surface favors the hydrophobic interaction with the nonpolar tail of cationic surfactants such as DTAB and renders the basal plane surface hydrophilic, in agreement with a previously reported decrease in contact angle when a cationic surfactant is present (Miller and Paruchuri 2004). However, electrostatic interaction originating from the point charge assigned to each atom in the talc edge surface and the DTAB surfactant dominates in the adsorption of DTAB at the talc edge surface, although complete water displacement does not occur.

MD simulation provides further information regarding the adsorption states of dextrin at talc surfaces. The dextrin molecule is attracted to and stabilizes at the talc basal plane surface, revealing a strong affinity for this surface. It appears that the adsorption involves hydrophobic attraction between the talc basal plane surface and the hydrophobic moieties of the dextrin molecule, as well as weak hydrogen bonding between negatively charged surface oxygen atoms and hydroxyl groups in the dextrin molecule. However, due to the strong interaction between the talc edge surface and water molecules, as well as the hydrogen bonding between the hydroxyl groups of the dextrin molecule with the surrounding water molecules, dextrin does not exhibit significant adsorption at the talc edge surface.

4.4 QUARTZ

Understanding the adsorption of long-chain alkyl amines and their surface structure is significant for improved nonsulfide flotation technology and the development of new flotation chemistry for the flotation of nonsulfide minerals. In the middle of the twentieth century, Gaudin and Fuerstenau first studied the use of long-chain alkyl amines for the flotation of oxide minerals, particularly the flotation of quartz with primary dodecylamine (DDA) (Gaudin and Fuerstenau 1955). Their studies showed that the collector adsorption density and zeta potential at the solid–water interface are directly correlated with wetting characteristics and appropriate conditions for flotation. For example, the flotation recovery and contact angle significantly increase with increasing pH in the range of pH 8 to pH 10 when 4×10^{-5} M DDA are used as a collector for the flotation of quartz (Fuerstenau 1957). However, at the same DDA concentration, further increase to pH values greater than pH 12 results in a decrease in flotation recovery and contact angle as the silica surface becomes hydrophilic.

A theory that is widely employed for interpretation of these experimental results is the hemi-micelle theory proposed by Gaudin, Fuerstenau, and Somasundaran (Fuerstenau 1956, 1957; Gaudin and Fuerstenau 1955; Somasundaran and Fuerstenau 1966). This theory suggests that amine adsorption initially occurs through columbic attraction of individual amine cations in the diffuse layer and in the Stern plane at the surface with their polar heads oriented toward the surface. At higher amine concentrations the adsorbed amine cations associate through interaction of their hydrocarbon chains, giving rise to hemi-micelles (i.e., half a micelle), a monolayer physically adsorbed at the oxide surface. For example, in the case of silica, for neutral and slightly acidic pH, DDA is almost completely ionized due to its pK_a of 10.6 (Somasundaran et al. 1988) while the silica surface is negatively charged, PZC at pH $\approx$ 2 (Fuerstenau 1956). Changing the pH significantly affects the amount of amine adsorbed because pH determines the magnitude of the surface potential. Because the silica surface potential becomes more negative with an increase in pH, adsorption and hydrophobicity increase up to about pH 10 (Castro et al. 1986). At higher pH values—above the pKa of 10.6 for DDA—the surface becomes hydrophilic even though amine adsorption is sustained. These results appear to be due to surface micelles or a bilayer of amine at the surface in order to account for the hydrophilic surface state at pH values exceeding pH 12 (Castro et al. 1986; Fuerstenau 1956; Fuerstenau and Renhe 2004).

More recent studies of amine adsorption mechanisms at surfaces have been carried out using atomic force microscopy (AFM), Fourier transform infrared spectroscopy (FTIR), x-ray photoelectron spectroscopy (XPS), and SFVS techniques (Bakshi et al. 2004; Castro et al. 1986; Chernyshova et al. 2000; Fuerstenau and Renhe 2004; Paruchuri et al. 2004; Schrodle and Richmond 2008; Subramanian and Ducker 2000; Velegol et al. 2000). AFM studies show that spherical micelles of cetyltrimethylammonium bromide/cetyltrimethylammonium chloride (CTAB/CTAC) surfactants form at silica surfaces near the critical micelle concentration (CMC) (Bakshi et al. 2004). Compared with tertiary and quaternary amines, primary amines, such as DDA, would have a different molecular structure at the silica surface. For example, AFM studies suggest that primary amine forms a featureless bilayer at mica surfaces.

In a preliminary MD simulation study (Wang et al. 2009), the adsorption behavior of DDA at a quartz surface was examined at different pH values. The simulation results show that at pH 6, the quartz surface is terminated mainly by hydroxyl groups so that the surface carries a slight negative charge. The initial configuration of DDA ions at the quartz surface is shown in Figure 4.21a, and a layer of DDA molecules is positioned in such a way that the polar head groups face the negatively charged quartz surface with their hydrophobic hydrocarbon tails in bulk solution. Figure 4.21b depicts the adsorption states of DDA at the quartz surface after 1ns of simulation. It is observed that at equilibrium the DDA molecules organize themselves into a micellar structure and that the main body of this micelle is stretched out into the bulk water with only several points attached to the quartz surface through the interaction between a number of DDA head groups and the quartz surface, suggesting a weak adsorption state. Figure 4.22 is a close-up snapshot of the DDA adsorption state, which clearly shows that the adsorption is achieved due to the hydrogen bonding between the DDA head group and quartz surface oxygen atoms.

At pH 12, the initial configuration of DDA molecules at the quartz surface was similar to that at pH 6, as shown in Figure 4.23a. Figure 4.23b shows the adsorption states after 1ns of simulation. Again, similar to what happens at pH 6, a micellar structure is formed, but this micelle apparently attached to the quartz surface more significantly, as suggested by the large number of attachment sites. A further examination of the micellar structure, as seen in Figure 4.24, reveals

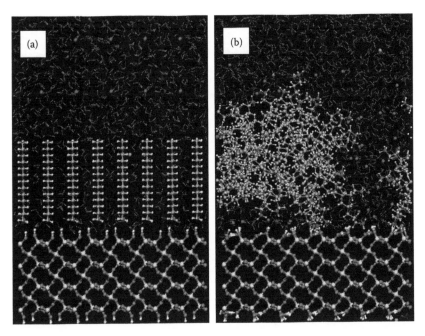

FIGURE 4.21 MDS snapshot of 40 DDA molecules near a quartz surface at pH equivalent to pH 6. The snapshots are taken at the initial state (a) and at 1ns (b). Color key: red, oxygen atom; white, hydrogen atoms; blue, nitrogen atoms; green, sodium ions; purple, chloride ions; light blue, carbon atoms. For clarity, only the skeletons of the water molecules are shown. (From Wang, X., Liu, J., Du, H., and Miller, J. D. *Langmuir,* 26: 3407, 2007. With permission.)

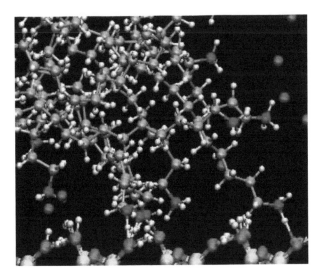

FIGURE 4.22 Close-up MDS snapshot of 40 DDA molecules near a quartz surface at pH equivalent to pH 6. The snapshots are taken at the equilibrium state (1ns For clarity, only the skeletons of the water molecules are shown. Color key: red, oxygen atom; white, hydrogen atoms; blue, nitrogen atoms; green, sodium ions; purple, chloride ions; light blue, carbon atoms. (From Wang, X., Liu, J., Du, H., and Miller, J. D. *Langmuir,* 26: 3407, 2007. With permission.)

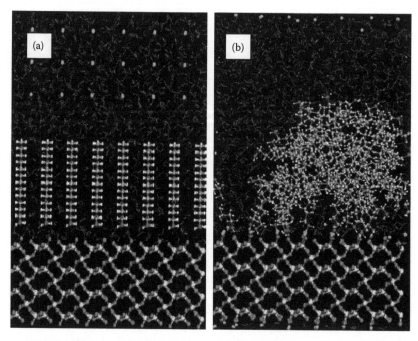

FIGURE 4.23   MDS snapshot of 40 DDA molecules near a quartz surface at pH equivalent to pH 12. The snapshots are taken at the initial state (a) and at 1ns (b). For clarity, only the skeletons of the water molecules are shown. Color key: red, oxygen atom; white, hydrogen atoms; blue, nitrogen atoms; green, sodium ions; purple, chloride ions; light blue, carbon atoms. (From Wang, X., Liu, J., Du, H., and Miller, J. D. *Langmuir,* 26: 3407, 2009. With permission.)

FIGURE 4.24   Close-up MDS snapshot of 40 DDA molecules near a quartz surface at pH equivalent to pH 12. The snapshots are taken at the equilibrium state (1ns). Color key: red, oxygen atom; white, hydrogen atoms; blue, nitrogen atoms; green, sodium ions; purple, chloride ions; light blue, carbon atoms. (From Wang, X., Liu, J., Du, H., and Miller, J. D. *Langmuir,* 26: 3407, 2009. With permission.)

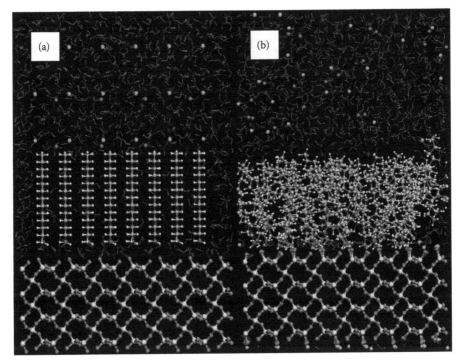

FIGURE 4.25 MDS snapshot of 40 DDA molecules near a quartz surface at pH equivalent to pH 10. The snapshots are taken at the initial state (a) and at 1 ns (b), respectively. Color key: Red, oxygen atoms; white, hydrogen atoms; blue, nitrogen atoms; green, sodium ions; purple, chloride ions; light blue, carbon atoms. For clarity, only the skeletons of the water molecules are shown. (From Wang, X., Liu, J., Du, H., and Miller, J. D. *Langmuir,* 26: 3407, 2009. With permission.)

that the adsorption is achieved due to the hydrogen bonding between the naked oxygen atoms at the quartz surface and hydrogen atom attached to the nitrogen atom in DDA. Even though positively charged sodium ions have a greater tendency to attach to the negatively charged quartz surface, there are still plenty of hydrogen bonding sites available to accommodate DDA head groups. These simulation results are in good agreement with the SFVS measurement results, that stable adsorption of DDA molecules at the negatively charged quartz surface is appreciable and that the organization of DDA molecules is in the form of a micellar structure (Wang et al. 2009).

At pH 10, as seen in Figure 4.25, the DDA adsorption state is totally different, and a well-shaped monolayer is formed at the quartz surface, producing a hydrophobic state. Figure 4.26 is the close-up snapshot of the adsorption state, which suggests that the adsorption is achieved through hydrogen bonding between the DDA head group and the naked oxygen at the quartz surface. Further, due to the strong adsorption of the DDA molecules and ions, most hydrogen-bonding sites at the quartz surface are taken, leaving no room to accommodate a large number of water molecules; thus a water-free surface is produced. These results complement the SFVS results nicely in both the DDA surface adsorption states and the water surface structures (Wang et al. 2009).

4.5 NATURALLY HYDROPHOBIC MINERALS

Naturally hydrophobic minerals are minerals of the surfaces which are of sufficiently low polarity that they are not well wetted by water. A selected list of naturally hydrophobic minerals is presented in Table 4.8 (Zettlemoyer 1969). In some instances, recovery of these minerals is desired, but in

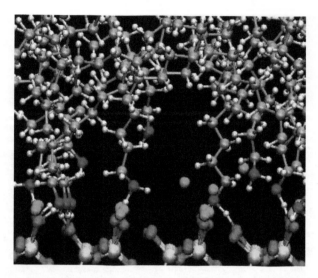

FIGURE 4.26   Close-up MDS snapshot of 40 DDA molecules near a quartz surface at pH equivalent to pH 10. The snapshots are taken at the equilibrium state (1ns). For clarity, only the skeletons of the water molecules are shown. Color key: red, oxygen atom; white, hydrogen atoms; blue, nitrogen atoms; green, sodium ions; purple, chloride ions; light blue, carbon atoms.

TABLE 4.8
Composition and Contact Angle of Selected Naturally Hydrophobic Minerals

Mineral	Composition	Contact Angle (Degree)
Graphite	C	86
Coal	Complex HC	20–60
Sulfur	S	85
Molybdenite	MoS_2	75
Talc	$Mg_3(Si_4O_{10})(OH)_2$	88
Iodyrite	AgI	20

Source: From Zettlemoyer, A. C., *Hydrophobic Surfaces*, Academic Press, New York, 1969. With permission.

other cases these minerals are of insufficient value and are considered as waste. Regardless of their value, the surface chemistry can be adjusted to either sustain their hydrophobic character or to create a hydrophilic surface state by the addition of appropriate flotation reagents.

Naturally hydrophobic minerals such as graphite and talc are common gangue minerals found in sulfide ores and are difficult to separate due to their tendency to float together with valuable sulfide minerals. Despite the relatively successful use of the polysaccharide group of chemicals (e.g., dextrin, guar gum, and carboxymethyl cellulose) as flotation depressants for these naturally hydrophobic minerals, the nature of the adsorption processes remains in debate. Consequently, the adsorption of amphipathic solutes at naturally hydrophobic minerals such as coal and graphite, talc, and sulfur is of interest to many researchers, and a substantial amount of research has been discussed.

Studies regarding the adsorption behavior of dextrin at molybdenite (Wie and Fuerstenau 1974), coal (Haung et al. 1978; Miller et al. 1983), and sulfur (Wisniewska 2005) surfaces have demonstrated the significance of hydrophobic interactions in adsorption processes. Chemical interactions related to the mineral surface metallic sites and consequent complexation effects are also considered as contributors to the adsorption mechanism (Liu and Laskowski 1989b).

To obtain further information regarding the adsorption states at naturally hydrophobic mineral surfaces, the interfacial phenomena and surfactant adsorption behavior at graphite and sulfur surfaces were studied using MD simulation, with some interesting results to report.

4.5.1 WATER AT A GRAPHITE SURFACE

An MD simulation snapshot of water molecules at a graphite (0001) surface is presented in Figure 4.27. An obvious gap exists between graphite and the aqueous phase, which is due to the natural hydrophobicity of graphite. It is evident that water has been excluded from the surface. A similar observation has been discussed in detail regarding water at the naturally hydrophobic talc basal plane surface in Section 4.7.1.1. The water density distribution along the graphite surface normal (Figure 4.27) further reveals a zero water density at the graphite surface.

Similar to previous reports in the literature (Wang et al. 2004a, 2004b), the primary water density peak is located at about 3.5 Å away from the surface, and this distance is larger than the distance between hydrogen bonded water–water molecules, which is approximately 2.8 Å (Dang and Pettitt 1990; Dang and Smith 1993; Lee and Rasaiah, 1996; Lynden-Bell and Rasaiah, 1997, 1988). This result demonstrates the weak interaction between water molecules and a graphite surface. In addition, a similar water dipole moment distribution at a graphite surface has been observed when compared with that at the talc basal plane surface. As seen in Figure 4.28, the water dipole angle distribution peak ranges from 50° to 130°, suggesting the random orientation of interfacial water molecules due to the absence of hydrogen-bonding donor and acceptor sites at the graphite surface.

4.5.2 DTAB ADSORPTION AT A GRAPHITE SURFACE

The DTAB adsorption process at the hydrophobic graphite surface (0001) is illustrated in Figure 4.29. Initially, 25 DTAB ions are sandwiched between graphite surfaces, with their polar head groups facing the substrates (Figure 4.29a). Similar to the DTAB–talc basal plane surface (see Section 4.7.2.1), due to the strong hydrophobic interaction between the substrates and the DTAB tails, spontaneous reorientation of surfactant molecules is observed, as seen in Figure 4.29b and 4.29c. The alkyl chain

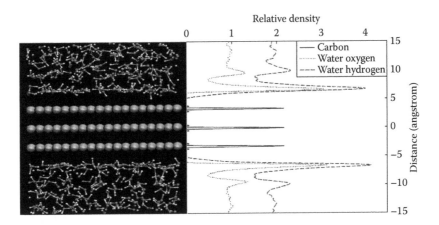

FIGURE 4.27 MD simulation snapshot of water at the graphite surface (left) and interfacial water density distribution at the graphite surface (right). For clarity, only the skeletons of the water molecules are shown. Color key: red, oxygen atom; white, hydrogen atoms; blue, carbon atoms.

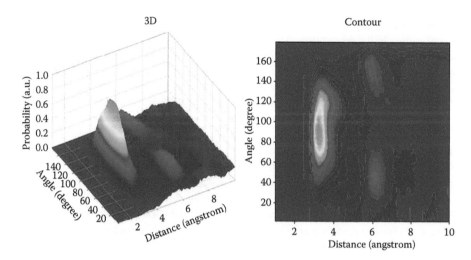

FIGURE 4.28 Water dipole moment distribution at the graphite surface. (The angle is between the water dipole moment and the talc basal plane surface normal.)

of the DTAB ion spreads at the graphite surface and the polar amine head groups stretch out into the aqueous phase, leading to a hydrophilic state at such graphite surfaces.

Also obvious is a spherical micelle structure in bulk water as observed early in the dynamic adsorption process (Figure 4.29b). Eventually, complete adsorption of the DTAB surfactant at the graphite surface is achieved (Figure 4.29d), revealing a strong affinity of the DTAB molecules for the graphite substrate.

The MD simulation and the state of organization of DTAB ions at the graphite surface suggest that hydrophobic interaction between the natural hydrophobic graphite surface and the hydrophobic tails of the DTAB ion accounts for the surfactant adsorption state. This observation is consistent with the results from previous studies regarding the significance of the hydrophobic interaction for the structurally similar surfactant, cetyltrimethylammonium bromide (CTAB), in adsorption at naturally hydrophobic mineral surfaces using surface characterization techniques such as contact angle measurements and atomic force microscope imaging (Bakshi et al. 2004; Paruchuri et al. 2006).

4.5.3 Dextrin Adsorption

4.5.3.1 Dextrin at a Graphite Surface

The behavior of the dextrin molecule at a graphite (0001) surface is presented in Figure 4.30. A couple of things are worth noting. First, the dextrin molecule migrates to and remains attached to the graphite surface at one end (Figure 4.30a and b), indicating its strong affinity for the graphite surface. Second, the dextrin molecule is unable to make full contact with the graphite surface during the entire simulation run (Figure 4.30b to d), suggesting that a complete spreading or occupation of the dextrin molecule at the graphite surface is not the most energetically stable state.

To understand dextrin organization at the graphite surface, the properties of the graphite surface, as well as the structure of the dextrin molecule, have to be considered. The graphite surface, which is composed of inactive neutral carbon atoms, is naturally hydrophobic and cannot form hydrogen bonds with water molecules. Consequently, water molecules are excluded from the graphite surface. However, the dextrin molecule shows a dual characteristic in that the presence of hydroxyl groups provides this molecule with plenty of hydrogen-bonding sites and enables the dextrin molecules to hydrogen-bond with surrounding water molecules. At the same time the low polarity of the methylene groups, as well as the C–H groups, in the dextrin molecule show a hydrophobic character,

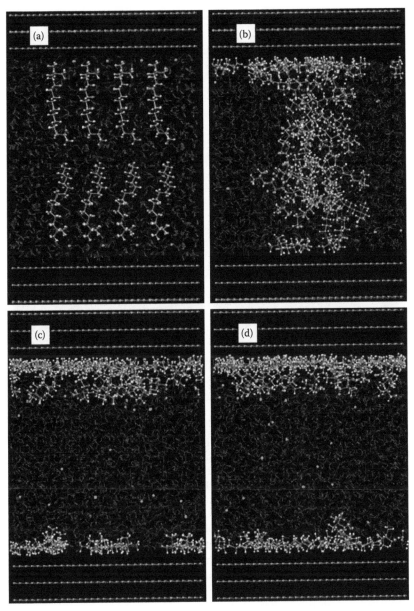

FIGURE 4.29 MDS snapshots of the adsorption states of 50 DTAB molecules at graphite surfaces for different simulation times. The snapshots are taken at 0 ps (a), 200ps (b), 500ps (c), and 1ns (d). For clarity, only the skeletons of the water molecules are shown. Color key: red, oxygen atom; white, hydrogen atoms; blue, nitrogen atoms; green, bromide atoms; light blue, carbon atoms. (From Du, H. and Miller, J. D., *Langmuir,* 23: 11587, 2007b. With permission.)

and these groups drive the molecule to escape from the water phase. When interacting with the hydrophobic graphite surface, the hydrophobic moieties of the dextrin molecule preferentially stay in contact with the graphite surface, as suggested by Figure 4.30. When compared with hydroxyl groups, the methylene and the C–H groups in the dextrin molecule are significantly closer to the graphite surface.

The hydrophilic hydroxyl groups favor hydrogen bonding with water molecules. As a compromise, the dextrin molecule reorients to expose as many hydrophobic moieties as possible to the

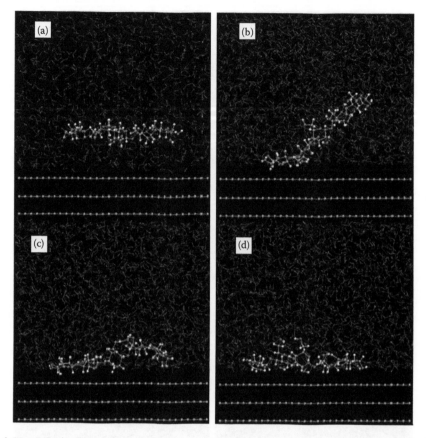

FIGURE 4.30 MD simulation snapshots of the dextrin molecule (five dextrose monomers) adsorption states at a graphite surface for different simulation times. The snapshots are taken at 0 ps (a), 200ps (b), 500ps (c), and 1ns (d). For clarity, only the skeletons of the water molecules are shown. Color key: red, oxygen atom; white, hydrogen atoms; light blue, carbon atoms. (Du, H. and Miller, J. D., *Langmuir,* 23: 11587, 2007b. With permission.)

hydrophobic graphite surface while the majority of the dextrin hydroxyl groups hydrogen-bond with water molecules. In conclusion, part of the dextrin molecule adsorbs at the graphite surface due to hydrophobic attraction, while the remainder of the molecule remains hydrated by the water phase through hydroxyl–water hydrogen bonding as observed in Figure 4.30.

4.5.3.2 Dextrin at a Sulfur Surface

The adsorption of dextrin molecules by sulfur has been reported to be similar to that observed for talc and coal. Hydrophobic interaction has been proposed to be the major adsorption mechanism (Brossard et al. 2008; Wisniewska 2005). In this study, a rhombic sulfur mineral was cut along the {111} plane to keep the integrity of each S_8 molecule. Our simulation results, as illustrated in Figure 4.31, suggest that the behavior of the dextrin molecule at the sulfur surface is almost identical to its behavior at the graphite surface.

The adsorption of the dextrin molecule at the sulfur surface is achieved by the attachment of a portion of the molecule. Yet the magnitude of the adsorption is weaker than at a graphite surface, as suggested by the significantly smaller portion of the dextrin molecule that sticks to the sulfur surface. Due to the similar hydrophobic character of the sulfur surface, which is composed of elemental sulfur atoms, the observed partial adsorption of the dextrin molecule at the sulfur surface is explained in the same way dextrin adsorption at the graphite surface is explained.

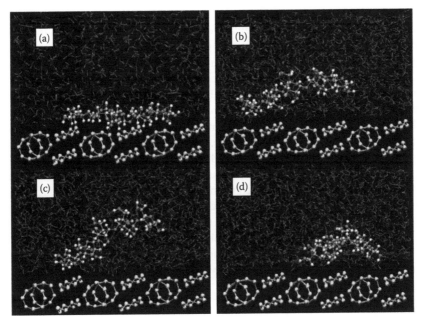

FIGURE 4.31 MD simulation snapshots of the adsorption states of the dextrin molecule (five dextrose monomers) at a sulfur {111} surface for different simulation times. The snapshots are taken at 0 ps (a), 200 ps (b), 500 ps (c), and 1 ns (d). For clarity, only the skeletons of the water molecules are shown. Color key: red, oxygen atom; white, hydrogen atoms; yellow, sulfur atoms; light blue, carbon atoms. (Du, H. and Miller, J. D., *Langmuir,* 23: 11587, 2007b. With permission.)

The significantly weakened adsorption state is understood by considering the structural difference between sulfur and graphite surfaces. It is known that rhombic sulfur is built up from cyclooctasulfur (S_8) molecules. Unlike graphite, which can be cleaved perfectly along its major cleavage plane {0001}, the cleavage of a rhombic sulfur crystal cannot produce an atomically flat surface due to its unique structure. A close examination of the sulfur surface {111} used in the simulation shows that the surface sulfur atoms are not positioned in the same plane. As a consequence, atomic notches occurred between each row of S_8 molecules, as illustrated in Figure 4.31. This atomic nanoroughness provides structural barriers to the adsorption of dextrin hydrophobic moieties at the surface while simultaneously remaining in an energetically stable molecular configuration. However, small water molecules are capable of dynamically occupying these surface voids, thus providing additional energy barriers for the dextrin adsorption.

4.5.4 Summary

MD simulation of water structures and cationic surfactant DTAB adsorption states at the graphite surface provide further information for the fundamental understanding of surfactant adsorption mechanisms at naturally hydrophobic surfaces. We have concluded from our MD simulations that, similar to MD simulation results obtained for the hydrophobic talc basal plane surface, water molecules are excluded from the graphite substrate due to the nonpolar features of the graphite surface, which favors the hydrophobic interaction with the nonpolar alkyl chains of cationic surfactant DTAB and renders the basal plane surface hydrophilic. These results are in agreement with previously reported results, which reveal a significant decrease in contact angle when cationic surfactants are present at sufficient concentration, and with the results from soft contact AFM imaging of surface micelle structures.

MD simulation that investigated the adsorption of dextrin at selected naturally hydrophobic surfaces such as graphite and sulfur suggests that hydrophobic interactions between the substrates and the hydrophobic moieties in the dextrin molecule play a significant role in the adsorption processes. At the graphite surface, which is composed of neutral elemental carbon atoms, the dextrin molecule reorients to expose as many hydrophobic moieties as possible to the hydrophobic graphite surface while the majority of the dextrin hydroxyl groups hydrogen bond with water molecules by stretching into the bulk aqueous phase.

The behavior of a dextrin molecule at the rhombic sulfur {111} surface suggests a similar adsorption state to that observed at the graphite surface. The dextrin molecule attaches to the sulfur substrate through hydrophobic interaction between the methyl group and the sulfur surface, while the remainder of the dextrin molecule stays in the aqueous phase. The decreased level of attachment of the dextrin molecule at the sulfur surface compared with that at the graphite surface may be due to the inherent atomic roughness of the sulfur surface, which allows instantaneous accommodation of water molecules within the sulfur surface region. Consequently, significant energetic and structural barriers are produced for all the hydrophobic moieties of the dextrin molecule, which inhibit interaction with the sulfur substrate. Nevertheless, the experimental dextrin adsorption density values for sulfur are equivalent to those observed for graphite.

4.6   FURTHER DISCUSSION

In this chapter the surface chemistry of selected nonsulfide flotation systems, including soluble alkali halide salts, phyllosilicates, quartz, and some naturally hydrophobic minerals, were studied using MD simulation. Issues such as water structure and dynamics, solution chemistry, interfacial water structure, and adsorption states for surfactants and macromolecules were examined. It is clear that MD simulation has been validated as a very useful tool to study the surface chemistry of certain flotation systems. As a complement to experimental studies, MD simulation analysis provides further information and understanding at the atomic level to issues such as water structure, particle dynamics, solution viscosities, mineral surface wetting characteristics, surface charge, and adsorption states. A wide application of MD simulation in the study of mineral surface chemistry is expected to have a significant impact on further advances in flotation technology.

It is also evident from MD simulation analysis that there are several limitations inherent in the empirical potential function, leading to some inaccuracies in computer simulation. One limitation is that once the force field is chosen, the potential parameters are fixed during the MD simulation. For example, carbon atoms in an alkyl chain and in a benzene ring are often treated using one set of potential parameters without the full consideration of their environments so that the number of atom types is minimized. This can lead to type-specific errors. Some atoms, such as aliphatic carbon or hydrogen atoms, are less sensitive to their surroundings and can be described by a single set of parameters. At the same time, other atoms such as oxygen and nitrogen are much more environmentally sensitive and therefore, require more types and parameters to account for the different bonding environments (Stote et al. 1999).

Another major limitation is the pairwise additive approximation, which is introduced to decrease the computational demand. In this approximation, the interaction energy between one atom and the rest of the atoms is calculated as a sum of pairwise (one atom to one atom) interactions; thus certain polarization effects are not explicitly included in the force field (Stote et al. 1999). This can lead to subtle differences between calculated and experimental results.

Further, an important point to consider is that the entropic effects are neglected in the potential energy function. Thus, when the system's energy reaches a minimum value as a sum of potential functions, it does not necessarily correspond to the equilibrium, or the most energy-stable state (Stote et al. 1999). The entropic effect can be evaluated from further analysis of the trajectories of particles in a simulation system (Schlitter 1993).

Last but definitely not least, due to the significance of water, proper selection of a water force field is critical for a realistic MD simulation. For example, when solution surface tension is determined from simulation, controversial results can be obtained with and without consideration of water molecule and solute atom polarity (Andersson et al. 2007; Gopalakrishnan et al. 2005; Jungwirth and Tobias 2002; Jungwirth et al. 2006; Thomas et al. 2007). In this regard, significant research has been carried out to develop water models that are both efficient computationally and realistic. Simulations reported in this paper used the SPC model; other popular water models include SPC/E, TIP3P, and TIP4P. Spoel et al. (1998) provides a good summarization of these water models.

4.7   ACKNOWLEDGMENTS

The authors are grateful to Professor Jayendran C. Rasaiah for engaging with us in numerous valuable discussions and to Professor Anh V. Nguyen for helping to initiate the MD simulation research. The financial support provided by the Department of Energy, Basic Science Division Grant No. DE-FG-03-93ER14315, National Natural Science Foundation of China under Grant No. 50904057 and No. 51090382, and National Basic Research Development Program of China (973 Program) under Grant No. 2007CB613501 are gratefully acknowledged. In addition, this study was prompted to some extent by collaborative research supported by NSF under grant Nos. INT-0227583 and INT-0352807.

REFERENCES

Abraham, F. F. 1978. The interfacial density profile of a Lennard–Jones fluid in contact with a (100) Lennard–Jones wall and its relationship to idealized fluid/wall systems: A Monte Carlo simulation. *J. Chem. Phys.* 68: 3713.

Allen, M. P., and Tildesley, D. J. 1987. Computer Simulation of Liquids. Oxford University Press, Oxford, UK 450.

Andersson, G., Morgner, H., Cwiklik, L., and Jungwirth, P. 2007. Anions of alkali halide salts at surfaces of formamide solutions: Concentration depth profiles and surface topography. *J. Phys. Chem. C* 111: 4379.

Arab, M., Bougeard, D., and Smirnov, K. S. 2003. Structure and dynamics of the interlayer water in an uncharged 2:1 clay. *Phys. Chem. Chem. Phys.* 5: 4699.

Bakshi, M. S., Kaura, A., Miller, J. D., and Paruchuri, V. K. 2004. Sodium dodecyl sulfate-poly(amidoamine) interactions studied by AFM imaging, conductivity, and Krafft temperature measurements. *J. Colloid Interface Sci.* 278: 472.

Balajee, S. R., and Iwasaki, I. 1969. Interaction of British gum and dodecylammonium chloride at quartz and hematite surfaces. *Trans. Am. Inst. Min. Metall. Pet. Eng.* 244: 407.

Bayerl, T. M., and Bloom, M. 1990. Physical properties of single phospholipid bilayers adsorbed to micro glass beads. A new vesicular model system studied by deuterium nuclear magnetic resonance. *Biophys. J.* 58: 357.

Berendsen, H. J. C., Grigera, J. R., and Straatsma, T. P. 1987. The missing term in effective pair potentials. *J. Phys. Chem.* 91: 6269.

Bridgeman, C. H., Buckingham, A. D., Skipper, N. T., and Payne, M. C. 1996. Ab-initio total energy study of uncharged 2:1 clays and their interaction with water. *Mol. Phys.* 89: 879.

Bridgeman, C. H., and Skipper, N. T. 1997. A Monte Carlo study of water at an uncharged clay surface. *J. Phys. Condens. Matter.* 9: 4081.

Brooks, B. R., Bruccoleri, R. E., Olafson, B. D., States, D. J., Swaminathan, S., and Karplus, M. 1983. CHARMM: A program for macromolecular energy, minimization, and dynamics calculations. *J. Comput. Chem.* 4: 187.

Brossard, S. K., Du, H., and Miller, J. D. 2008. Characteristics of dextrin adsorption by elemental sulfur. *J. Colloid Interface Sci.* 317: 18.

Bryant, R. G. 1996. The dynamics of water–protein interactions. *Ann. Rev. Biophys. Biomol. Struct.* 25: 29.

Cacelli, I., Cinacchi, G., Prampolini, G., and Tani, A. 2004. Computer simulation of solid and liquid benzene with an atomistic interaction potential derived from ab initio calculations. *J. Am. Chem. Soc.* 126: 14278.

Cao, Q., Du, H., Miller, J. D., Wang, X., and Cheng, F. 2010. Surface chemistry features in the flotation of KCl. *Miner. Eng.* 23: 365.

Castro, S. H., Vurdela, R. M., and Laskowski, J. S. 1986. The surface association and precipitation of surfactant species in alkaline dodecylamine hydrochloride solutions. *Colloids Surf.* 21: 87.

Chernyshova, I. V., Rao, K. H., Vidyadhar, A., and Shchukarev, A. V. 2000. Mechanism of adsorption of long-chain alkylamines on silicates. A spectroscopic study. 1. Quartz. *Langmuir* 16: 8071.

Cho, K.-H., No, K. T., and Scheraga, H. A. 2002. A polarizable force field for water using an artificial neural network. *J. Mol. Struct.* 641: 77.

Chowdhuri, S., and Chandra, A. 2001. Molecular dynamics simulations of aqueous NaCl and KCl solutions: Effects of ion concentration on the single-particle, pair, and collective dynamical properties of ions and water molecules. *J. Chem. Phys.* 115: 3732.

Chowdhuri, S., and Chandra, A. 2003. Hydration structure and diffusion of ions in supercooled water: Ion size effects. *J. Chem. Phys.* 118: 9719.

Cramer, C. J. 2002. *Essentials of Computational Chemistry: Theories and Models.* John Wiley & Sons, Hoboken New Jersey, 562.

Cygan, R. T., Liang, J.-J., and Kalinichev, A. G. 2004. Molecular models of hydroxide, oxyhydroxide, and clay phases and the development of a general force field. *J. Phys. Chem. B.* 108: 1255.

Daivis, P. J., Travis, K. P., and Todd, B. D. 1996. A technique for the calculation of mass, energy, and momentum densities at planes in molecular dynamics simulations. *J. Chem. Phys.* 104: 9651.

Dang, L. X. 1992. Development of nonadditive intermolecular potentials using molecular dynamics: Solvation of lithium(1^+) and fluoride ions in polarizable water. [Erratum to document cited in CA117(10):98630n]. *J. Chem. Phys.* 97: 1614.

Dang, L. X. 1994. Free energies for association of Cs^+ to 18-crown-6 in water. A molecular dynamics study including counter ions. *Chem. Phys. Lett.* 227: 211.

Dang, L. X. 1995. Mechanism and thermodynamics of ion selectivity in aqueous solutions of 18-crown-6 ether: A molecular dynamics study. *J. Am. Chem. Soc.* 117: 6954.

Dang, L. X. 1997. Simulations of water and water-chlorinated hydrocarbon liquid/liquid interfaces with polarizable potential models. Book of Abstracts, 214th ACS National Meeting, Las Vegas, NV, September 7–11, p. PHYS-134.

Dang, L. X., and Smith, D. E. 1993. Molecular dynamics simulations of aqueous ionic clusters using polarizable water. *J. Chem. Phys.* 99: 6950.

Dang, L. X., and Pettitt, B.M. 1990. A theoretical study of like ion pairs in solution. *J. Phys. Chem.* 94: 4303.

Dillon, S. R., and Dougherty, R. C. 2002. Raman studies of the solution structure of univalent electrolytes in water. *J. Phys. Chem. A* 106: 7647.

Du, H., and Miller, J. D. 2007a. A molecular dynamics simulation study of water structure and adsorption states at talc surfaces. *Int. J. Miner. Process.* 84: 172.

Du, H., and Miller, J. D. 2007b. Adsorption states of amphipatic solutes at the surface of naturally hydrophobic minerals: A molecular dynamics simulation study. *Langmuir* 23: 11587.

Du, H., and Miller, J. D. 2007c. Interfacial water structure and surface charge of selected alkali chloride salt crystals in saturated solutions: A molecular dynamics modeling study. *J. Phys. Chem. C* 111: 10013.

Du, H., Liu, J., Ozdemir, O., Nguyen, A. V., and Miller, J. D. 2008. Molecular features of the air/carbonate solution interface. *J. Colloid Interface Sci.* 318: 271.

Du, H., Rasaiah, J. C., and Miller, J. D. 2007. Structural and dynamic properties of concentrated alkali halide solutions: A molecular dynamics simulation study. *J. Phys. Chem. B* 111: 209.

Engkvist, O., and Stone, A. J. 2000. Adsorption of water on the NaCl(001) surface. III. Monte Carlo simulations at ambient temperatures. *J. Chem. Phys.* 112: 6827.

Forester, T. T., and Smith, W. 1995. *DL_POLY User Manual*, CCLRC Daresbury Laboratory, Daresbury, Warrington, Cheshire, UK.

Fuerstenau, D. W., and Huang, P. 2003. Interfacial phenomena involved in talc flotation and depression. *XXII International Mineral Processing Congress.* South African Institute of Mining and Metallurgy in Marshalltown, Cape Town, South Africa. 1034.

Fuerstenau, D. W. 1956. Streaming-potential studies on quartz in solutions of ammonium acetates in relation to the formation of hemi-micelles at the quartz-solution interface. *J. Phys. Chem.* 60: 981.

Fuerstenau, D. W. 1957. Correlation of contact angles, adsorption density, zeta potentials, and flotation rate. *AIME Trans.* 208: 1365.

Fuerstenau, D. W., and Renhe, J. 2004. The adsorption of alkylpyridinium chlorides and their effect on the interfacial behavior of quartz. *Colloids Surf. A* 250: 223.

Fuerstenau, M. C., Jameson, G., and Yoon, R.-H. 2007. *Froth Flotation: A Century of Innovation*. Society for Mining Metallurgy & Exploration, Englewood, Colorado, 904.

Gaudin, A. M., and Fuerstenau, D. W. 1955. Quartz flotation with cationic collectors. *Trans. Am. Inst. Min. Metall. Pet. Eng.* 202: Tech. Pub. 4104-B.

Gopalakrishnan, S., Jungwirth, P., Tobias, D. J., and Allen, H. C. 2005. Air-liquid interfaces of aqueous solutions containing ammonium and sulfate: Spectroscopic and molecular dynamics studies. *J. Phys. Chem. B* 109: 886.

Haile, J. M. 1997. *Molecular Dynamics Simulation: Elementary Methods*. Wiley, New York, 489.

Hancer, M., Celik, M. S., and Miller, J. D. 2001. The significance of interfacial water structure in soluble salt flotation systems. *J. Colloid Interface Sci.* 235: 150.

Haung, H., Calara, J. V., Bauer, D. L., and Miller, J. D. 1978. Adsorption reactions in the depression of coal by organic colloids. *Recent Dev. Sep. Sci.* 4: 115.

Healy, T. W. 1974. Principles of adsorption of organics at solid–solution interfaces. *J. Macromol. Sci. Chem.* 8: 603.

Heinz, H., Castelijns, H. J., and Suter, U. W. 2003. Structure and phase transitions of alkyl chains on mica. *J. Am. Chem. Soc.* 125: 9500.

Heinz, H., Koerner, H., Anderson, K. L., Vaia, R. A., and Farmer, B. L. 2005. Force field for mica-type silicates and dynamics of octadecylammonium chains grafted to montmorillonite. *Chem. Mater.* 17: 5658.

Heinz, H., Vaia, R. A., Krishnamoorti, R., and Farmer, B. L. 2006. Self-assembly of alkylammonium chains on montmorillonite: Effect of chain length, head group structure, and cation exchange capacity. *Chem. Mater.* 19: 59.

Hunt, J. P. 1963. *Metal Ions in Aqueous Solution*. W. A. Benjamin, New York, 124.

Jacobs, J. D., Koerner, H., Heinz, H., Farmer, B. L., Mirau, P., Garrett, P. H., and Vaia, R. A. 2006. Dynamics of alkyl ammonium intercalants within organically modified montmorillonite: Dielectric relaxation and ionic conductivity. *J. Phys. Chem. B* 110: 20143.

Jenkins, P., and Ralston, J. 1998. The adsorption of a polysaccharide at the talc-aqueous solution interface. *Colloids Surf. A.* 139: 27.

Jiang, J., and Sandler, S. I. 2003. A new model for the viscosity of electrolyte solutions. *Ind. Eng. Chem. Res.* 42: 6267.

Jungwirth, P., and Tobias, D. J. 2002. Ions at the air/water interface. *J. Phys. Chem. B.* 106: 6361.

Jungwirth, P., Finlayson-Pitts, B. J., and Tobias, D. J. 2006. Introduction: Structure and chemistry at aqueous interfaces. *Chem. Rev.* 106: 1137.

Kalinichev, A. G., and Kirkpatrick, R. J. 2002. Molecular dynamics modeling of chloride binding to the surfaces of calcium hydroxide, hydrated calcium aluminate, and calcium silicate phases. *Chem. Mater.* 14: 3539.

Kaminsky, M. 1957. The concentration and temperature dependence of the viscosity of aqueous solutions of strong electrolytes. III. KCl, K_2SO_4, $MgCl_2$, $BeSO_4$, and $MgSO_4$ solutions. *Zeitschrift fuer Physikalische Chemie.* 12: 206.

Kamiya, N., Watanabe, Y. S., Ono, S., and Higo, J. 2005. AMBER-based hybrid force field for conformational sampling of polypeptides. *Chem. Phys. Lett.* 401: 312.

Kara, M., Yuzer, H., Sabah, E., and Celik, M. S. 2003. Adsorption of cobalt from aqueous solutions onto sepiolite. *Water Res.* 37: 224.

Kim, J., Kim, G., and Cremer, P. S. 2001. Investigations of water structure at the solid/liquid interface in the presence of supported lipid bilayers by vibrational sum frequency spectroscopy. *Langmuir* 17: 7255.

Koenig, B. W., Krueger, S., Orts, W. J., Majkrzak, C. F., Berk, N. F., Silverton, J. V., and Gawrisch, K. 1996. Neutron reflectivity and atomic force microscopy studies of a lipid bilayer in water adsorbed to the surface of a silicon single crystal. *Langmuir* 12: 1343.

Koneshan, S., and Rasaiah, J. C. 2002. Computer simulation studies of aqueous sodium chloride solutions at 298 K and 683 K. *J. Chem. Phys.* 113: 8125.

Koneshan, S., Lynden-Bell, R. M., and Rasaiah, J. C. 1998. Solvent structure, dynamics, and ion mobility in aqueous solutions at 25°C. *J. Phys. Chem. B* 102: 4193.

Koneshan, S., Lynden-Bell, R. M., and Rasaiah, J. C. 1998. Friction coefficients of ions in aqueous solution at 25 C. *J. Am. Chem. Soc.* 120: 12041.

Larentzos, J. P., Greathouse, J. A., and Cygan, R. T. 2007. An ab initio and classical molecular dynamics investigation of the structural and vibrational properties of talc and pyrophyllite. *J. Phys. Chem. C* 111: 12752.

Lazarevic, S., Jankovic-Castvan, I., Jovanovic, D., Milonjic, S., Janackovic, D., and Petrovic, R. 2007. Adsorption of $Pb2^+$, $Cd2^+$ and $Sr2^+$ ions onto natural and acid-activated sepiolites. *Appl. Clay Sci.* 37: 47.

Leach, A. 2001. *Molecular Modelling: Principles and Applications,* 2nd ed. Prentice Hall, Upper Saddle River, New Jersey, 784.

Lee, S. H., and Rasaiah, J. C. 1994. Molecular dynamics simulation of ionic mobility.1. Alkali metal cations in water at 25°C. *J. Chem. Phys.* 101: 6964.

Lee, S. H., and Rasaiah, J. C. 1996. Molecular dynamics simulation of ion mobility. 2. Alkali metal and halide ions using the SPC/E model for water with simple truncation of ion–water potential. *J. Phys. Chem.* 100: 1420.

Leng, D., Keffer, J., and Cummings, P. T. 2003. Structure and dynamics of a benzenedithiol monolayer on a Au(111) surface. *J. Phys. Chem. B* 107: 11940.

Liu, Q., and Laskowski, J. S. 1989a. The interactions between dextrin and metal hydroxides in aqueous solutions. *J. Colloid Interface Sci.* 130: 101.

Liu, Q., and Laskowski, J. S. 1989b. The role of metal hydroxides at mineral surfaces in dextrin adsorption, II. Chalcopyrite-galena separations in the presence of dextrin. *Int. J. Miner. Process.* 27: 147.

Liu, Q., and Laskowski, J. S. 1989c. The role of metal hydroxides at mineral surfaces in dextrin adsorption, I. Studies on modified quartz samples. *Int. J. Miner. Process.* 26: 297.

Liu, X., Zhou, G., Zhang, S., Wu, G., and Yu, G. 2007. Molecular simulation of guanidinium-based ionic liquids. *J. Phys. Chem. B* 111: 5658.

Luck, W. A. P. 1973. *Water: A Comprehensive Treatise*, Ed. Felix Franks. Plenum, New York. 3, 211.

Luck, W. A. P., and Schioeberg, D. 1979. Spectroscopic investigations of the structure of liquid water and aqueous solutions. *Adv. Mol. Relax. Interact. Processes* 14: 277.

Lynden-Bell, R. M., and Rasaiah, J. C. 1996. Mobility and solvation of ions in channels. *J. Chem. Phys.* 105: 9266.

Lynden-Bell, R. M., and Rasaiah, J. C. 1982. From hydrophobic to hydrophilic behaviour: A simulation study of solvation entropy and free energy of simple solutes. *J. Chem. Phys.* 107: 1982.

Max, J.-J., and Chapados, C. 2000. Infrared spectra of cesium chloride aqueous solutions. *J. Chem. Phys.* 113: 6803.

Max, J.-J., and Chapados, C. 2001. IR spectroscopy of aqueous alkali halide solutions: Pure salt-solvated water spectra and hydration numbers. *J. Chem. Phys.* 115: 2664.

Mayo, S. L., Olafson, B. D., Goddard, W. A. III. 1990. DREIDING: A generic force field for molecular simulations. *J. Phys. Chem.* 94: 8897.

McNamara, J. P., Muslim, A., Abdel-Aal, H., Wang, H., Mohr, M., Hillier, I. H. and Bryce, R. A. 2004. Towards a quantum mechanical force field for carbohydrates: A reparametrized semi-empirical MO approach. *Chem. Phys. Lett.* 394: 429.

Miller, J. D., and Paruchuri, V. K. 2004. Surface micelles as revealed by soft contact atomic force microscopy imaging. *Recent Res. Dev. Surf. Colloids* 1: 205.

Miller, J. D., Nalaskowski, J., Abdul, B., and Du, H., 2007. Surface characteristics of kaolinite and other selected two layer silicate minerals. *Can. J. Chem. Eng.* 85: 617.

Miller, J. D., Laskowski, J. S., and Chang, S. S. 1983. Dextrin adsorption by oxidized coal. *Colloids and Surfaces* 8: 137.

Miller, J. D., Yalamanchili, M. R., and Kellar, J. J. 1992. Surface charge of alkali halide particles as determined by laser-Doppler electrophoresis. *Langmuir* 8: 1464.

Miranda, P. B., and Shen, Y. R. 1999. Liquid interfaces: A study by sum-frequency vibrational spectroscopy. *J. Phys. Chem. B* 103: 3292.

Mondello, M., and Grest, G. S. 1997. Viscosity calculations of n-alkanes by equilibrium molecular dynamics. *J. Chem. Phys.* 106: 9327.

Morris, G. E., Fornasiero, D., and Ralston, J. 2002. Polymer depressants at the talc–water interface: Adsorption isotherm, microflotation and electrokinetic studies. *Int. J. Miner. Process.* 67: 211.

Nalaskowski, J., Drelich, J., Hupka, J., and Miller, J. D. 2003. Adhesion between hydrocarbon particles and silica surfaces with different degrees of hydration as determined by the AFM colloidal probe technique. *Langmuir* 19: 5311.

Nalaskowski, J., Abdul, B., Du, H., and Miller, J. D. 2007. Anisotropic character of talc surfaces as revealed by streaming potential measurements, atomic force microscopy, molecular dynamics simulations and contact angle measurements. *Can. Metall. Q.* 46: 227.

Nickolov, Z. S., and Miller, J. D. 2005. Water structure in aqueous solutions of alkali halide salts: FTIR spectroscopy of the OD stretching band. *J. Colloid Interface Sci.* 287: 572.

Nickolov, Z. S., Wang, X., and Miller, J. D. 2004. Liquid/air interfacial structure of alcohol-octyl hydroxamic acid mixtures: A study by sum-frequency spectroscopy. *Spectrochim. Acta, Part A.* 60A: 2711.

Park, J. M., Cho, J.-H., and Kim, K. S. 2004. Atomic structure and energetics of adsorbed water on the NaCl(001) surface. *Phys. Rev. B: Condens. Matter* 69: 233403.

Paruchuri, V. K., Nguyen, A.V., and Miller, J. D. 2004. Zeta potentials of self-assembled surface micelles of ionic surfactants adsorbed at hydrophobic graphite surfaces. *Colloids Surf. A* 250: 519.

Paruchuri, V. K., Nalaskowski, J., Shah, D. O., and Miller, J. D., 2006. The effect of cosurfactants on sodium dodecyl sulfate micellar structures at a graphite surface. *Colloids Surf. A* 272: 157.

Pearlman, D. A., Case, D. A., Caldwell, J. W., Ross, W. S., Cheatham, T. E., DeBolt, S., et al. 1995. "AMBER", a package of computer programs for applying molecular mechanics, normal mode analysis, molecular dynamics and free energy calculations to stimulate the structural and energetic properties of molecules. *Comput. Phys. Commun.* 91: 1.

Ponder, J. W., Case, D. A., and Valerie, D. 2003. Force Fields for Protein Simulations, in Advances in Protein Chemistry. Academic Press, Maryland Heights, Missouri, 27.

Pramanik, A., Kalagi, R. P., Barge, V. J., and Gadre, S. R., 2005. Adsorption of water on sodium chloride surfaces: electrostatics—guided ab initio studies. *Theor. Chem. Acc.* 114: 129.

Pugh, R. J. 1989. Macromolecular organic depressants in sulfide flotation—A review, 2. Theoretical analysis of the forces involved in the depressant action. *Int. J. Miner. Process.* 25: 131.

Rasaiah, J. C. 1988. Theories of electrolyte solutions. *NATO ASI Ser., Ser. B* 193: 89.

Rath, R. K., Subramanian, S., and Laskowski, J. S. 1997. Adsorption of dextrin and guar gum onto talc. A comparative study. *Langmuir* 13: 6260.

Ruckenstein, E., and Churaev, N. V. 1991. A possible hydrodynamic origin of the forces of hydrophobic attraction. *J. Colloid Interface Sci.* 147: 535.

Schlitter, J. 1993. Estimation of absolute and relative entropies of macromolecules using the covariance matrix. *Chem. Phys. Lett.* 215: 617.

Schrodle, S., and Richmond, G. L. 2008. In situ non-linear spectroscopic approaches to understanding adsorption at mineral-water interfaces. *J. Phys. D: Appl. Phys.* 41: 033001.

Smith, D. E., and Dang, L. X. 1994a. Computer simulations of cesium–water clusters: Do ion-water clusters form gas-phase clathrates? *J. Chem. Phys.* 101: 7873.

Smith, D. E., and Dang, L. X. 1994b. Computer simulations of NaCl association in polarizable water. *J. Chem. Phys.* 100: 3757.

Smith, D. E., and Dang, L. X. 1994c. Interionic potentials of mean force for $SrCl_2$ in polarizable water. A computer simulation study. *Chem. Phys. Lett.* 230: 209.

Somasundaran, P., and Fuerstenau, D. W. 1966. Mechanisms of alkyl sulfonate adsorption at the alumina–water interface. *J. Phys. Chem.* 70: 90.

Somasundaran, P., and Moudgil, B. M. 1988. *Surfactant Science Series 27: Reagents in Mineral Technology*, CRC Press, Boca Raton, FL, 755.

Stote, R., Dejaegere, A., Kuznetsov, D., and Falquet, L., 1999. Tutrial of molecular dynamics simulations, CHARMM. http://www.ch.embnet.org/MD_tutorial/

Subramanian, V., and Ducker, W. A. 2000. Counterion effects on adsorbed micellar shape: Experimental study of the role of polarizability and charge. *Langmuir* 16: 4447.

Terpstra, P., Combes, D., and Zwick, A. 1990. Effect of salts on dynamics of water: A Raman spectroscopy study. *J. Chem. Phys.* 92: 65.

Thomas, J. L., Roeselova, M., Dang, L. X., and Tobias, D., 2007. Molecular dynamics simulations of the solution–air interface of aqueous sodium nitrate. *J. Phys. Chem. A* 111: 3091.

Uchida, H., and Matsuoka, M. 2004. Molecular dynamics simulation of solution structure and dynamics of aqueous sodium chloride solutions from dilute to supersaturated concentration. *Fluid Phase Equilib.* 219: 49.

van der Spoel, D., van Maaren, P. J., and Berendsen, H. J. C. 1998. A systematic study of water models for molecular simulation: Derivation of water models optimized for use with a reaction field. *J. Chem. Phys.* 108: 10.

van Gunsteren, W. F., Billeter, S. R., Eising, A. A., Hünenberger, P. H., Krüger, P., Mark, A. E., et al. 1996. *Biomolecular Simulation: The GROMOS96 Manual and User Guide*.

Veeramasuneni, S., Hu, Y., and Miller, J. D. 1997. The surface charge of alkali halides—Consideration of the partial hydration of surface lattice ions. *Surf. Sci.* 382: 127.

Velegol, S. B., Fleming, B. D., Biggs, S., Wanless, E, J., and Tilton, R. D. 2000. Counterion effects on hexadecyltrimethylammonium surfactant adsorption and self-assembly on silica. *Langmuir* 16: 2548.

Wang, J., Kalinichev, A. G., Kirkpatrick, R. J., and Cygan, R. T., 2005. Structure, energetics, and dynamics of water adsorbed on the muscovite (001) surface: A molecular dynamics simulation. *J. Phys. Chem. B* 109: 15893.

Wang, J., Kalinichev, A. G., and Kirkpatrick, R. J. 2004a. Molecular dynamics modeling of the 10-A phase at subduction zone conditions. *Earth Planet. Sci. Lett.* 222: 517.

Wang, J., Kalinichev, A. G., and Kirkpatrick, R. J. 2004b. Molecular modeling of water structure in nano-pores between brucite (001) surfaces. *Geochim. Cosmochim. Acta.* 68: 3351.

Wang, X., Liu, J., Du, H., and Miller, J. D. 2009. States of adsorbed dodecyl amine and water at a silica surface as revealed by vibrational spectroscopy. *Langmuir* 26: 3407.

Wie, J. M., and Fuerstenau, D.W. 1974. Effect of dextrin on surface properties and the flotation of molybdenite. *Int. J. Miner. Process.* 1: 17.

Wisniewska, S. K. 2005. Surface analysis of selected hydrophobic materials. Thesis (PhD), The University of Utah, 235.

Yalamanchili, M. R., Kellar, J. J., and Miller, J. D. 1993. Adsorption of collector colloids in the flotation of alkali halide particles. *Int. J. Miner. Process.* 39: 137.

Yu, C. J., Richter, A. G., Datta, A., Durbin, M. K., and Dutta, P. 1999. Observation of molecular layering in thin liquid films using x-ray reflectivity. *Phys. Rev. Lett.* 82: 2326.

Zettlemoyer, A. C. 1969. *Hydrophobic Surfaces,* Ed. F. M., Fowkes. Academic Press, New York.

Zhou, G., Liu, X., Zhang, S., Yu, G., and He, H. 2007. A force field for molecular simulation of tetrabutylphosphonium amino acid ionic liquids. *J. Phys. Chem. B* 111: 7078.

Zhu, Y., Su, Y., Li, X., Wang, Y., and Chen, G. 2008. Evaluation of Amber force field parameters for copper (II) with pyridylmethyl-amine and benzimidazolylmethyl-amine ligands: A quantum chemical study. *Chem. Phys. Lett.* 455: 354.

5 Application of Molecular Modeling in Pharmaceutical Crystallization and Formulation

Sendhil K. Poornachary, Pui Shan Chow, and Reginald B. H. Tan

CONTENTS

5.1 INTRODUCTION

Crystallization is essentially a molecular recognition process occurring on a grand scale that allows separation and purification of the desired compound to produce high-purity products. The ways in which a molecule is packed in the crystal structure depends on the balance of the intermolecular interactions that it can achieve for a given conformation (Price 2009). Variations in the molecular packing arrangement in the crystal structure of a drug (polymorphic forms) can directly influence its physical properties such as dissolution rate and stability (Bernstein 2002; Datta and Grant 2004). It is therefore important to understand the relationship between the crystal structure of a drug and the properties of pharmaceutical solids. This will allow for selection of the most suitable crystal form for development into a drug product. Crystal growth morphology (i.e., shape or habit) of crystalline drugs can influence fundamental properties of the material such as rate of dissolution, wettability, and compressibility (Yu and York 2000). Crystal shape is also critical to downstream processing of solid drugs, including filtration, drying, particle flow, and milling (Davey and Garside 2000; Wood 2001). The extent of expression of different crystal planes, and therefore functional groups, has been attributed to the dependence of these pharmaceutical properties on crystal growth morphology (Danesh et al. 2000; Heng and Williams 2006).

For a given drug molecule, the attributes of product crystals depend on various operating conditions, including supersaturation (vis-à-vis solute concentration), solvent medium, presence of impurities and additives, crystallization temperature, and hydrodynamics (Black et al. 2004; Blagden et al. 1998; Chew et al. 2004; Gong et al. 2008; Mukuta et al. 2005; Poornachary et al. 2008a; Spruijtenburg 2000; Xie et al. 2010; Yu et al. 2005). Therefore, a holistic understanding of the mechanisms of crystal growth and the influence of the various operating conditions is a prerequisite for the design and development of a robust crystallization process (Yu et al. 2007). Besides, understanding the molecular and crystal chemistry of the material is equally important because they dictate the influence of the macroscopic factors on crystallization and crystal properties.

With the goal of achieving a better understanding of pharmaceutical crystallization that leads to optimization of particle properties, molecular modeling (MM) and computational chemistry methods can be applied at different levels (Docherty and Meenan 1999). First, MM can be used in the examination of the chemistry of surfaces and solid-state structures. Second, in conjunction with computational chemistry, MM allows prediction of crystal properties, and thereafter, to link the calculated properties to performance characteristics. Third, the application of MM can complement experimental studies—for instance, in the design of novel solid-state structures with improved properties and in morphological engineering of crystalline drug products.

The aim of this chapter is to provide an overview of the theories and methodologies governing MM of crystal structure and surfaces, growth morphology, and the effects of impurities and additives on crystallization. Different computational methods are used in these modelings, including *ab initio* or molecular orbital calculations, semiempirical methods, molecular mechanics, molecular dynamics, and Monte Carlo (MC) simulation. The reader is recommended to some excellent reference textbooks on the principles of MM and simulation (Hinchliffe 2003; Leach 2001; Myerson 1999).

5.2   INTERMOLECULAR INTERACTIONS IN THE CRYSTALLINE SOLID-STATE

The cohesive forces that hold the crystalline structure of organic molecular crystals encompass varied intermolecular interactions, including dispersive (van der Waals attractive and repulsive) interactions, electrostatic (coulombic) forces, dipole–dipole interactions, hydrogen bonds, and inter-ring (π-π stacking) interactions (Aakeröy and Boyett 1999). These nonbonded interactions within the crystal structure—and the intramolecular interactions relating to all internal degrees of freedom in a gas-phase molecule such as the deformation of bond lengths, bond angles, and internal rotations—can be described by a potential energy function known as *force field*. Originally, force fields were developed to reproduce the spectroscopic, thermochemical, or structural data for various sets of chemical compounds. These are parameterized to calculate energy values corresponding to those of a vibrationally averaged molecule existing in the gas phase. However, with the advancement in computational chemistry of crystalline materials, some modern force fields were developed not just to examine molecular properties but also to deal explicitly with various intermolecular interactions in the crystal structure. The basic assumption of the force field, or atom–atom, method, is that the interaction between two molecules can be approximated by the sum of the interactions between the constituent atom pairs.

The crystal binding, or cohesive energy, of single component molecular materials is often referred to as the *lattice energy* (E_{latt}) and can be calculated by summing all the interactions between a central molecule and all the surrounding molecules in the crystal lattice (Docherty and Meenan 1999). Each atom–atom interaction pair consists of a van der Waals attractive and repulsive interaction, an electrostatic interaction, and in some cases, a hydrogen-bonding potential. The van der Waals interactions are commonly described using the Lennard–Jones 6-12 potential function or the Buckingham potentials. The electrostatic interaction is described by a coulombic term $(q_i q_j r^{-1})$, in which a fractional charge $\pm q$ is assigned to each atom, the sign determining whether the atom in the molecular arrangement has an excess or deficiency of electrons compared with the neutral atom. These charges are usually determined from molecular orbital calculations. Fractional charges can also be represented by higher-order moments derived from experimental electrostatic density distributions (Berkovitch-Yellin 1985). However, in the case of a crystalline solid composed of polar molecules, it is essential that both the magnitude and the inherent anisotropy of the charge distribution are described accurately. In such cases, often an improved description of the charge distribution is obtained by replacing fractional point charges with a more realistic distributed multipole model (Day et al. 2005). The distributed multipole model is used to fit the output of an *ab initio* calculation and, in turn, obtain the asymmetric distribution of charge around a polar molecule or moiety for the molecule of interest. Hydrogen-bonding interactions are essentially special van der Waals interactions calculated using modified versions of the Lennard–Jones potential functions, which account for some of the important structural features of hydrogen bonds.

The lattice energy is an important parameter in evaluating the thermodynamic properties of crystals. The calculated value of lattice energy can be compared with the experimental sublimation enthalpy (ΔH_{sub}) as a check of the description of the intermolecular interactions in the crystal structure by a chosen force field (Clydesdale et al. 2005). However, there could be many corrections that must be applied to characterize the lattice energy more precisely—for example, the energy associated with conformation, polarization, charge transfer, vibrational modes, and intramolecular H-bonding. One particular example is the case of molecules that crystallize as zwitterions (e.g., glycine) for which the proton-transfer energy (ΔE_{pt}) must be included in the calculation of the lattice energy (Bisker-Leib and Doherty 2001; Poornachary 2008b). ΔE_{pt} is the energy associated with the transfer of a proton from the NH_3^+ group to the COO^- group of the glycine zwitterion molecule ($NH_3^+ CH_2 COO^-$) to form the neutral molecule ($NH_2 CH_2 COOH$). The relationship between the lattice energy, sublimation enthalpy, and proton-transfer energy can be described as follows:

$$E_{latt} = -\Delta H_{sub} - 2RT + \Delta E_{pt} \qquad (5.1)$$

where $2RT$ is the factor that accounts for the difference between the gas-phase enthalpy and the vibrational contribution to crystal enthalpy.

The selection of a suitable force field for crystal calculations can also be based on the root mean square (RMS) deviation of the atomic positions used to characterize the deviation of the energetically minimized crystal structure from the initial structure, which is determined through the use of x-ray diffraction data. The smaller the RMS value, the more the minimized structure resembles the experimental structure, and hence the more accurate the atomic charge description and governing energy expression are considered (Givand et al. 1998).

The Cambridge Structural Database (CSD) contains crystal structures of organic small molecules and metal–organic compounds maintained and distributed by the Cambridge Crystallographic Data Centre (CCDC) (http://www.ccdc.cam.ac.uk/). It provides information on molecules and crystals, which is crucial for MM studies, as well as force field generation and evaluation. The CSD is distributed along with a number of surrounding programs for search and information retrieval (ConQuest), structure visualization (Mercury), numerical postsearch analysis (Vista), and database creation (PreQuest). The CSD is used in applications that cover a wide range of topics, including structure correlation, conformational analysis, hydrogen bonding and other intermolecular interactions, studies of crystal packing, extended structural motifs, crystal engineering, and polymorphism (Allen and Motherwell 2002; Childs et al. 2009; Sarma and Desiraju 1999).

5.3   CRYSTAL STRUCTURE PREDICTION

The crystal structure of a substance can be determined experimentally by conventional single-crystal x-ray diffraction. However, when the single-crystal dimension is smaller than 0.05 mm (along each axis of the crystal), synchrotron and neutron scattering techniques can be used (Datta and Grant 2004). Sometimes it may be difficult to obtain suitable single crystals (of organic molecules with a complex chemical structure) for structure determination using any of these methods. In such cases, computational techniques can potentially be used to predict crystal structures. Besides, with the issue of crystal polymorphism, it may be a challenging task to explore experimentally all the crystal structures that are important practically. Crystal structure prediction using computational methods can thus be used as a complimentary technique to solid form screening (Datta and Grant 2004; Price 2009).

5.3.1   METHODOLOGY

The initial step in crystal structure prediction involves generation of an optimized molecular conformation and atomic point charges that adequately describe the electrostatic potential of the isolated molecule in the gas phase. Subsequently, energetically feasible crystal structures for a given optimized molecular conformer are generated, typically by applying an MC search algorithm. To this end, lattice energies of many "trial" molecular crystal structures are computed using a suitable force field method. This procedure can generate many thousands of crude crystal structures, of which a significant number can be similar in packing arrangement. The structures of very similar energy minima are clustered by applying a clustering process. Subsequently, the resulting crystal structure is minimized by allowing the degrees of freedom, including the conformation of the molecule and the unit cell parameters, to relax. The symmetry elements of the chosen space group are retained during the minimization step. Consequently, a crystal energy landscape—lattice (potential) energy versus cell volume per molecule—is constructed for the molecule packed in different space groups (Ouvrard and Price 2004; Payne et al. 1999; Price 2009).

This procedure is illustrated by using the case studies of aspirin (nonpolymorphic system) and paracetamol (polymorphic system), both widely used as antipyretic (fever suppressant) and analgesic (painkiller) drugs.

5.3.2   CONFORMATIONAL MAPPING

The molecular structure of aspirin has three bonds (denoted by the dihedral angles τ_1, τ_2, and τ_3 in Figure 5.1a). Rotation is possible around these bonds, and thus aspirin is an ideal model compound for studying the effects of conformational flexibility on crystal structure prediction. The dihedral angles τ_1 and τ_2 were only considered for calculating the conformational maps because the carboxylic acid group is usually in the same plane as the phenyl ring. The gas-phase conformational maps were obtained by using different computational approaches (Payne et al. 1999): (1) the AM1 semiempirical quantum mechanics method as implemented in the MOPAC program (Stewart 1990), (2) the Dreiding 2.21 force field (Mayo et al. 1990) with electrostatic (ESP) charges obtained from 6-31G** molecular orbital calculations in the Cerius^2 program (1995), and (3) *ab initio* calculations using restricted Hartree-Fock formalism at the 6-31G** level. While method (1) suggested that a planar conformer ($\tau_1 = \pm180$, $\tau_2 = 0$) would be less stable (7.1 kJ/mol) than that of a conformer extracted from the crystal structure (CSD Ref. code ACSALA01, with $\tau_1 = -81.4$, $\tau_2 = -5.7$), method (2) indicated that the planar conformer is more stable than the crystal structure conformer by 13.0 kJ/mol. Method (3) suggested that the relative stability of conformers would follow the data predicted by the AM1 method.

Ouvrard and Price (2004) also calculated the potential energy surface of aspirin at the AM1, MP3 (semiempirical), and B3LYP/6-31G(d) (density functional) levels and found that the molecular

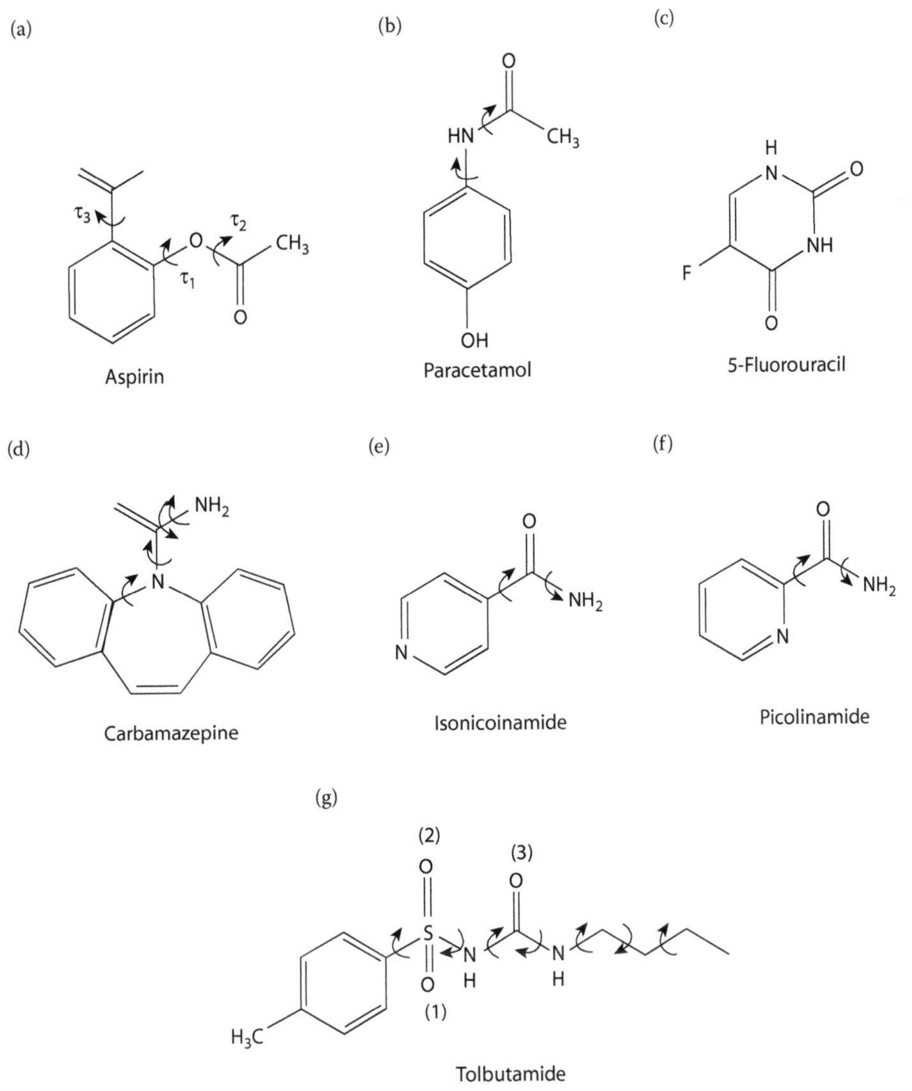

(a)

(b)

(c)

Aspirin

Paracetamol

5-Fluorouracil

(d)

(e)

(f)

Carbamazepine

Isonicoinamide

Picolinamide

(g)

Tolbutamide

FIGURE 5.1   Molecular structures.

conformation adopted in the crystal structure was close to a local minimum predicted using the density functional method. The analysis revealed that none of the low-energy conformers was planar, despite the fact that planar molecules could have an advantage in producing dense and energetically favorable crystal packing. The gas phase conformer that was most similar to that found in the crystal structure was about 3.5 kJ/mol less stable than the global minimum.

These sets of calculations highlight some important points with respect to crystal structure prediction. First, various high levels of *ab initio* theory are required to estimate the gas-phase conformations and energy differences. Second, the molecular conformation adopted in the crystal structure could be close to a local minimum (not the global minimum) found in the gas phase *ab initio* energy calculation. Hence, it is necessary to rigorously explore potential conformers of a particular molecule before structure prediction. Otherwise, it is likely that potential polymorphs will be missed by the prediction process because energy barriers may not be surmounted between conformations.

5.3.3 CRYSTAL ENERGY LANDSCAPES

For conformationally flexible molecules, both the lattice energy and the total energy of the predicted crystal structures are to be considered in evaluating the stability order. The most stable crystal structure at 0 K is obtained by optimizing $E_{tot} = E_{latt} + \Delta E_{intra}$, a balance of the intermolecular and intramolecular forces. Initially, the accuracy of the potential set for crystal structure prediction is validated by reproducing an experimental (if known) crystal structure of the molecule. To this end, the conformer extracted from the known crystal structure is optimized using *ab initio* methods, followed by structure prediction using force field methods.

The search for minima in the lattice energy for aspirin (Payne et al. 1999) was carried out using the "Polymorph Predictor" module in Cerius^2, with the molecule held in low-energy conformations in commonly occurring space groups (i.e., $P\bar{1}$, $P2_1$, $P2_1/c$, $P2_12_12_1$, and $C2/c$; these five space groups account for approximately 75% of the structures in the CSD). In this program, the similarity between a predicted crystal structure and an experimental structure is calculated based on the partial radial distribution functions (RDFs) between pairs of force field atom types of the two structures being compared. Smaller the value of this similarity measure, higher the similarity between the structures being compared. Despite using the same set of user defined parameters during replicate runs of polymorph prediction (for the MC and Simulated Annealing procedure), there were a significant number of structures unique to each run, suggesting that it was unlikely to exhaustively locate all possible structures near the predicted global minimum in a single run. The calculation results indicated that the lattice energy of the experimentally known crystal structure was 5.6 kJ/mol lower than any of the potential crystal structures, even though the total energies (lattice + intramolecular) of a number of predicted structures were lower.

Ouvrard and Price (2004) used the MOLPAK program (Holden et al. 1993) to generate hypothetical packing structures of aspirin, with the molecule held rigidly in the B3LYP-optimized low-energy conformations (Figure 5.2). The molecular charge distribution of each conformer was obtained by a distributed multipole analysis (DMA) of the charge density calculated using the GDMA program (Stone 1999) on the Gaussian98 (Frisch et al. 1998) charge density file. Lattice

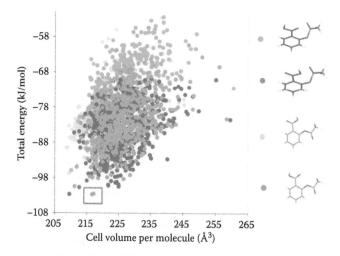

FIGURE 5.2 Lattice energy landscape of the hypothetical crystal structures of aspirin generated using different gas phase conformations. The legend shows the *ab initio* optimized molecular conformers used in the lattice energy calculations. There are two unique structures close to the global minimum in lattice energy (region highlighted), one of them corresponding to the experimental crystal structure. The other structure seems unlikely to be an observable polymorph as it has a very small shear elastic constant, implying that it can readily deform. (Adapted from Ouvrard, C. and Price, S. L., *Cryst. Growth Des.*, 4: 1119, 2004.)

energy minimization of the hypothetical crystal structures was performed with the DMAREL (Willock et al. 1995) program using the DMA-based model potential. The calculations indicated that the possibility of a planar conformer of aspirin in a crystal structure appeared to be unlikely because of a higher intramolecular energy penalty (about 12 kJ/mol). Although the experimentally observed crystal structure was predicted as one of the most thermodynamically stable structures, the calculated lattice energies were sensitive to small distortions of the molecular conformation. Hence, small differences in molecular conformation and the presence of intramolecular hydrogen bonding can affect the hydrogen bonding motifs in the predicted crystal structures. By selecting a large number of molecular conformations that are sufficiently low in energy, it is plausible that any loss in conformational energy could be balanced by improvements in the lattice energy. Therefore, conformational energies calculated at high levels of *ab initio* methods will need to be combined with an accurate calculation of lattice energies and the highest quality intermolecular force fields to obtain sufficiently accurate total crystal energies.

5.3.4 ENTROPIC AND KINETIC FACTORS

The methods for polymorph prediction just described were based on the static lattice energies of different crystal structures, which approximate the relative stability at 0 K. This lattice energy landscape is usually adequate for predicting thermodynamically feasible structures (e.g., within 10 kJ mol^{-1} of the most stable structure) because differences in the thermal contributions (e.g., entropy and zero-point energy) generally do not exceed a few kilojoules per mole at room temperature between real and hypothetical structures (Price 2009). Nevertheless, to distinguish between thermodynamically stable and kinetically preferred crystal polymorphs, it would be more ideal to compute the crystal energy landscapes by locating the minima in the free energy at the temperatures and pressures under which crystallization takes place. One way of addressing this is to additionally estimate the dynamical contributions to free energy—for example, on the basis of rigid body harmonic estimates of the infrared, Raman, and terahertz phonon frequencies (Day et al. 2003) and elastic constants (Day et al. 2001), which can reorder the relative energies with temperature. However, this method will only probe the curvature of the static energy surface in the region around the minimum and ignore the real shape of the free energy landscape, such as the barrier separating adjacent basins and their dependence on the thermodynamic conditions. Molecular dynamics (MD) simulations (Karamertzanis et al. 2008) can help assess the effect of temperature on the energetic barriers to transformation from a metastable to a thermodynamically stable crystal structure. In addition, when combined with crystal structure prediction methodologies, the modeling of thermal expansion can assist the determination of crystal structures from nonindexable powder diffraction data by establishing the variable dependence of peaks due to the anisotropy of interactions in the lattice.

It is commonly found that many unobserved hypothetical crystal structures are predicted close to the global minimum in lattice energy. In such cases, kinetic factors such as ease of nucleation, barriers to transformation to more thermodynamically stable forms, and rates of crystal growth may influence the crystallization of metastable polymorphs. An assessment of the relative growth volumes of the different polymorphs can serve as an additional factor to aid the prediction of a kinetically favored form (Coombes et al. 2005). Towards testing this proposition, the morphologies of a number of observed and hypothetical low-energy crystal structures of paracetamol (Figure 5.1b)—generated based on searches for the global minima in the lattice energy (Beyer et al. 2001)— were calculated using the attachment energy model. The attachment energy model (Hartman and Bennema 1980) assumes that growth of a crystal face is proportional to the absolute value of the attachment energy—that is, the energy released when a stoichiometric growth layer of material is added to the surface. Accordingly, faces with low absolute attachment energies grow most slowly and thus have the most morphological importance. More discussion on the application of attachment energy model to predict growth morphologies of organic crystals can be found later in this chapter.

The attachment energy calculations were performed using the GULP program (Gale and Rohl 2003) with the same force field potential sets used for the crystal structure prediction (Beyer et al. 2001). The calculation results showed that both the observed forms of paracetamol (form I and form II) grow relatively fast, with the metastable form II growing the fastest (Figure 5.3). In the case of those hypothetical crystal structures that yielded platelike growth morphologies (and a correspondingly larger aspect ratio), the attachment energy of the most dominant face was much lower than that for the other faces, and hence the relative volume growth rate was notably smaller. The calculations suggested that a single face dominates the overall crystal growth and, in turn, predicted which thermodynamically feasible crystal structures could have a kinetic advantage in crystal growth. In particular, the role of kinetics in crystal structure prediction was highlighted for one of the hypothetical crystal structures (AK6) found in the polymorph prediction study (Beyer et al. 2001). This low-energy structure was suggested as a possible structure of form III, the third highly metastable form of paracetamol obtained by crystallization from the melt (Peterson et al. 2002). Further analysis helped to understand how the degree of variation in growth volumes depends on the variation of crystal packings within the low-energy range of different polymorphs. For instance, all the predicted crystal structures of paracetamol that contained chains of dimers had low relative growth volumes compared with structures that had other hydrogen-bonding motifs. This was reasoned on the following basis. A slice with strong interactions within the growth units results in higher value of slice energy and hence lower attachment energy, since the lattice energy is approximately the same for all polymorphs.

Another important kinetic factor that can influence the polymorphic outcome of a crystallization process is solvent–solute interactions because they can significantly affect the initial association of solute molecules en route to crystal nucleation (Davey et al. 2001; Parveen et al. 2005). This has been highlighted by MD simulation studies (Hamad et al. 2006) on the effects of solvents on the polymorphism of 5-fluorouracil (Figure 5.1c), a drug used for the treatment of solid tumors. The RDFs were calculated using the DL_POLY program (Smith and Forester, 1996) by evaluating a histogram of the distances between every pair of atoms of defined atomic types. Analysis of the simulation results showed that the polar water molecules form strong hydrogen bonds with the $C = O$ groups of 5-fluorouracil, producing a strong hydration sphere around most of the molecule,

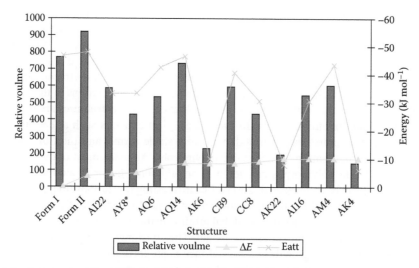

FIGURE 5.3   Relative growth volume, attachment energy (Eatt) for the dominant face, and energy above the global minimum (ΔE) for predicted crystal structures of paracetamol. The growth volume is plotted relative to a cube where each face has an Eatt 10 kJ mol^{-1}. The lattice energies and crystal structure identifiers were obtained from references (From Beyer, T., Day, G. M., and Price, S. L., *J. Am. Chem. Soc.* 123: 5086, 2001; Coombes, D. S., et al. *Cryst. Growth Des.* 5: 879, 2005. With permission.)

except the hydrophobic F region. As a result, formation of the doubly hydrogen-bonded dimer is hindered in aqueous solution and, by default, close hydrophobic F···F interactions are promoted. In conjunction with this solution chemistry, the polymorphic form I (thermodynamically stable form) is crystallized from aqueous solution. In form I crystal structure, the molecular-packing arrangement results in regions where four fluorine atoms are in close proximity, approaching within 3.2 Å. This crystal packing is unusual considering the highly electronegative nature of fluorine atom. The simulations strongly suggested that the $C=O···H-N$ double hydrogen bonds present in the crystal structure must be formed later during the self-assembling process in water, with the initial interactions more likely leading to formation of close F···F contacts and single hydrogen bonds. In contrast, doubly hydrogen-bonded dimers were formed in dry nitromethane solution, consistent with the crystallization of the doubly hydrogen-bonded ribbon structure of form II from this solvent. Interestingly, this (metastable) polymorph corresponded to the global energy minimum structure from the computational search (Hulme et al. 2005). When nitromethane is doped with water, the water forms hydrogen bonds to the solute and interferes with the formation of doubly hydrogen-bonded dimers. This simulation result is consistent with the crystallization of form I from this hygroscopic solvent when it is not dried.

5.3.5 CRYSTAL STRUCTURE SOLUTION FROM POWDER X-RAY DIFFRACTION DATA

Sometimes it is difficult to obtain a good quality single crystal of sufficient size for structure determination using the single crystal X-ray diffraction method. In that case, the crystal structure of the material of interest can be determined from its powder x-ray diffraction data using the Rietveld refinement method (Rietveld 1969). The methodology involves four stages (Docherty and Meenan 1999):

1. The experimental powder pattern is indexed, and the unit cell dimensions and space group are determined.
2. An initial molecular structure is built, either using semiempirical or *ab initio* calculations; in certain cases the molecular structure is extracted from a known crystal polymorph.
3. The molecular structure is introduced into the unit cell and the potential packing motifs are examined for the definite space group symmetry.
4. The Rietveld refinement is applied, in which the packing arrangement and molecular structure are optimized so that the simulated pattern of the proposed structure "fits" the experimental pattern.

MC simulation and lattice energy minimization calculations are used to generate trial packing structures. This methodology is illustrated using the case study of tolbutamide, an antidiabetic drug.

The crystal structure of tolbutamide form IV was solved (Thirunahari et al. 2010) from its powder diffraction data using the "Reflex Plus" module in Materials Studio (2005). Initially, the unit cell parameters and space group were determined using an indexing algorithm known as X-Cell (Neumann 2003). Pawley refinement of the trial unit cells against the experimental data was performed using the "Powder Refinement" tool in Reflex. At this stage, the proposed unit cell and the function profile parameters, including background coefficients, zero-point shift, and peak width parameters, were refined over a 2θ range (5°–50°). The initial molecular model of tolbutamide was taken from the form II crystal structure (CSD Ref. code ZZZPUS03) and placed in the refined unit cell. The structure solution search was performed by an MC parallel tempering method using the "Powder Solve" module (Engel et al. 1999) in Materials Studio. A total of 13 degrees of freedom, including the position of the molecule defined by three coordinates and the orientation of the molecule defined by three rotation angles and seven torsion angles (Figure 5.1g), was assigned for an asymmetric unit. The March-Dollase method (Dollase 1986) was used to model the preferred orientation effects in the powder pattern because this crystal form grows as thin needles. Hydrogen atoms were included only in the final Rietveld refinement. The feasibility of close contacts and hydrogen bonding was verified before accepting the final solution. The final weighted profile R-factor and profile R-factor for this structure were found to be 5.38 and 3.96%, respectively.

The advantage of the crystal structure solution from powder diffraction data over the *ab initio* crystal structure prediction method is that in the former approach the accuracy of prediction can be validated by matching the simulated powder pattern with the experimental powder diffraction pattern. In retrospect, concerns over global minimization algorithms, force field, and charge distribution accuracies are reduced with this method.

5.3.6 Predicting the Formation of Co-crystals

Co-crystals are crystals consisting of two or more neutral compounds (as opposed to salts) that are in their pure form and are solid at room temperature (as opposed to solvates) (Aakeroy 2005). Co-crystals of an active pharmaceutical ingredient (API) and a co-crystal former (or co-former) can have different properties when compared with crystals of a pure API (e.g., improved bioavailability and reduced hygroscopicity) (Almarsson and Zawarotko 2004; Trask et al. 2005). Crystal engineering approaches based on different hydrogen-bonding synthons have been proposed to guide the search for co-crystals (Walsh et al. 2003). These guidelines suggest possible co-formers that can be tested experimentally for co-crystal formation with the API. Some insights into the tendency of co-crystals formation can be gained (He et al. 2008) through the characterization of intermolecular pair interactions between the API and the co-former using pulsed gradient spin-echo nuclear magnetic resonance (PGSE NMR). The pathways for solution co-crystallization can be designed by understanding the ternary phase diagram of API, co-former, and solvent (Habgood et al. 2010). The range of co-crystals generated by various experimental screening methods shows that many co-crystals do not have the hydrogen-bonding motifs expected from crystal engineering guidelines. Therefore, more advanced techniques based on computational chemistry are required for more reliable predictions of co-crystal formation.

The methodology of crystal energy landscapes developed for prediction of single compound crystal structures can be extended for successful prediction of co-crystal formation (Karamertzanis et al. 2009). However, the search for co-crystals using this computational technique is more challenging than it is for pure component crystals because of the additional number of degrees of freedom of a multiphase system. With an objective to simplify the computational effort, a three-stage methodology has been proposed:

1. Generate a co-crystal energy landscape with lattice energy minimization for a rigid molecule, performed using an isotropic atom–atom potential.
2. Refine the predicted crystal structures using a more realistic electrostatic model.
3. Relax the flexible degrees of freedom of the API and co-former by optimizing their intramolecular interactions, followed by cell refinement for a selection of likely structures.

The final step gives a more accurate calculation of the lattice energy of the predicted crystal structures by optimizing the molecular conformation so as to improve the intermolecular interactions. The reference lattice energies for single component crystals are calculated at this stage. The lattice energies of each co-crystal are compared with the weighted sum of the lattice energies of the pure components to determine the co-crystal's stability.

The aforesaid methodology was used (Habgood et al. 2010) to compute crystal energy landscapes and consequently predict the co-crystallization of carbamazepine (an anticonvulsant drug used in the treatment of epilepsy) with pyridine carboxamides (Figure 5.1d to f). Experimentally, while isonicotinamide (INA) co-crystallizes with carbamazepine (CBZ), picolinamide (PA)—despite its similarity to other pyridine carboxamides in the homologous series—does not appear to form a co-crystal with CBZ. This discrepancy in the co-crystallization behavior cannot be explained from the viewpoint of traditional crystal engineering principles. For example, synthon design would suggest the formation of an $R_2^2(8)$ dimer between the CBZ and the carboxamide amide groups of the co-formers. Instead, a CBZ = CBZ $R_2^2(8)$ dimer is found in all the crystal structures for this series and other co-formers containing the amide group. Based on this observation, it is expected that very similar synthons would be formed in a CBZ-PA system.

From the crystal energy calculations, two 1:1 co-crystal structures of CBZ and INA were predicted to have lower or comparable lattice energies than the sum of the pure component lattice energies, which corresponded to the known co-crystal structures. However, lattice energies of predicted CBZ-PA co-crystal structures were less stable than the pure component lattice energies, implying that CBZ and PA would not form a co-crystal. These simulation results were consistent with the experimental results and can be further explained in terms of intermolecular hydrogen-bonding capability within the co-crystals. Fewer hydrogen bonds were observed in the predicted CBZ-PA structures compared with the CBZ-INA structures because the amide hydrogen atom in PA did not take part in hydrogen bonding, whereas the analogous atom in INA contributed to hydrogen bonding. The reason for this was explained on the basis of the electrostatic potential surface of the PA and INA molecules. In PA, the electrostatic potentials usually associated with an amide and cyclic nitrogen's hydrogen-bonding abilities are in close proximity and so are cancelled out to a large extent. Thus the strength of intermolecular interactions, including hydrogen bonds to CBZ, would be much weaker for PA than INA, despite the similarities in their steric interactions.

5.4   MODELING CRYSTAL GROWTH MORPHOLOGY

Modeling the growth morphology of organic crystals from the internal crystal structure is governed by the molecular structure of the solute, crystal geometry, intermolecular interactions in the crystal structure, and the growth mechanism. In addition, environmental factors such as the nature of solvent, the degree of supersaturation, and the presence of impurities can profoundly affect crystal shape. In this section, models that simulate the shape of organic crystals using geometrical rules and intermolecular interaction energies between the building blocks of the crystal (i.e., internal factors) are discussed. Then, models that account for the effects of external factors such as solvent type and supersaturation are reviewed. The influence of impurities on crystal growth morphology is discussed in the next section.

5.4.1   THE BRAVAIS–FRIEDEL–DONNAY–HARKER RULE AND ATTACHMENT ENERGY MODELS

The pioneering works on modeling crystal growth morphology by Bravais, Friedel, Donnay, and Harker were based on crystallographic and geometric considerations. The *Bravais–Friedel–Donnay–Harker* (BFDH) *rule* states that a larger interplanar spacing in the crystal lattice (d_{hkl}) leads to a more prominent face being expressed along that orientation in the external crystal morphology. The BFDH morphology of the α-glycine crystal (Figure 5.4b) was derived from its crystal structure (CSD Ref. code GLYCIN03) using the "Morphology" module in Materials Studio. Although this model can adequately describe all the possible growth faces of the crystal habit, it does not accurately predict the morphological importance and the aspect ratios when compared with the solution grown crystal habit (Figure 5.4a).

Hartman and Bennema (1980) proposed the attachment energy model, which relates the energies of the bonds between the crystal-building units to its external shape. Two important concepts were developed as part of this model—the periodic bond chains (PBCs) that define periodically repeated chains of strong intermolecular interactions and the attachment energy (E_{att}). Based on the number of PBCs that are parallel to a face, crystal faces can be classified as flat (F) face with 2 PBCs, step (S) face with 1 PBC, or kink (K) face with 0 PBCs. The morphological importance of the faces follows

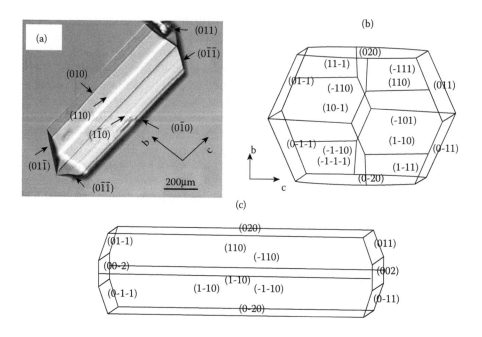

FIGURE 5.4   The growth morphology of α-glycine crystal: (a) Experimental habit as grown from pure aqueous solution, (b) BFDH model, and (c) attachment energy model.

the order: F (slow growth) and S (intermediate growth) and K (rapid growth). The energy required to form a growth layer in the crystal bulk of the thickness of d_{hkl} is defined as the *slice energy* (E_{sl}), and the energy released upon the attachment of the slice to a (hkl) face of a growing crystal face is the *attachment energy*. The sum of the slice energy and attachment energy of any given orientation (hkl) is the lattice energy and thus a constant. The morphological importance of an F-face is estimated by assuming that growth velocity normal to the face is proportional to its attachment energy. In line with this model, crystal morphology is bounded by the slow-growing faces that have lower E_{att} values.

The framework of the attachment energy model has been used by different research groups to simulate the habit of organic molecular crystals. Clydesdale et al. (1996) developed the program HABIT, a tool that uses force field methods to calculate the interaction energy within organic crystals, and thereafter, predicts the growth morphology. The attachment energy model is also implemented in the "Morphology" module of Materials Studio and provides a fast and easy tool for crystal habit prediction. This program was used to predict the α-glycine crystal habit (Poornachary et al. 2008b) with the COMPASS force field (Sun 1998). The atomic charges of a glycine zwitterion molecule were obtained from *ab initio* calculations performed using the Dmol³ program (Delly 2000) in Materials Studio. The simulated growth morphology (Figure 5.4c) was in good agreement with the experimental crystal habit with respect to the dominant faces, however, the (b/c) aspect ratio (ca. 7.0) was much higher as compared with the experimental habit (2.62 ± 0.5). These discrepancies between the simulated and experimental habits can be attributed to strong solvent interactions with the {011} faces that are highly polar in nature (Berkovitch-Yellin 1985; Gnanasambandam and Rajagopalan 2010).

The attachment energy model in general provides good predictions of vapor-grown crystals or crystals grown in systems in which the solvent does not interact strongly with the solute. However, in many instances, the simulated crystal shapes differ from experimental because the kinetic effects due to supersaturation, solvent, and impurities dominate the crystal growth process.

5.4.2 THE EFFECT OF SOLVENT

To simulate the habit of solution-grown crystals, the interactions of solvent molecules at the crystal–solution interface could be considered. In most cases, it is assumed that the solvent affects crystal habit through preferential adsorption of solvent molecules on specific faces and that removal of solvent molecules before the deposition of oncoming solute molecules causes retardation of crystal growth. The extent of solvation of a crystal face could be qualitatively understood from the relative polarities of the various crystal faces, which can be obtained from electrostatic potential maps calculated at closest approach distances (Berkovitch-Yellin 1985).

Modeling crystal growth from solutions through a fully simulation-based approach with an all-atom description is restricted by computational resource limitations. MD could only simulate very short time events (typically picoseconds to nanoseconds) and therefore, multiscale methods are required to completely explain solvent effects on crystal growth at the molecular level. One of the early works (Liu et al. 1995) proposed a method that involves an "interface structure analysis" to estimate the concentration of "growth units"—that is, solute molecules that have orientations appropriate for adsorbing onto the crystal surface. The free energy barrier associated with the incorporation of a growth unit from the solution to a kink site on the crystal surface is calculated and, in turn, used to obtain detailed growth kinetics through consideration of the theories of Burton-Cabrera-Frank (BCF) growth mechanism (Davey and Garside 2000), which is applicable for crystal growth at low to moderate supersaturations. This method was further extended (Gnanasambandam and Rajagopalan 2010) to account for all possible orientations that a solute molecule could undergo in the solution phase before incorporating into the crystalline phase. The solvent-dependent properties such as the concentration of adsorbed growth units on the individual crystal faces were obtained from MD calculations and used alongside crystallographic properties such as the interplanar distance and the anisotropy (intermolecular interaction energy) factor to determine the relative growth rates and morphology. By this method, the relative growth rates for the morphologically important (010) and (011) faces of α-glycine were calculated, and the resultant aspect ratios of the crystal habit (2.88 ± 0.5) were found to be comparable with the experiments (2.62 ± 0.5) (Poornachary et al. 2007).

Piana et al. (2005) used a multiscale modeling approach involving MD and kinetic MC (KMC) simulation to simulate the effects of solvent, supersaturation, and surface defects (such as screw dislocations) on crystal growth. Both dissolution and growth rates of urea crystals were simulated at the nanometer scale using MD calculations. The rates for all possible molecular transitions between unique sites that might occur on any surface were provided as input to KMC calculations to simulate three-dimensional crystal growth in the millimeter scale. In this way, the needlelike morphology characteristic of urea crystals growing from water solution and the polar morphology of urea crystals growing from a methanol solution were successfully simulated. However, this approach calls for lengthy simulation time to calculate reliable rates for molecular crystal growth and limits its utility for process engineering applications.

Win and Doherty (1998, 2000) proposed a model for simulating crystal shape under solution growth that requires knowledge of the intermolecular interactions (obtained from attachment energy calculations) and the surface free energy of the pure solvent. The crystal–solvent interfacial free energies were calculated from the pure component energies via a geometric mean approximation developed previously for estimating the free energy of adhesion (Girifalco and Good 1957). Solvent surface free energies can be either measured for solvents in contact with air, or for certain classes of solvents, calculated from correlations based on molar volume and bulk solubility parameters. The kink free energy (γ^{kink}) of the most stable edge of a likely crystal face (as derived from the BFDH model) was used as a criterion to identify the crystal growth mechanism (i.e. the two-dimensional nucleation model or the screw dislocation growth model). Consequently, the relative growth rate of the face was calculated from the growth models, with the values of kink and edge free energies provided as input. This method underpins the idea that a strong intermolecular interaction in the crystal structure contributes to nearest-neighbor bonds, and therefore stable edges on the crystal lattice with lower kink (edge) free energies. For instance, kink structures involving hydrogen bonds can have lower free energies and hence faster growth. One of the limitations of this approach is that it applies to cases where there are mainly dispersive forces of interaction between the solvent and the crystal.

The application of this method was illustrated for ibuprofen crystals grown from polar (ethanol) and nonpolar (hexane) solvents. Experimentally, it was found that ibuprofen grows as elongated needles along the *b*-axis from hexane while equant low aspect ratio crystals are formed from ethanol; this solvent-mediated change in habit has been patented as a process improvement. The BFDH model and attachment energy model (simulated assuming ibuprofen dimer as the growth unit) are shown in Figure 5.5a and b. Upon estimating the interface properties for ibuprofen crystals in hexane and ethanol, a two-dimensional nucleation model was applied for growth from hexane because of the low values of kink free energies ($\gamma^{kink} < RT$). The model predicted $\Delta G_c < 3RT$ for the {011} face at moderate supersaturations, and therefore, a rough fast growth could be expected along the *b*-axis, resulting in a needlelike crystal habit. The growth morphology was hence derived (Figure 5.5c) by assuming the growth rate of {011} face to be an order of magnitude (10 times) greater than the average growth rates of the unroughened {100} and {002} faces. For growth from ethanol, higher values of kink free energies were calculated for the crystal faces, and hence the screw dislocation model was applied. An equant crystal shape with an aspect ratio of about two was predicted (Figure 5.5d).

Hammond et al. (2007) suggested a surface-specific, grid-based search method for estimating the crystal-solution interfacial energies and using it to predict solvent-mediated crystal morphology. In this approach, the specific surface energies were computed from the surface attachment energies of the crystal faces, similar to the method of Win and Doherty (2000) discussed earlier. Then the binding energies of the solute and solvent molecules with the crystal surface were calculated using an atom–atom potential energy approach, via a systematic grid-based search procedure by treating the interfacial probe molecules as rigid bodies. The degrees of freedom in this grid-based search were the location of the center of the coordinates of the probe molecule with respect to the surface cleaved from the bulk crystallographic crystal structure and the orientation of the probe molecule at that location. Each configuration was subjected to energy minimization, and the minimum

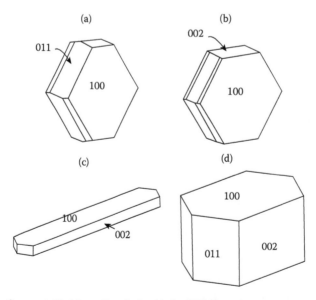

FIGURE 5.5 Ibuprofen crystal habit predicted via: (a) the BFDH model, (b) the attachment energy model, (c) and (d) as grown from hexane and ethanol, respectively. (From Win, D. and Doherty, M. F., *AIChE J*. 46: 1348, 2000. With permission.)

interaction energy identified in this way was used to calculate the solvent-mediated attachment energy for the (hkl) crystal face. For a mixture of solvents, the specific interaction energy of a crystal surface with its saturated solution was calculated via a simple mass balance derived from the solvent composition and the solute equilibrium solubility. This method was applied to predict the morphology of aspirin crystallized from ethanol–water solution (Figure 5.6). Although the attachment energy model provided a good general match to the experimental morphology, the predicted habit was much thicker along the a-axis when compared with the experimentally observed shape. In contrast, the aspirin crystal habit predicted after accounting for the effect of surface wetting by the solvent produced a thinner morphology, providing a better match with the experimental habit.

5.4.3 The Effect of Supersaturation

Crystal habits can be influenced by several physical parameters that are not reflected by the attachment energy. The most important parameter is the driving force for crystallization, often expressed in terms of the supersaturation. Boerrigter et al. (2002, 2004) introduced a new model based on discrete MC simulation to account for the effects of supersaturation on crystal growth. The approach was applied to predict the supersaturation-dependent morphology of paracetamol grown from aqueous solution. The first step involved construction of the "crystal graph", which represents the mutual bonds (interactions) between all the growth units (paracetamol molecules) making up the crystal structure. To this end, the crystal structure of paracetamol (monoclinic form) was minimized using the Dreiding force field in Cerius^2 with ESP-derived charges. In the resulting crystal graph, every growth unit was connected to 16 neighboring growth units. The bond energies were all scaled down to fit the total crystal energy to the dissolution enthalpy in water so that the supersaturation parameter could be compared with the experimental values.

The second step involved a connected net analysis, which identifies all possible F faces for paracetamol that contain one or more connected nets. The connected nets were determined for the {hkl} forms using the program FACELIFT (Boerrigter et al. 2001) by applying the selection rules according to the BFDH rules. The third step involved determination of edge free energies of the growth steps and two-dimensional nuclei on the crystal surface on the basis of the connected nets. The edge free energy–calculated from the difference in broken-bond energy between an S interface and an F interface–was used as a parameter to simulate the growth mechanisms under low or

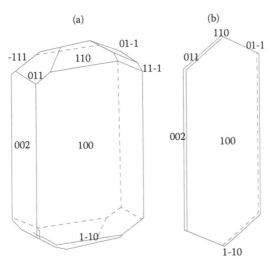

FIGURE 5.6   Aspirin crystal habit predicted via: (a) the attachment energy model and (b) as derived from solution surface energies (38% ethanol–water, soluble at 50°C). (From Hammond, R. B., et al. *Cryst. Growth Des.*, 7: 1571, 2007. With permission.)

medium supersaturation conditions. Note that the classical attachment energy model usually adopts the broken-bond energy associated with the strongest connected net on a flat interface to compute the slice and attachment energies, respectively.

In this approach, the concept of a *roughening transition temperature* is used to relate the thermodynamics of crystal growth interface to the bond energies within the connected nets. The roughening transition temperature of a face (hkl) is determined by the edge free energy of the two-dimensional nuclei on that face. Below this transition temperature, the F orientation of a face is maintained and above this temperature, the barrier for the formation of the two-dimensional nuclei vanishes as the free energy of formation is zero. This phenomenon is called *thermal roughening*. Above its roughening temperature, a face will start to grow faster, either disappearing from the morphology or appearing as a rounded-off face. Below its roughening temperature, a face can still lose its flat appearance beyond a certain driving force, a phenomenon called *kinetic roughening*. Therefore, faces with relatively small edge free energies for two-dimensional nucleation are expected to kinetically roughen at relatively small driving force.

MC simulations were run using the program MONTY (Boerrigter et al. 2004) to determine the growth rates of the various faces of paracetamol from its crystal graph. As mentioned before, the crystal graph defines the number of possible configurations of the growth units at the crystal surface. The growth (attachment) and etch (detachment) probabilities of the growth units were calculated as a function of the driving force for crystallization. While the energy parameters for the crystal phase used in the MC simulation were obtained from the crystal graph, the solution phase parameters were obtained from simulation or experimental data. For instance, the dissolution enthalpy and entropy can be derived from solubility data and in turn can be related to Gibbs free energy. The linear growth rate of a face (hkl) is determined using the relationship:

$$R_{hkl} = (d_{hkl}/N_{slice}) * (N_{growth} - N_{etch})/t_N$$

(5.2)

where d_{hkl} is the interplanar distance between growth slices, N_{growth} and N_{etch} are the total number of growth and etch events during the simulation, N_{slice} is the total number of growth units in the growth slice, and t_N is the total simulation time passed after N events.

On the basis of experimental observations (Ristic et al. 2001) concerning the growth mechanisms for paracetamol, a spiral growth mechanism was adopted for all faces except the {110} faces, for which a two-dimensional nucleation mechanism was used. The simulated growth rates as a function of supersaturation and the predicted growth morphologies showed a good correspondence with the experimental crystal habits (Figure 5.7).

5.5   MODELING THE EFFECT OF IMPURITIES AND ADDITIVES ON CRYSTALLIZATION

A key issue in pharmaceutical manufacturing is the reproducibility of solid-state attributes of the crystalline product. Batch-to-batch variation in the crystal size, habit, polymorphism, or chemical purity could be caused when certain impurities—usually, reaction by-products from the upstream processes—are present in the crystallizing solution. The resulting modification of crystal properties can affect downstream processing and formulation of the final product. Sometimes tailor-made and polymeric additives are used to tune the crystallization process and, hence, crystal properties at the molecular level. For instance, a good crystal habit can facilitate better filtration and separation. A stable (or metastable) polymorph can influence the stability and formulation behavior of the drug product.

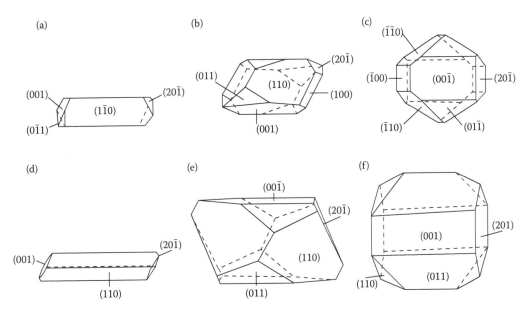

FIGURE 5.7 Growth morphology of monoclinic paracetamol based on the MONTY simulation results for supersaturation of (a) just below, (b) just above, and (c) well above the nucleation barrier for the {110} face. Experimental morphologies of paracetamol at (d) low, (e) medium, and (f) high supersaturation in aqueous solution. (Redrawn from Ristic, R. I., et al. *J. Phys. Chem. B*, 105: 9057, 2001; Boerrigter, S. X. M., et al. *Cryst. Growth Des.*, 2: 357, 2002. With permission.)

5.5.1 IMPURITY INTERACTIONS AT THE CRYSTAL SURFACE

In most cases, separation and purification via crystallization are highly selective due to the fact that molecular recognition process at the crystal-solution interface acts in such a way as to select the host molecules and reject impurities. However, sometimes the solute and impurity molecules are not discriminated at certain crystal faces, especially when the impurity has many of the structural and chemical characteristics of the primary solute but differs only in some specific way. A systematic approach toward understanding the effects of such impurities on crystal growth has been developed using the concept of tailor-made additives (Weissbuch et al. 2003). These additives are structurally similar to the solute molecules and are basically composed of two moieties. The first, known as the *binder,* has a similar structure (and stereochemistry) to that of the substrate molecule on the crystal surface where it adsorbs. The second, referred to as the *perturber,* is modified when compared with the substrate molecule and thus hinders the attachment of the oncoming solute molecules to the crystal surface. Several classic examples in the literature highlight this type of interaction mechanism in molecular crystals.

The pioneering work of Weissbuch et al. (2003) has elegantly shown that naturally occurring α-amino acids operate as tailor-made additives and cause habit modification along the enantiopolar *b*-axis of α-glycine crystals. This effect is stereospecific in nature because the L- and D-amino acids cause habit modification along the −*b*- and +*b*-axes of the α-glycine crystal, respectively. Poornachary et al. (2007) observed additional habit modification along the *c*-axis of α-glycine crystal in the presence of aspartic acid (impurity) in the crystallization medium (Figure 5.8a and b). Taking into account the known solution chemistry of the impurity and building on the stereoselective interaction mechanism, it was proposed that the two differently charged impurity species—zwitterions and anions—interact preferentially with the {010} and {011} faces of α-glycine, respectively. The molecular differentiation between the impurities zwitterion and anion was analyzed based on

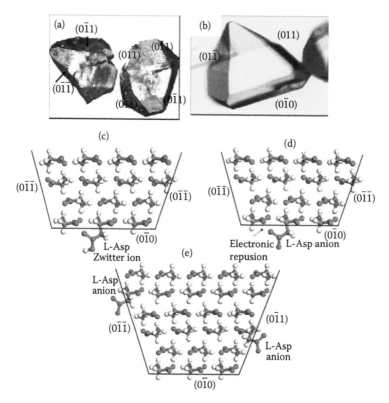

FIGURE 5.8 Habit modification in α-glycine crystallized in the presence of L-asp acid: (a) 0.5 wt% (w/w of glycine), (b) 1.5 wt%. MM of the interaction of impurity species with α-glycine: (c) L-Asp zwitterions and (d) L-Asp anion on the $(0\bar{1}0)$ face, and (e) L-Asp anion on the $(0\bar{1}1)$ and $(0\bar{1}\bar{1})$ faces. (From Poornachary, S. K., et al. *Cryst. Growth Des.,* 7: 254, 2007. With permission.)

intermolecular interactions at the crystal surface. In accordance with the stereoselective interaction mechanism, an L-aspartic (L-asp) acid zwitterion will adsorb preferentially on the (010) face (Figure 5.8c). However, the anion would incur electronic repulsive effects on the (010) face due to close contacts between the negatively charged carboxylate group of the anion and of glycine molecules at an adjacent site on the surface (Figure 5.8d). But at the same time, the anion may not incur such electrostatic repulsive forces of interaction on the {011} faces on account of the oblique nature of the crystal faces delineating the glycine dimer motifs along the *c*-axis (Figure 5.8e). These specific interactions of the impurity species with the α-glycine crystal are further substantiated using binding energy calculations and explained in the ensuing section. Upon strong adsorption on the crystal surface, the L-asp zwitterion and anion cause subsequent growth inhibition along the *b*-axis and *c*-axis of α-glycine, respectively, and result in habit modification.

Morphological changes in paracetamol crystals by the impurities *p*-acetoxyanilide and acetanilide have been explained based on host–additive interactions (Hendriksen 1998). The crystal structure of paracetamol is predominantly hydrogen bonded with molecular chains packed in a "herring bone" fashion (Figure 5.9). Both *p*-acetoxyanilide and acetanilide, commensurate with the molecular textures at the crystal interfaces, incorporate into the host crystal lattice. Consequently, *p*-acetoxyanilide, whose ester group in the *p*-position is larger than the –OH group of the paracetamol molecule, can provide steric hindrance to solute molecules approaching the growth surface. Besides, the absence of a proton donor in *p*-acetoxyanilide to contribute to the existing hydrogen-bonding network in the host crystal structure results in a blocking effect. However, acetanilide has no group

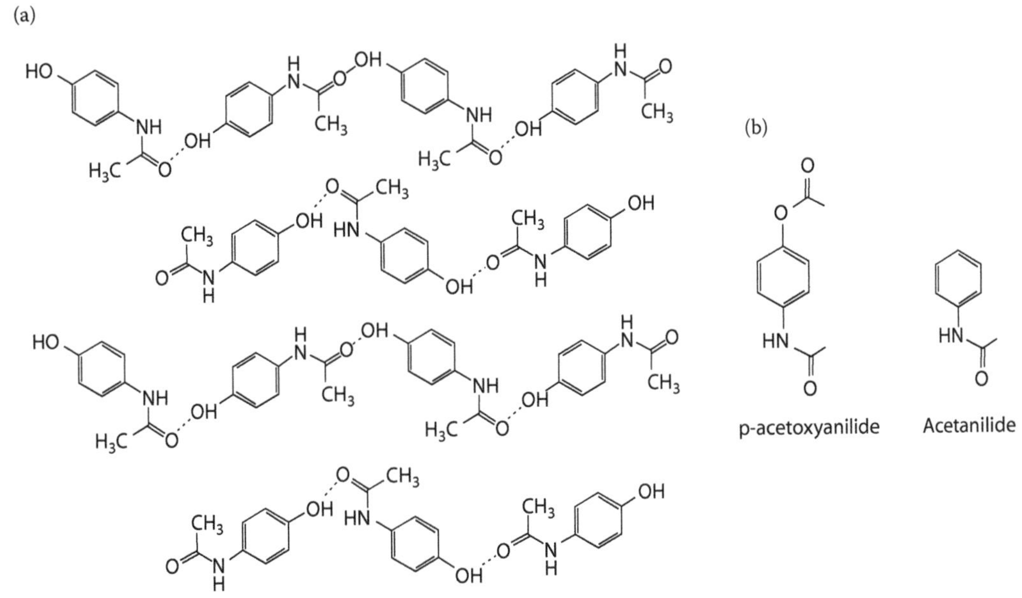

(a)

(b)

p-acetoxyanilide   Acetanilide

FIGURE 5.9 (a) Molecular packing arrangement in paracetamol crystal structure and (b) molecular structures of the impurities.

in the *p*-position, so there is no steric hindrance; but because it has no proton donor, it remains a strong blocker.

5.5.2 BINDING ENERGY CALCULATIONS

While examples just discussed help illustrate the structural relationship between an impurity molecule and the host crystal surface, computational chemistry involving binding energy calculations can aid in quantifying the crystal–impurity intermolecular interactions. In this approach, the attachment energy method for crystal habit modeling is modified to be able to treat the effects of impurity species on the growth morphology. For this, an additional factor, the *differential binding energy*, is introduced to assess the likelihood of impurity incorporation into the host crystal lattice (Clydesdale et al. 2005). Differential binding energy is defined as the difference between the incorporation energies of solute (E_b) and the impurity molecule ($E_{b'}$):

$$\Delta b = E_{b'} - E_b = (E_{sl'} + E_{att'}) - (E_{sl} + E_{att})$$

(5.3)

In this equation, $E_{sl'}$ is the slice energy calculated with an impurity molecule at the center of the (hkl) crystal surface and $E_{att'}$ is the attachment energy of a growth slice containing an impurity onto a pure surface. This equation shows that the impurities are likely to incorporate on a crystal face where there is a minimum change in the binding energy. If Δb is strongly dependent upon crystal orientation, then the incorporation will be specific to one crystal face and vice versa.

The MM approaches used for prediction of impurity incorporation into crystal lattice can be grouped into two major categories: surface docking and built-in models. Both approaches are based on the classic attachment energy model. In the built-in approach, an impurity molecule is substituted successively in each symmetry position of host molecules in the crystal unit cell and its orientation is optimized through energy minimization and MD. Then the crystal morphology under the influence of the impurity is obtained by averaging the modified attachment energy values from all the calculations. In the surface-docking approach, the individual faces are cleaved from the crystal (pure) structure and the binding energy is calculated for an impurity docked on the surface. The magnitude of the binding energy is related to an effective reduction in the attachment energy and, hence, to the growth rate of a crystal face. The interaction forces among the host molecules play an important role in the determination of a suitable modeling approach (Lu and Ulrich 2004). If the major contribution to the lattice energy of organic molecule is from hydrogen bonds (anisotropic), both the built-in and surface-docking approaches can be applied. When the hydrogen bond energy contribution is minimum (isotropic), only the surface-docking approach is effective in simulating the impurity-mediated habit modification. It is supposed that if a crystal has an isotropic intermolecular bonding, its intermolecular forces do not change much when a host molecule is substituted by an impurity molecule in the built-in approach. Examples illustrating the use of both approaches for binding energy calculations are reviewed.

The surface docking approach was employed to calculate binding energies of aspartic acid adsorbed on α-glycine crystal faces (Poornachary et al. 2008c). The atomistic potential energy calculations were performed using the "Forcite" module in Materials Studio. The net atomic charges of glycine and aspartic acid molecules were computed from *ab initio* single-point energy calculations and subsequent Mulliken population analysis using the Dmol^3 module (Delley 2000) in Materials Studio. The COMPASS force field potential (Sun 1998) was used to calculate intermolecular interaction energies within the crystal lattice because this force field was validated as adequately describing the intermolecular interactions in the glycine crystal. Incorporation of the impurity zwitterion into the α-glycine crystal lattice is modeled by substituting the C–H bond (one that is normal to the *b*-axis) of a glycine molecule on the (010) face by the side chain moiety of the impurity molecule. An aspartic acid anion is created by deprotonating the side chain carboxylic acid group. The interaction energy of the impurity species is calculated on a (010) F face, (010) S face, and (011) face through an energy minimization procedure. The calculation results show preferential interaction of aspartic acid zwitterion on the (010) F face (more negative value of differential binding energy) and of the aspartic acid anion either at a kink site on the (010) step or on the (011) surface (Table 5.1). Further analysis of the individual terms contributing to the binding energy show that the anions incur electrostatic repulsive effects on incorporation on the F (010) surface and, in contrast, gain attractive coulombic interactions on the (011) face. These results, in turn, corroborate with the growth inhibition observed experimentally along the *b*-axis and *c*-axis of α-glycine crystals at different impurity concentrations in the crystallizing medium (Poornachary et al. 2007).

On investigating the interaction of impurities at the glycine crystal surfaces, the hydration effect of impurity ions was not quantitatively treated. In principle, the solvation effect would dampen the electrostatic interactions as computed from the Coulomb's law equation using the permittivity

TABLE 5.1

Binding Energies of L-Aspartic Acid Molecular Species on the α-Glycine Crystal Faces

	Binding Energy (kcal/mol)			Binding Energy Difference (kcal/mol)	
Crystal Face	Glycine	L-Asp Zwitterion	L-Asp Anion	L-Asp Zwitterion	L-Asp Anion
(010) Flat face	−50.0	−59.3	−55.2	−9.3	−5.2
(010) Step face	−43.4	−54.6	−61.9	−11.2	−18.5
(011) Face	−39.5	−47.6	−52.9	−8.1	−13.4

L-Asp, L-aspartic acid

of free space. The effect of hydration of ions on the intermolecular interactions can be accounted either through explicit solvent models or implicit dielectric models. For instance, from distance-dependent dielectric models, it is suggested that the effective dielectric constant can deviate from the vacuum value by a factor of 3–5 (Leach 2001). Nevertheless, considering the fact that in the proposed molecular models the $O \cdots O$ contact distances between the residual (carboxylate) group of aspartic acid anion molecule adsorbed on the crystal surface and of a glycine molecule on the crystal lattice is much smaller (ca. 2.0 Å), the repulsive (or attractive) effects operating at the crystal interfaces could be expected to be significantly stronger, even in a solution environment (Poornachary et al. 2008c).

Myerson and Jang (1995) applied binding energy calculations to screen impurities (alkanoic acids) as potential nucleation and growth inhibitors during the crystallization of adipic acid (AA) ($C_4H_8(CO_2H)_2$)—an excipient used in the pharmaceutical industry as tablet lubricant, acidulant in effervescent tablets, and a constituent of tablet coating films. The calculations were done using Cerius^2 software (1995), and the procedure involved MD and molecular mechanics (MM) simulation. MD simulation, performed under constant NVE (moles (N), volume (V), energy (E)) ensemble and anneal dynamics, helped to arrive at a reasonable initial position/conformation of the additive molecule on the crystal surface and also to relax initial strain in the additive molecule. After this, the additive-crystal interactions were optimized using an MM energy minimization procedure. MD and MM calculations were done using the Dreiding force field (Mayo et al. 1990), with the atom partial charges calculated using the charge equilibration method. The binding energies were found to increase with increasing carbon number to C_{14} (myristic acid—$CH_3(CH_2)_{12}COOH$), to decrease from C_{14} to C_{16} (palmitic acid—$CH_3(CH_2)_{14}COOH$), and then to increase again above C_{16} (stearic acid—$CH_3(CH_2)_{16}COOH$). This interesting result was rationalized based on the conformation of the alkanoic acids and the resulting binding forces between the additives and the crystal surface. The alkanoic acids with less than 14 carbon atoms assumed a long straight-chain on the crystal surface and therefore their binding energies—characterized by the sum of electrostatic, hydrogen bonds and van der Waals forces—tended to increase with the molecular weight. However, between C_{14} and C_{16}, the alkanoic acids assumed a coiled structure, resulting in more self-interaction than interaction with the crystal surface. As a result, their binding energies dropped, but increased again with molecular weight. Interestingly, these computational results correlated well with the experimental metastable zone widths measured for nucleation of AA from ethanol solution in the presence of the alkanoic acids. The effects of additives on the nucleation rate of AA were related to the strength of the intermolecular bonds that form during the adsorption process.

Clydesdale et al. (2005) used the built-in approach for binding energy calculations to investigate the habit modification of AA mediated by the action of homologous impurities (caproic acid (CA) – $C_5H_{11}CO_2H$, glutaric acid (GA) – $C_3H_6(CO_2H)_2$, and succinic acid (SA) – $C_2H_4(CO_2H)_2$). The computer program HABIT98 was used to calculate modified slice and attachment energy values associated with the impurity adsorption process. The results showed that the impurity-mediated morphology is less needlelike, with the higher index {302} faces becoming more important due to favorable binding of the impurity. The AA molecules were oriented perpendicular to this face, giving rise to larger voids on the surface where impurity incorporation is particularly favored. By comparing the binding energy values of each impurity on each of the morphologically important faces, it was postulated that GA has a higher affinity to AA crystal. The extent of habit modification was explained on the basis of intermolecular bonding interactions that are disrupted in the presence of the impurities. On incorporation into the AA crystal, SA (dicarboxylic acid with a chain length two atoms shorter than AA) results in the loss of the main hydrogen bonds due to the shortening of the alkyl chain along the packing direction. With GA (dicarboxylic acid with one carbon atom shorter than AA chain length), because of the shortening of the alkyl chain and the cis orientation of the carboxylic acid, one of the bonds is disrupted along the main hydrogen-bonding chain but another hydrogen bond is formed normal to the axis. With CA, due to the replacement of one carboxylic acid group with a methyl group, the main hydrogen bonding chain is disrupted.

Clydesdale et al. (2003) used binding energy calculations as a measure of the equilibrium impurity segregation coefficient (K) for solids crystallized from mother liquor in the presence of impurities ($K = C_{solid}/C_{liquid}$, where C_{solid} and C_{liquid} represent the impurity concentration in the crystallized solid and mother liquor, respectively). If the differential binding energy values are low for the crystallized material, it implies favorable impurity incorporation, and the material would be expected to have $K > 1$. On application of this method, the previously developed morphological modeling techniques were improved upon by allowing the orientation of the adsorbing impurity molecule to be relaxed with respect to the intermolecular forces of the crystal bulk, thus achieving improved binding energy values.

In the examples just discussed, usually a single value of binding energy is calculated for adsorption of impurity on a specific crystal face and, in turn, related to modified-growth rates. Fiebig et al. (2007) developed a new model to predict crystal morphology that depends on the concentration of impurities in the bulk liquid. The approach is based upon the relation proposed by Davey and Mullin (1974) relating S velocities and impurity concentration to fractional surface coverage. The growth rate ratios of the morphologically important faces as modified by the impurity were predicted by approximating the adsorption energy in the Langmuir isotherm model with the binding energy difference. Using the open force field module of Cerius², MD simulations were performed in a vacuum for each face of the phenyl salicylate (salol) crystal with one adsorption site occupied by the impurity. Charge distributions and optimum molecular conformation of salol and the impurities (Figure 5.10 inset) were determined by semiempirical quantum mechanical methods. The resulting molecular geometries were evaluated based on experimental heats of formation. The Dreiding force field and semiempirical PM5 partial charges were validated against the experimental sublimation energy. Intramolecular interactions were eliminated during the calculation by applying the "rigid body" feature of Cerius². The strain on the molecular conformation of the adsorbed molecule was calculated as the change in conformation energy and summed to the nonbonded interaction energies

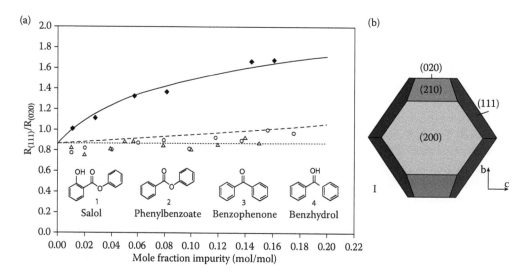

FIGURE 5.10   (a) The ratios of center-to-face distance of the faces (020) and (111) of salol predicted using the model equations relating the step velocities and surface coverage to impurity concentration. The distance ratios calculated in the presence of impurities are represented by (—) phenyl benzoate, (---) benzophenone, and (···) benzhydrol, respectively. The experimental values are represented by (♦) phenyl benzoate, (o) benzophenone, and (Δ) benzhydrol, respectively. (b) Growth morphology of salol predicted using the attachment energy model. (From Fiebig, A., Jones, M. J., and Ulrich, J., *Cryst. Growth Des.*, 7:1623, 2007. With permission.)

to obtain the binding energy. The calculations predicted a higher affinity of phenyl benzoate, and the extent of growth inhibition was in good agreement with the melt crystallization data in the range of the impurity concentration studied (Figure 5.10).

While the effect of impurity concentration on crystal growth can be predicted to a certain extent using this approach, it does not fully account for the combined effects of impurity concentration and supersaturation. A more rigorous approach may involve kinetic MC simulations (van Enckevort and Los 2008), which is applied to simulate the impact of tailor-made inhibitors on the growth of {001} faces of the simple cubic Kossel crystal. In this theoretical approach, a solid-on-solid model is considered for simulating crystal growth. Growth and dissolution rates, impurity surface coverage, and impurity incorporation into the bulk lattice are measured from the probabilities for the addition and removal of growth (and inhibitor) units. The sticking fraction of growth units, which is defined as the total number of additions (and removals) of growth units as a function of time, is used as a measure of the growth rate. The growth rates are expressed as a function of;

- The driving force [supersaturation or undersaturation, expressed as the difference in chemical potential per growth unit in the fluid ($\Delta\mu/kT$)]
- The bond strength (φ) of the solute (impurity) in the solid bulk and at the solid–liquid interface
- The impurity concentration in the mother phase

Tailor-made impurities are defined as additives with horizontal and downward bonds similar to the growth units but with weaker or repelling bonds upward.

Several important and general conclusions were drawn from the MC simulations. The {001} surfaces of the Kossel crystal grow layer-wise through steps if the bond energies between neighboring solid units are relatively stronger ($\varphi/kT \geq 0.8$) and the driving force is less than the transition value for kinetic roughening. The tailor-made inhibitors, in this case, result in no or very slow growth at the lowest supersaturation values, despite the presence of growth steps. On increasing the supersaturation beyond the dead supersaturation zone, the sticking fraction is gradually increased and eventually becomes linear with supersaturation. The concentration of impurities incorporated into the crystal increases for increasing supersaturations up to a maximum value near the dead supersaturation $(\Delta\mu/kT)_d$ and beyond that decreases at the same rate as the surface coverage. The explanation for this is that, as the progress of steps is completely blocked by the adsorbed impurities and only minimum growth occurs, only a few impurity molecules will be incorporated. At higher supersaturation, the steps can pass the blocking centers and a large part of the adsorbed impurities are readily built into the crystal lattice. For $\Delta\mu/kT > (\Delta\mu/kT)^*$, the surface and bulk fraction of inhibitors decrease because the probability ratio of inhibitor attachment to growth unit attachment on the crystal surface decreases with increasing supersaturation. These results are in agreement with the Cabrera and Vermilyea's theory (1958) of step pinning by adsorbed impurities.

5.6   APPLICATIONS IN FORMULATION

Understanding the surface properties of pharmaceutical materials is important to many applications associated with the crystallization and formulation of a drug product. The surface free energy of APIs and solid-state pharmaceutical excipients is an important physicochemical property that governs interfacial interactions during particle processing, including milling, granulation, and powder compaction. Quantifying the surface properties can prove to be extremely useful in the more accurate prediction of processing and formulation behavior—for example, the performance of drug aerosols in inhaler formulations, the wettability and dispersibility in suspensions, and the dissolution behavior.

5.6.1 Surface Properties of Pharmaceutical Materials

Inverse gas chromatography (IGC) is a useful tool for analyzing the surface properties and surface energetics of pharmaceutical powders. IGC uses a combination of nonpolar (n-alkanes) and polar (water, acetone, chloroform, tetrahydrofuran (THF)) vapor probes to characterize the dispersive surface energy and surface acid–base functionality. MM can help in the interpretation and correlation of the IGC data to the precise chemical nature of the crystal surface. In this methodology (Grimsey et al. 2002), the interaction energies among individual atoms of the probe molecule and atoms of the test molecule oriented as in the crystal surface are calculated to simulate surface adsorption. Initially, the molecular structure was extracted from its crystal structure and the electrostatic potential around the isolated molecule was determined by *ab initio* calculations performed at the Hartree–Fock 6-31G** level using the GAMESS *ab initio* quantum chemistry package (Schmidt et al. 1993). Point charges were assigned to each atom by fitting the electrostatic potential distribution around each molecule. The van der Waals and electrostatic interactions between this molecule representing the crystal surface and a probe molecule was calculated using the GRID program (Molecular Discovery, Oxford) (Goodford 1985) by evaluating the interaction energies throughout and around the test molecule. The probe molecules were modeled as follows: acetone using carbonyl oxygen, THF using the oxygen atom of the furan ring, and water treated as an electrically neutral sphere with no dipole moment but capable of donating or accepting up to two hydrogen bonds. Contours of equal energy were plotted around the test molecule, with negative interaction energies denoting regions of favorable interaction. The preferential interaction sites for each probe could thus be predicted from the position and orientation of these regions.

These calculations helped in understanding the effects of milling on the surface properties of paracetamol crystals. The surface energy data of milled paracetamol crystals revealed that the dispersive component (γ_s^d) and the specific component for an acidic probe (chloroform) increased, while the specific component for a basic probe (THF) decreased. This implies that the overall surface showed an increase in both apolar and basic properties, with a corresponding decrease in its acidic component. It is assumed that paracetamol will preferentially split along the (010) cleavage plane during milling, which is found to be the crystal face with the smallest attachment energy. The (010) plane exposes hydrophobic methyl groups and pockets containing benzene ring moieties together with the carbonyl functionality. The OH moiety is sterically hindered and inaccessible to any adsorbing probes. As a result, the surface is overall apolar and basic in nature. Thus the measured changes in surface energy using IGC mirror the exposure of particular chemical groups at that surface.

The calculations also provided insights into the effect of relative humidity (RH) on the surface energetics of paracetamol. At 0% RH (dry) and 45% RH (ambient) conditions, no changes were observed in the dispersive surface energy (γ_s^d) of paracetamol. However, the values for the specific interactions with acetone and THF were observed to decrease. In order to understand this behavior, the preferential interaction sites for the different probe molecules as predicted using this method were analyzed. The predicted interaction regions for water were around the terminal hydroxyl group, the amine, and the carbonyl chain. THF showed similar interaction areas around the hydroxyl and amine groups while the alkyl probes interacted most strongly in a direction perpendicular to the aromatic ring. By comparing these results, it was concluded that in the case where interaction sites for water showed considerable differences to those for probe molecules (alkanes), no changes in the surface energetic parameters were measured with relative humidity. In the case where the preferential interaction sites for water corresponded with those for the probes (THF), this was reflected by a decrease in the measured specific interactions. This correspondence, in turn, suggests that the water is blocking these interaction sites from the vapor probes.

5.6.2 Stabilization of Crystalline Drug Nanosuspensions

The formulation of a poorly water-soluble drug into crystalline nanosuspensions is one of the means of achieving greater bioavailability of a drug (Rabinow 2004). However, the nanosuspensions have a tendency to aggregate, making the formulation unstable. Surfactants can help stabilize crystalline nanosuspensions, and, to this end, MM can provide a useful tool. Konkel and Myerson (2008) have used an empirical MM approach to predict the stabilization of crystalline nanosuspensions of model pharmaceutical drugs by the surfactant polysorbate 80. The approach involves calculation of the ratio of binding energy of the surfactant to the binding energy of the drug to the surface of the crystal. The growth morphology for each drug was calculated from its crystal structure using the attachment energy method as implemented in the Morphology module of Materials Studio. The faces with significant facet area were first selected for binding-energy calculations because they were most likely to be observed in the experimental crystal habit. The energy of a molecule of polysorbate 80 and the drug were minimized using the same energy parameters as the drug crystal and subsequently positioned over the crystal surface so that the surfactant molecule could interact with the surface (hydrogen bonding was monitored to aid in the initial positioning). The lowest energy configurations for the interaction of the surfactant (or drug molecule) with the crystal surface were obtained from MD simulation performed using the "Discover" module in Materials Studio.

From the results obtained, the ratio of the polysorbate 80 binding energy to that of the drug correlated well with the rank order stabilization of the crystalline suspension. A ratio of approximately 3 (for celecoxib and fluorometholone drugs) correlated to moderate stability; a ratio larger than 3 (nabumetone) correlated to excellent stabilization. For the drugs with poor stabilization (carbamazepine), the ratio was around 1. It was also observed that the faces with the greatest interaction with the surfactant were the ones with the most polar groups on the surface that could form hydrogen bond with the oxyethylene arms of the polysorbate 80 molecule. The effect of the aqueous environment was ignored in this adsorption model and so were the effects of various possible conformations of the oxyethylene side chains of the surfactant. Nevertheless, these simulations gave a good prediction of the relative stability of the suspensions in a limited computational time, and thus provide a rapid screening tool to aid in the design of empirical formulation studies.

5.7   CONCLUDING REMARKS

Toward a greater understanding of molecule to solid-state structure relationships and the designing of pharmaceutical crystals with desired properties, MM and computational chemistry go hand in hand with experimental analysis. The application of MM can be associated with the various stages during the development of solid drug products. First, the *ab initio* prediction of crystal structures from the molecular structure using crystal energy landscapes has a great potential in understanding the polymorphism of drug crystals and can aid in screening the stable crystal forms for drug development. This methodology has been further extended to predict the formation of multicomponent crystals such as co-crystals, salts, and solvates. The challenges posed in all these applications are associated with the high conformational flexibility of the drug molecules contributing to a large number of degrees of freedom that need to be optimized during the lattice energy calculations. In turn, this calls for more accurate force fields and charge description to characterize the intermolecular (and intramolecular) interactions between the drug molecules in the predicted crystal structures.

Second, MM methods can be applied to predict the growth morphology of crystals by considering the packing arrangement within the crystal structure (internal factors) and the environmental growth factors such as solvent, supersaturation, and impurities. The crystal shape can be linked to the internal factors through crystallographic (geometric) analysis and by relating the growth rates of crystal faces to the attachment energy, which is a measure of intermolecular interactions between growth units. While these techniques for morphology prediction have been implemented in commercial software packages and thus provide a fast and easy tool, the influence of various environmental growth factors on crystal shape has not been built in. The effects of impurities (mostly reaction by-products) and some solvents on crystal habit modification are interpreted as a form of "molecular trickery," wherein the impurity, because of its similarity with solute molecule structure, preferentially adsorbs on certain crystal faces by mimicking the solute molecule. When incorporated, the impurity disrupts growth in the direction normal to the face by interfering with the normal intermolecular bonding pattern. Binding energy calculations, using the molecular mechanics approach, have been employed to determine the ease of incorporation of the impurity and its subsequent disruption of growth (Poornachary et al. 2011). Recent studies are geared toward simulating the effects of solvent, supersaturation, and impurities on crystal growth using different computational approaches, including MD and KMC calculations.

Finally, MM can enhance our understanding of surface properties of pharmaceutical materials, proving to be extremely useful in prediction of processing and formulation behavior. For instance, the surface functionalities present on different faces of crystal morphology can be modeled and linked with experimental characterization techniques (using atomic force microscopy, X-ray photoelectron spectroscopy, and inverse gas chromatography). Such information can provide valuable insights into the performance of drug aerosols in inhaler formulations, the wettabilty and dispersibility of crystalline drugs in suspensions, and the powder compaction and dissolution behaviors.

REFERENCES

Aakeröy, C. B., and Boyett, R. E. 1999. Molecular mechanics and crystal engineering. In *Crystal Engineering: The Design and Application of Functional Solids*. Eds. K. R. Seddon and M. Zaworotko. Kluwer Academic Publishers, Dordrecht, The Netherlands, 69–82.

Aakeroy, C. B., and Salmon, D. J. 2005. Building co-crystals with molecular sense and supramolecular sensibility. *Cryst. Eng Comm.* 7: 439.

Allen, F. H., and Motherwell, W. D. S. 2002. Applications of the Cambridge Structural Database in organic chemistry and crystal chemistry. *Acta Crystallogr., Sect. B.* 58: 407.

Almarsson, O., and Zawarotko, M. J. 2004. Crystal engineering of the composition of pharmaceutical phases. Do pharmaceutical co-crystals represent a new path to improved medicines? *Chem. Commun.* 17: 1889.

Berkovitch-Yellin, Z. 1985. Toward an *ab initio* derivation of crystal morphology. *J. Am. Chem. Soc.* 107: 8239.

Bernstein, J. 2002. *Polymorphism in Molecular Crystals.* Oxford University Press, UK.

Beyer, T., Day, G. M., and Price, S. L. 2001. The prediction, morphology, and mechanical properties of the polymorphs of paracetamol. *J. Am. Chem. Soc.* 123: 5086.

Bisker-Leib, V., and Doherty, M. F. 2001. Modeling the crystal shape of polar organic materials: Prediction of urea crystals grown from polar and nonpolar solvents. *Cryst. Growth Des.* 1: 455.

Black, S. N., Cuthbert, M. W., Roberts, R. J., and Stensland, B. 2004. Increased chemical purity using a hydrate. *Cryst. Growth Des.* 4: 539.

Blagden, N., Davey, R. J. Rowe, R. and Roberts R. 1998. Disappearing polymorphs and the role of reaction by-products: The case of sulphathiazole. *Int. J. Pharm.* 172: 169.

Boerrigter, S. X. M., Cuppen, H. M., Ristic, R. I., Sherwood, J. N., Bennema, P., and Meekes, H. 2002. Explanation for the supersaturation-dependent morphology of monoclinic paracetamol. *Cryst. Growth Des.* 2: 357.

Boerrigter, S. X. M., Grimbergen, R. F. P., and Meekes, H. 2001. FACELIFT-2.50, a program for connected net analysis: Department of solid state chemistry. University of Nijmegen, The Netherlands.

Boerrigter, S. X. M., Josten, G. P. H., van de Streek, J., Hollander, F. F. A., Los, J., Cuppen, H. M., et al. 2004. MONTY: Monte Carlo crystal growth on any crystal structure in any crystallographic orientation: Application to fats. *J. Phys. Chem. A* 108: 5894.

Cabrera, N., and Vermilyea, D. 1958. The growth of crystals from solution. In *Growth and Perfection of Crystals*. Eds. R. H. Doremus, B. W. Roberts, and D. Turnbull, John Wiley & Sons. Inc., New York, 393-410.

Cerius² (Release 2.0), 1995. Molecular Simulations Inc., San Diego, CA.

Chew, C. M., Ristic, R. I., Dennehy, R. D., and De Yoreo, J. J. 2004. Crystallization of paracetamol under oscillatory flow mixing conditions. *Cryst. Growth Des.* 4: 1045.

Childs, S. L., Wood, P. A., Rodríguez-Hornedo, N., Reddy, L. S., and HardCastle, K. I. 2009. Analysis of 50 crystal structures containing carbamazepine using the *Materials* module of *Mercury* CSD. *Cryst. Growth Des.* 9 :1869.

Clydesdale, G., Hammond, R. B., and Roberts, K. J. 2003. Molecular modeling of bulk impurity segregation and impurity-mediated crystal habit modification of naphthalene and phenanthrene in the presence of heteroimpurity species. *J. Phys. Chem. B* 107: 4826.

Clydesdale, G., Roberts, K. J., and Docherty, R. 1996. HABIT95—A program for predicting the morphology of molecular crystals as a function of the growth environment. *J. Cryst. Growth* 166: 78.

Clydesdale, G., Thomson, G. B., Walker, E. M., Roberts, K. J., Meenan, P., and Docherty, R. 2005. A molecular modeling study of the crystal morphology of adipic acid and its habit modification by homologous impurities. *Cryst. Growth Des.* 5: 2154.

Coombes, D. S., Catlow, C. R. A., Gale, J. D., Rohl, A. L., and Price, S. L. 2005. Calculation of attachment energies and relative volume growth rates as an aid to polymorph prediction. *Cryst. Growth Des.* 5: 879.

Danesh, A., Davies, M. C., Hinder, S. J., Roberts, C. J., Tendler, S. J. B., Williams, P. M., and Wilkins, M. J. 2000. Surface characterization of aspirin crystal planes by dynamical chemical force microscopy. *Anal. Chem.* 72: 3419.

Datta, S., and Grant, D. J. W. 2004. Crystal structures of drugs: Advances in determination, prediction and engineering. *Nature Reviews* 3: 42.

Davey, R. J., Blagden, N., Righini, S., Alison, H., Quayle, M. J., and Fuller, S. 2001. Crystal polymorphism as a probe for molecular self-assembly during nucleation from solutions: The case of 2,6-dihydroxybenzoic acid. *Cryst. Growth. Des.* 1: 59.

Davey, R. J., and Garside, J. 2000. *From Molecules to Crystallizers*. Oxford Chemistry Primer, 86, Oxford University Press, UK.

Davey, R. J., and Mullin, J. W. 1974. Growth of the {100} faces of ammonium dihydrogen phosphate crystals in the presence of ionic species. *J. Cryst. Growth* 26: 45.

Day, G. M., Motherwell, W. D. S., and Jones, W. 2005. Beyond the isotropic atom model in crystal structure prediction of rigid molecules: Atomic multipoles versus point charges. *Cryst. Growth Des.* 5: 1023.

Day, G. M., Price, S. L., and Leslie, M. 2001. Elastic constant calculations for molecular organic crystals. *Cryst. Growth Des.* 1: 13.

Day, G. M., Price, S. L., and Leslie, M. 2003. Atomistic calculations of phonon frequencies and thermodynamic quantities for crystals of rigid organic molecules. *J. Phys. Chem. B* 107: 10919.

Delley, B. 2000. From molecules to solids with the $Dmol^3$ approach. *J. Chem. Phys.* 113: 7756.

Docherty, R., and Meenan, P. 1999. The study of molecular materials using computational chemistry. In *Molecular Modeling Applications in Crystallization*. Ed. A. S. Myerson, Cambridge University Press, New York 106–165.

Dollase, W. A. 1986. Correction of intensities for preferred orientation in powder diffractometry: Application of the March model. *J. Appl. Crystallogr.* 19: 267.

Engel, G. E., Wilke, S., Konig, O., Harris, K. D. M., and Leusen, F. J. J. 1999. PowderSolve—A complete package for crystal structure solution from powder diffraction patterns. *J. Appl. Crystallogr.* 32: 1169.

Fiebig, A., Jones, M. J., and Ulrich, J. 2007. Predicting the effect of impurity adsorption on crystal morphology. *Cryst. Growth Des.* 7:1623.

Frisch, M. J., Trucks, G. W., Schlegel, H. B., Scuseria, G. E., Robb, M. A., Cheeseman, J. R., et al. 1998. GAUSSIAN 98 (A6 ed.). Gaussian Inc.: Pittsburgh, Pa.

Gale, J. D., and Rohl, A. L. 2003. The General Utility Lattice Program (GULP). *Mol. Simul.* 29: 291.

Girifalco, L. A., and Good, R. J. 1957. A theory for the estimation of surface and interfacial energies: I. Derivation and application to interfacial tension. *J. Phys. Chem.* 61: 904.

Givand, J., Rousseau, R. W., and Ludovice, P. J. 1998. Characterization of L-isoleucine crystal morphology from molecular modeling. *J. Cryst. Growth*, 194: 228.

Gnanasambandam, S., and Rajagopalan, R. 2010. Growth morphology of-glycine crystals in solution environments: An extended interface structure analysis. *CrystEngComm.* 12: 1740.

Gong, Y., Colliman, B. M., Mehrens, S. M., Lu, E., Miller, J. M., Blackburn, A., et al. 2008. Stable-form screening: Overcoming trace impurities that inhibit solution-mediated phase transformation to the stable polymorph of sulfamerazine. *J. Pharm. Sci.* 97: 2130.

Goodford, P. J. 1985. A computational procedure for determining energetically favorable binding sites on biologically important macromolecules. *J. Med. Chem.* 28: 849.

Grimsey, I. M., Osborn, J. C., Doughty, S. W., York, P., and Rowe, R. C. 2002. The application of molecular modeling to the interpretation of inverse gas chromatography data. *J. Chromatogr. A* 969: 49.

Habgood, M., Deij, M. A., Mazurek, J., Price, S. L., and ter Horst, J. H. 2010. Carbamazepine co-crystallization with pyridine carboxamides: Rationalization by complementary phase diagrams and crystal energy landscapes. *Cryst. Growth Des.* 10: 903.

Hamad, S., Moon, C., Catlow, C. R. A., Hulme, A. T., and Price, S. L. 2006. Kinetic insights into the role of solvent in the polymorphism of 5-fluorouracil from molecular dynamics simulations. *J. Phys. Chem. B* 110: 3323.

Hammond, R. B., Pencheva, K., Ramachandran, V., and Roberts, K. J. 2007. Application of grid-based molecular methods for modeling solvent-dependent crystal growth morphology: Aspirin crystallized from aqueous ethanolic solution. *Cryst. Growth Des.* 7: 1571.

Hartman, P., and Bennema, P. 1980. The attachment energy as a habit controlling factor. I. Theoretical considerations. *J. Cryst. Growth* 49: 145.

He, G., Jacob, C., Guo, L. F., Chow, P. S., and Tan, R. B. H. 2008. Screening for cocrystallization tendency: The role of intermolecular interactions. *J. Phys. Chem. B.* 112: 9890.

Hendriksen, B. A., Grant, D. J. W., Meenan, P., and Green, D. A. 1998. Crystallization of paracetamol (acetaminophen) in the presence of structurally related substances. *J. Cryst. Growth* 183: 629.

Heng, J. Y. Y., and Williams, D. R. 2006. Wettability of paracetamol polymorphic forms I and II. *Langmuir* 22: 6905.

Hinchliffe, A. 2003. *Molecular modeling for beginners.* John Wiley & Sons, Chichester, West Sussex, England.

Holden, J. R., Du, Z. Y., and Ammon, H. L. 1993. Prediction of possible crystal structures for C-, H-, N-, O-, and F-containing organic compounds. *J. Comput. Chem.* 14: 422.

Hulme, A. T., Price, S. L., and Tocher, D. A. 2005. A new polymorph of 5-fluorouracil found following computational crystal structure predictions. *J. Am. Chem. Soc.* 127: 1116.

Karamertzanis, P. G., Kazantsev, A. G., Issa, N., Welch, G. W. A., Adjiman, C. S., Pantelides, C. C., et al. 2009. Can the formation of pharmaceutical cocrystals be computationally predicted? 2. Crystal structure prediction. *J. Chem. Theory Comput.* 5: 1432.

Karamertzanis, P. G., Raiteri, P., Parrinello, M., Leslie, M., and Price, S. L. 2008. The thermal stability of lattice energy minima of 5-fluorouracil: Metadynamics as an aid to polymorph prediction. *J. Phys. Chem. B* 112: 4298.

Konkel, J. T., and Myerson, A. S. 2008. Empirical molecular modeling of suspension stabilization with polysorbate 80. *Mol. Simul.* 34: 1353.

Leach, A. R. 2001. *Molecular Modeling Principles and Applications.* Pearson Prentice Hall, New York.

Liu, X. Y., Boek, E. S., Briels, W. J., and Bennema, P. 1995. Prediction of crystal growth morphology based on structural analysis of the solid fluid interface. *Nature* 374: 342.

Lu, J. J., and Ulrich, J. 2004. Improved understanding of molecular modeling—the importance of additive incorporation. *J. Cryst. Growth* 270: 203.

Materials Studio (Version 4.0), 2005. Accelrys Software Inc., San Diego, CA.

Mayo, S. L., Olafson, B. D., and Goddard, W. A. III. 1990. DREIDING: A generic force field for molecular simulations. *J. Phys. Chem.* 94: 8897.

Mukuta, T., Lee, A. Y., Kawakami, T., and Myerson, A. S. 2005. Influence of impurities on the solution-mediated phase transformation of an active pharmaceutical ingredient. *Cryst. Growth Des.* 5: 1429.

Myerson, A. S. (ed.). 1999. *Molecular Modeling Applications in Crystallization*, Cambridge University Press, New York.

Myerson, A. S., and Jang, S. M. 1995. A comparison of binding energy and metastable zone width for adipic acid with various additives. *J. Cryst. Growth* 156: 459.

Neumann, M. A. 2003. X-cell: A novel indexing algorithm for routine tasks and difficult cases. *J. Appl. Crystallogr.* 36: 356.

Ouvrard, C., and Price, S. L. 2004. Toward crystal structure prediction for conformationally flexible molecules: The headaches illustrated by aspirin. *Cryst. Growth Des.* 4: 1119.

Parveen, S., Davey, R. J., Dent, G., and Pritchard, R. G. 2005. Linking solution chemistry to crystal nucleation: The case of tetrolic acid. *Chem. Commun.* 12: 1531.

Payne, R. S., Rowe, R. C., Roberts, R. J., Charlton, M. H., and Docherty, R. 1999. Potential polymorphs of aspirin. *J. Comput. Chem.* 20: 262.

Peterson, M. L., Morissette, S. L., McNulty, C., Goldweig, A., Shaw, P., LeQuesne, M., et al. 2002. Iterative high-throughput polymorphism studies on acetaminophen and an experimentally derived structure for form III. *J. Am Chem. Soc.* 124: 10958.

Piana, S., Reyhani, M., and Gale, J. D. 2005. Simulating micrometer-scale crystal growth from solution. *Nature* 438: 70.

Poornachary, S. K., Chow, P. S., and Tan, R. B. H. 2008a. Influence of solution speciation of impurities on polymorphic nucleation in glycine. *Cryst. Growth Des.* 8: 179.

Poornachary, S. K., Chow, P. S., and Tan, R. B. H. 2008b. Impurity effects on the growth of molecular crystals: Experiments and modeling. *Adv. Powder Technol.* 19: 459.

Poornachary, S. K., Chow, P. S., and Tan, R. B. H. 2008c. Effect of solution speciation of impurities on -glycine crystal habit: A molecular modeling study. *J. Cryst. Growth* 310: 3034.

Poornachary, S. K., Chow, P. S., Tan, R. B. H. and Davey, R. J. 2007. Molecular speciation controlling stereoselectivity of additives: Impact on the habit modification in-glycine crystals. *Cryst. Growth Des.* 7: 254.

Poornachary, S. K., Lau, G., Chow, P. S., Tan, R. B. H., and George, N. 2011. The effect and counter-effect of impurities on crystallization of an agrochemical active ingredient: stereochemical rationalization and nanoscale crystal growth visualization. *Cryst. Growth Des.* 11: 492.

Price, S. L. 2009. Computed crystal energy landscapes for understanding and predicting organic crystal structures and polymorphism. *Acc. Chem. Res.* 42: 117.

Rabinow, B. E. 2004. Nanosuspensions in drug delivery. *Nature Rev.* 3:785–796.

Rietveld, H. M. A profile refinement method for nuclear and magnetic structures. 1969. *J. Applied Crystallogr.* 2: 65.

Ristic, R. I., Finnie, S., Sheen, D. B., and Sherwood, J. N. 2001. Macro- and micromorphology of monoclinic paracetamol grown from pure aqueous solution. *J. Phys. Chem. B* 105: 9057.

Sarma, J. A. R. P., and Desiraju, G. R. 1999. Polymorphism and pseudopolymorphism in organic crystals: A Cambridge Structural Database study. In *Crystal Engineering: The Design and Application of Functional Solids*. Eds. K. R. Seddon and M. Zaworotko. Kluwer Academic Publishers, Dordrecht, The Netherlands, 325–356.

Schmidt, M. W., Baldridge, K. K., Boatz, J. A., Elbert, S. T., Gordon, M. S., Jensen, J. H. et al. 1993. General atomic and molecular electronic structure system. *J. Comp. Chem.* 14: 1347.

Smith, W., and Forester, T. R.. 1996. DL_POLY_2.0: A general purpose parallel molecular dynamics simulation package. *J. Mol. Graphics* 14: 136.

Spruijtenburg, R. 2000. Examples of the selective preparation of a desired crystal modification by an appropriate choice of operating parameters. *Org. Process Res. Dev.* 4: 403.

Stewart, J. P. P. 1990. MOPAC: A semiempirical molecular orbital program. *J. Comput.-Aided Mol. Des.* 4: 1.

Stone, A. J. (ed.). 1999. *GDMA: A program for performing distributed multipole analysis of wave functions calculated using the Gaussian program system*. University of Cambridge, UK.

Sun, H. 1998. COMPASS: An *ab initio* force-field optimized for condensed-phase applications—Overview with details on alkane and benzene compounds. *J. Phys. Chem. B*, 102: 7338.

Thirunahari, S., Aitapamula, S. S., Chow, P. S., and Tan, R. B. H. 2010. Conformational polymorphism of tolbutamide: A structural, spectroscopic, and thermodynamic characterization of Burger's forms I IV. *J. Pharm. Sci.* 99: 2975.

Trask, A. V., Motherwell, W. D. S., and Jones, W.. 2005. Pharmaceutical co-crystallization: Engineering a remedy for caffeine hydration. *Cryst. Growth Des.* 5: 1013.

van Enckevort, W. J. P., and Los, J. H. 2008. "Tailor-made" inhibitors in crystal growth: A Monte Carlo simulation study. *J. Phys. Chem. C* 112: 6380.

Walsh, R. D. B., Bradner, M. W., Fleischman, S., Morales, L. A., Moulton, B., Rodriguez-Hornedo, N., et al. 2003. Crystal engineering of the composition of pharmaceutical phases. *Chem. Commun.* 2: 186.

Weissbuch, I., Lahav, M., and Leiserowitz, L. 2003. Toward stereochemical control, monitoring, and understanding of crystal nucleation. *Cryst. Growth Des.* 3: 125.

Willock, D. J., Price, S. L., Leslie, M., and Catlow, C. R. A. 1995. The relaxation of molecular crystal structures using a distributed multipole electrostatic model. *J. Comput. Chem.* 16: 628.

Win, D., and Doherty, M. F. 1998. A New technique for predicting the shape of solution-grown organic crystals. *AIChE J.* 44: 2501.

Win, D., and Doherty, M. F. 2000. Modeling crystal shapes of organic materials grown from solution. *AIChE J.* 46: 1348.

Wood, W. M. L. 2001. A bad (crystal) habit—And how it was overcome. *Powder Tech.* 121: 53.

Xie, S., Poornachary, S. K., Chow, P. S., and Tan, R. B. H. 2010. Direct precipitation of micron-size salbutamol sulphate—New insights into the action of surfactants and polymeric additives. *Cryst. Growth Des.* 10: 3363.

Yu, Z. Q., Chew, J. W., Chow, P. S., and Tan, R. B. H. 2007. Recent advances in crystallization control: an industrial perspective. *Chem. Eng. Res. Design* 85: 893.

Yu, Z. Q., Tan, R. B. H., and Chow, P. S.. 2005. Effects of operating conditions on agglomeration and habit of paracetamol crystals in anti-solvent crystallization. *J. Cryst. Growth* 279: 477.

Yu, S. B., and York, P. 2000. Crystallization processes in pharmaceutical technology and drug delivery design. *J. Cryst. Growth* 211: 122.

6 Studies on the Microstructure in Water–Surfactant Systems Using Atomistic and Mesoscale Simulations

K. Ganapathy Ayappa and Foram M. Thakkar

CONTENTS

6.1   INTRODUCTION

Amphiphilic molecules consist of both hydrophobic and hydrophilic parts. Because of their industrial importance, the most commonly studied amphiphiles are surfactant molecules, which are widely used in detergency, enhanced oil recovery, mineral beneficiation, and a range of interfacial phenomenon. Oil–water–surfactant systems display a wide variety of interesting phase behavior; some of the microstructures encountered in these systems, such as reverse micelles, micelles, and the lamellar phase, are illustrated in Figure 6.1. Hydrocarbons (oils) associate with the hydrophobic tail regions of the microstructure, giving rise to swollen micelles or reverse micelles, in which water is confined as nanodroplets, with the hydrocarbon tails associated with the oil phase. Many of the generic features observed in the oil–water–surfactant phase diagram are preserved, as long as the surfactant is replaced by molecules that contain hydrophobic and hydrophilic entities. Hence, in addition to surfactants, which are the most commonly used and studied amphiphiles, lipid molecules and block copolymers can also associate into various microstructures by virtue of their amphiphilic nature.

The thermodynamics of self-assembly has been widely investigated using simple phenomenological models, mean field–based lattice models, and more sophisticated self-consistent field theory

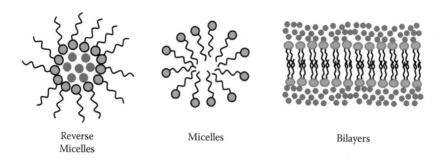

Reverse
Micelles

Micelles

Bilayers

FIGURE 6.1   Schematic representation of microstructure commonly found in oil–water–surfactant systems.

methods. Supported by a wide range of experimental techniques, the existence of a given phase and the factors that control the formation of a particular phase in both ternary and binary systems are reasonably well understood.

In addition to identifying the microstructure related to a particular phase, studying the microscopic features associated with a given structure is of interest. In this regard, molecular simulations have played an important role in elucidating detailed structural and dynamic events associated with a given microstructure and have become an indispensable tool in developing our understanding of various systems ranging from simple monatomic fluids to more complex fluids made up of polymers, surfactants, and proteins. From the early Monte Carlo (MC) (Metropolis 1953) and molecular dynamics (MD) (Alder and Wainwright 1957) simulations of hard sphere systems and later on soft sphere fluids (Rahman 1964; Wood and Parker 1957), simulation techniques have advanced significantly, and today a variety of techniques are available for evaluating thermodynamic, structural, and dynamic properties. Because of the many degrees of freedom involved, the processes that take place in physical systems occur over a wide range of time and length scales, as illustrated in Figure 6.2. Physicochemical processes of engineering relevance require understanding ranging from the electronic structure of materials to events that govern phenomena at the macroscopic or continuum level. The typical time and length scales of the processes under investigation pose limitations on the level of molecular detail that one chooses to use, as well as the choice of a given method. Figure 6.2 illustrates the regime of length and time scales probed using classical MD simulations. In this regime, the length scales range from 1 to 10 nm and the time scales are typically in the nanosecond (ns) regime. With the advances in parallel computing, a wide variety of self-assembled structures can now be investigated using classical MD, also referred to as all-atom simulations. Because water (solvent) and ions are treated explicitly, these simulations are restricted to smaller system sizes and simulation times, which typically range from 10 to 100 ns. The systems that fall within the purview of length and time scales studied with classical MD simulations are the structure and dynamics of self-assembled monolayers in which long-chain alkane molecules organize into ordered two-dimensional structures on a solid surface, reverse micelles, surfactant bilayers, and protein dynamics. Many of these structures also have dynamics at different length and time scales that are not accessible to classical atomistic simulations.

In the mesoscale regime, the interest is in probing systems on the time scale of a few μs and length scales of 100 to 1000 nm. These systems include the lamellar phase, structure and dynamics of phases with complex microstructures that form in oil–water–surfactant systems, dynamics of colloidal suspensions, protein folding, and aggregation pathways. A common feature in these systems is the coupling between the microscopic (molecular) and macroscopic structures. For example, in the case of self-assembling systems, the aggregation behavior arises from the competing volumes of hydrophilic and hydrophobic entities giving rise to structures such as micelles and bilayers. On macroscopic length scales, these systems are quite well characterized. However, the formal connection between the two length scales and the intervening dynamics are not well established. This chapter focuses on the reverse micellar and the lamellar phases and summarizes recent and ongoing efforts

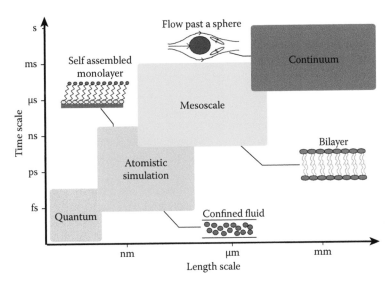

FIGURE 6.2 Schematic representation of the various simulation methods and the corresponding length and time scales typically accessible to each method. Quantum methods include density functional, *ab initio* and Carr-Parinello molecular dynamics. Atomistic methods include molecular dynamics and Monte Carlo simulations. Mesoscale methods include coarse-grained molecular dynamics, dissipative particle dynamics, Brownian dynamics, and lattice Boltzmann simulations. Continuum methods involve solution of heat, mass, and momentum transport equations using a variety of numerical methods such as the finite element and finite difference methods. Methods such as coarse-grained molecular dynamics include the overlap region between atomistic and mesoscale simulations, while dissipative particle dynamics and lattice Boltzmann methods include the overlap between mesoscale and continuum regimes.

in our laboratory to understand the microstructure of these systems at the molecular and mesoscale levels of description.

6.2   ION EXCHANGE IN REVERSE MICELLES

The microstructure of reverse micelles is characterized by surfactant head groups enclosing a core of water, thereby separating water from the nonpolar solvent, which is the continuous phase. Reverse micelles are also referred to as the L2 phase in the ternary oil–water–surfactant phase diagram. The diameter of the reverse micelle, which typically lies between 20 and 100 Å, is characterized by the molar ratio of water to surfactant (w_o), which typically ranges from 0 to 45. An understanding of the properties of water in these systems is important for a number of technologically relevant processes such as nanoparticle synthesis (Bandyopadhyaya et al. 2000) and protein solubilization (Bratko et al. 1988). Water confined in the reverse micelles presents a unique situation in contrast to bulk or free water (Bhattacharyya and Bagchi 2000). Layering of water adjacent to the surfactant head groups occurs as a result of the strong hydrophilic nature of the surfactant–water interface and water in the central regions of the reverse micelle possesses bulklike properties. This dual environment has triggered interest in studying the effects of "crowding" on protein folding pathways because the hydrophobic interactions are likely to be different from those observed in bulk water. Molecular simulations are particularly powerful in studying structured inhomogeneous fluids, because detailed molecular information that might not be accessible experimentally is available in such studies. Here, we focus on two phenomena that can be elucidated only with the detailed atomistic models that have been developed to study reverse micelles. The first phenomenon is related to the distribution and solvation of ions inside the aqueous core of the reverse micelle, and the second aspect is concerned with the influence of confinement on the hydrophobic effect.

6.2.1   INTERACTION POTENTIALS AND SIMULATION PROCEDURE

The Na-AOT reverse micelle is a widely investigated reverse micelle system made up of the sodium salt of a two-tailed anionic surfactant, sodium di(2-ethylhexyl) sulfosuccinate. The interior of the aqueous reverse micelle is modeled as a rigid cavity, with a united atom representation for the sulfonate head group (Faeder and Ladanyi 2000; Pal et al. 2005). The head groups protrude from the cavity boundary and are tethered only in the radial direction by means of a harmonic potential. Interactions between reverse micelles are neglected in the model; hence periodic boundary conditions and Ewald summations for the electrostatics are not required. Water is treated using the extended simple point charge, or SPC/E, model and the potential parameters for all the species are listed in Table 6.1.

The interaction potential, U_{ij}, between charged species i and j consists of both Coulombic and van der Waals interactions treated using a 12-6 LJ potential. Hence

$$U_{ij} = \frac{q_i q_j}{4\pi\epsilon_0 r_{ij}} + 4\epsilon_{ij}\left[(\sigma_{ij} / r_{ij})^{12} - (\sigma_{ij} / r_{ij})^6\right] \tag{6.1}$$

where q_i and q_j are the electrostatic charges, ϵ_0 is the permittivity of vacuum, r_{ij} the distance between the atomic centers, ϵ_{ij} the LJ interaction energy, and σ_{ij} the LJ diameter. In the SPC/E model for water, charges are situated on atomic sites located on a rigid water molecule. Interaction of hydrogen atoms between water molecules and with other charged species is only Coulombic in nature.

For unlike species the LJ potential parameters were obtained using the Lorentz–Berthelot mixture rules

$$\sigma_{ij} = (\sigma_i + \sigma_j) / 2 \qquad \text{and} \qquad \epsilon_{ij} = \sqrt{\epsilon_i \epsilon_j}$$

where ϵ_i and ϵ_j are the self-interaction energy parameters ϵ_{ii} and ϵ_{jj}, respectively.

The interaction potential between a site in the interior of the reverse micelle (excluding the head groups) and the region external to the head groups is obtained using a mean field approximation, by

TABLE 6.1

Potential Parameters Used for Various Species in the Atomistic Model for the Reverse Micelle*

Ions	σ (Å)	ε (kJ/mol)	q (e)
SO$_3^-$	6.0	2.0907	−1
Na$^+$	2.275	0.48208	+1
Hydrocarbon medium	2.5	1.92422	
O (SPC/E)	3.169	0.6502	−0.8476
H (SPC/E)	—	—	0.4238
Li$^+$	1.51	0.6904	+1
K$^+$	3.331	0.4184	+1
Cs$^+$	3.883	0.4184	+1
Cl$^-$	4.409	0.4184	−1

*The sulfonate group is treated as a united atom.

Source: From Pal, S., Vishal, G., Gandhi, K.S., and Ayappa, K.G., *Langmuir*, 21, 767, 2005. Copyright 2005 American Chemical Society.

integrating the LJ interactions between an atom within the reverse micelle and the external region represented by a continuum of sites. The resulting potential represents only the van der Waals contribution from the external hydrocarbon region, and electrostatic contributions from the water continuum are neglected. In spherical polar coordinates, this procedure results in the following expression for the interaction potential

$$u_{iw}(z) = 8\pi\rho_{hc}\epsilon_{iw}\sigma_{iw}^3 \left[(\sigma_{iw}/R)^9 F(z,6) - (\sigma_{iw}/R)^3 F(z,3) \right] \tag{6.2}$$

where $z = r/R$, R is the radius of the reverse micelle cavity, r is the distance of the site from the center of the reverse micelle, ρ_{hc} is the density of sites in the hydrocarbon exterior, and ϵ_{iw} and σ_{iw} are the LJ interaction parameters between an atomic species and the hydrocarbon exterior.

The functions $F(z,3)$ and $F(z,6)$ are:

$$F(z,3) = \frac{2}{3(1-z^2)}$$

and

$$F(z,6) = \frac{2(5+45z^2+63z^4+15z^6)}{45(1-z^2)^9} \tag{6.3}$$

respectively. In Equation 6.2, ϵ_{iw} and σ_{iw} for oxygen, sodium counterion and the added salt species are computed using the Lorentz–Berthelot mixture rules with σ_w and ϵ_w taken to represent the parameters for the hydrocarbon exterior (Table 6.1). The oxygen atoms of the water molecule, added cations and anions, and sodium counterions interact with the region external to the head groups with the interaction potential given in Equation 6.2. The hydrocarbon density, $\rho_{hc} = 0.0241$ Å^{-3}.

The head groups Z$^-$ are held at the interface of the cavity wall and the inner core of the reverse micelle by a harmonic potential of the form (Faeder and Ladanyi 2000)

$$u(r') = \frac{1}{2}k(r'-r_e')^2 \tag{6.4}$$

with $r_e' \equiv R - r_e = 2.5$ Å, where r_e is the equilibrium radial distance of the center of the head groups, $r' = R - r$ where r is the radial location of the head group and the spring constant, $k = 600$ kcal/molÅ^{-2}.

The density distribution of the various atoms inside the reverse micelle is calculated by counting the number of atoms located in a spherical shell of thickness Δr. The density distribution is evaluated using,

$$\rho(r) = \frac{\langle N(r,r+\Delta r) \rangle}{4\pi r^2 \Delta r} \tag{6.5}$$

For a particular species i, the running integral

$$RI_i(r) = \int_0^r \rho_i(r')4\pi r'^2 dr' \tag{6.6}$$

yields the number of particles inside a radial location r.

Pair distribution functions $g_{A-B}(r)$ at a given distance r are computed using

$$g_{A-B}(r) = \frac{\langle N_B(r, r+\Delta r)\rangle}{4\pi r^2 \Delta r} \tag{6.7}$$

where N_B is the number of particles of species B located at a distance r from species A. The number of B atoms that surround the A atoms within a given distance r from the reference atom A is obtained by evaluating the integral, $\int_0^r g_{AB}(r')4\pi r'^2 dr'$.

The $w_o = 4$ reverse micelle corresponds to 140 water molecules and 35 sodium counterions. MC simulations were carried out using the standard Metropolis sampling procedure. In addition to the displacement moves, rotation of the water molecules involved a rigid body rotation around one of the three coordinate axes chosen at random (Allen and Tildesley 1987). MC simulations typically consist of 160 to 200 million equilibration moves followed by 60 to 80 million moves, during which system properties were accumulated. MD was implemented using the SHAKE algorithm (Ryckaert et al. 1977), which is a constraint dynamics method in which the bonds in the water molecules are held rigid.

6.2.2   Density Distributions

Results are presented for the $w_o = 4$ reverse micelle corresponding to a radius of 14.1 Å. Simulations are carried out by introducing alkali salts of Li^+, K^+, Na^+, and Cs^+ into the aqueous core of the reverse micelle. These studies were carried out for low loading (1 MCl molecule, L1) and high loading (3 MCl molecules, L2) salt concentrations. Figure 6.3 illustrates the density distribution of oxygen, Na^+ counterion, and the added cations Li^+, K^+, Na^+, and Cs^+. The water density distribution illustrates the influence of the hydrophilic interior on the density of water in the vicinity of the head groups. A well-defined water peak is observed at a radius of 12 Å. Water in this layer partially hydrates the counterions and forms hydrogen bonds with the negatively charged head groups. The density distributions reveal distinctly different locations of the added cations, with the Li^+ and Na^+ (Figure 6.3a and b) cations preferring to remain in the aqueous core of the reverse micelle. On the other hand, K^+ (Figure 6.3c) and Cs^+ (Figure 6.3d) prefer to reside with the Na^+ counterions associated with the anionic head groups. The added ion pair has a negligible effect on the density distributions for oxygen compared with the Na–AOT reverse micelle in the absence of added salt. This relatively small influence of added ions on water density distribution has also been observed in our recent work for ions confined between two extended hydrophobic surfaces such as graphite (Malani et al. 2006). In the case of K^+ and Cs^+, the changes to the Na^+ counterion density distributions are the greatest. Figure 6.3c and d, illustrates a small decrease (compared with Figure 6.3a and b) in the first Na^+ counterion density peak located at 12 Å and an increase in the counterion density within the core of the reverse micelle. Similar preferential distribution of added cations is observed when three cations are added into the system.

6.2.3   Ion Hydration

In Figure 6.4, the Li^+–O and Na^+–O pair density distributions are shown for the ion added into the aqueous core of the reverse micelle. When NaCl was the added salt (Figure 6.4), the identity of the added cation was kept separate from the Na^+ counterions. The value of the running integrals up to the first minima in the pair distributions yields the number of oxygen nearest neighbors, also known as the average hydration number.

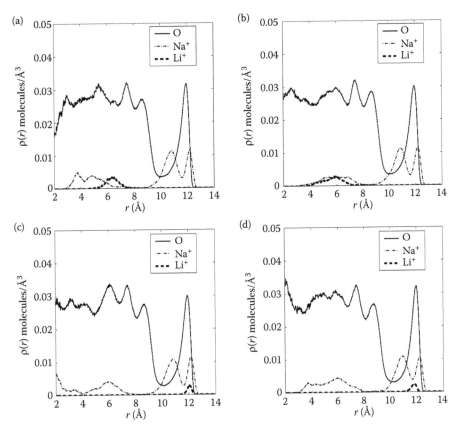

FIGURE 6.3 Density distributions inside the reverse micelle with one molecule of added salt (dashed lines). The added Li^+ ion prefers to remain in the aqueous core region of the reverse micelle, however, Cs^+ (c) and K^+ (d) are associated with the head groups of the reverse micelle. (From Pal, S., Vishal, G., Gandhi, K.S., and Ayappa, K.G., *Langmuir*, 21, 767, 2005. Copyright 2005 American Chemical Society.)

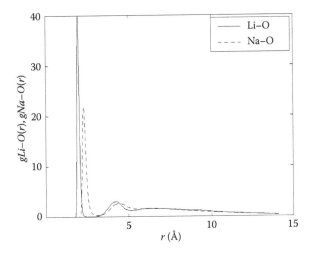

FIGURE 6.4 Ion–water (oxygen) pair distribution functions reveal the formation of a strong first hydration shell for both Li and Na (added) cations. (From Pal, S., Vishal, G., Gandhi, K.S., and Ayappa, K.G., *Langmuir*, 21, 767, 2005. Copyright 2005 American Chemical Society.)

Hydration numbers for the added cations in the reverse micelle are a useful structural indicator of the degree to which water in the reverse micelle differs from free water. For the Li^+ cation, which remains in the aqueous core, the hydration number of 4.03 is very similar to the hydration number of 4.1 reported for ions at infinite dilution in SPC/E water at 298 K (Lee and Rasaiah 1996). However, for Na^+, the hydration numbers are 5.06 and 5.9 in the reverse micelle and free water, respectively. The differences for Na^+ could partly be attributed to the difference in the potential parameters used for Na^+ in our study with those used in the free water. Despite the limited amount of available free water within the reverse micelle, the Li^+–O and Na^+–O (Figure 6.4) pair distributions reveal the presence of strongly bound first hydration shells as seen by the zero intensity in the region between the first and second peaks. The first and second peak positions are also similar to those observed for ions in free water. Examination of the K^+–O and Cs^+–O distributions did not show the formation of a distinct peak as in the case of Li and Na cations, and the hydration numbers were less than 0.4, indicating significantly reduced hydration in the head group region compared with the corresponding bulk values of 7.2 and 9.6 for K and Cs cations, respectively.

In situations where the ions remain solvated (Li^+, Na^+) in the aqueous core of the reverse micelle, we compare the ion–oxygen pair distributions with those reported in bulk water (Lee and Rasaiah 1996) at infinite dilution. Comparison of the peak positions reveals that the positions for the Li^+–O and Na^+–O are similar to those observed in the bulk fluid. It is interesting to note that these positions are unaltered at high salt concentrations. On comparing the peak intensities for Li–O and Na–O between the low and high loading salt concentration situations, we observe a lowering of the peak intensities at high salt concentrations. This lowering is greater in the case of Na^+ than in the case of Li^+, indicating that the smaller ion with the lower hydration number still has sufficient water for solvation at the higher loadings (3 molecules of added salt). Similar to the low loading situation (Figure 6.4), the Li^+–O distributions reveal the presence of a strongly bound first hydration shell as seen by the zero intensity in the region between the first and second peaks. The locations of the peaks for the M^+–O (where M^+ is the added cation) distributions at high loading are, however, similar to the ones observed at low loadings.

6.2.4 ION EXCHANGE

For both ion loadings studied, we investigate whether an equivalent number of counterions are displaced when the added cation is found near the head group region. The running integrals, which are evaluated till $r = 10$ Å for the Na^+ counterions, are illustrated for both loadings in Figure 6.5. The running integrals are evaluated between $0 \leq r \geq 10$Å, because this region represents water that is not bound to the head group region. Further, at $r = 10$ Å, a minimum in the oxygen density distribution is observed and the counterion density peaks have nearly decayed to zero. These trends can be observed in any of the density distributions illustrated in Figure 6.3. Because a total of 35 sodium counterions are present in the system, the number of sodium counterions present in the head group region ($r > 10$ Å) is obtained as a difference between the total number of counterions and that obtained from the running integrals evaluated up to $r = 10$ Å. From such an analysis, the average number of counterions present, in the absence of added cations, in the head group region is 30.13. From the running integrals, at L1 loadings, the number of counterions in the head group region is 29.85, 29.64, 29.29, and 29.13 for Li^+, Na^+, K^+, and Cs^+, respectively. The corresponding numbers for the L2 loadings are 28.42, 29.95, 27.46, and 27.17, respectively. These numbers clearly indicate that the presence of added cations tends to reduce the number of sodium counter ions present in the head group region. The decrease in the number of counterions is greater for the L2 loading, as one would expect. To quantify the decrease in the number of sodium cations in the head group region we define an ion-exchange efficiency.

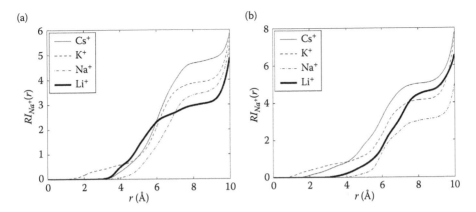

FIGURE 6.5 Running integrals of the sodium counterions for the (a) low loading and (b) high loading situations. (From Pal, S., Vishal, G., Gandhi, K.S., and Ayappa, K. G., Langmuir, 21, 767, 2005. Copyright 2005 American Chemical Society.)

$$\eta_{IE} = \frac{N_{Na^+}^d}{N_{M^+}^a} \qquad (6.8)$$

where $N_{Na^+}^d$ is the number of counterions displaced from the head group region and $N_{M^+}^a$ is the number of added cations into the reverse micelle.

The values of η_{IE} are 0.28/0.57 (L1/L2) for Li$^+$, 0.84/0.89 for K$^+$, and 1.0/0.99 for Cs$^+$. A value of $\eta = 1$ indicates that a complete exchange occurs, and this is observed for the case of Cs$^+$ for both loadings. The exchange efficiency for K$^+$ is also high, above 84%. The data indicate that the lowest propensity for exchange occurs for Li$^+$. However, for the L2 loading, η_{IE} for Li$^+$ increases to 57% from 28%, which is observed at lower loadings. The density distributions also reveal the presence of a small Li$^+$ peak in the region of the head group.

A simple model, based on enthalpic contributions is able to explain the ion-exchange trends observed in the reverse micelle. We present the salient aspects of the model, and the reader is referred to the work by Pal et al. (2005) for the complete treatment. We assume, based on previous studies (Lynden-Bell and Rasaiah 1997), that entropic contributions play a small role, and the process is dominated by the enthalpic effects. In Figure 6.6, configuration 1 illustrates a situation in which all the counterions M$_1^+$ are located near the head groups and M$_2^+$, the cation corresponding to the added salt, is within the free water region of the reverse micelle. In configuration 2 shown in Figure 6.6, the added cation M$_2^+$ replaces the counterion by an ion-exchange process. For further discussion, M$_1^+$ and M$_2^+$ will be denoted as M$_1$ and M$_2$, respectively. Let $U_{M_1^S-M_2^F}^{(1)}$ and $U_{M_2^S-M_1^F}^{(2)}$ represent the potential energy for configurations 1 and 2, respectively. The superscript S denotes species located at the surface (head group region), and the superscript F denotes species in the free water region located in the aqueous core of the reverse micelle. Considering only the potential energy contributions that involve ions M$_1$ and M$_2$, and neglecting interactions with the anion of the added salt, the interaction potential for configuration 1 is

$$U_{M_1^S-M_2^F}^{(1)} = U_{M_1-M_1}^S + U_{M_1-Z^-}^S + U_{M_1-O}^S + U_{M_1-H}^S + U_{M_2-O}^F \qquad (6.9)$$

where $U_{M_1-M_1}^S$ is the interaction energy between counterions at the surface, $U_{M_1-Z^-}^S$ is the interaction energy between M$_1$ and the head groups, $U_{M_1-O}^S$ is the energy of M$_1$ with bound water at the surface,

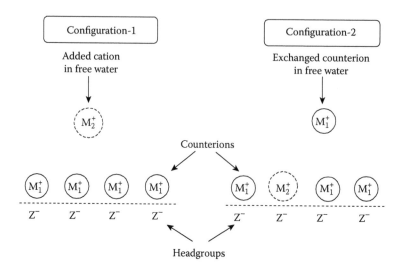

FIGURE 6.6 Schematic illustration of cation exchange near the head groups of reverse micelle. (From Pal, S., Vishal, G., Gandhi, K.S., and Ayappa, K.G., *Langmuir*, 21, 767, 2005. Copyright 2005 American Chemical Society.)

$U^S_{M_1-H}$ is the energy of M_1 with the hydrocarbon exterior (Equation 6.2), and $U^F_{M_2-O}$ is the interaction energy of M_2 with the free water in the aqueous core of the reverse micelle. In a similar manner, the potential energy contribution for configuration 2 is

$$U^{(2)}_{M_2^S-M_1^F} = U^S_{M_2-M_1} + U^S_{M_2-Z^-} + U^S_{M_2-O} + U^S_{M_2-H} + U^F_{M_1-O} \qquad (6.10)$$

If ΔU represents the difference in energies between the two configurations, then

$$\Delta U = U^{(2)}_{M_2^S-M_1^F} - U^{(1)}_{M_1^S-M_2^F} \qquad (6.11)$$

Substituting Equations 6.9 and 6.10 into Equation 6.11 and assuming that the ratio of the solvation energy of the ion i at the surface to that in the free water region is α_i ($0 < \alpha_i < 1$), which implies $U^S_{M_1-O} = \alpha_1 U^F_{M_1-O}$ and $U^S_{M_2-O} = \alpha_2 U^F_{M_2-O}$, the expression for ΔU reduces to

$$\Delta U = \Delta U_a + \Delta U_b + \Delta U_s + \Delta U_h \qquad (6.12)$$

where $\Delta U_a = U^s_{M_2-M_1} - U^s_{M_1-M_1}$, $\Delta U_b = U^s_{M_2-Z^-} - U^s_{M_1-Z^-}$, $\Delta U_s = (1-\alpha_1)\, U^F_{M_1-O} + (\alpha_2-1)\, U^F_{M_2-O}$, is the difference in solvation energies resulting from the exchange of cations M_2 with M_1 and $\Delta U_h = U^s_{M_2-H} - U^s_{M_1-H}$

The various energetic contributions to ΔU are summarized in Table 6.2. In arriving at these numbers, the change in solvation energy from surface to free water is estimated from previous simulation studies, yielding $\alpha_1 = \alpha_2 = 0.234$. In agreement with our observations in the simulations, the exchange of K^+ and Cs^+ with the Na^+ counterion is energetically favorable because $\Delta U < 0$ and for the case of Li^+, $\Delta U > 0$, indicating that Li^+ would prefer to remain solvated in the aqueous core of the reverse micelle, as observed in this study.

TABLE 6.2

Model Estimates for the Change in Energy Associated with Replacing the Cation of the Added Salt with the Counterion*

Ion	ΔU_a (kJ/mole)	ΔU_b(kJ/mole)	ΔU_s (kJ/mole)	$\Delta U_a + \Delta U_b + \Delta U_s$ (kJ/mole)
Cs$^+$	−60	112	−195	−143
K$^+$	−38	92	−139	−85
Li$^+$	25	−61	94.2	58.2

*$\Delta U < 0$ indicates that an ion-exchange is favorable (Cs$^+$ and K$^+$), and $\Delta U > 0$ implies that an ion-exchange is unfavorable (Li$^+$).

Source: From Pal, S., Vishal, G., Gandhi, K.S., and Ayappa, K.G., *Langmuir*, 21, 767, 2005. Copyright 2005 American Chemical Society.

6.3 HYDROPHOBIC EFFECT IN REVERSE MICELLES

The hydrophobic effect, which is used to describe interactions between nonpolar entities, has widespread implications in bilayer formation, protein folding, protein–protein aggregation, and growth. While studying the hydrophobic effect, efforts have been directed toward understanding the thermodynamic origins that govern interactions and forces that exist between nonpolar or hydrophobic entities immersed in water. Interactions between hydrophobic solutes in water have been investigated extensively using a combination of theory, free energy calculations, and molecular simulations. The Helmoltz free energy or, equivalently, the potential of mean force (PMF), calculated from the pair distribution function, between two hydrophobic solutes present in water shows the presence of two minima, indicating two stable configurations for a system in equilibrium (Pangali et al. 1979; Wallqvist 1991). The global minima, which occurs at smaller distances, corresponds to the configuration in which the solutes are in contact, giving rise to the contact pair. The second local minimum, which occurs at larger distances, corresponds to the water separated pair. When a nonpolar entity is placed in bulk water, the water molecules surrounding that entity will attempt to readjust themselves to retain their hydrogen-bonding network. To minimize this loss of hydrogen bonds, the most favorable (thermodynamically) configuration is the contact pair. In the contact pair configuration, nonpolar solutes create the least amount of exposed nonpolar surface, thereby minimizing the loss of the hydrogen bonds. This configuration also minimizes the entropy loss that would occur if water were to completely surround both nonpolar entities. The schematic in Figure 6.7 illustrates the trends associated with the PMF that exists between two hydrophobic entities surrounded by water. The PMF,

$$w(r) = -k_B T \ln g_{ss}(r) \tag{6.13}$$

where k_B is Boltzmann constant, T is temperature, and g_{ss} is the solute–solute pair distribution function. Figure 6.8 illustrates the pair correlation function and the PMF for two methane particles inside the reverse micelle. In comparison with the double minima associated with the PMF for hydrophobic solutes in bulk water, the PMF within the reverse micelle consists of a single minima. In these simulations, the PMF was obtained by averaging the pair correlation function from five independent simulation runs with different initial configurations and the uncertainty in the well depth of the PMF is about 1 kJ/mol. Thus for confinement within the reverse micelle only, the contact pair is stabilized and the solvent-separated pair is not observed. The position of the contact pair is similar to that observed for methane in bulk water; however, the well depth is

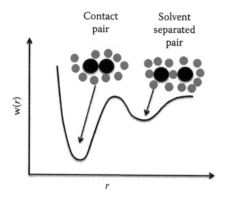

FIGURE 6.7 Schematic representation of the potential of mean force $w(r)$, as a function of the distance r between solute pairs. The first minima corresponds to the contact pair and the second to the solvent-separated pair.

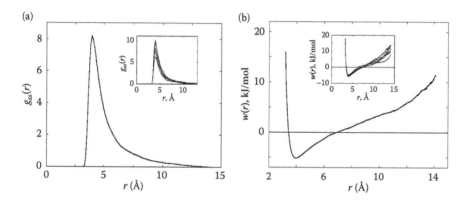

FIGURE 6.8 (a) Pair distribution function and (b) potential of mean force, $w(r)$ for two methane molecules in the reverse micelle. In contrast to the potential of mean force observed in free water where two minima are observed, here only a single minima associated with the contact pair is observed. (Reproduced with permission from Rao, P. V. G., Gandhi, K. S., and Ayappa, K. G., *Langmuir*, 23, 12795, 2007. Copyright 2007 American Chemical Society.)

5.2 kJ/mol inside the reverse micelle compared with about 3 kJ/mol for methane in bulk SPC/E water. Hence, the contact pair is significantly more stable in the environment of the reverse micelle. The presence of the single minimum is also observed for other LJ solutes with diameters in the range of 2.5 to 4.0 Å, and the well depth at the minimum of the PMF is found to increase with the solute diameter.

The absence of the secondary minima associated with the solvent-separated pair has been attributed to insufficient water in the reverse micelle to stabilize this configuration. A simple geometric argument to support this hypothesis is given in the following section. Consider two solute particles, in the solvent-separated configuration schematically represented in Figure 6.9.

For water to form the solvent-separated pair, let l_m be the minimum distance to be accommodated in the aqueous core region of the reverse micelle. From the pair correlation function between methane and the oxygen, the location of the first minima is observed (Rao et al. 2007) at a distance of $l_{so} = 5.6$ Å. Let l_r be the distance between atomic centers of the solute atoms in the solvent-separated configuration, hence the minimum distance, $l_m = l_r + 2l_{so}$. For the case of methane in the solvent-separated configuration in free water, $l_r = 7$ Å and $l_m = 18.2$ Å. For the contact pair, $l_r = 3.95$ Å and the corresponding value of $l_m = 15.15$ Å. From the water density distributions in the reverse micelle,

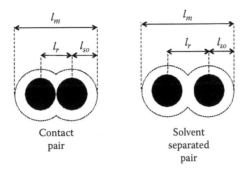

FIGURE 6.9 Representation of relevant distances required to evaluate the minimum diameter of available free water in the reverse micelle.

the available free water lies within a diameter of 20 Å if the distance between the first density minima (from the walls of the reverse micelle) is used to define the boundaries of free water and a diameter of 18 Å if the second maxima in the density distribution is used. From an examination of the density distributions, 18 Å provides a more reasonable estimate of the free water regions. Clearly, this diameter is sufficient to accommodate the contact pair, but not the solvent-separated pair. We note that water molecules that surround the methane particles must have sufficient free water to complete their hydrogen bond network. Thus, if the solvent-separated pair was to be formed in the center of the reverse micelle, there would be insufficient water to complete the hydrogen bond network purely from steric considerations.

The reverse micelle and the associated structural properties that were investigated in this section, require the detailed level of description to describe phenomena such as ion-exchange or the hydrophobic effect. Thus, water is treated explicitly in these systems to capture effects associated with hydrogen bonding and solvation. In what follows, we will illustrate phenomena in which this level of atomistic detail is not required and the solvent (water) is treated as a "bead" that represents a group of water molecules and hydrocarbon chains treated as a collection of beads to represent lipid molecules. These reduced descriptions, which form the basis for mesoscale descriptions, are able to probe phenomena on larger length and time scales compared with the fully atomistic descriptions.

6.4 POLYMER-GRAFTED MEMBRANES: MELTING AND MECHANICAL PROPERTIES

In this section, we turn our attention to systems at the mesoscale, with particular emphasis on the melting characteristics and mechanical properties of polymer-grafted bilayer membranes. The lamellar phase consists of alternating bilayer stacks with intervening water, as illustrated in Figure 6.10. The bilayers are composed of amphiphilic molecules, and bilayers composed of lipids assume special significance because of their biological relevance. Polymer-grafted membranes are formed when polymers are chemically grafted to the head group of the lipid molecule. Polymer grafting is usually carried out to modify the interfacial properties of the membrane and have applications (Lasic and Needham 1995; Tanaka and Sackmann 2005) in the stabilization of liposomes designed for targeted drug delivery, synthesis of supported bilayers for biomaterial applications, surface modification of implanted medical devices to prevent biological fouling, and stabilization of colloids. Grafted polymers are found to sterically stabilize liposomes and provide protection against attack from the immune system, thereby increasing their *in vivo* lifetimes (Lasic and Needham 1995; Marsh et al. 2003). Polymers play an important role in the synthesis of supported membranes where the bilayer rests on a polymer cushion tethered to a solid substrate (Tanaka and Sackmann 2005). These polymer-supported membranes have shown

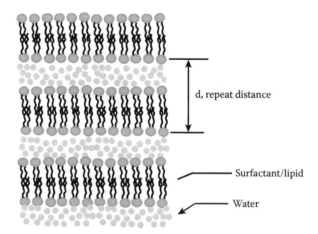

FIGURE 6.10   Schematic of the lamellar phase, which consists of stacks of bilayers with intervening water. The lamellar phase is characterized by the repeat distance, d, as illustrated above.

remarkable stability in air and have the potential to be used for *in vitro* biosensor applications. A commonly used polymer in grafting studies is polyethylene glycol (PEG), which is hydrophilic and biocompatible. At low grafting densities, polymers retain their mushroom configuration, and at high grafting densities the polymers extend away from the surface, giving rise to the brush regime in good solvent conditions. In addition to the steric interactions, grafted polymer brushes impart lateral interactions, which are of fundamental importance in membrane biophysics. Experimental and theoretical studies on the changes in the structural and mechanical properties of the lipid bilayers in the presence of grafted polymers have been reviewed recently (Tribet and Vial 2008). In general, the presence of grafted polymer decreases the main gel to liquid-crystalline transition temperature and increases the bending rigidity at low grafting fractions. In a spin-labeled electron spin resonance (ESR) study (Montesano et al. 2001), the pretransition temperature associated with the rippled phase was found to increase with grafting density up to a threshold grafting fraction and signatures of a pretransition could not be detected above a threshold grafting fraction. X-ray diffraction and micropipette experiments have also been performed to study the influence of grafted PEG on the lipid–cholesterol bilayer system (Needham et al. 1992). At low PEG concentrations, the presence of PEG in the brush regime increases the interbilayer repulsion (Needham et al. 1992). Using small-angle x-ray and neutron scattering, thinning and lateral stretching of polymer-grafted bilayer is observed with increasing polymer concentration (Castro-Roman and Ligoure 2001). Theories for polymer-grafted membranes are primarily based on the early mean field theories developed for polymers grafted on a hard surface (Alexander 1977; de Gennes 1980). Marsh et al. (2003) reviewed the extension of these theories to polymer-grafted lipid membranes and included the predictions based on self-consistent mean field theories (Milner and Witten 1988). The scaling relations for polymer brush height, area expansion, shift in chain-melting transition, and elastic constants for polymers in the brush regime as a function of the grafting density and chain length are discussed. Models for the intermembrane interactions, both in brush and mushroom regime and the bilayer micelle transition are found to agree well with available experimental data. In an earlier work, the influence of grafted polymers on membrane elastic constants was studied (Hristova and Needham 1994).

Bilayer systems with grafted polymers represent a mesoscopic system, and an appropriate simulation technique must be employed to study these systems. A successful mesoscopic simulation method bridges the connection between relevant microscopic and macroscopic phenomena, making it possible to observe mesoscopic system properties in a computationally efficient manner. In this regard, dissipative particle dynamics (DPD) has evolved as a promising method and has been applied to study a variety of systems (Ayappa et al. 2007). Although DPD has been widely used to study the behavior of the lamellar phase, the influence of grafted polymers on the structure and in particular the melting transition of polymer-grafted bilayers has only been recently studied (Thakkar and Ayappa 2010a).

6.4.1 DISSIPATIVE PARTICLE DYNAMICS

Dissipative Particle Dynamics (DPD) is a particle-based mesoscopic simulation method. The basic strategy is to modify the Langevin method, leading to a momentum-conserving algorithm and thereby producing correct hydrodynamics. DPD was formulated (Hoogerbrugge and Koelman 1992; Koelman and Hoogerbrugge 1993) primarily to study hydrodynamics. In a DPD simulation, particles interact through conservative, dissipative, and random forces, and similar to conventional MD the forces are central and pairwise additive in nature. This simple construction makes DPD more versatile and easily extendable compared to other methods. The relationship of the original DPD algorithm and its equivalent Fokker–Planck representation was established by Español and Warren (1995) to derive a fluctuation dissipation theorem relating the random and dissipative forces. Unlike other mesoscopic simulation techniques, such as lattice gas and lattice Boltzmann simulations, DPD is performed in continuous space and time, yielding an algorithm that conserves momentum and mass in a constant temperature ensemble.

In a DPD simulation (Español and Warren 1995; Groot and Warren 1997), the interactions are assumed to be pairwise additive, much like a standard MD simulation. However, in addition to the conservative force, dissipative and random forces are also present. The force acting between a pair of particles i and j is

$$\mathbf{F}_{ij} = \mathbf{F}_{ij}^C + \mathbf{F}_{ij}^D + \mathbf{F}_{ij}^R \qquad (6.14)$$

where the dissipative force is

$$\mathbf{F}_{ij}^D = -\gamma w_D(r_{ij})[\mathbf{e}_{ij} \cdot \mathbf{v}_{ij}]\mathbf{e}_{ij} \qquad (6.15)$$

and the random force is

$$\mathbf{F}_{ij}^R = \sigma w_R(r_{ij})\mathbf{e}_{ij}\theta_{ij} \qquad (6.16)$$

In these equations $\mathbf{r}_{ij} = \mathbf{r}_i - \mathbf{r}_j$ is the distance between two particles, $r_{ij} = |\mathbf{r}_{ij}|$. The unit vector, $\mathbf{e}_{ij} = \mathbf{r}_{ij}/r_{ij}$, and the relative velocity, $\mathbf{v}_{ij} = \mathbf{v}_i - \mathbf{v}_j$. γ and σ, are the strengths of dissipative and random forces, respectively, and θ_{ij} the Gaussian white noise term satisfies the following conditions:

$$\theta_{ij} = \theta_{ji}$$

$$\langle \theta_{ij}(t) \rangle = 0$$

and

$$\langle \theta_{ij}(t)\theta_{kl}(\tau) \rangle = (\delta_{ik}\delta_{jl} + \delta_{il}\delta_{jk})\delta(t - \tau) \qquad (6.17)$$

To preserve Boltzmann statistics in the canonical ensemble, the relationship between the weight functions, $w_D(r) = [w_R(r)]^2$ results in the following fluctuation–dissipation relation (Español and Warren 1995):

$$k_B T = \frac{\sigma^2}{2\gamma} \qquad (6.18)$$

where T is the temperature and k_B the Boltzmann constant. The fluctuation–dissipation relation, which expresses a balance between the strength of the dissipative and random forces, is derived from the equivalent representation of the dynamic equation as a Fokker–Planck equation.

Following the original prescription of the method, the weight functions in Equations 6.15 and 6.16 used are (Español and Warren 1995; Groot and Warren 1997)

$$w_R = \sqrt{w_D} = \begin{cases} 1-r, & r \leq 1 \\ 0, & r > 1 \end{cases} \tag{6.19}$$

where r is scaled by r_c, which is the cutoff radius for the interaction and defines an appropriate length scale for the DPD simulation. The conservative force between DPD beads is soft repulsive, and

$$\mathbf{F}_{ij}^C = \begin{cases} a_{ij}(1-r)\mathbf{e}_{ij}, & r \leq 1 \\ 0, & r > 1 \end{cases} \tag{6.20}$$

The corresponding potential to $\mathbf{F}_{ij}^C$ is

$$u(r_{ij}) = \begin{cases} \dfrac{a_{ij}}{2}(1-r)^2, & r \leq 1 \\ 0, & r > 1 \end{cases} \tag{6.21}$$

In this formulation and in the rest of this chapter all units are reduced with respect to the energy $\varepsilon = 1$ and size $r_c = 1$.

DPD does not restrict the form of the conservative force. Hence, it is easy to include various bonded and nonbonded interactions as conservative force for complex molecules. For the case of the chain-type molecule (e.g., polymer), a simple harmonic potential is used:

$$U_{\text{bond}} = \frac{k_s}{2}(r_{ij} - r_0)^2 \tag{6.22}$$

where k_s is the spring constant and r_0 denotes the equilibrium distance between the two beads. To control the chain stiffness, the bond bending potential introduced between two adjacent bonds (connecting three consecutive beads) is

$$U_\theta = \frac{k_\theta}{2}(\theta - \theta_0)^2 \tag{6.23}$$

where k_θ is the bending spring constant and θ_0 is the equilibrium angle between two adjacent bonds.

One of the central issues concerning the applicability of DPD to real systems lies in obtaining appropriate parameters. To relate the DPD system to the system of interest, the parameters a_{ij}, r_c, and system density (which determines the level of coarse graining) have to be specified. Groot and Warren (1997) established a method to estimate the repulsion parameter, a_{ij} (Equation 6.23), using appropriate thermodynamic properties. In the case of a pure fluid the repulsion parameter is determined from the compressibility of the system of interest. This involves generating the pressure–density relationship for the coarse-grained system evolving in a DPD framework, from which the compressibility is obtained. While fixing the repulsion parameter, the density of the DPD system, which is a free parameter, has to be judiciously chosen (Keaveny et al. 2005; Venturoli et al. 2005). In the case of a binary (AB) liquid interface the repulsion parameters (a_{AB}) are related to the Flory–Huggins parameter of the solute. The variation of the Flory–Huggins and repulsion parameters as a function of bead size, which determines the level of coarse graining, has also been investigated with the aim of computing interfacial tensions in a phase-separated binary mixture (Maiti and McGrother 2004).

The parameter σ in Equation 6.16 determines the strength of the random force and γ, the strength of the dissipative force, is related to σ (Equation 6.18). σ is usually fixed with trade-off between a sufficiently large time step, and accuracy as measured by the ability of the simulation to maintain a constant desired temperature. Higher values of σ lead to an artificial temperature rise. Because the dissipative force depends on the velocity, the numerical integration scheme requires some attention.

The present study focuses on studying the influence of grafted polymers on the lipid bilayer membrane. It is generally accepted that the tensionless state corresponds to the minimum free energy state of an infinite bilayer. A tensionless state can be achieved if the corresponding area per head group, a_h, for the membrane is known. This information can be obtained by carrying out systematic simulations at different a_h values and determining the a_h corresponding to zero tension.

In the present study, a modified Andersen barostat is proposed and tested to achieve the tensionless state. The implementation of the Andersen barostat (Anderson 1980) within the framework of a DPD simulation has been recently demonstrated for bulk fluids (Trofimov et al. 2005). The Andersen barostat is an extended system in which one solves additional equations for the volume fluctuations. Standard equations for volume fluctuations are

$$\dot{V} = \Pi / M \tag{6.24}$$

and

$$\dot{\Pi} = p(t) - p_{ext} \tag{6.25}$$

where V is the volume and Π is the momentum conjugate of the volume. M is the mass of the piston associated with the barostat, $p(t)$ is the instantaneous pressure, and p_{ext} is the specified external pressure. Solving these equations with the equations of motion for the particles yields the constant pressure ensemble. However, solution of the barostat equations along with the DPD equations was found to lead to decaying volume oscillations during the equilibration phase (Trofimov et al. 2005). To minimize these oscillations, an additional damping term was added to the equation for the momentum conjugate of volume (Equation 6.28), resulting in

$$\dot{\Pi} = p(t) - p_{ext} - \Gamma\Pi \tag{6.26}$$

The interfacial tension across the bilayer is

$$\gamma = \frac{1}{2}\int_{-\frac{L}{2}}^{\frac{L}{2}} [p_N(z) - p_T(z)]dz \tag{6.27}$$

where the normal pressure

$$p_N(z) = p_{zz}(z) \tag{6.28}$$

and the tangential pressure

$$p_T(z) = \frac{p_{xx}(z) + p_{yy}(z)}{2} \tag{6.29}$$

The factor of half in Equation 6.30 is due to the presence of two water–lipid interfaces. Equation 6.30 can be rewritten as

$$\gamma = \frac{L_z}{2}(\bar{p}_N - \bar{p}_T) \tag{6.30}$$

where the overbar denotes the average values of the pressure tensor component. We drop the overbar for further discussion. The equations of motion for the extended system are

$$\dot{L}_\alpha = \Pi_\alpha / M \tag{6.31}$$

and

$$\dot{\Pi}_\alpha = p_{\alpha\alpha}(t) - p_{\alpha\alpha,e} - \Gamma\Pi_\alpha \tag{6.32}$$

where $\alpha = x, y, z$. L is the dimension of the simulation box in the α direction, Π_α is the associated momentum coordinate, and M is the piston mass. In order to maintain a constant interfacial tension, we specify the tangential pressure components $p_{xx,e}$ and $p_{yy,e}$. During the simulation $p_{zz,e}$ is evaluated using

$$p_{zz,e} = \frac{2}{L_z}\gamma_e + \frac{p_{xx} + p_{yy}}{2} \tag{6.33}$$

where p_{xx} and p_{yy} are the instantaneous pressures and $\gamma_e = 0$ to achieve a tensionless state. While implementing the barostat, the tangential pressure components, the piston mass, and the dissipation strength were chosen as $M = 0.01$ and $\Gamma = 5.0$, respectively. We tested this implementation of the barostat for a bulk isotropic fluid consisting of 3000 water beads at various pressures. The resulting pressure–density relationship is shown in Figure 6.11. The results are in excellent agreement with that obtained from the canonical ensemble, NVT (N, number of particles, V, volume and T, temperature) simulations (Groot and Warren 1997).

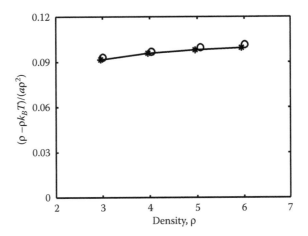

FIGURE 6.11   The excess pressure divided by $a\rho^2$ as a function of density $a_{ij} = 25.0$. $*$ indicates the data from our simulation with barostat, whereas $\circ$ indicates the NVT simulation data. (Taken from Groot, R. D. and Warren, P. B. J. *Chem. Phys.* 107: 4423, 1997.)

TABLE 6.3

Repulsion Parameter a_{ij} of Various Beads*

a_{ij}	W	H	T	A	B
W	25.0	15.0	80.0	25.0	15.0
H	15.0	15.0	80.0	−15.0	25.0
T	80.0	80.0	25.0	80.0	80.0
A	25.0	−15.0	80.0	80.0	80.0
B	15.0	25.0	80.0	80.0	80.0

$*$ W represents the water bead. H and T represent the hydrophilic head and hydrophobic tail bead of the lipid, respectively. A and B represent the polymer beads. The lipid parameters are similar to those used in earlier studies without polymer grafting Kranenburg, M., Venturoli, M., and Smit, B., *J. Phys. Chem. B,* 107: 11491, 2003.

6.4.2   MELTING TRANSITION OF POLYMER-GRAFTED BILAYERS

The DPD simulations were carried out with 800 lipid molecules and 54,224 total particles in a tensionless state. The repulsion parameters used for the results reported here are shown in Table 6.3, and the bonded interaction parameters are shown in Table 6.4. Because the area per head group, a_h, changes as a function of temperature, the amount of polymer in the system is measured in terms of the grafting fraction,

$$G_f = \frac{\text{Number of polymer molecules}}{\text{Number of lipid molecules}} \tag{6.34}$$

The grafting fraction was varied from 0 to 0.16 for studying the melting transition. A preassembled bilayer, with an equal amount of polymer on both the leaflets was chosen as the initial configuration for all simulations, and the first bead of the polymer chain is chemically grafted to the

TABLE 6.4

Bonded Interaction Parameters Used for Polymer-Grafted Membranes

Constants	Lipid	Polymer
k_s	100.0	20.0
r_o	0.7	0.5
k_θ	10.0	—
θ_o	180°	—

head groups of the lipid using a harmonic potential. In all studies, the polymer chain consisted of 20 beads and the lipids consist of 1 head bead and 6 tail beads (HT_6). The equations of motion are integrated using the modified velocity–Verlet algorithm (Groot and Warren 1997) with $\Delta t = 0.01$. Higher value of time step is avoided to prevent artifacts in the pressure profile across the bilayers (Jakobsen et al. 2005).

To determine the structure of the bilayer, the orientational order parameter was defined as

$$s = \frac{1}{2}\langle 3\cos^2\theta - 1\rangle \tag{6.35}$$

where θ is the angle between the vector obtained by joining the first and last bead of the lipid with the bilayer normal. The order parameter has the value 1 if all the lipids are aligned along the bilayer normal. It has a value of –0.5 if the lipids are aligned perpendicularly to the bilayer normal. The value 0 indicates a completely random orientation of the lipid tails. The order parameter defined earlier can be related to the experimentally measured order parameter using NMR spectroscopy with deuterium substitution. The area per head group is the total ensemble average x-y plane area divided by the number of lipids in a leaflet.

To compute the density profile, the computation cell was divided into 200 bins parallel to the bilayer surface. The density of a particular species i is

$$\rho_i(z) = \left\langle \frac{N_i(z + \Delta z/2, z - \Delta z/2)}{A\Delta z} \right\rangle \tag{6.36}$$

where N_i is the number of particles of species i, A is the area of the bin of thickness z, and the brackets denote a time average. The density profiles for each bead in the lipid, polymer, and solvent are computed.

The influence of polymer grafting on the melting transition of lipid bilayers is illustrated in Figure 6.12, in which the variations of the order parameter, s, membrane thickness d, and area per head group, a_h, are plotted as a function of reduced temperature. The order parameter is a sensitive measure of the degree of order within the lipid tail beads. Because the melting of bilayer membranes is predominantly driven by chain melting, the tilt order parameter, s, is an appropriate indicator of the melting transition. At low temperatures, the order parameter is close to unity, indicating the formation of the L_β phase, which is characterized by the lipid tails oriented parallel to the bilayer normal. This low temperature phase is illustrated in the snapshots at two different grafting fractions (Figures 6.13 and 6.14). As the temperature is raised, the increased thermal energy increases the entropy in the chains, and this has two major consequences on the integrity of the membrane. First, the increased entropy leads to an increase in the area per head group, a_h, as illustrated in Figure 6.12b. To keep the microstructure within the realm of the lamellar phase, the membrane thickness, d, decreases to prevent water from penetrating the bilayer phase. Thus melting is accompanied by

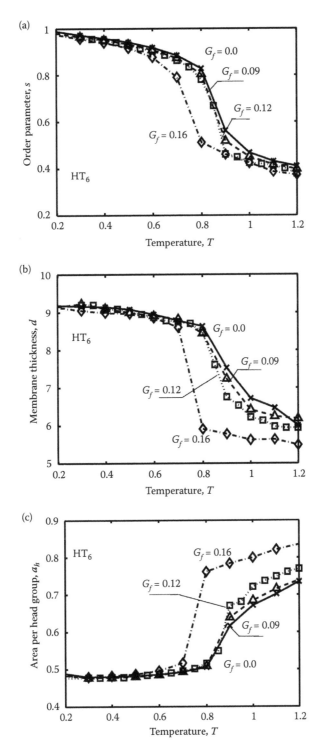

FIGURE 6.12 Order parameter, membrane thickness, and area per head group for HT_6 membranes as a function of temperature. Increasing the grafting fraction, G_f to 0.16 is seen to lower the transition temperature of the membrane. At a low grafting fraction of 0.09, the melting transition is unaffected compared with the polymer free membrane ($G_f = 0.0$). (Thakkar, F. M., and Ayappa, K. G., *Biomicrofluidics*, 4, 032203, 2010b. Copyright 2010 American Institute of Physics (AIP).)

(a)   (b)

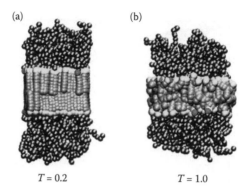

$T = 0.2$ $T = 1.0$

FIGURE 6.13 Snapshots for HT_6 with $G_f = 0.09$ illustrating the L_β phase at $T = 0.2$ and the L_α phase at $T = 1.0$. In the L_β phase the tails are oriented along the bilayer normal. The decrease in the membrane thickness is clearly observed across the melting transition. Color key: Pink, lipid tails; yellow, lipid heads; green, grafted polymer; blue, grafted polymer bead attached to the lipid head group. For clarity the solvent beads are excluded.

(a)   (b)

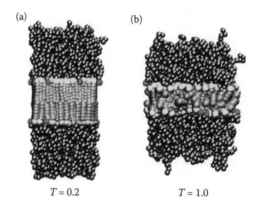

$T = 0.2$ $T = 1.0$

FIGURE 6.14 Snapshots for HT_6 with $G_f = 0.16$ illustrating the L_β phase at $T = 0.2$ and the L_α phase at $T = 1.0$. In the L_β phase the tails are oriented along the bilayer normal. The decrease in the membrane thickness when compared with the snapshot at $G_f = 0.09$ (Figure 6.13) is clearly observed. Color key: Pink, lipid tails; yellow, lipid heads; green, grafted polymer; blue, grafted polymer bead attached to the lipid head group. For clarity the solvent beads are excluded.

an increased a_h and a decrease in d. As the grafting fraction is increased these qualitative features remain; however, the melting shifts to lower temperatures as the polymer exerts a lateral pressure on the membrane, and the a_h changes occur at smaller temperatures. At the lower values of $G_f = 0.09$, the changes are minor and the melting characteristics are unaltered compared with the membrane in the absence of polymer ($G_f = 0.0$). At $G_f = 0.16$ the changes are quite significant. The higher temperature structures correspond to the L_α phase and are illustrated in the snapshots of Figures 6.13 and 6.14. The greater reduction in membrane thickness for the higher temperature L_α phase at the higher G_f can also be clearly observed. The changes in the membrane thickness and area per head group can also be rationalized based on arguments of excluded volume (Israelachvili 1985). The lamellar phase is found to form when the ratio $0.5 < (\eta = a_h l/V) < 1$, where l is the length and V the volume associated with the hydrophobic tail of the lipid molecule. Thus η represents a match between the projected head group area, a_h, and the projected area of the hydrocarbon tail. Because we are dealing with single-tail lipids, this ratio is expected to be closer to unity. Thus, if the volume V does not change appreciably across the transition, an increase in a_h must be accompanied by a decrease in l, as reflected in the decrease in membrane thickness d. Assuming a value of $\eta = 1$, the value of V evaluated in the L_β and L_α phase from the corresponding a_h and d data is invariant at a value of about 2.3. This indicates that changes in the a_h are predominantly compensated by changes in the membrane thickness d across the transition.

6.4.3 ELASTIC PROPERTIES OF POLYMER-GRAFTED BILAYERS

The mechanical properties of a lipid bilayer can be described by the Helfrich theory (Safran 2003), which treats the bilayer as a smooth, undulating surface. The free energy F of a lipid bilayer is given by

$$F = \int \left(\gamma + \frac{\kappa}{2}(c_1 + c_2 - c_0)^2 + \bar{\kappa}c_1c_2 \right) dA \tag{6.37}$$

where γ is the interfacial tension, κ is the bending modulus, and $\bar{\kappa}$ is the saddle-splay modulus. c_1 and c_2 represent the instantaneous principal curvature, and c_0 is the spontaneous curvature. Spontaneous curvature is the preferred curvature of the surface. For a flat symmetric bilayer the spontaneous curvature is zero.

The energy of a bilayer is given by Equation 6.40. Because the membrane topology is unchanged, integration over the Gaussian curvature is a constant, which means that the fluctuations of the energy are governed only by the bending modulus κ and the interfacial tension γ:

$$E = \int \gamma + \frac{\kappa}{2}(c_1 + c_2)^2 \, dA \tag{6.38}$$

The local deviation of the interface $h(x,y)$ is defined as

$$h(x, y) = z(x, y) - z_0 \tag{6.39}$$

where z_0 is the mean position of the interface. For small curvatures the principle curvatures can be written as a function of $h(x,y)$,

$$c_1 + c_2 = \nabla^2 h \tag{6.40}$$

and the elemental area,

$$dA = \sqrt{1 + (\nabla h)^2} \, dx dy. \tag{6.41}$$

Substituting Equations 6.40 and 6.41 into Equation 6.38 we get,

$$E = \int \left(\gamma + \frac{\kappa}{2}(\nabla^2 h)^2 \right) dx dy \sqrt{1 + (\nabla h)^2} \tag{6.42}$$

Expanding the square root term and neglecting higher order terms,

$$E = \int \left(\gamma + \frac{\kappa}{2}(\nabla^2 h)^2 \right) \left(1 + \frac{1}{2}(\nabla h)^2 \right) dx dy \tag{6.43}$$

Collecting terms and neglecting terms higher than quadratic,

$$E = \gamma A_p + \frac{\gamma}{2}(\nabla h)^2 A_p + \frac{\kappa}{2}(\nabla^2 h)^2 A_p \tag{6.44}$$

where A_p is the simulation box area. In this equation, the first term is the reference state energy, $E(0)$, which is independent of the curvature term. Expressing the energy as a deviation from the reference energy,

$$e(h(x,y)) = \frac{(E(h) - E(0))}{A_p} \tag{6.45}$$

$$e(h) = \frac{\gamma}{2}(\nabla h)^2 + \frac{\kappa}{2}(\nabla^2 h)^2 \tag{6.46}$$

The Fourier transform of the this equation yields

$$\tilde{e}(h(q)) = \frac{\gamma}{2}q^2 h(q) + \frac{\kappa}{2}q^4 h(q) \tag{6.47}$$

where $q = (2\pi / L)k$ and k is the wave number ($k = 1,2,3,\dots$). According to the equipartition theorem,

$$\langle \tilde{e}(h(q)) \rangle = \frac{1}{A_p} \frac{k_B T}{2} \tag{6.48}$$

Considering the ensemble average of Equation 6.47 and using Equation 6.48 yields (Rekvig et al. 2004) the intensity of Fourier modes,

$$\langle |h(q)|^2 \rangle = \frac{k_B T}{A_p} \frac{1}{\kappa q^4 + \gamma q^2} \tag{6.49}$$

Defining $S(q) = \langle |h(q)|^2 A \rangle$ the equation reduces to

$$S(q) = \frac{k_B T}{\kappa q^4 + \gamma q^2} \tag{6.50}$$

This equation indicates that the undulatory modes will dominate over stretching modes in the small q limit, provided γ is sufficiently small. For frequencies smaller than $q = q_c = \sqrt{\gamma/\kappa}$, the stretching modes become important. Hence, it is important to have a tensionless bilayer in order to be able to extract the bending modulus. For frequencies smaller than $q_o = 2\pi / d$, where d is the membrane thickness, the continuum theory is no longer valid and for zero tension membranes, protrusion modes begin to dominate over the undulatory modes (Marrink and Mark 2001). For evaluating the bending modulus, simulations were carried out with a preassembled bilayer consisting of 3200 lipid molecules in a tensionless state. All simulations were carried out with a total of 56,000 particles corresponding to an initial reduced particle density of 3.0. The simulation temperature was chosen as 1.5, ensuring that all the systems are in the L_α phase. Equilibration consisted of 50,000 steps, and particle coordinates were collected every 50 steps over the next 950,000 steps for computing time averaged properties.

While computing the bending modulus it is important to have an accurate representation of the surface. In the previous methods used in the literature, the surface representation contains fitting parameters that affect the mapping of the head group positions onto the surface $h(x,y)$. Although the roughness of the local surface at the level of the head group is not likely to influence the small q behavior, which dictates the bending modulus, it influences the large q behavior of the system. The high q modes are important while computing Bragg peaks (Fasolino et al. 2007) and in understanding the relaxation dynamics over the entire spectral range. In the present study, the surface $h(x,y)$ is represented using a Delaunay triangulation method, in which the position of each of the head group atoms is assigned to the vertices of a triangle. The method is simple to implement and does not involve any surface fitting parameters. For the purpose of completion, the procedure used for computing the bending modulus is outlined later and the schematic in Figure 6.15 illustrates the various quantities involved in determining the surface $h(x,y)$.

For every lipid molecule the tail-to-head vector for lipid, i, $\mathbf{r}_{th,i}$ is defined as the coordinates

$$\mathbf{r}_{th,i} = \mathbf{r}_{h,i} - \mathbf{r}_{t,i}$$

where $\mathbf{r}_{h,i}$ and $\mathbf{r}_{t,i}$ are the vector positions of the head and tail bead, respectively.

The dot product of the vector, $\mathbf{r}_{th,i}$, with the bilayer normal, $\mathbf{n}$, is used to determine whether the lipid molecule belongs to upper or lower leaflets denoted by L_u and L_l, respectively.

Hence

$$\mathbf{n}\cdot\mathbf{r}_{th,i} > 0 \Rightarrow i \in L_u$$

$$\mathbf{n}\cdot\mathbf{r}_{th,i} < 0 \Rightarrow i \in L_l$$

Delaunay triangulation was performed on sets L_u and L_l separately using the (x,y) coordinates of the head groups during this operation. Interpolation of this triangulated surface on the regular $n_x \times n_y$ grid was carried out to obtain the surfaces $z_u(x,y)$ and $z_l(x,y)$ corresponding to the sets L_u and L_l, respectively. The schematic shown in Figure 6.15 indicates the procedure by which the upper and lower leaflets are defined. Figure 6.16 shows the interfaces $z_u(x,y)$ and $z_l(x,y)$ obtained using the previously described procedure for a given configuration of lipid molecules.

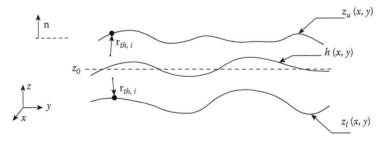

FIGURE 6.15 Schematic of the upper ($z_u(x, y)$) and lower ($z_l(x, y)$) surfaces. $\mathbf{r}_{th,i}$ is the tail-to-head vector for lipid i. Both leaflets are obtained by a dot product of $\mathbf{r}_{th,i}$ with the outward bilayer normal $\mathbf{n}$. The midplane surface $h(x, y)$ is obtained from upper and lower leaflets.

The surface $h(x,y)$ is obtained from the average of the two surfaces,

$$h(x,y) = \frac{z_u(x,y) + z_l(x,y)}{2} - z_0$$

where z_0 is the mean position of the interface for a given configuration defined as

$$z_0 = \frac{1}{N}\sum_{i=1}^{N} z_{h,i}$$

where N is the total number of lipid molecules in both leaflets (L_u and L_l) and $z_{h,i}$ is the z coordinate of the head group of the i-th lipid. Fast Fourier transforms were performed on $h(x,y)$ to get the spectral intensities $h_{\text{grid}}(q)$.

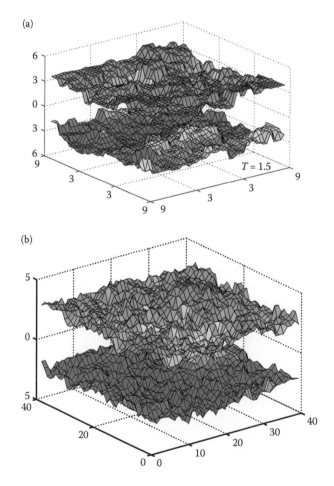

FIGURE 6.16   Configuration of the upper and lower leaflets for the HT_6 bilayer for (a) $G_f = 0.0$ and (b) $G_f = 0.2$. Both membranes are in the L_α phase at $T = 1.5$. The intensity of undulations is reduced in the presence of grafted polymer. This reduction in the intensity leads to a stiffening of the membrane and consequently an increase in the bending rigidity with grafted polymer.

The actual spectrum $h(q)$, is obtained by dividing the grid-based spectrum with the spectral damping factor (Cooke and Deserno 2005; Müller et al. 2006)

$$|h(q)|^2 = \frac{|h_{grid}(q)|^2}{sinc^2\left(\dfrac{i}{n_x}\right)sinc^2\left(\dfrac{j}{n_y}\right)}$$

where $sinc(x) = sin(\pi x)/(\pi x)$, i and j indicate the position of the point on the grid, and n_x and n_y are the total number of grid points in x and y directions, respectively. The origin of this factor lies in discretization artifacts and has been discussed by Cooke et al. (2005).

Time average of the magnitudes of these intensities along with the ensemble averaged box area, $\langle |h(q)|^2 A \rangle$, is used to obtain the bending modulus from Equation 6.50. As an illustration, the surfaces corresponding to the head groups of the upper and lower leaflets for HT_6 with $G_f = 0.2$ at $T = 1.5$ are shown in Figure 6.16.

Polymer grafting on the bilayer induces significant changes in the bilayer properties, and these changes on the transition temperature, lateral area expansion, and membrane thickness were discussed earlier in the text. Apart from various structural changes, the elastic properties also get altered because of the presence of grafted polymer. The presence of grafted polymer increases the area per head group and decreases the membrane thickness. Both these changes reduce the bending modulus of a given membrane. However, the polymer contribution increases the overall bending modulus of the membrane. Table 6.5 shows the bending modulus κ as a function of grafting fraction G_f for the HT_6 system, where κ is seen to increase with grafting fraction.

Figure 6.17 illustrates the density profiles for the HT_6 bilayer at various grafting fractions corresponding to a temperature of $T = 1.5$. As the grafting fraction increases, the degree of overlap of the tail beads is seen to increase in the core region of the bilayer, leading to a decrease in the membrane thickness and causing κ to decrease (Equation 6.51). The effect of polymers on the elastic properties of lipid membranes has been analyzed using mean field and scaling theories (Marsh 2001; Milner and Witten 1988). The presence of polymer increases the area per head group, a_h, which decreases the area stretch modulus k_A, and the membrane thickness, d. The combined effect of the polymer on the bending modulus of the membrane can be expressed as

---

TABLE 6.5

The Bending Modulus and Other Structural Properties as a Function of Grafting Fraction (G_f) for the HT_6 Lipid Bilayer*

G_f	κ	d	a_h	$\dfrac{\Delta\kappa}{\kappa_0}$	$\Delta\kappa_I$	$\Delta\kappa_{II}$
0.0	2.35	5.91	0.7704	0.0	0.0	0.0
0.04	3.43	5.87	0.7709	0.46	−0.013	0.47
0.09	5.36	5.76	0.7953	1.28	−0.049	1.33
0.16	5.37	5.40	0.8824	1.91	−0.164	2.00
0.20	7.99	5.11	0.9475	2.40	−0.252	2.65

*d is the membrane thickness, a_h is the area per head group. $\Delta\kappa = \kappa - \kappa_0$, where κ_0 is the bending modulus in the absence of polymer ($G_f = 0$). $\Delta\kappa_I$ and $\Delta\kappa_{II}$ represent contributions to changes in the bending modulus arising from membrane thinning and contributions from the polymer, respectively (Equation 6.53).

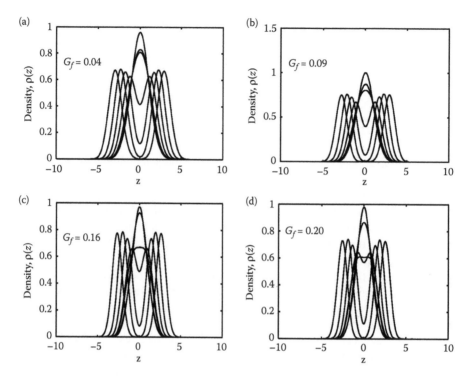

FIGURE 6.17   Density profiles of HT_6 with various values of grafting fraction G_f. With increase in the grafting fraction, the overlap in the core region increases and the membrane thickness decreases.

$$\kappa = \frac{k_A(G_f, a_h)}{\alpha} d^2 + \beta \left(\frac{G_f}{a_h} \right)^p \qquad (6.51)$$

In this expression, the area stretch modulus, k_A, is a function of both G_f and a_h, d is the membrane thickness, α is a proportionality constant, and $p = 5/2$ in the mean field theory and $7/3$ in the self-consistent field theory. The temperature, polymer length, and monomer size are held fixed in our simulations, so β is a constant for a given value of p. The first term on the right-hand side (RHS) of Equation 6.54 represents the influence of the bending modulus from a continuum analysis (Goetz et al. 1999). An increase in G_f increases k_A because of the elastic contribution from the brush alone and the increase in a_h as a result of the presence of the polymer lowers k_A. These opposing effects have been shown to partially compensate each other to some extent, resulting in a weak dependence of k_A because of the grafted polymer (Marsh 2001). The changes in k_A further diminish when the degree of polymerization is small and G_f is low; as is the case for our simulations. In what follows, we are only interested in estimating the relative contributions from each of the two terms on the RHS of Equation 6.54. for convenience of notation we set $k_A = \epsilon k_{A0}$, where k_{A0} is the bare membrane expansion modulus, and $0 < \epsilon$ represents the change resulting from grafting. For the low values of G_f and degree of polymerization, $n_p = 20$ is used in our study, ϵ is estimated to range between 0.8 and 1.2 from the analysis for monolayers and bilayers by Marsh (2001). Noting that the bare membrane modulus,

$$\kappa_0 = k_{A0} d_0^2 / \alpha \qquad (6.52)$$

Equation 6.54 can be rewritten as

$$\frac{\Delta\kappa}{\kappa_0} = \left(\frac{\epsilon d^2}{d_0^2} - 1\right) + \frac{\beta}{\kappa_0}\left(\frac{G_f}{a_h}\right)^p \tag{6.53}$$

where $\Delta\kappa = \kappa - \kappa_0$, and the two terms on the RHS of Equation 6.56 will henceforth be referred to as $\Delta\kappa_I$ and $\Delta\kappa_{II}$, respectively. Because we have evaluated the membrane thickness d as a function of G_f, Equation 6.56 can be used to estimate the relative contribution to changes in κ from changes in d by explicitly evaluating $\Delta\kappa_I$. This term is always negative when $0 < \epsilon \leq 1$ and corresponds to a net decrease in the bending modulus contribution from $\Delta\kappa_I$, because of both a decrease in d and a decrease in k_A over the polymer-free bilayer. If $\epsilon > 1$, the first term can be either positive or negative depending on the ratio of d/d_0. Because $\Delta\kappa/\kappa_0$ (Equation 6.56) is known from the simulation, $\Delta\kappa_{II}$, which represents the effect of polymer contributions explicitly to the changes in the bending modulus, can be obtained as a difference from Equation 6.53.

Using the values of d and the values of $\Delta\kappa/\kappa_0$ obtained from the Fourier analysis, we evaluated the contribution from $\Delta\kappa_I$ and $\Delta\kappa_{II}$. In the first part of the analysis, we set $\epsilon = 1.0$. In this case, membrane thinning by a reduction in d has a negative contribution to $\Delta\kappa$ and opposes the increase in κ from the grafted polymer, as reflected in $\Delta\kappa_{II}$ in Equation 6.53. Table 6.5 lists the contributions from $\Delta\kappa_I$ and $\Delta\kappa_{II}$ to the overall change in the bending modulus as a function of G_f. The contributions from changes in d lower the overall bending modulus of the membrane as the grafting fraction is increased. The polymer contribution to the relative change in the bending modulus is reflected in $\Delta\kappa_{II}$. The results indicate that the dominant contribution to changes in κ is from the grafted polymer. As one would expect from physical considerations, the largest contributions from changes in d are observed at the highest values of G_f. Hence for moderate grafting densities, the mechanical properties of the grafted membrane are only weakly influenced by changes induced in the membrane itself and the dominant contributions arise from the presence of polymer. Note that this effect would be further weighted toward the polymer as the length of the polymer chain is increased provided ϵ remained close to unity. In this analysis, we assumed that the stretch modulus is relatively unaffected by the presence of the polymer, resulting in $\epsilon = 1$. We examined this assumption (Thakkar and Ayappa 2010b) in a recent study and found that including variations in k_A because of the presence of the polymer did not change the qualitative trends in the contributions from $\Delta\kappa_I$ and $\Delta\kappa_{II}$ toward the overall change in the bending modulus.

6.5 CONCLUSION

Using two examples of commonly found microstructure in water–surfactant systems, we illustrated the role that molecular simulations play in developing our understanding of these complex systems. In the case of reverse micelles, a fully atomistic model is required to elucidate the nature of solvation within the aqueous regions and the effects associated with ion-exchange that is observed in these systems. Confinement within the reverse micelle has a strong influence on the PMF associated with the hydrophobic effect. As a result of the lack of sufficient water, only minimum contact is formed between two methane particles confined within the reverse micelle. In the latter part of the chapter, we studied the influence of polymer grafting on the melting behavior and the mechanical properties of lipid bilayers. In contrast to the detailed atomistic models used for the reverse micelle study, a coarse-grained description based on DPD is used for the polymer-grafted bilayers. In this description, discrete atoms are replaced with beads, with each bead typically representing 2 or 3 atomic units. This permits studying systems at larger length scales, which are particularly significant while evaluating mechanical properties such as the bending modulus. Although favorable from this point of view, coarse-grained methods must be systematically parameterized to make quantitative comparison with experiments. We finally point out that the choice of a given method is decided on the level of information required from a particular study. Hence, a coarse-grained description would be inappropriate to study phenomena observed in the reverse micelle, where atomistic level details are required to capture hydrogen bonding and ion solvation effects.

REFERENCES

Alder, B. J., and Wainwright, T. E. 1957. Phase transition for a hard sphere system. *J. Chem. Phys* 27: 1208.

Alexander, S. 1977. Adsorption of chain molecules with a polar head a scaling description. *J. Physique* 38: 983

Allen, M. P., and Tildesley, D. J. 1987. *Computer Simulation of Liquids.* Clarendon Press, Oxford.

Andersen, H. C. 1980. Molecular dynamics simulations at constant pressure and/or temperature. *J. Chem. Phys.* 72: 2384.

Ayappa, K. G., Malani, A., Patil, K., and Thakkar, F. 2007. Molecular simulations: Probing systems from the nanoscale to mesoscale. *J. Indian I. Sci.* 87: 35.

Bandyopadhyaya, R., Kumar, R., and Gandhi, K. S. 2000. Simulation of precipitation reactions in reverse micelles. *Langmuir* 16: 7139.

Bhattacharyya, K., and Bagchi, B. 2000. Slow dynamics of constrained water in complex geomentries. *J. Phys. Chem.* 104: 10603.

Bratko, D., Luzar, A., Chen, S. H. 1988. Electrostatic model for protein/reverse micelle complexation. *J. Chem. Phys.* 89: 545.

Castro-Roman, F., and Ligoure, C. 2001. Lateral stretching of fluid membranes by grafted polymer brushes. *Europhys. Lett.* 53, 483–489.

Cooke, I. R., and Deserno, M. 2005. Solvent-free model for self-assembling fluid bilayer membranes: Stabilization of the fluid phase based on broad attractive tail potentials. *J. Chem. Phys.* 123: 224710.

de Gennes, P. G. 1980. Conformations of polymers attached to an interface. *Macromolecules* 13: 1069.

Español, P., and Warren, P. 1995. Statistical mechanics of dissipative particle dynamics. *Europhys. Lett.* 30: 191.

Faeder, J., and Ladanyi, B. M. 2000. Molecular dynamics simulations of the interior of aqueous reverse micelles. *J. Phys. Chem. B* 104: 1033.

Fasolino, A., Los, J., and Katsnelson, M. 2007. Intrinsic ripples in graphene. *Nat. Mater. (UK)* 6: 858.

Goetz, R., Gompper, G., and Lipowsky, R. 1999. Mobility and elasticity of self-assembled membranes. *Phys. Rev. Lett.* 82: 221.

Groot, R. D., and Warren, P. B. 1997. Dissipative particle dynamics: Bridging the gap between atomistic and mesoscopic simulation. *J. Chem. Phys.* 107: 4423.

Hoogerbrugge, P. J., and Koelman, J. M. V. A. 1992. Simulating microscopic hydrodynamic phenomena with dissipative particle dynamics. *Europhys. Lett.* 19: 155.

Hristova, K., and Needham, D. 1994. The influence of polymer-grafted lipids on the physical properties of lipid bilayers: A theoretical study. *J. Colloid. Int. Sci.* 168: 302.

Israelachvili, J. N. 1985. *Intermolecular and Surface Forces: With Applications to Colloidal and Biological Systems.* Academic, London.

Jakobsen, A. F., Mouritsen, O. G., and Besold, G. 2005. Artifacts in dynamical simulations of coarse-grained model lipid bilayers. *J. Chem. Phys.* 122: 204901.

Keaveny, E. E., Pivkin, I. V., Maxey, M., and Karniadakis, G. E. 2005. A comparative study between dissipative particle dynamics and molecular dynamics for simple-and complex-geometry flows. *J. Chem. Phys.* 123: 104107.

Koelman, J., and Hoogerbrugge, P. 1993. Dynamic simulations of hard sphere suspensions under steady shear. *Europhys. Lett.* 21: 363.

Kranenburg, M., Venturoli, M., and Smit, B. 2003. Phase behavior and induced interdigitation in bilayers studied with dissipative particle dynamics. *J. Phys. Chem. B* 107: 11491.

Lasic, D. D., and Needham, D. 1995. The "stealth" liposome: A prototypical biomaterial. *Chem. Rev.* 95: 2601.

Lee, S. H., and Rasaiah, J. C. 1996. Molecular dynamics simulation of ion mobility. 2. Alkali metal and halide ions using SPC/E model for water at 250C. *J. Phys. Chem.* 100: 1420.

Lynden-Bell, R. M., and Rasaiah, J. 1997. From hydrophobic to hydrophilic behaviour. *J. Chem. Phys.* 107: 1981.

Maiti, A., and McGrother, S. 2004. Bead–bead interaction parameters in dissipative particle dynamics: Relation to bead-size, solubility parameter, and surface tension. *J. Chem. Phys.* 120: 1594.

Malani, A., Ayappa, K., and Murad, S. 2006. Effect of confinement on the hydration and solubility of NaCl in water. *Chem. Phys. Lett.* 431: 88.

Marrink, S. J., and Mark, A. E. 2001. Effect of undulations on surface tension in simulated bilayers. *J. Phys. Chem. B* 105: 6122.

Marsh, D. 2001. Elastic constants of polymer-grafted lipid membranes. *Biophys. J.* 81: 2154.

Marsh, D., Bartucci, R., and Sportelli, L. 2003. Lipid membranes with grafted polymers: Physicochemical aspects. *Biochim. Biophys. Acta* 1615: 33.

Metropolis, N., Rosenbluth, A. W., Rosenbluth, M. N., and Teller, A. H. 1953. Equation of state calculations by fast computing machines. *J. Chem. Phys.* 21: 1087.

Milner, S., and Witten, T. 1988. Theory of grafted polymer brush. *Macromolecules* 21: 2610.

Montesano, G., Bartucci, R., Belsito, S., Marsh, D., and Sportelli, L. 2001. Lipid membrane expansion and micelle formation by polymer-grafted lipids: Scaling with polymer length studied by spin-label electron spin resonance. *Biophys. J.* 80: 1372.

Müller, M., Katsov, K., and Schick, M. 2006. Biological and synthetic membranes: What can be learned from a coarse-grained description? *Phys. Rep.* 434: 113.

Needham, D., McIntosh, T., and Lasic, D. 1992. Repulsive interactions and mechanical stability of polymer-grafted lipid membranes. *Biochim. Biophys. Acta* 1108: 40.

Pal, S., Vishal, G., Gandhi, K. S., and Ayappa, K. G. 2005. Ion-exchange in reverse micelles. *Langmuir* 21: 767.

Pangali, C., Rao, M., and Berne, B. J. 1979. A Monte Carlo simulation of the hydrophobic interaction. *J. Chem. Phys.* 71: 2975.

Rahman, A. 1964. Correlations in the motion of atoms in liquid argon. *Phys. Rev.* 136: A405.

Rao, P. V. G., Gandhi, K. S., and Ayappa, K. G. 2007. Enhancing the hydrophobic effect in confined water nanodrops. *Langmuir* 23: 12795.

Rekvig, L., Hafskjold, B., and Smit, B. 2004. Simulating the effect of surfactant structure on bending moduli of monolayers. *J. Chem. Phys.* 120: 4897.

Ryckaert, J.-P., Ciccotti, G., and Berendsen, H. 1977. Numerical integration of the Cartesian equations of motion of a system with constraints: Molecular dynamics of n-alkanes. *J. Comput. Phys.* 23: 327.

Safran, S. A. 2003. *Statistical Thermodynamics of Surfaces, Interfaces, and Membranes.* West-view, Boulder, CO.

Tanaka, H., and Koga, K. 2005. Formation of ice nanotube with hydrophobic guests inside carbon nanotube. *J. Chem. Phys.* 123: 094706.

Tanaka, M., and Sackmann, E. 2005. Polymer-supported membranes as models of the cell surface. *Nature* 437: 656.

Thakkar, F. M., and Ayappa, K. G. 2010a. Effect of polymer grafting on the bilayer gel to liquid-crystalline transition. *J. Phys. Chem. B* 114: 2738.

Thakkar, F. M., and Ayappa, K. G. 2010b. Investigations on the melting and bending modulus of polymer grafted bilayers using dissipative particle dynamics. *Biomicrofluidics* 4: 032203.

Tribet, C., and Vial, F. 2008. Flexible macromolecules attached to lipid bilayers: Impact on fluidity, curvature, permeability and stability of the membranes. *Soft Matter* 4: 68.

Trofimov, S. Y., Nies, E. L. F., and Michels, M. A. J. 2005. Constant pressure simulations with dissipative particle dynamics. *J. Chem. Phys.* 123: 144102.

Venturoli, M., Smit, B., and Sperotto, M. M. 2005. Simulation studies of protein-induced bilayer deformations, and lipid-induced protein tilting, on a mesoscopic model for lipid bilayers with embedded proteins. *Biophys. J.* 88: 1778.

Wallqvist, A. 1991. Molecular dynamics study of a hydrophobic aggregate in an aqueous solution of methane. *J. Phys. Chem.* 95: 8921.

Wood, W. W., and Parker, F. R. 1957. Monte Carlo equation of state of molecules interacting with the Lennard-Jones potential. I. A supercritical isotherm at about twice the critical temperature. *J. Chem. Phys.* 27: 720.

7 Molecular Simulation of Wetting Transitions on Novel Materials

Sandip Khan and Jayant K. Singh

CONTENTS

7.1 INTRODUCTION

Surface wettability plays an important role in nature and numerous industrial applications such as coatings, paintings, adhesives, microfluidic technology, microelectronics, textiles, and so on. Many examples of super-hydrophobicity are found in nature, especially in plants and insects. For example, lotus leaves (Barthlott and Neinhuis 1997) are super-hydrophobic because of their rough-surface microstructure and presence of wax coating. Self-cleaning occurs as water droplets remove surface particles as they roll off the leaves. Super-hydrophobicity also helps many plants and insects float on water because of the buoyancy force provided by the trapped air in their rough surface (Kim et al. 2010). Lady's mantle leaf, water strider feet, rice leaves, and fisher spiders are the most common examples present in the nature. Super-hydrophobicity mainly results from a combination of low surface energy and high surface roughness. Numerous methods to attain these two requirements have been reported experimentally (Heslot et al. 1990; Modaressi and

Garnier 2002; Ran et al. 2008; Shirtcliffe et al. 2004; Spori et al. 2008; Yoshimitsu et al. 2002) and computationally (Berim and Ruckenstein 2009; Brandon et al. 2003; Bulnes et al. 2007; Grest et al. 2006; Hirvi and Pakkanen 2007, 2008; Lundgren et al. 2007; Patankar 2003; Shi and Dhira 2008; Yaneva et al. 2004; Yang et al. 2008). Understanding of surface wettability can lead to novel materials for different purposes, for example, the super-oleophobic surface, anti-frosting surface, anti-fouling surface, and so forth.

Several approaches have been reported for combining materials of low surface energy with suitable surface roughness. Two methods are generally employed in experiment; one method is to roughen a normally smooth hydrophobic surface and another method is to roughen a hydrophilic surface followed by hydrophobic coating. These two techniques are well reviewed by Bhushan and Jung (2010). Different experimental fabrication techniques are used for creating micro and nano rough structures, such as lithography (photo, x-ray, and soft), etching (plasma, laser, and chemical), deposition (chemical vapor decomposition, adsorption, and self-assembly).

Molecular modeling is another attractive approach that can provide necessary insight to the phenomena near surfaces. In this chapter, we illustrate methods that are commonly used for the study of wettability on solid surfaces. We begin with the thermodynamics of liquid–solid interface in the next section followed by simulation techniques and some illustrative examples.

7.2 THERMODYNAMICS OF LIQUID DROP ON A SURFACE

Consider a drop on a surface, as illustrated in Figure 7.1. The angle between the tangent drawn at the contact line between the three phases (solid, liquid, and vapor) and the substrate surface is known as the *contact angle*. Under equilibrium, the contact angle depends on the surface and interfacial energies. The relation between contact angle and these interfacial tensions is given by the Young's equation:

$$\gamma_{sv} = \gamma_{sl} + \gamma_{lv}\cos\theta \tag{7.1}$$

where γ is the interfacial tension and subscript s, l, and v represent solid, liquid, and vapor, respectively. The contact angle of the droplet on the surface is represented by θ. Equation 7.1 relates the interfacial tensions across the contact line of the three-phase system.

Wetting behavior of a surface is characterized typically in terms of the contact angle. When the sum of the solid–liquid and liquid–vapor surface tension is equal to the solid–vapor surface tension, the three-phase contact line vanishes ($\theta = 0$) and liquid film spreads over the whole surface. This is known as *complete wetting state*, as shown in Figure 7.1b, and the corresponding temperature is the wetting temperature. Below wetting temperature, liquid may form a droplet on the surface, which is known as the *partially wetting state* (Figure 7.1a). The contact angle can vary from 0° to 180° according to the characteristic of the surface. When the contact angle is less than 90°, liquid prefers the surface, recognized as hydrophilic in nature. On the other hand, the contact angle is above 150° for super-hydrophobic surface. When the contact angle increases to 180° for which the sum of solid–vapor and the liquid–vapor is equal to the solid–liquid interfacial tension, the liquid droplet forms complete spherical shape on the surface. The situation is depicted in Figure 7.1c, and is known as the *complete drying state*.

(a)   (b)   (c)

FIGURE 7.1   Wetting of a liquid drop on a flat substrate: (a) partial wetting state, (b) complete wetting state, (c) complete drying state.

Young's equation did not account for the third-phase interaction in the vicinity of each interface that may change the interfacial energies. To correct this equation, line tension was introduced. Line tension is defined as the interfacial energy per unit length across the three-phase contact line. It is found that line tension is more important when radius of droplet is less than a micron (Marmur 1996), which can be found in microelectronic systems and microfluidic devices (Heine et al. 2005). Thus, modified Young's equation can be used for microscopic contact angle calculation, described as

$$\gamma_{sv} = \gamma_{sl} + \gamma_{lv}\cos\theta + \left(\frac{\tau}{r_B}\right)$$ (7.2)

where τ and r_B are line tension and radius of the droplet, respectively.

Macroscopic contact angle can be derived at a infinitely large drop, that is, $1/r_B \to 0$, which yields a well-known Young's equation (Equation 7.1):

$$\cos\theta_\infty = \left(\frac{\gamma_{sv} - \gamma_{sl}}{\gamma_{lv}}\right)$$ (7.3)

Equation (7.2) can be rewritten in terms of macroscopic contact angle and line tension, as shown:

$$\cos\theta = \cos\theta_\infty - \left(\frac{\tau}{\gamma_{lv}}\right)\frac{1}{r_B}$$ (7.4)

Line tension may be important for rough and heterogeneous surfaces. However, the previous relation is applicable for only smooth surfaces and needs further correction for accounting the surface roughness or heterogeneity. Line tension affects the stability of emulsion and foams and would play an important role in microfluidic and nanofluidic devices. As summarized by Amirfazli and Neumann (2004), line tension has been found to be negative as well as positive.

It is known that wetting is greatly influenced by surface roughness. Wenzel (1936) modified the Young's equation for rough surface as follows:

$$\cos\theta' = \frac{r(\gamma_{lv} - \gamma_{sl})}{\gamma_{lv}} = r\cos\theta$$ (7.5)

where r is the roughness factor, which was defined as the ratio of the actual area of the solid surface to the projected one. It is assumed that liquid completely penetrates the groves as shown in Figure 7.2a. This model is based on the equilibrium state of the droplet and does not include the contact angle hysteresis, but it has been observed that liquid droplets in Wenzel state undergo large contact angle hysteresis (Dettre and Johnson 1963). Contact angle hysteresis is defined as the difference between advancing and receding contact angles.

In contrast to the Wenzel model, if the droplet penetrates the grooves of the surface and air is trapped in between the droplet and the rough surface, the contact angle of the droplet is greatly

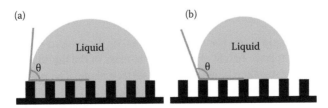

FIGURE 7.2 Wetting of a liquid drop on a textured surface. (a) Wenzel regime, (b) Cassie–Baxter regime.

influenced by the surface fraction, wetted by the liquid. Cassie and Baxter (1944) proposed an equation for such a heterogeneous system:

$$\cos\theta'' = rf\cos\theta' + f - 1 \tag{7.6}$$

where f is the fraction of surface wetted by liquid.

7.3   CONTACT ANGLE MEASUREMENT

Experimentally, contact angle is measured based on geometry of the droplet or the interfacial tension across the contact line using various techniques such as tilting plate, sessile bubble and drop, captive bubble, the Wilhelmy method, and so on. Similarly, in simulation, three methods are generally used to find the contact angle.

7.3.1   Fitting Method

MD simulations, in general, employ this technique to calculate the contact angle droplet of liquids on surfaces. In MD simulation, molecular trajectories are captured in the interval of 1 to 2 ps after equilibration period. The equilibration period may vary from 1 to 2 ns for simple systems such as water droplets on a smooth graphite surface. After equilibration period droplet stops spreading further on the surface and is usually static on the surface. However, in some cases droplets can move on the surface, such as water droplets on asymmetric surfaces (Halverson et al. 2008). From the MD simulation trajectories, fluid isochore profiles are obtained by introducing a cylindrical binning normal to the surface, keeping the center of mass of the droplet as a reference axis and horizontal binning parallel to the surface. Averages are taken over all configurations. All bins are assumed to be of equal volume. Generally, bins are considered to be of the same height in z direction and the surface area of horizontal bins is varied, to keep the volume fixed. To extract the contact angle from such a profile, a two-step procedure is adopted, as described by de Ruijter et al. (1999). From the isochore profile one can find the location of the equimolar dividing surface in the horizontal layer of the binned drop. Subsequently, a circular best fit through these points is extrapolated to the surface, where the contact angle θ is measured as shown in Figure 7.3. The boundary between equilibrated liquid and vapor interface for a given droplet is determined at the position where the density is half of bulk solvent density and is modeled using the following relation:

$$\rho(r) = \frac{1}{2}(\rho^L + \rho^V) - \frac{1}{2}(\rho^L - \rho^V)\tanh\left(\frac{2(r - r_e)}{d}\right) \tag{7.7}$$

where ρ^L and ρ^V are liquid and vapor densities, respectively, r is the distance from origin to the droplet surface, r_e is the center of the interface region, and d is the interface thickness. Isochore profile is considered from 3 to 4 molecular diameters from the surface to avoid the influence of density fluctuations at the liquid–solid interface.

7.3.2   Center of Mass Method

In this method (Fan and Cagin 1995; Hautman and Klein 1991), a microscopic contact angle is calculated by comparing the average height of the center of mass of the liquid cluster to that of an ideal sessile drop in the shape of a sphere intersecting the surface plane. The position of the sphere relative to the plane is determined by the center-of-mass position and the condition that the volume of the sphere in the half space above the surface plane contains the correct number of liquid molecules (assuming a uniform density in the idealized drop equal to that of bulk). The equation employed is

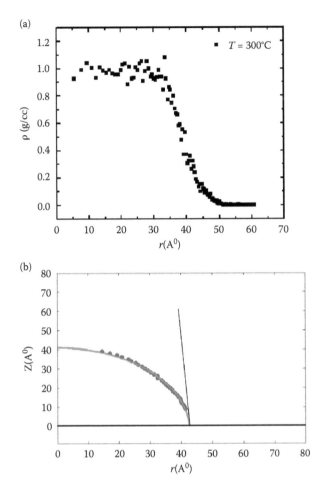

FIGURE 7.3   (a) Density profile of water droplet across x-axis from center of mass of droplet. (b) Schematic for the computation of contact angle for graphite–water system at 300 K.

$$\langle z_{cm}\rangle = (2)^{-\frac{4}{3}} R_0 \left(\frac{1-\cos\theta}{2+\cos\theta}\right)^{\frac{1}{3}} \left(\frac{3+\cos\theta}{2+\cos\theta}\right) \tag{7.8}$$

where θ is the contact angle; $\langle z_{cm}\rangle$, the average height of center of mass, is measured relative to the planar surface; and $R_0 = (3N/4\pi\rho_0)^{1/3}$ is the radius of a free spherical drop of N liquid molecules at uniform density ρ_0.

7.3.3   GC-TMMC APPROACH

We briefly describe the usage of grand canonical transition matrix Monte Carlo simulation (GC-TMMC) for obtaining the interfacial tension along the contact line. GC-TMMC simulations are conducted in a grand canonical ensemble at a constant chemical potential μ, volume V, and temperature T. Microstate probability in this ensemble is represented as

$$\Pi_s = \frac{1}{\Xi} \frac{V^{N_s}}{\Lambda^{3N_s} N_s!} \exp[-\beta(U_s - \mu N_s)] \tag{7.9}$$

where $\beta = 1/k_B T$ is the inverse temperature, k_B is the Boltzmann's constant, Ξ is the grand partition function, U_s is the interaction energies of particles of microstate s, and Λ is the de Broglie wavelength. GC-TMMC simulation stores the transition probability of one microstate to another microstate for every particle state that allows calculating the probability of single-particle state.

$$\Pi_N = \sum_{N_s=N} \Pi_s \qquad (7.10)$$

Once a probability distribution has been collected at a given value of chemical potential, histogram reweighting (Ferrenberg and Swendsen 1988) is used to shift the probability distribution to other values of the chemical potential. If coexistence chemical potential is not known *a priori,* it can be calculated by recursively applying histogram reweighting technique using the recorded distribution until probability distribution, Π_N^{coex}, is obtained such that areas under the vapor and liquid regions in the probability distribution are equal. Densities of phases are calculated from the first moment of Π_N^{coex} distribution. This approach is used to obtain vapor–liquid coexistence conditions. However, for obtaining the contact angle on a surface, two independent simulations are carried out. A separate simulation is conducted to obtain the bulk vapor–liquid surface tension. The other simulation is conducted at the coexistence bulk chemical potential but in the presence of the surface. Bulk simulation is employed to find out the liquid–vapor surface tension, γ_{lv}. The interfacial energy for a finite-size system with box length, L, is determined from the maximum likelihood in the liquid phase Π_{max}^{liquid}, thin film regions Π_{max}^{vapor}, and minimum likelihood in the interface region Π_{min}.

$$\beta F_L = \frac{1}{2}(\ln \Pi_{max}^{vapor} + \ln \Pi_{max}^{liquid}) - \ln \Pi_{min} \qquad (7.11)$$

Binder's formalism (1982) can be used to calculate true interfacial tension as follows:

$$\beta \gamma_L = \frac{\beta F_L}{2L^2} = C_1 \frac{1}{L^2} + C_2 \frac{\ln(L)}{L^2} + \beta \gamma \qquad (7.12)$$

where γ_L is the interfacial tension for a system of box length L, γ is the interfacial tension for infinite system, C_1 and C_2 are constants, and F_L represents the free energy of the vapor–liquid interface for a finite system size L.

In the presence of the surface in one end and hard wall on the other end (this is included mainly to enhance the simulation sampling and reduce the run length of the simulation) probability distribution is not bimodal anymore, as shown in Figure 7.4. The low density peak corresponds to the free energy associated with the solid–vapor interface, and the plateau corresponds to the free energy associated with solid–liquid and vapor–liquid interface in the saturated condition. The ratio of probability of these two particle states can be written as:

$$\beta S = -[\ln \Pi_{vap} - \ln \Pi_{plateau}] = \beta A[\gamma_{sv} - (\gamma_{sl} + \gamma_{lv})] \qquad (7.13)$$

where γ_{sl} and γ_{sv} are the interfacial tension of solid–liquid and solid–vapor, respectively, and S is the spreading coefficient. This equation can be combined with Young's equation to get the following:

$$\cos\theta = 1 + \frac{S}{A\gamma_{lv}} \qquad (7.14)$$

Errington et al. (2005) recently used this method to obtain the contact angle of Lennard-Jones (LJ) fluids on 9-3 wall. A typical probability distribution of LJ on 9-3 wall is shown in Figure 7.4.

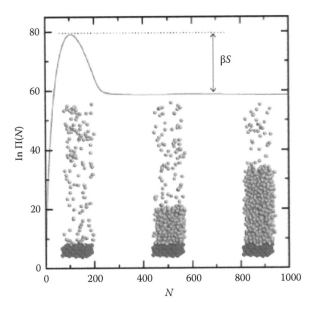

FIGURE 7.4 A typical particle number probability distribution of LJ fluid in the presence of a 9-3 surface generated from a one-wall simulation conducted at the bulk liquid–vapor saturation chemical potential and at a temperature below a wetting point. (Adapted from Grzelak, E. M. and Errington, J. R., *J. Chem. Phys.* 128: 014710, 2008. With permission.)

However, this method requires a separate simulation to obtain liquid–vapor surface tension values and is therefore computationally intensive. More description of this method can be found elsewhere (Grzelak and Errington 2008).

7.4 PHASE TRANSITION ON SURFACES

7.4.1 PREWETTING TRANSITION

Wetting transition is closely associated with a temperature called *wetting temperature*, T_w, at which the adsorption state transforms from *partial wetting* to *complete wetting*. Below the wetting temperature, the thickness of the film adsorbed on a surface remains finite at all sub-saturation pressures. Above wetting temperature prewetting transition, a first-order transition, might be observed between two surface phase states differing by thickness of the adsorbed film. This prewetting transition stems from the saturation curve at the wetting temperature and terminates at prewetting critical point, T_{pwc} (shown in Figure 7.5a), where thin and thick films (surface phase states) become indistinguishable. In 1977, Cahn predicted the existence of wetting transition for a two-phase mixture of fluids near a third phase surface. Independently, Ebner and Saam (1977) also predicted wetting and prewetting transition of argon film adsorbed onto a weakly attractive solid carbon dioxide surface. The authors used density functional theory (DFT) for the prediction of wetting transition. Since then, DFT has been widely used to study the wetting phenomena. Experimental evidence, which came much later, supports the prediction of prewetting transitions. Examples for helium adsorption can be found on Cs (Hallock 1995) and Rb (Phillips et al. 1998), liquid hydrogen on various substrates (Cheng et al. 1993), and acetone on graphite (Kruchten and Knorr 2003).

The prewetting line can be located by observing the adsorption isotherm with increasing density with increase in the pressure or chemical potential of the system at different temperatures. Prewetting transition can be observed as a first-order transition before saturation pressure or chemical potential in between T_w and T_{pwc}. At T_{pwc} the thin and thick films are indistinguishable, that is,

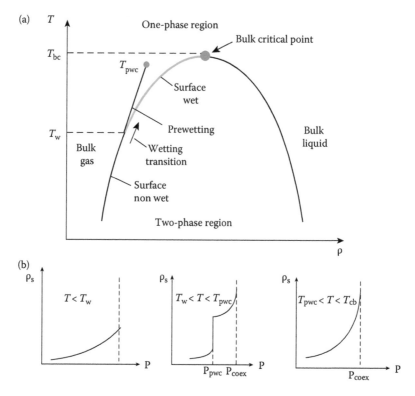

FIGURE 7.5 (a) Typical phase diagram of fluid near the surface. (b) Adsorption isotherms at different temperature.

line tension between two phases vanishes. With decreasing temperature, the prewetting pressure or chemical potential shifts toward the bulk saturation pressure or chemical potential. Moreover, at T_w, no more first-order transition is seen because prewetting pressure and chemical potential are the same as that of the bulk saturation values. Prewetting transition can be investigated via MD or MC slab-based techniques as shown in our earlier work (Saha et al. 2009; Singh et al. 2008). Such investigation is not most suited for phase transition near the surface, because appropriate gas density should be known beforehand to obtain surface phase transitions using slab-based technique. Without proper density information, it is usually a time-consuming investigation. In addition, it is rather tricky to distinguish between quasi–two dimensional vapor–liquid transitions and prewetting transitions without extensive study of adsorption isotherms with precise control of chemical potential. GC-TMMC along with histogram reweighting (Ferrenberg and Swendsen 1988), on the other hand, is more suitable method for such investigation. Details of GC-TMMC simulation techniques are given elsewhere (Errington 2004; Errington and Wilbert 2005). In our recent work (Khan and Singh 2010), we studied the prewetting transition of associating fluid for different associating strength and reported the wetting temperatures and prewetting critical temperatures using GC-TMMC. To evaluate the wetting temperature, first the difference between the bulk saturation chemical potential and prewetting chemical potential, $\Delta\mu$, is recoded for a series of temperatures. Subsequently, using the series of $\Delta\mu(T)$, we extrapolated $\Delta\mu$ to zero to determine the wetting temperature. Theoretically (Ancilotto and Toigo 1999), it is argued that $\Delta\mu$ scales as $(T - T_w)^{3/2}$ for surface potential with van der Waals tail approximately $1/z^3$. Hence, for the given set of interaction parameters with van der Waals tail for the surface potential, a linear fit for $\Delta\mu^{2/3}$ against T can be used to obtain the wetting temperature, as shown in Figure 7.6. Wetting temperature is found to increase with an increase in associating strength. Prewetting critical temperature calculation is relatively trickier for thin–thick films because simple approaches widely used for bulk vapor–liquid

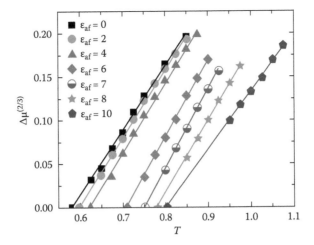

FIGURE 7.6 Difference of bulk saturation chemical potential and prewetting chemical potential versus temperature for one-site associating fluids. (Adapted from Khan, S. and Singh, J. K., *J. Chem. Phys.* 132: 144501, 2010. With permission.)

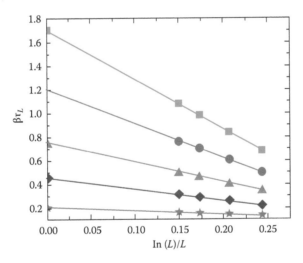

FIGURE 7.7 Typical system size dependence of the boundary tension for model one-site associating fluid. Data points from top to bottom are for increasing temperatures. Lines provide an extrapolation to the infinite system size. (Adapted from Khan, S. and Singh, J. K., *J. Chem. Phys.* 132: 144501, 2010. With permission.)

phase transitions are not applicable (Singh et al. 2003). In our work on associating fluids, we used the fact that at prewetting critical temperature, boundary tension or free-energy barrier between thick and thin films should vanish. True or infinite size boundary tension is obtained by the finite size scaling approach, as shown Figure 7.7. Prewetting critical temperature can be obtained from extrapolating the series of true boundary tension as a function of temperature to zero by fitting a second-order polynomial function, as shown in Figure 7.8. The results for model associating fluids from molecular simulations are in agreement with the DFT's results at lower association strength. At higher association, we observed a complex behavior that has not been seen or predicted earlier by the DFT. The length of the prewetting line (difference in T_{pwc} and T_w) is seen to decrease with increasing association strength, until a certain associating strength, and subsequently it increases substantially. This behavior is in disagreement with the DFT study and also raises a question of second-order transition (i.e., zero prewetting length) at higher association strength as predicted by the DFT, which is yet to seen by molecular simulation.

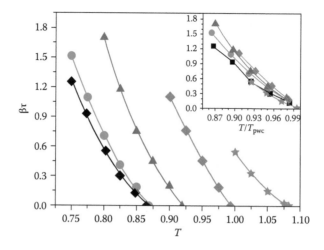

FIGURE 7.8 Infinite system size boundary tension versus temperature for various association strengths. Curves serve as a guide to the eye. Inset: Infinite system size boundary tension versus reduced temperature for different associating strengths. Symbol square, circle, triangle, diamond, and star represent boundary tension for increasing associating strengths. (Adapted from Khan, S. and Singh, J. K., *J. Chem. Phys.* 132: 144501, 2010. With permission.)

7.4.2 Wetting on Smooth Surface

As described in the method, contact angle relation with the system size can be used for obtaining the line tension. In this section, we describe in detail the calculations of contact angles of water on two different surfaces, graphite and boron–nitride. Typically, to start a contact angle simulation, one needs to construct the surface at the bottom of the simulation box and an amorphous cell of water on the top of surface. For example, for a water–graphite system, one can start with a unit cube of graphite to make a large surface according to the system size by using a supercell in the lateral direction, as shown in Figure 7.9. A vacuum cell is created above the surface with the lateral dimension the same as that of the surface and the height large enough to avoid the periodic interaction of surface on the droplet. An amorphous cell of water is usually constructed by randomly placing water molecules into the cell to maintain the density of the cell, as shown in Figure 7.10.

The knowledge of the surface properties of various materials and their temperature dependence allows one to use them in a more efficient way. One of the methods frequently used to obtain information about surface properties is via the contact angle a drop makes with a solid surface. Various methods can be employed to calculate the contact angle (Bernardin et al. 1997; Jones and Adamson 1968; Klier et al. 1995; Suzuki and Koboki 1999). The contact angle of a nano droplet can be obtained directly by molecular simulation (de Ruijter et al. 1999; Giovambattista et al. 2007).

The existing experimental data provide examples of different temperature behaviors of the contact angle for different materials. For example, the contact angle almost linearly decreases with temperature in the case of a water droplet on a crystalline naphthalene surface (Jones and Adamson 1968). A nonlinear decrease in the contact angle of water on polymer surfaces with increasing temperature was also observed (Petke and Ray 1969). The opposite behavior, an increase in contact angle with temperature was also found for water on perchloroethylene, glycerol, ethylene glycol, and, diethylene glycol on elastomers, mixtures (Budziak et al. 1991), ^{4}He, and ^{3}He on Cs (Klier and Wyatt 1996).

Dutta et al. (2010) studied the wetting temperature and line tension for a water–graphite system using MD simulations. Water–water interaction is described by the SPC/E model

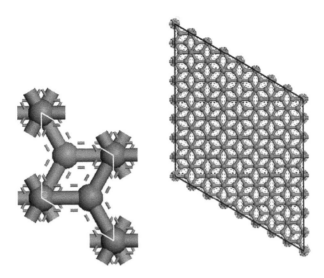

FIGURE 7.9   Schematic diagram of unit cell of graphite and a graphite surface.

FIGURE 7.10   Schematic diagram of water droplet on a graphite surface.

(Berendsen et al. 1987), and surface–water interaction is represented by the Lennard–Jones potential. The finite size scaling approach can be used for obtaining the macroscopic contact angle. In this approach, series of simulations are conducted using different system sizes. For example, Dutta et al. (2010) performed four sets of simulations, with 4000, 5000, 6000, and 7000 water molecules at different temperatures. Microscopic contact angle for each system size is evaluated as shown in the discussion of the contact angle method. These microscopic contact angles are then extrapolated to get macroscopic contact angle as shown in Figure 7.11. Wetting temperature, T_w, is evaluated from a series of contact angles as a function of temperature. The data are extrapolated linearly to obtain the temperature at which the contact angle becomes zero, as described in Figure 7.12. Using the above approach, Dutta et al. (2010) calculated T_w for the graphite–water (470 ± 5 K) and boron–nitride (438 ± 5 K) systems. Their estimate of the wetting temperature of water on the graphite surface is in agreement with that obtained from the GCMC simulation of Zhao (2007).

Line tensions for the water on graphite and boron–nitride surfaces are also calculated for various temperatures by Dutta et al. (2010). To calculate the line tension, the authors took the advantage of Equation 7.4. System size analysis data can be used to obtain the slope, $-\tau/\gamma_{lv}$, as shown in

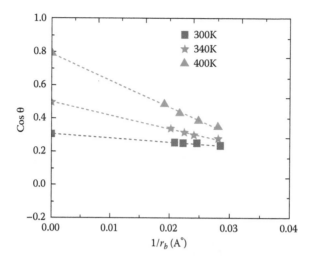

FIGURE 7.11 Dependence of the system size on contact angle of water droplet on the Boron Nitride (BN) surface at different temperatures. Symbols square, star, triangle represent contact angles at 300K, 340K, and 400K, respectively. Data points, from right to left, represent droplet with 2000, 3000, 4000, 5000 water molecules. (Adapted from Dutta, R. C., Khan, S., and Singh, J. K., *Fluid Phase Equilibria*, 302, 310, 2011.)

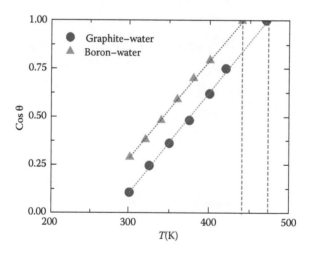

FIGURE 7.12 Dependency of cosine of macroscopic contact angle θ against temperature for graphite–water and BN–water systems. Dashed line along the data points are guide to the eye. Dashed vertical lines represent the estimated wetting temperatures. (Adapted from Dutta, R. C., Khan, S., and Singh, J. K., *Fluid Phase Equilibria*, 302, 310, 2011.)

Figure 7.12. The numerical values for vapor–liquid surface tension, γ_{lv}, were taken from the study by Vega and de Miguel (2007) for SPC/E water at different temperatures. The calculated values of line tensions are on the order of 10^{-10} N, which are in good agreement with the order of the line tension values reported in the literature (Mugele et al. 2002; Rowlinson and Widom 1982). Dutta et al. (2010) obtained 3.06×10^{-10} N as the line tension value at 300 K, which is in good agreement with the previously reported values on the graphite–water system at the same temperature by Werder et al. (2003). In a recent work, Zangi and Berne (2008) assumed a constant line tension for different temperatures for their study of water on a hydrophobic plate. However, Dutta et al. (2010) found that the line tension substantially increases with increasing temperature. The variation of the line tension for water on graphite and Boron Nitride (BN) surfaces is shown in Figure 7.13.

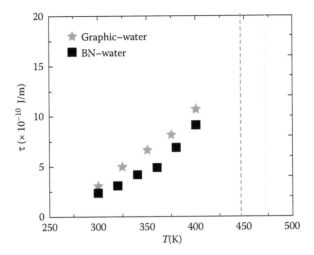

FIGURE 7.13 Line tension versus temperature for graphite–water and BN–water systems. (Adapted from Dutta, R. C., Khan, S., and Singh, J. K., *Fluid Phase Equilibria*, 302, 310, 2011.)

In general, there is a consensus of positive line tension near first-order wetting (Indekeu 1994). At wetting, short-ranged interactions are characterized by a finite τ with finite slope, whereas the retarded van der Waals interactions exhibits finite τ with diverging slope, and the longer ranged interactions (e.g., nonretarded van der Waals) exhibit diverging τ. However, which of the two forces (long or short ranged) dominates cannot be ascertained from the calculated line tension data.

7.4.2.1 Effect of Composition on the Contact Angle

The composition of liquid mixtures can also affect the contact angle. Lundgren et al. (2002) studied the contact angle of water and ethanol mixture with different compositions on a graphite surface. With the addition of ethanol, the wetting properties of the droplet increase and the contact angles decrease. The ethanol molecules are concentrated close to the hydrophobic solid surface and at the top of the droplet, at the water–vapor interface with the hydroxyl group pointing toward the center of the droplet and the alkyl chain pointing toward the vacuum.

7.4.3 WETTING ON CHEMICAL PATTERNED SURFACE

It is now possible, experimentally, to tune the surface property as per the desired applications by different methods of surface treatment such as microcontact printing, plasma treatment, and so forth. Wetting of substrate mainly depends on the surface energy of the substrate. By introducing different lyophilic (–COOH–, –OH–, –NH2–) or lyophobic (–CH2–, –CF2) groups to the surface, one can change the surface energy of substrate and thus the wetting behavior of the surface. For example, Chai et al. (2009) investigated the wetting behavior of the silica surface with different degrees of surface hydroxylation and silanization via MD simulation. They observed complete transformation from hydrophilicity to hydrophobicity with an increase in silanization percentage. In their study, the contact angle varied from 20° to 120°. Another way to modify the surface is via a self-assembled monolayer (SAM) without the need for sophisticated analytical equipment that allows precise control of liquid affinity to the surface. SAM is discussed in detail in the next section.

7.4.3.1 Self-Assembled Monolayer

Amphiphilic molecules have a tendency to organize themselves to form different types of morphologies such as bilayer, multilamellar, vesicles, spherical and cylindrical micelles, and so on in bulk solution and are well understood experimentally (Gruner 1989) or by simulation. However, the

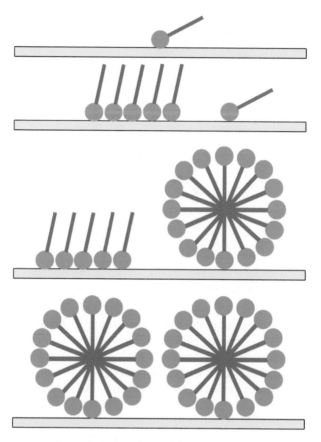

FIGURE 7.14   Different types of morphologies of amphiphilic molecules on the surface.

self-assembly can be different at the solid–fluid interface. A typical set of structures near the surface is shown in Figure 7.14.

Monolayers of the self-assembled amphiphilic molecules on the solid have gained more attention because of their high selectivity, stability, sensitivity, and potential applications in biosensors (Cheng et al. 2008; Jung et al. 2006; Kim et al. 2008), biomedical devices (Fragoso et al. 2008; Park et al. 2007; Zhang et al. 2007), and lubrication. Self-assembled monolayer (SAM) can be formed on various substrates (gold, silver, copper, platinum, palladium, silicon, and silicon dioxide) by chemisorptions of organic molecules containing groups such as thiols, disulfides, amines, acids, or silanes. Typically, an SAM comprises three significant parts: the surface active head groups that bind to the substrate, the alkyl chains that provide stability to the monolayer, and terminal functional groups that could be used for selectivity, catalytic action, and so forth, per the applications. For example, alkanethiol molecules are mostly used to form SAMs on gold surface (Azzam et al. 2006; Birss et al. 2003; Noh et al. 2006) because the gold atoms have strong interaction toward the thiol group and lower the free energy of the interface between metal and the ambient environment. Organization of the fluid molecules on the substrate depends on the relative interactions of fluid–fluid and fluid–substrate. Thus, properties of SAMs can be tuned by modifying the substrate and fluid molecules. An extensive research has been carried out experimentally (Bohmer et al. 1992; Ducker and Wanless 1999; Manne et al. 1994), but very little effort has been directed toward simulation because of the requirement of heavy computations for the accurate representation of real systems. However, recent advancements in computational power and molecular simulation methodologies have enabled us to simulate relatively large systems to understand their behavior from molecular point of view. In this section, we will discuss future trends of simulations by critically evaluating experimental results and how they can be related to theoretical simulations.

In many applications, it is necessary to prepare SAMs of appropriate orientation, packing density, and stability. Typically, these can be achieved by physical adsorption, chemical activators, chemical cross-linking, or exchange process. Physical adsorption is based on the charge–charge interaction of polar molecules with the surface. Chemical activators are used to modify the chemical activity of the binding groups. The exchange process is used to modify the monolayer by replacing it with another monolayer; these mixed monolayers are more oriented and stable. Pan and Rothberg (2005) showed that mixed SAMs of small and large thiol molecules on a gold surface respond better than a homogenous system.

An extensive research has been done experimentally for the adsorption of amphiphilic surfactants on various surfaces such as silica (Bohmer et al. 1992; Lee et al. 1989; Levitz et al. 1984), rutile, graphite (Ducker and Grant 1996; Manne et al. 1994), mica (Ducker and Wanless 1999; Pashley and Israelachvii 1981), and gold (Jaschke et al. 1997) using different experimental techniques such as atomic force spectroscopy (AFM), surface force apparatus, and neutron scattering. Most of the studies were directed toward understanding the factors affecting the surfactant morphologies (hemimicelles, monolayers, bilayers, spherical, or cylindrical) at the solid–fluid interface. It is noteworthy that the concentration of surfactant plays a major role in defining the structure of an adsorbed monolayer. Grant et al. (1998) studied the effect of alkyl chain of alkyl-poly(ethylene oxide) on silica and graphite surfaces. They found a critical chain length at which a homogeneous layer transforms to the hemicylindrical structure. However, structural transition of surfactant is still controversial because of experimental difficulties. On the other hand, molecular modeling has been used by a few groups to clear some of the arguments as well provide insight to the self-assembly behavior of surfactant near surfaces. Srinivas et al. (2006) studied self-organization of aqueous surfactant n-alkyl-poly(ethylene oxide) on a graphite surface by coarse-grain MD simulation. They found that short-chain surfactants form a nearly perfect monolayer on the graphite surface, whereas continuous hemicylinders were found for long-chain surfactants. The diameter of a hemicylinder is found to be in good agreement with an earlier AFM study (Manne and Gaub 1995). Wanless et al. (1996) studied the effect of NaCl concentration on the adsorption of sodium dodecylsulfate (SDS) surfactant on graphite surface. They proposed that with increase of NaCl concentration there is decrease in the interaggregate spacing rather than aggregate size of surfactants. Similarly, Jaschke et al. (1997) observed from AFM images that in the absence of halide counterion, surfactant (hexadecyltrimethylammonium hydroxide [C_{16}TAOH] and SDS) adsorption is driven by the tail group— surface interaction leading to a hemicylindrical structure on the surface while in the presence of halide counterion, because of the charged surface, surfactants adsorb preferentially on the surface leading to the full-cylindrical structure.

Zehl et al. (2009) studied the adsorption behavior of small amphiphilic surfactants on solid surface by means of MC simulation. The authors modeled amphiphilic molecules as a chain of hydrophobic and hydrophilic sphere with square well potential for pair interaction. They found different types of structure, like two-dimensional cluster phases, monolayers, cluster networks, bilayers, and wormlike admicelles, depending on the ratio between the strength of chain–chain hydrophobic interactions and adsorption energy of surfactant molecules at the surface. In spite of extensive study of large surfactant on different surfaces, a little effort has been made for small amphiphilic organic molecules. Wang et al. (2004) studied the adsorption behavior of methanol, ethanol, n-butanol, n-hexanol, and n-octanol on mica surface by the AFM technique. They observed vertically oriented bilayer for methanol and ethanol and tilted monolayer for n-butanol, n-hexanol, and n-octanol. Khan and Singh (2011) also studied the same system with the help of MD and obtained good agreement with the experimental results.

Water interaction has significant effect on the selectivity or stability of the SAM. Few studies have been reported on the structural and dynamic properties of water (layering of water molecules) (Gordillo and Martí 2002; Lee and Rossky 1993; Raghavan et al. 1991) or the behavior of the contact angle (Hautman and Klein 1991; Hirvi and Pakkanena 2006; Lundgren and Allan 2002) near hydrophobic and hydrophilic surfaces. Yang and Weng (2008) performed a series of molecular dynamics to study the structural and dynamic properties of a water layer on alkanethiol SAM adsorbed on the gold surface with different end functional groups, that is, methyl, carboxyl, and hydroxyl. The authors found existence of two well-defined water layers on the virgin gold surface, whereas in the presence of adsorbed SAMs, layering disappeared. The orientation of water O–H groups was parallel to the SAM surface near the water–methyl interface, whereas it was uniform near the water–carboxyl and water–hydroxyl interfaces. They also studied the mobility of water near SAMs and found that at the water–methyl interface water had more lateral mobility compared to the water–carboxyl and water–hydroxyl interfaces but a little less than that of bulk. Similarly, Shenogina et al. (2009) studied the contact angle of water on SAMs with different head groups with varying hydrophobicity by means of MD simulation. They got excellent agreement with experimental results. They also related the interfacial thermal conductance with the contact angle. Chai et al. (2009) used SAMs to modulate the surface hydrophilicity and hydrophobicity. They investigated the wetting properties for hydroxylated and silanized amorphous silica surfaces in terms of contact angles. The hydroxylated silica surface is hydrophilic in nature, but by replacing the –OH group with $-Si(CH_3)_3$ group it becomes hydrophobic, depending on the percentage of $-Si(CH_3)_3$ group. Halversona et al. (2008) found imbalance force at the contact line between water drop on inhomogeneous methyl- and hydroxyl-terminated alkanethiol chains on gold surface by mean of MD simulation. This imbalance force, at contact line, drives the water droplet to move on the surface. Halversona et al. (2009) also studied the effect of surfactant addition on the wetting characteristics of a water droplet on a graphite surface. They found that the addition of alkyl polyethoxylate in water drop has more effect on wetting compared to the trisiloxylate surfactant. In the case of polysiloxylate surfactant, the contact angle reduces from 110° to 80°, whereas for polyethoxylate it can reduce to 55°.

7.4.3.2   Effect of Surface Polarity on Contact Angle

Giovambattista et al. (2007) showed that surface polarity can also play a role in wetting phenomena. They demonstrated the idea by using MD simulation on water on hydroxylated SiO_2 surface. The authors performed hydroxylation of the surface by attaching one hydrogen (H) atom per surface oxygen atom. To control the polarity of the surface the authors introduced a parameter in the columbic interactions:

$$q_i = K * q_{0,i} \quad (0 \leq K \leq 1) \tag{7.15}$$

where $q_{0,i}$ represents the charges of the atoms when the surface is fully hydroxylated. K is the scaling parameter by which the surface can be tuned from a nonhydroxylated ($K = 0$) to a hydroxylated surface ($K = 1$).

The authors conducted a detailed study of the orientation of the water molecules adjacent to the surface while tuning the columbic interaction. They observed one hydrogen bond (HB–OH) vector preferentially normal to the surface in the case of an idealized flat surface ($K = 0$) and the remaining three HB vectors associated with other water molecules in the bulk. In contrast, with an ideal hydroxylated hydrophilic silica surface, they exhibited one HB from an oxygen atom of water molecules with the surface while the remaining three HB contributed to the water molecules in the bulk. They found the extreme value of K, below which the surface behaves as a hydrophobic surface, and beyond this, it behaves as hydrophilic surface. This illustrates that a polar surface can also behave as a hydrophobic surface. Similarly, Ohlar and Langel (2009) performed MD simulation of water droplets on titanium oxide (TiO_2) and found that the reduction of the partial charges of TiO_2 mimicked a hydrophobic surface.

7.4.4 WETTING ON A PHYSICALLY PATTERNED SURFACE

7.4.4.1 Super-Hydrophobic and Super-Hydrophilic Surface

It is well known that roughness plays an important role over wetting phenomena. Both experiments and simulations find significant effects of surface roughness on the surface wetting properties, which can be explained by two main hypotheses attributed to Wenzel (1936) and Cassie and Baxter (1944). Numerous fractal structures have been reported experimentally as able to significantly improve the surface property of a solid surface (Extrand 2002; He et al. 2003; Lenz and Lipowsky 1998; Patankar 2004; Ran et al. 2008; Shirtcliffe et al. 2005; Yoshimitsu et al. 2002). Spoor et al. (2008) studied various surfaces with different textures. Surfaces examined were sandblasted glass microscope slides (SBG), as well as replicas of sandblasted (large grit) acid-etched titanium (SLA); lotus leaves (LLR); and golf tee–shaped micropillars of photoresist (GTM) on a silicon wafer (as presented in Figure 7.15). The contact angles for these surfaces ranged from 20° and 105°.

Yoshimitsu et al. (2002) also studied the contact angle for pillar surfaces systematically on silicon wafers and observed that pillar dimension has a significant effect on the contact angle, as illustrated in Figure 7.16.

Many experimental results have been reported (Patankar 2004; Ran et al. 2008; Shirtcliffe et al. 2005) on the transition of the Wenzel state to the Cassie state. For example, Shirtcliffe et al. (2005) observed the transition of water from the Wenzel state to the Cassie state on patterned copper surfaces with different roughness. Similarly, Ran et al. (2008) studied the surface wettability of nanoporous alumina substrates and were able to transform from the Wenzel state to the Cassie state by systematically changing the hole diameter and hole depth.

Along with experiments, much progress has been reported in simulation regarding the effect of the surface texture on wetting (Hirvi and Pakkanen 2007, 2008; Lundgren et al. 2007; Yang et al.

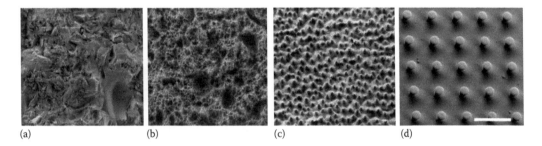

(a)   (b)   (c)   (d)

FIGURE 7.15 SEM images for (a) SBG, (b) SLA, (c) LLR, and (d) GTM. (Adapted from Ran, C. et al. *Langmuir* 24: 9952, 2008. With permission.)

Contact angle of water θ' [deg]	114	138	155	151	153
Roughness factor r	1.0	1.1	1.2	2.0	3.1
Pillar height c [μm]	0	10	36	148	282

FIGURE 7.16 Water droplet on different patterned silicon wafer. (Adapted from He, B., Patankar, N. A., and Lee, J., *Langmuir* 19: 4999, 2003. With permission.)

2008). Lundgren et al. (2007) studied systematically the pillar surface with a wide variation of pillar height, pillar width, and pillar gap and reported the effect of these parameters on the transition from the Wenzel state to the Cassie state.

Yang et al. (2008) performed MD simulations to study contact angle and the contact angle hysteresis on hydrophobic and hydrophilic surfaces. They showed the contact angle on hydrophobic surfaces depends strongly on roughness rather than fractal dimension. For hydrophobic surfaces, they did not observe any contact angle hysteresis resulting in thermal fluctuation; however, for hydrophilic surfaces, strong contact angle hysteresis was observed because of the higher energy barrier. Hirvi and Pakkanen (2008) studied the impact and sliding of a nanosized water droplet on nanostructured hydrophobic polyethylene (PE) and hydrophilic polyvinyl chloride (PVC) polymer surfaces. The initial velocities of droplets (v) varied from 10 m/s to 500 m/s. At higher impact velocity on collision with pillar surfaces leads to deformation of the water droplet and penetrate into the grooved surfaces. However, the final equilibrium shape was unchanged. For the hydrophilic surface they observed an asymmetric Wenzel state, and for hydrophobic surface the water droplet bounced on the textured surface. They also calculated friction during sliding of the water droplet on the textured surface.

We recently studied the effect of roughness on wetting behavior of a grooved, patterned graphite surface (Dutta et al. 2011) by examining the change in contact angle of a water droplet, as shown in Figure 7.17.

The grooves are defined by groove width a, height b, and spacing c. Roughness factor r was computed using the following formula given by Wenzel (1936):

$$r = \frac{\text{actual surface}}{\text{geometric surface}} \tag{7.16}$$

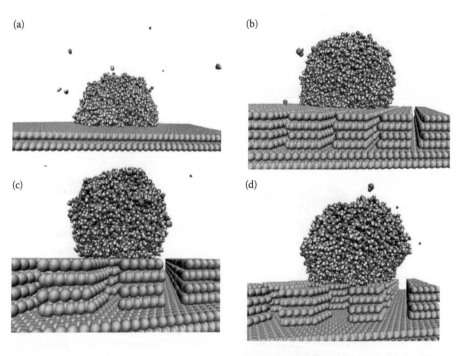

FIGURE 7.17 Representative snapshots of structure of the simulated water droplets on pillar surfaces of varying surface fraction and roughness: (a) $f = 1$, $r = 1$; (b) $f = 0.7146$, $r = 2.1$; (c) $f = 0.6253$, $r = 2.1$; (d) $f = 0.5$, $r = 2.1$.

So, r is expressed as

$$r = \frac{(a+c)^2 + 4bc}{(a+c)^2} \qquad (7.17)$$

By varying pillar height, b, we can vary the roughness of the surface.

The macroscopic contact angle is computed by the finite size scaling method, as described in an earlier section. To study the effect of surface fraction (the fraction of the solid on the interface) on the contact angle, we varied the ratio of groove width to groove spacing of the pillar (i.e., varying surface fraction, f) from 0 to 0.5. This study is performed at a constant roughness factor ($r = 2.1$). Using MD and system size analysis, we demonstrated that contact angle increases with decrease in the surface fraction, which is in agreement with Cassie and Baxter's predictions (1944) qualitatively.

We also studied the effect of pillar height by varying the pillar height from 3 Å to 27.5 Å at constant surface fraction ($f = 0.5$, i.e., at constant pillar width and gap between the pillars). It is seen that with increase in the roughness factor from 1 to 3.3, the contact angle increases from 83.3° to 130°. In particular, we observed that by controlling the roughness factor the system can be changed from the Wenzel regime to the Cassie-Baxter regime.

7.4.4.2 Super Oleophobic Surface

It is of practical importance to find a low-energy surface with suitable texture that can repel oil-like fluids (oleophobic surface). Experimentally, several papers have been published in this field (Han and Steckl 2009; Hoefnagels et al. 2007; Hsieha et al. 2005; Shibuichi et al. 1998; Steele et al. 2009). Recently, an excellent work was published on oleophobic fiber (Han and Steckl 2009). In this study, the authors successfully produced hydrophobic and oleophobic electrospun fibers using coaxial electrospinning with Teflon AF sheath and poly(ε-caprolactone) core materials. The core material was used to generate surface roughness and was coated with Teflon AF sheath to reduce surface energy. The combination of these two properties produces super-hydrophobicity as well as oleophobicity. Similarly, Steele et al. (2009) described a technique to fabricate super oleophobic coatings by spray casting ZnO nanoparticles blended with perfluroacrylic polymer suspensions using a cosolvent. This coating can be easily applied to almost all surfaces.

7.4.4.3 Anti-Frosting Surface

Frosting is a common phenomenon in refrigeration when moist air contacts the cold surface below its triple point. Initially, it enhances heat transfer, but as the ice layer thickness increases, it acts as an insulator and subsequent increases in pressure loss finally block the airflow. The system needs frequent defrosting to allow continuous operation. Frost formation on different parts (wings, propeller, etc.) of aircraft creates problems in maintaining desired altitude by disturbing the smooth flow of air and increasing drag. Sometimes it is not possible to break up ice on parts of an aircraft, which increases weight and interferes with flight control, a major problem for small aircrafts. Accumulation of ice in parts of the engine can force shutdown of the engine. Thus, it is desirable to develop smart material that can prevent ice formation on the surface. Many attempts have been made to find out appropriate surface that can reduce frost formation (Liu et al. 2008). It is understood that a low-energy surface or hydrophobic surface can retard ice formation; however, once ice forms, its thickness increases rapidly. Several low-energy substance polymer coatings have been tried to reduce frost formation. Liu et al. (2006) using polymeric materials and a hydrophilic agent developed a new material that could delay frost nucleation and minimize the frost growth on cold surfaces. Many anti-frost surfaces have been reported experimentally (Liu et al. 2006; Wang et al. 2007), but molecular simulation efforts in this direction are limited.

7.4.4.4   Anti-Fouling Surface

Anti-fouling surfaces find numerous applications such as in blood-compatible materials, biosensor coatings, membrane coating, bioreactor surface coating, and marine surface coating. Numerous investigations have been carried out in this field (Reisch et al. 2009; Xua et al. 2008). Fouling of heat transfer equipment is a universal problem in the industry. A fouling layer increases resistance to heat transfer, leading to a decrease in operating efficiency and increase in heat loss. These fouling materials can undergo cracks, which could lead to serious damage to the equipment. In addition, a fouling layer decreases the cross-sectional area, which increases pressure drop across the apparatus. Many surface treatment technologies (Bornhorst et al. 1999; Yang et al. 2000) have been reported to counter the fouling problem. The Ni–P deposits can inhabit the growth of ice formation on surface and can be used in heat transfer equipment (Cheng et al. 2009).

Protein fouling on biocompatible surfaces is of great concern to medical science. Polyethylene glycol (PEG) has strong resistance to protein adsorption, which has led to development of PEG-modified surfaces (Reisch et al. 2009).

Biofouling (Callow and Callow 2002) is common phenomenon generally found in marine systems and caused by deposition of microorganism on a surface. Tri-n-butyltin (TBT)-containing paints can be used to coat the surface for controlling biofouling; however, because of its acute toxicity toward nontarget marine species, TBT recently was banned. Hence, it is of urgent need to identify appropriate surface coatings that strongly resist the attachment and growth of microorganisms and are stable in marine environment. Not much work has been done to understand the mechanisms of biofouling and develop suitable coatings using molecular simulation techniques. This is an area in which computational approaches such as molecular simulation can make significant contributions in development and guidance for producing suitable surfaces.

7.8   SUMMARY

In this chapter, we summarized molecular simulation methods for investigating wetting transitions particularly contact angle computations. These aspects about surfaces are regularly studied in experiments; however, this is not the case for simulations, although there is no reason these studies cannot be done in coming years. The major bottleneck in the utility of molecular modeling tools is the availability of correct force field parameters. In the case of organic fluids, force fields are well known and are of reasonable quality. However, for other systems, such as metals and mineral surfaces, force field information is not widely available, leading to lack of confidence in routine use of these tools. Nevertheless, with an increasing number of scientists working on molecular simulations and developing methods and force fields, use of molecular modeling simulations will see rapid expansion.

REFERENCES

Amirfazli, A., and Neumann, A. W. 2004. Status of the three-phase line tension, *Adv. Colloid Interface Sci.* 110: 121.

Ancilotto, F., and Toigo, F., 1999. Prewetting transitions of Ar and Ne on alkali-metal surfaces. *Phys. Rev. B* 60: 9019.

Azzam, W., Bashir, A., Terfort, A., Strunskus, T., and Wöll, C. 2006. Combined STM and FTIR Characterization of Terphenylalkanethiol Monolayers on Au(111): Effect of Alkyl Chain Length and Deposition Temperature, *Langmuir* 22: 3647.

Barthlott, W, and Neinhuis, C. 1997. Purity of the sacred lotus, or escape from contamination in biological surfaces, Planta 202: 1.

Berendsen, H. J. C., Grigera, J. R., and Straatsma, T. P. J. 1987. The missing term in effective pair potentials. *Phys. Chem.* 91: 6269.

Berim, G. O., and Ruckenstein, E. J. 2009. Simple expression for the dependence of the nanodrop contact angle on liquid-solid interactions and temperature. *Chem. Phys.* 130: 184712.

Bernardin, J. D., Mudawar, I., Walsh, C. B., and Franses, E. I. 1997. Film boiling heat transfer of droplet streams and sprays. *Int. J. Heat Mass Transfer* 40: 2579.

Bhushan, B., and Jung, Y. C. 2010. Natural and biomimetic artificial surfaces for superhydrophobicity, self-cleaning, low adhesion, and drag reduction. *Prog. Mater. Sci.* 56: 1.

Binder, K. 1982. monte carlo simulation of surface tension for two and three-dimensional lattice gas model. *Phys. Rev. A* 25: 1699.

Birss, V., Dang, K., Wong, J. E., and Wong, R. P. C. 2003. Adsorption of quaternary pyridinium compounds at Pt electrodes in neutral and weakly alkaline solutions. *J. Electroanal. Chem.* 550: 67.

Bohmer, M. R., Koopal, L. K., and Janssen, R. 1992. Adsorption of Nonionic Surfactants on Hydrophilic Association in the Adsorbed Layer Surfaces. An Experimental and Theoretical Study on association in the adsorbed layer. *Langmuir* 8: 2228.

Bornhorst, A., Müller-Steinhagen, H., and Zhao, Q. 1999. Reduction of Scale Formation Under Pool Boiling Conditions by Ion Implantation and Magnetron Sputtering on Heat Transfer Surfaces. *Heat Transf. Eng.* 20: 6.

Brandon, S., Haimovich, N., Yeger, E., and Marmur, A. 2003. Partial wetting of chemically patterned surfaces: The effect of drop size. *J. Colloid Interface Sci.* 263: 237.

Budziak, C. J., Vargha-Butler, E. I., and Neumann, A. W. 1991. Temperature Dependence of Contact Angles on Elastomers. *J. Appl. Polym. Sci.* 42: 1959.

Bulnes, F., Ramirez-Pastor, A. J., and Zgrablich, G. 2007. Scaling Behavior of Adsorption on Patchwise Bivariate Surfaces Revisited. *Langmuir* 23: 1264.

Cahn, J. W. J. 1977. Critical point wetting. *Chem. Phys.* 66: 3667.

Callow, M. E., and Callow, J. A. 2002. marine biofouling : a sticky problem. *Biologist* 49: 10.

Cassie, A. B. D., and Baxter, S. 1944. Wettability of porous surfaces. *Trans. Faraday Soc.* 40: 546.

Chai, J., Liu, S., and Yang, X. 2009. Molecular dynamics simulation of wetting on modified amorphous silica surface. *App. Surf. Sci.* 255: 9078.

Cheng, E., Mistura, G., Lee, H. C., Chan, M. H. W., Cole, M. W., Carraro, C., et al. 1993. Wetting transitions of liquid hydrogen films. *Phys. Rev. Lett.* 70: 1854.

Cheng, F., Gamble, L. J., and Castner, D. G. 2008. XPS, TOF-SIMS, NEXAFS, and SPR Characterization of Nitrilotriacetic Acid-Terminated Self-Assembled Monolayers for Controllable Immobilization of Proteins. *Anal. Chem.* 80: 2564.

Cheng, Y. H., Zou, Y., Cheng, L., and Liu, W. 2009. Effect of the microstructure on the properties of Ni–P deposits on heat transfer surface. *Surf. Coating Tech.* 203: 1559.

De Ruijter, M. J. D., Blake, T. D., and Coninck, J. D. 1999. Dynamic Wetting Studied by Molecular Modeling Simulations of Droplet Spreading. *Langmuir* 15: 7836.

Dettre, R. H., and Johnson, R. E. Jr. 1963. Contact Angle Hysteresis II. Contact Angle Measurements on Rough Surfaces. *Adv. Chem. Ser.* 43: 136.

Ducker, W. A., and Grant, L. M. 1996. Effect of Substrate Hydrophobicity on Surfactant Surface-Aggregate Geometry. *J. Phys. Chem.* 100: 11507.

Ducker, W. A., and Wanless, E. J. 1999. Adsorption of Hexadecyltrimethylammonium Bromide to Mica: Nanometer-Scale Study of Binding-Site Competition Effects. *Langmuir* 15: 160.

Dutta, R. C., Khan, S., and Singh, J. K. 2011. Wetting transition of Water on Graphite and Boron-Nitride Surfaces: A Molecular Dynamics Study. *Fluid Phase Equilibria* 302: 310.

Dutta, R. C., Khan, S., and Singh, J. K. 2011. Unpublished work.

Ebner, C., and Saam, W. F. 1977. New Phase-Transition Phenomena in Thin Argon Films. *Phys. Rev. Lett.* 38: 1486.

Errington, J. R. 2004. Prewetting Transitions for a Model Argon on Solid Carbon Dioxide System. *Langmuir* 20: 3798.

Errington, J. R., and Wilbert, D. W. 2005. Prewetting Boundary Tensions from Monte Carlo Simulation. *Phys. Rev. Lett.* 95: 226107.

Extrand, C. W. 2002. Model for Contact Angles and Hysteresis on Rough and Ultraphobic Surfaces. *Langmuir* 18: 7991.

Fan, C. F., and Cagin, T. J. 1995. Wetting of crystalline polymer surfaces: A molecular dynamics simulation. *Chem. Phys.* 103: 9053.

Ferrenberg, A. M., and Swendsen, R. H. 1988. New Monte Carlo technique for studying phase transitions. *Phys. Rev. Lett.* 61: 2635.

Fragoso, A., Laboria, N., Latta, D., and O'Sullivan, C. K. 2008. Electron Permeable Self-Assembled Monolayers of Dithiolated Aromatic Scaffolds on Gold for Biosensor Applications. *Anal. Chem.* 80: 2556.

Giovambattista, N., Rossky, P. J., and Debenedetti, P. G., 2007. Effect of Surface Polarity on Water Contact Angle and Interfacial Hydration Structure. *J. Phys. Chem. B* 111: 9581.

Gordillo, M. C., and Martí, J. 2002. Molecular dynamics description of a layer of water molecules on a hydrophobic surface. *J. Chem. Phys.* 117: 3425.

Grant, L. M., Tiberg, F., and Ducker, W. A. *J. Phys. Chem. B* 1998. Nanometer-Scale Organization of Ethylene Oxide Surfactants on Graphite, Hydrophilic Silica, and Hydrophobic Silica. 102: 4288.

Grest, G. S., Heine, D. R., and Webb, E. B. 2006. Liquid Nanodroplets Spreading on Chemically Patterned Surfaces. *Langmuir* 22: 4745.

Gruner, S. M. 1989. Stability of Lyotropic Phases wlth Curved Interfaces. *J. Phys. Chem.* 93: 7562.

Grzelak, E. M., and Errington, J. R. 2008. Computation of interfacial properties via grand canonical transition matrix Monte Carlo simulation. *J. Chem. Phys.* 128: 014710.

Hallock, R. B. J. 1995. Review of some of the experimental evidence for the novel wetting of He on alkali metals. *Low Temp Phys.* 101: 31.

Halverson, J. D., Maldarelli, C., Couzis, A., and Koplik, J. 2008. A molecular dynamics study of the motion of a nanodroplet of pure liquid on a wetting gradient. *J. Chem. Phys.* 129: 164708.

Halverson, J. D., Maldarelli, C., Couzis, A., and Koplik, J. 2009. Wetting of hydrophobic substrates by nanodroplets of aqueous trisiloxane and alkyl polyethoxylate surfactant solutions. *Chem. Eng. Sci.* 64: 4657.

Han, D., and Steckl, A. J. 2009. Superhydrophobic and Oleophobic Fibers by Coaxial Electrospinning. *Langmuir* 25: 9454.

Hautman, J., and Klein, M. L. 1991. Microscopic wetting phenomena. *Phys. Rev. Lett.* 67: 1763.

He, B., Patankar, N. A., and Lee, J. 2003. Multiple Equilibrium Droplet Shapes and Design Criterion for Rough Hydrophobic Surfaces. *Langmuir* 19: 4999.

Heine, D. R., Grest, G. S., and Webb, E. B. 2005. Surface Wetting of Liquid Nanodroplets: Droplet-Size Effects. *Phys. Rev. Lett.* 95: 107801.

Heslot, F., Cazabat, A. M., Levinson, P., and Fraysse, N. 1990. experiments on wetting on the scale of nanometer: influence surface energy. *Phys. Rev. Lett.* 65: 599.

Hirvi, J. T., and Pakkanen, T. A. 2007. Wetting of Nanogrooved Polymer Surfaces. *Langmuir* 23: 7724.

Hirvi, J. T., and Pakkanen, T. A. 2008. Nanodroplet impact and sliding on structured polymer surfaces. *Surf. Sci.* 602: 1810.

Hirvi, J. T., and Pakkanen, T. A. 2006. molecular dynamics simulations of water droplets on polymer surfaces. *J. Chem. Phys.* 125: 144712.

Hoefnagels, H. F., Wu, D., De, G., and Ming, W. 2007. Biomimetic Superhydrophobic and Highly Oleophobic Cotton Textiles. *Langmuir* 23: 13158.

Hsieha, C.-T., Chena, J.-M., Kuoa, R.-R., Linb, T.-S., and Wuc, C.-F. 2005. Influence of surface roughness on water- and oil-repellent surfaces coated with nanoparticles. *App. Surf. Sci.* 240: 318.

Indekeu, J. O. 1994. Line tension at wetting. *J. Mod. Phys. B* 8: 309.

Jaschke, M., Butt, H. J., Gaub, H. E., and Manne, S. 1997. Surfactant Aggregates at a Metal Surface. *Langmuir* 13: 1381.

Jones, J. B., and Adamson, A. W. 1968. Temperature dependence of contact angle and of interfacial free energies in the naphthalene-water-air system. *J. Phys. Chem.* 72: 646.

Jung, S.-H., Son, H.-Y., Yuk, J. S., Jung, J.-W., Kim, K. H., Lee, C.-H., et al. 2006. Oriented immobilization of antibodies by a self-assembled monolayer of 2-(biotinamido)ethanethiol for immunoarray preparation. *Colloids Surf. B Biointerface* 47: 107.

Khan, S., and Singh, J. K. 2010. prewetting transition of one site associating fluids. *J. Chem. Phys.* 132: 144501.

Khan, S., and Singh, J. K. 2011. Unpublished work.

Kim, H., Choi, D., Kim, Y., Baik, S., and Moon, H. J. 2010. Shape Optimization of Symmetric Cylinder Shape on Buoyancy Using Fourier Series Approximation. *Fluids Eng.* 132: 051206.

Kim, S. J., Gobi, K. V., Tanaka, H., Shoyama, Y., and Miura, N. 2008. A simple and versatile self-assembled monolayer based surface plasmon resonance immunosensor for highly sensitive detection of 2,4-D from natural water resources. *Sensors Actuators B* 130: 281.

Klier, J., and Wyatt, A. F. G. 1996. Contact angle of liquid He mixtures on Cs: Evidence for ^{3}He at the He-Cs interface. *Phys. Rev. B* 54: 7350.

Klier, J., Stefanyi, P., and Wyat, A. F. G. 1995. Heat Capacity Measurements of ^{3}He-^{4}He Mixtures in Aerogel. *Phys. Rev. Lett.* 75: 3709.

Kruchten, F., and Knorr, K. 2003. Multilayer Adsorption andWetting of Acetone on Graphite. *Phys. Rev. Lett.* 91: 085502.

Lee, E. M., Thomas, R. K., Cummins, P. G., Staples, E. J., and Penfold, J. 1989. Determination of the structure of a surfactant layer adsorbed at the silica/water interface by neutron reflection. *Chem. Phys. Lett.* 162: 196.

Lee, S. H., and Rossky, P. J. 1993. A comparison of the structure and dynamics of liquid water at hydrophobic and hydrophilic surfaces-a molecular dynamics simulation study. *J. Chem. Phys.* 100: 3334.

Lenz, P. and Lipowsky, R. 1998. Morphological Transitions of Wetting Layers on Structured Surfaces. *Phys. Rev. Lett.* 80: 1920.

Levitz, P., Miri, A. E., Keravis, D., and Damme, H. V. 1984. Adsorption of Nonionic Surfactants at the Solid-Solution Interface and Micellization: A Comparative Fluorescence Decay Study. *J. Colloid Interface Sci.* 99: 484.

Liu, Z., Gou, Y., and Cheng, J. W. S. 2008. Frost formation on a super-hydrophobic surface under natural convection conditions. *Int. J. Heat Mass Transfer* 51: 5975.

Liu, Z., Wang, H., Zhang, X., Meng, S., and Ma, C. 2006. An experimental study on minimizing frost deposition on a cold surface under natural convection conditions by use of a novel anti-frosting paint. Part I. Anti-frosting performance and comparison with the uncoated metallic surface. *Int. J. Refrigeration* 29: 229.

Lundgren, M., Allan, N. L., and Cosgrove, T. 2002. Wetting of Water and Water/Ethanol Droplets on a Non-Polar Surface: A Molecular Dynamics Study. *Langmuir* 18: 10462.

Lundgren, M., Allan, N. L., and Cosgrove, T. 2007. Modeling of Wetting: A Study of Nanowetting at Rough and Heterogeneous Surfaces. *Langmuir* 23: 1187.

Manne, S., and Gaub, H. E. 1995. molecular organization of surfactants at solid-liquid interfaces. *Science* 270: 1480.

Manne, S., Cleveland, J. P., Gaub, H. E., Stucky, G. D., and Hansma, P. K. 1994. Direct Visualization of Surfactant Hemimicelles by Force Microscopy of the Electrical Double Layer. *Langmuir* 10: 4409.

Marmur, A. 1996. Line Tension and the Intrinsic Contact Angle in Solid–Liquid–Fluid Systems. *J. Colloid Interface Sci.* 186: 462.

Modaressi, H., and Garnier, G. 2002. Mechanism of Wetting and Absorption of Water Droplets on Sized Paper: Effects of Chemical and Physical Heterogeneity. *Langmuir* 18: 642.

Mugele, F., Becker, T., Nikopoulos, R., Kohonen, M., and Herminghaus. S. 2002. Capillarity at the Nanoscale: an AFM view. *J. Adhes. Sci. Tech.* 16: 9771.

Noh, J., Kato, H. S., Kawai, M., and Hara, M. 2006. Surface Structure and Interface Dynamics of Alkanethiol Self-Assembled Monolayers on Au(111). *J. Phys. Chem. B* 110: 2793.

Ohler, B., and Langel, W. 2009. Molecular Dynamics Simulations on the Interface between Titanium Dioxide and Water Droplets: A New Model for the Contact Angle. *J. Phys. Chem. C* 113: 10189.

Pan, S., and Rothberg, L. 2005. Chemical Control of Electrode Functionalization for Detection of DNA Hybridization by Electrochemical Impedance Spectroscopy. *Langmuir* 21: 1022.

Park, S., Lee, K.-B., Choi, I. S., Langer, R., and Jon, S. 2007. Dual Functional, Polymeric Self-Assembled Monolayers as a Facile Platform for Construction of Patterns of Biomolecules. *Langmuir* 23: 10902.

Pashley, R. M., and Israelachvii, J. N. 1981. A comparison of surface forcces and interfacial properties of mica in purified surfactant solutions. *Colloids Surf.* 2: 169.

Patankar, N. A. 2003. On the Modeling of Hydrophobic Contact Angles on Rough Surfaces. *Langmuir* 19: 1249.

Patankar, N. A. 2004. Transition between Superhydrophobic States on Rough Surfaces. *Langmuir* 20: 7097.

Petke, D., and Ray, B. R. 1969. Temperature dependence of contact angles of liquids on polymeric solids. *J. Colloid Interface Sci.* 31: 216.

Phillips, J. A., Ross, D., Taborek, P., and Rutledge, J. E. 1998. Superfluid onset and prewetting of 4He on rubidium. *Phys. Rev. B* 58: 3361.

Raghavan, K., Foster, K., and Berkowitz, M. 1991. Comparison of the structure and dynamics of water at the Pt(111) and Pt(100) interfaces: molecular dynamics study . *Chem. Phys. Lett.* 177: 426.

Ran, C., Ding, G., Liu, W., Deng, Y., and Hou, W. 2008. Wetting on Nanoporous Alumina Surface: Transition between Wenzel and Cassie States Controlled by Surface Structure. *Langmuir* 24: 9952.

Reisch, A., Voegel, J.-C., Gonthier, E., Decher, G., Senger, B., Schaaf, P., and Mesini, P. J. 2009. Polyelectrolyte Multilayers Capped with Polyelectrolytes Bearing Phosphorylcholine and Triethylene Glycol Groups: Parameters Influencing Antifouling Properties. *Langmuir* 25: 3610.

Rowlinson, J. S., and Widom, B. W. 1982. *Molecular Theory of Capillarity.* Clarendon Press, Oxford.

Saha, A., Singh, S. P., Singh, J. K., and Kwak, S. K. 2009. Quasi-2D and prewetting transitions of square-well fluids on a square-well substrate. *Mol. Phys.* 107: 2189.

Shenogina, N., Godawat, R., Keblinski, P., and Garde, S. 2009. How Wetting and Adhesion Affect Thermal Conductance of a Range of Hydrophobic to Hydrophilic Aqueous Interfaces. *Phys. Rev. Lett.* 102: 156101.

Shi, B., and Dhira, V. K. 2008. Molecular dynamics simulation of the contact angle of liquids on solid surfaces. *J. Chem. Phys.* 130: 034705.

Shibuichi, S., Yamamoto, T., Onda, T., and Tsujii, K. 1998. Super Water- and Oil-Repellent Surfaces Resulting from Fractal Structure. *J. Colloid Interface Sci.* 208: 287.

Shirtcliffe, N. J., McHale, G., Newton, M. I., and Perry, C. C. 2004. Wetting and Wetting Transitions on Copper-Based Super-Hydrophobic Surfaces. *Langmuir* 20: 10146.

Shirtcliffe, N. J., McHale, G., Newton, M. I., and Perry, C. C. 2005. Wetting and Wetting Transitions on Copper-Based Super-Hydrophobic Surfaces. *Langmuir* 21: 937.

Singh, J. K., Kofke, D. A., and Errington, J. R. 2003. Surface tension and vapor–liquid phase coexistence of the square-well fluid . *J. Chem. Phys.* 119: 3405.

Singh, J. K., Sarma, G., and Kwak, S. K. 2008. Thin-thick surface phase coexistence and boundary tension of the square-well fluid on a weak attractive surface. *J. Chem. Phys.* 128: 044708.

Spori, D. M., Drobek, T., Zurcher, S., Ochsner, M., Sprecher, C., Muhlebach, A., and Spencer, N. D. 2008. Beyond the Lotus Effect: Roughness Influences on Wetting over a Wide Surface-Energy Range. *Langmuir* 24: 5411.

Srinivas, G., Nielsen, S. O., Moore, P. B., and Klein, M. L. 2006. Molecular Dynamics Simulations of Surfactant Self-Organization at a Solid–Liquid Interface. *J. Am. Chem. Soc.* 128: 848.

Steele, A., Bayer, I., and Loth, E. 2009. Inherently Superoleophobic Nanocomposite Coatings by Spray Atomization. *Nano Lett.* 9: 501.

Suzuki, A., and Koboki, Y. J. 1999. Static contact angle of sessile air bubbles on polymer Gel surfaces in water. *Appl. Phys.* 38: 2910.

Vega, C., and de Miguel, E. 2007. Surface tension of the most popular models of water by using the test-area simulation method. *J. Chem. Phys.* 126: 154707.

Wang, H., Tang, L., Wu, X., Dai, W., and Qiu, Y. 2007. Fabrication and anti-frosting performance of super hydrophobic coating based on modified nano-sized calcium carbonate and ordinary polyacrylate. *App. Surf. Sci.* 253: 8818.

Wang, L., Song, Y., Zhang, B., and Wang, E. 2004. Adsorption behaviors of methanol, ethanol, n-butanol, n-hexanol and n-octanol on mica surface studied by atomic force microscopy. *Thin Solid Film* 458: 197.

Wanless, E. J., and Ducker, W. A. 1996. Organization of Sodium Dodecyl Sulfate at the Graphite-Solution Interface. *J. Phys. Chem.* 100: 3207.

Wenzel, R. N. 1936. Resistance of solid surfaces to wetting by water. *Ind. Eng. Chem. Res.* 28: 988.

Werder, T., Walther, J. H., Jaffe, R. L., Halicioglu, T., and Koumoutsakos, P. 2003. On the Water–Carbon Interaction for Use in Molecular Dynamics Simulations of Graphite and Carbon Nanotubes. *J. Phys. Chem. B* 107: 1345.

Xua, F. J., Li, H. Z., Lib, J., Eric, Y. H., T., Zhud, C. X., Kanga, E. T., and Neoha, K. G. 2008. Spatially well-defined binary brushes of poly(ethylene glycol)s for micropatterning of active proteins on anti-fouling surfaces. *Biosensors Bioelectron* 24: 773.

Yaneva, J., Milchev, A., and Binder, K. 2004. Polymer nanodroplets forming liquid bridges in chemically structured slit pores: A computer simulation. *J. Chem. Phys.* 121: 12632.

Yang, A., and Weng, C. 2008. Influence of alkanethiol self-assembled monolayers with various tail groups on structural and dynamic properties of water films. *J. Chem. Phys.* 129: 154710.

Yang, C., Tartaglinoa, U., and Persson, B. N. J. 2008. Nanodroplets on rough hydrophilic and hydrophobic surfaces. *Eur. Phys. J.* 25: 139.

Yang, Q. F., Ding, J., and Shen, Z. Q. 2000. Investigation on fouling behaviors of low-energy surface and fouling fractal characteristics. *Chem. Eng. Sci.* 55: 797.

Yoshimitsu, Z., Nakajima, A., Watanabe, T., and Hashimoto, K. 2002. Effects of Surface Structure on the Hydrophobicity and Sliding Behavior of Water Droplets. *Langmuir* 18: 5818.

Zangi, R., and Berne, B. R. 2008. Temperature Dependence of Dimerization and Dewetting of Large-Scale Hydrophobes: A Molecular Dynamics Study. *J. Phys. Chem. B* 112: 8634.

Zehl, T., Wahab, M., Schiller, P., and Mogel, H. J. 2009. Monte Carlo Simulation of Surfactant Adsorption on Hydrophilic Surfaces. *Langmuir* 25: 2090.

Zhang, S., Huang, F., Liu, B., Ding, J., Xu, X., and Kong, J. 2007. A sensitive impedance immunosensor based on functionalized gold nanoparticle–protein composite films for probing apolipoprotein A-I. *Talanta* 71: 874.

Zhao, X. 2007. Wetting transition of water on graphite: Monte Carlo simulations. *Phys. Rev. B* 76: 144301.

8 Molecular Modeling of Capillary Condensation in Porous Materials

Sudhir K. Singh and Jayant K. Singh

CONTENTS

8.1 INTRODUCTION

Knowledge of phase diagrams and thermodynamic properties of a material system is crucial for improving existing materials and developing new materials. Particularly, condensation in capillaries plays a direct role in various industries such as membrane separation (separation of olefin and paraffin gas mixtures using adsorptive separation), purification (water purification using nanoporous membranes), catalysis (nanoporous catalysts used for fluid catalytic cracking processes), adsorption/gas storage (e.g., it has been observed that some complex carbon nanoporous materials are capable of storing natural gas at unprecedented density and comparatively at very low pressure of conventional natural gas tanks, which in fact could increase the viability of adsorbed gas-fueled vehicles in the near future), drying, enhanced oil recovery (structure of porous rocks can affect fluid saturation and interfacial properties and hence relative permeability), sensors (detection of gas leakage or detection of multiple ions simultaneously in the solution phase of liquids), and lubrication and moisture transport in soil. For these reasons, knowledge of capillary condensation at different state points is necessary for efficiently designed novel porous materials and hence improved industrial processes. The usefulness of molecular modeling and simulations can be exploited as a screening tool to identify new and better candidates. Therefore, phase transitions in porous media have been an active research area since the introduction of high-speed computers.

The capillary condensation phenomenon often involves vapor and liquid phases in equilibrium. Hence, vapor–liquid and liquid–liquid equilibrium studies become extremely important and essential not only from a scientific point of view but also for the efficient design and improved operation of various industrial processes (Altwasser et al. 2005; Bhatia 1998; Gupta et al. 1997; Jackson and McKenn 1990; Matranga et al. 1992). Several studies done over the last decade using theoretical approaches such as density functional theory (Evans 1992) and molecular simulation (Evans 1990; Koga and Tanaka 2005; Rivera et al. 2002; Zangi 2004), as well as experimental studies (Thommes and Findenegg 1994) on well-characterized porous materials, have provided considerable insight into the nature of these transitions and capillary condensation phenomenon. In this chapter, our aim is to provide an introduction to phase transitions under confinement, capillary condensation, hysteresis phenomenon, critical properties, and crossover behavior from 3D to 2D fluids in porous materials. In addition, we summarize molecular modeling techniques to study these phenomena.

8.2   PHASE DIAGRAM CALCULATIONS: VAPOR–LIQUID AND LIQUID–LIQUID EQUILIBRIA

Some questions that are of interest in regard to vapor–liquid and liquid–liquid phase transitions in confined geometry are:

1. What are the available techniques to evaluate the phase diagram in confined geometry?
2. How does confinement affect the phase diagram; that is, what is the influence of confinement on the vapor–liquid and liquid–liquid phase transitions, critical properties, interfacial properties, and critical mixing point of a solute in a solvent?
3. How are these properties affected by pore morphology such as pore size, shape, and material?

To address the aforementioned questions, various investigations have been undertaken, resulting in breakthroughs in this field. In the following discussion, we summarize some of the techniques used for investigating properties of confined fluids within the framework of experiment, theory, and simulation.

8.3   EXPERIMENTAL TECHNIQUE

We briefly describe the techniques used to obtain capillary condensation and evaluate the associated properties. Typically, in experiments, volumetric adsorption apparatus made of stainless steel components and fittings is used. The typical schematic of the apparatus is shown in Figure 8.1. The pressure sensor, dosing volume, and associated valves are kept at a constant temperature, and a variable temperature bath surrounding the sample fits immediately below the constant temperature box. Temperatures are measured with a platinum resistance thermometer. Blank and sample cells have the same volume and are kept at the same operating temperatures. The blank cell is used to estimate the excess adsorption in the sample cell. At low temperatures, at which the density of the adsorbed fluid is much greater than that of the bulk vapor, the excess amount adsorbed is determined with accuracy; at higher temperatures and pressures, however, the error may be comparatively larger. Using this apparatus, adsorption isotherms are determined as follows: A given amount of adsorptive is added to the system, and the temperature is then increased stepwise from the lowest temperature to the highest. At each step, temperature, pressure, and the excess amount adsorbed (per unit mass of the adsorbent) are measured and adsorption isotherms are generated. Similarly, the process is reversed with temperature decreasing stepwise. If hysteresis is absent, the points obtained while decreasing temperature coincide with those obtained with increasing temperature, whereas if hysteresis is present, this is not the case. A more detailed description of the apparatus and operating procedure is given elsewhere (Machin 1999).

Ordered mesoporous materials are of significant industrial interest, especially in fields of catalysis, novel adsorbents, nanoparticles/nanocomposites, and sensor applications. Hence, it becomes necessary to understand the various fundamental thermodynamic aspects related to the mesoporous material, which in turn will facilitate better design of new material for intended industrial application.

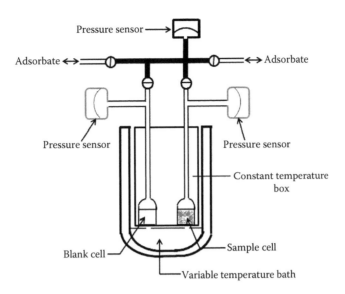

FIGURE 8.1   Typical schematic of a volumetric adsorption apparatus.

Figure 8.2 presents a typical adsorption isotherm of hexafluoroethane (HFE) on an ordered mesoporous silica (MCM-41) (Machin 2003). In this figure, the ordinate denotes the amount of fluid adsorbed as a function of relative fugacity (same as relative vapor pressure for a single component) at various sets of temperatures. The total amount adsorbed, n_t, is given by the following expression:

$$n_t = n_e + V_p \rho_{vap} \tag{8.1}$$

where n_e is the excess adsorption, V_p is the pore volume, and ρ_{vap} is the bulk vapor density. At lower temperatures and pressures, $n_t \approx n_e$, but may be up to 10% larger at higher temperatures. Typical isotherms are shown in Figure 8.2. At lower temperatures, isotherms exhibit features and characteristics of numerous adsorbates (of different size and shapes) on ordered mesoporous materials with initial monolayer–multilayer growth at low relative fugacities, leading to an abrupt, nearly vertical jump, followed by a nearly horizontal plateau at higher relative fugacities. The height of the step and its slope both decrease at higher temperatures, with the step becoming indistinct/absent at a sufficiently high temperature, with the point at which this occurs depending on the adsorbate as well as the adsorbent characteristics such as shape, size of the adsorbate and interaction parameters, and pore size distribution of the adsorbent. The standard Brunauer, Emmett, and Teller (BET) equation (Gregg and Sing 1982) provides an excellent description of each isotherm up to the onset of the step or well into the multilayer region when the step is absent, that is:

$$n_t/n_m = C(f/f^\circ) / \{(1 - f/f^\circ)[1 + (C-1)(f/f^\circ)]\} \tag{8.2}$$

where n_t is the total amount adsorbed at relative fugacity f/f°, n_m is the monolayer capacity, and C is a BET parameter. The solid lines in Figure 8.2 illustrate the goodness of fit of Equation (8.2) to the experimental data.

The BET parameters C and n_m are given by some empirical relations which, for HFE in MCM-41 material, are as follows (Machin 2003)

$$C = \exp(-0.01234T + 5.3414) \tag{8.3}$$

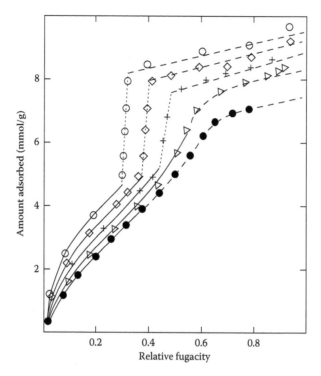

FIGURE 8.2 Adsorption isotherms of HFE on MCM-41 at different temperatures. Symbols open circle, diamond, cross, triangle, and closed circle represent the temperatures 196 K, 216 K, 236 K, 256 K, and 276 K, respectively. (Adapted from Machin, W. D., *Phys. Chem. Chem. Phys.*, 5: 203, 2003. With permission.)

$$n_m = 7.321 - 2.761 \times 10^{-2}T + 4.379 \times 10^{-5}T^2 \tag{8.4}$$

For temperature $T < 255$ K (for the considered system), the relative fugacity corresponding to the lower position of the step (f_L) is given by

$$f_L/f^\circ = 3.516 \times 10^{-3}T - 0.3877 \tag{8.5}$$

and at the top of the step is given by:

$$f_U/f^\circ = 4.090 \times 10^{-3}T - 0.4799 \tag{8.6}$$

The amounts adsorbed at saturation n_s are obtained by extrapolation of the linear plateau region (represented by the dashed line in Figure 8.2) to $f/f^\circ = 1$, and these values for the system are given by:

$$n_s = 11.132(1 - T/T_{cb})^{0.14} \tag{8.7}$$

where T_{cb} is the bulk critical temperature of the fluid.

The abrupt step in adsorption isotherms, resulting from reversible capillary condensation of the adsorbate, is shown as a tie-line (represented by a dotted line in Figure 8.2) between a less dense capillary gas phase and a dense capillary liquid phase. As temperature increases, the length of the tie-line (i.e., step height) decreases and eventually vanishes at the capillary critical temperature. Further, with the help of isotherms at different temperatures, the amount adsorbed in the capillary gas and liquid phases can be estimated and a vapor–liquid phase diagram constructed. Using this technique, a typical capillary vapor–liquid phase diagram is shown in Figure 8.3 for HFE in MCM-41 (Machin 2003).

The average density along each tie-line (the rectilinear diameter), $(n_U + n_L)/2$, decreases slowly as shown by the following relation obtained by fitting an empirical form to the experimental data:

$$(n_U + n_L)/2 = -2.441 \times 10^{-3} T + 6.994 \tag{8.8}$$

The above average density line intersects the phase boundary line (shown by the solid curve in Figure 8.3) at the capillary critical temperature (T_{cc}). The density difference, $n_U - n_L$, obeys the scaling law typical of bulk fluids, given by:

$$(n_U - n_L) = k(1 - T/T_{cc})^\beta \tag{8.9}$$

where k is a constant and β is the critical scaling parameter, which has the value of 0.326 for bulk 3D fluid (Fisher 1964). For the HFE in MCM-41 system, the values of β and k are estimated as 0.37 and 5.967, respectively.

In experiments, because of the possibility of the appearance of a metastable state(s), it is often difficult to know if true thermodynamic equilibrium has been achieved in any reasonable experimental time scale. Some common experimental difficulties include: (1) the precise control on operating parameters, (2) the possibility that the surfaces and pore structure may change with temperature or pressure, and (3) trace amounts of impurities in the adsorbate may preferentially adsorb on the pore walls, leading to spurious results. Hence, with the available experimental techniques it becomes quite difficult to predict the trends of the phase behavior and related thermodynamic properties with reasonable accuracy, especially at extremely small nanopores.

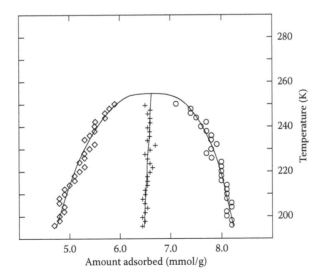

FIGURE 8.3 The capillary coexistence curve of HFE in MCM-41 for the first-order phase transition. Symbols diamond and circle represent the amount adsorb at bottom and top of each step (shown by dotted line in Figure 8.2), respectively, at $T < 255$ K. The rectilinear diameter (average density along each tie-line i.e., $(n_U + n_L)/2$) is represented by the cross symbol. (Adapted from Machin, W. D., *Phys. Chem. Chem. Phys.* 5: 203, 2003. With permission.)

8.4 THEORETICAL TECHNIQUE

With regard to the theoretical calculation of the phase diagram of confined fluid, density functional theory (DFT) has found great utility (Evans 1992). The key step in DFT is to establish the Helmholtz free energy as a functional of density distribution. Early attempts to accomplish this failed because of the neglect of structural correlations in the Helmholtz free energy functional (Yu et al. 2006). However, since then, several attempts have been made to include the structure correlations of inhomogeneous fluid (Choudhury and Ghosh 2003; Kim and Lee 2004; Lutsko 2008; Muller et al. 2003; Tang and Wu 2003) in the confined geometry.

In the DFT approach, the grand potential (Ω) of the fluid in a pore at a given temperature, T, and chemical potential, μ, is a function of the local fluid density $\rho(r)$ (Ravikovitch et al. 2001), where r is the magnitude of the vector joining the center of one particle to other:

$$\Omega[\rho(r)] = F[\rho(r)] + \int [v(r) - \mu]\rho(r)dr \tag{8.10}$$

where $F[\rho(r)]$ the Helmholtz is free energy functional and $v(r)$ is the potential of the pore walls. The criterion for equilibrium is that the grand potential, $\Omega[\rho(r)]$, is a minimum; that is, the density profile $\rho(r)$ of the adsorbate within the pore satisfies the condition:

$$\frac{d\Omega[\rho(r)]}{d\rho(r)} = 0 \tag{8.11}$$

In a perturbation expression, the Helmholtz free energy functional is the sum of the contribution from a reference system of hard spheres and the attractive contributions; that is

$$F[\rho(r)] = F_h[\rho(r)] + \frac{1}{2}\iint \rho(r)\rho(r')\phi_{attr}(|r-r'|)drdr' \tag{8.12}$$

where F_h is the hard sphere Helmholtz free energy functional and in the right-hand side second term of Equation 8.12, the mean field approximation is used. The hard sphere term can be further split into an ideal gas component and an excess component:

$$F_h[\rho(r)] = \int \rho(r)\left[kT(\ln(\rho(r))-1) + f_{ex}(\overline{\rho}(r))\right] \tag{8.13}$$

where the excess Helmholtz free energy per particle, f_{ex}, can be described by the Carnahan–Starling equation of state for a system of hard spheres (Carnahan and Starling 1969) using a smoothed density $\overline{\rho}$. The smoothed density $\overline{\rho}(r)$, is defined as:

$$\overline{\rho}(r) = \int w(|r-r'\overline{\rho}(r))\rho(r')dr' \tag{8.14}$$

where $w(|r-r'|)$ is a weighted function (Lastoskie et al. 1993). Therefore, the grand potential under the confined system can be expressed as:

$$\Omega[\rho(r)] = \int \rho(r)[kT(\ln(\Lambda^3\rho(r))-1) + f_{ex}(\overline{\rho}(r))]dr + \int \rho(r)[v(r)-\mu]dr$$
$$+ \frac{1}{2}\iint \rho(r)\rho(r')\phi_{attr}(|r-r'|)drdr' \tag{8.15}$$

The attractive contributions for, for example, a Lennard–Jones (LJ) fluid, were modeled with the cut-shifted Lennard–Jones (CSLJ) potential, split according to the Weeks–Chandler–Anderson (WCA) prescription (Weeks et al. 1971) at its minimum $r_m = 2^{1/6} \sigma_{ff}$, where, σ_{ff} is the diameter of the fluid particle.

Combining Equations 8.10 to 8.15, one can obtain:

$$\rho(r) = \exp\left[-\frac{1}{kT}\left[f_{ex}(\bar{\rho}(r')) + \int \rho(r') f'_{ex}(\bar{\rho}(r'))dr' \times \frac{w(|r - r'|\bar{\rho}(r'))}{1 - \bar{\rho}_1(r') - 2\bar{\rho}_2(r')\bar{\rho}(r')} + v(r) - \mu \right. \right.$$
$$\left. \left. + \iint \rho(r')\phi_{attr}(|r - r'|)dr' \right]\right] \quad (8.16)$$

Consequently, density profiles may be obtained by iteration of Equation 8.16, enabling the calculation of other properties of the fluid such as capillary condensation, adsorption, hysteresis, and phase transition.

It has been shown that the mean-field weight function DFT (MFWDFT) (Peng and Yu 2008) can successfully predict the density profile, adsorption isotherms, and vapor–liquid phase transitions in slit-like pores for model fluids. Further, it has been confirmed that the results of these calculations are in good agreement with molecular simulation results (Hamada et al. 2007; Sweatman 2001; Vishnyakov et al. 2001). Evidence of this may be seen in Figure 8.4, which shows a typical comparison of the phase diagram of methane (modeled by an LJ potential) in a smooth graphite slit pore as calculated using theoretical and molecular simulation methodologies. From Figure 8.4, one can see that the mean field theory (MFT) (Katsov and Weeks 2001) dramatically overestimates the vapor–liquid critical points of the confined fluid, exceeding even the bulk critical point, which cannot be true. On the other hand, MFWDFT predicts a slightly lower critical temperature, as expected, and is in reasonably good agreement with the simulation data. Moreover, DFT results predict that under confinement, the vapor–liquid coexistence curve shrinks and the critical temperature decreases, similar to the experimental observations.

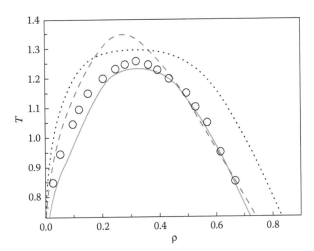

FIGURE 8.4 Vapor–liquid phase diagrams of methane in a graphite pore with fluid–solid interaction of $\varepsilon_{sf}/k_B = 21.5$ K and pore width of $H = 10\sigma$, where σ is diameter of methane. Temperature, T, and density, ρ, are in adimensional form in the plot. The open circles and dashed and solid lines represent the results from the simulations, MFT, and MFWDFT, respectively. The dotted curve represents the bulk vapor-liquid equilibrium obtained from simulation. (From Vishnyakov, A., et al. *Langmuir* 17: 4451, 2001; Adapted from Peng, B. and Yu, Y.-X., *J. Phys. Chem. B.* 112: 15407, 2008.)

Using the DFT approach, Zhang and Wang (2006) investigated the effect of the surface–fluid interaction strength on the vapor–liquid phase diagram and critical point shift in a single cylindrical pore for a model (square-well) fluid. The authors noticed that the dependence of the critical temperature shift on the surface–fluid interaction does not follow a simple monotonic behavior; rather, it increases with the strength of the surface–fluid interaction for weak surfaces initially and subsequently decreases for strong surfaces. A similar trend is reported with different fluid–fluid interactions and different pore widths. On the other hand, the critical density of square-well fluids in a confined space increases monotonically as the surface–fluid interaction becomes more attractive.

8.5 MOLECULAR SIMULATION APPROACH

8.5.1 VAPOR–LIQUID PHASE TRANSITION AND CRITICAL POINT

Molecular simulation techniques are used to bridge the gap between experimental outcomes and related theoretical investigations, with increased computational power and new, efficient algorithms being added advantages in this bridging. Vapor–liquid phase equilibria are invariably related to the phenomenon of adsorption and capillary condensation. Adsorption of a gas by a porous material is described quantitatively by an adsorption isotherm, the amount of gas adsorbed by the material at a fixed temperature as a function of pressure. The uptake of a fluid into a porous material could be intuitively viewed simply as the filling of an existing vacuum, but adsorption has long been recognized as a far more subtle phenomenon. Gibbs expressed the concept of adsorption on a general thermodynamic basis as follows. For the system of a fluid in contact with an adsorbent, the amount adsorbed as the quantity of fluid that is in *excess* (see Equation 8.1) of that which would be present if the adsorbent had no influence on the behavior of the fluid (Nicholson and Parsonage 1982).

In capillary condensation phenomena, multilayer adsorption from the vapor phase into a porous medium proceeds until the pores are filled with condensed liquid from the vapor phase. In this process, vapor condensation occurs below the bulk saturation pressure of the liquid because of the increased number of interactions between vapor phase molecules inside the porous media. Once the condensation process is completed, a vapor–liquid interface is formed in the pores with an equilibrium vapor pressure, P, which is below the bulk saturation pressure, P_S, of the fluid. For very large pores (i.e., macropores), the relation between P and P_S, can be well described with the macroscopic form of the Kelvin equation (Melrose 1966):

$$N_A kT \ln \frac{P}{P_S} = V\left(P - P_S + \frac{2\gamma \cos\theta}{r}\right) \tag{8.17}$$

where θ is the contact angle, V is the molar liquid volume, γ is the vapor–liquid surface tension, N_A is the Avogadro number, k is Boltzmann constant, and T is the absolute temperature. In the case of smaller size range mesoporous materials, certain empirical modifications depending on the shape of the adsorbate meniscus have been done (Kruk et al. 1997). By incorporating the statistical film thickness, pore-wall potential, and curvature-dependent surface tension, it has been shown that results predicted by the molecular models of porous materials gave fairly good quantitative agreement with experiments (Casanova et al. 2008; Kruk et al. 1997; Miyahara et al. 2000; Morishige and Tateishi 2003). Fisher and Nakanishi (1981) predicted that when the width of a pore approaches a few molecular diameters, the macroscopic concept of the Kelvin equation fails, and consideration of the role of solid surfaces on fluid properties becomes important. Also, the gas and liquid in a pore can no longer be treated as uniform phases separated by a well-defined interface. Further, the vapor phase cannot be treated as an ideal gas and the vapor–liquid surface tension can no longer be assumed the same as the bulk. Consequently, the Kelvin equation has been modified to take into account gas nonideality, adsorbate layers, and adsorption field. Chen et al. (2006) proposed a microscopic form of the Kelvin equation that takes into account the two necessary aspects, that is, Laplace's pressure and oscillatory compression pressure, that become increasingly important in micropores and small mesopores. The modified equation is:

$$N_A kT \ln \frac{p}{p_S} = V_{L0}\left(\frac{2\gamma\cos\theta}{r} + \Delta p_c\right)\left(1 - \frac{\beta}{2}\left(\frac{2\gamma\cos\theta}{r} + \Delta p_c\right)\right) \tag{8.18}$$

where β is the compressibility coefficient of the liquid and ΔP_c is the pressure change caused by the adsorption compression (Aranovich and Donohue 2003). Chen et al. (2006), with a typical example of n-hexadecane, concluded that capillary vapor pressure shows oscillating behavior with a pore width similar to that seen in experiments (Fisher and Israelachvili 1981). The comparison is shown in Figure 8.5. Moreover, such oscillating behavior is also observed in the spreading pressure of the various model fluids (Singh et al. 2009, 2010a) confined in the small nanopores.

Similar to capillary condensation, another interesting phenomenon related to porous material is capillary evaporation, or desorption. It has been observed that capillary condensation (adsorption) and capillary evaporation (desorption) do not take place at the same pressure and result in a an interesting phenomenon known as *hysteresis*. A typical illustration of the phenomenon is shown in Figure 8.6 with a model fluid using model slit pore geometry, calculated using grand canonical Monte Carlo (GCMC) and molecular dynamics. In Figure 8.6, adsorbed fluid density versus relative vapor pressure, P/P_0, is plotted. Several important things should be noted. There is excellent agreement between molecular dynamics and Monte Carlo results. Both methods predict the same sequence of states for the system on adsorption and desorption. Figure 8.7 presents the corresponding configurations of key states of the system going through the adsorption and desorption process. In adsorption the system goes through several layering stages and then undergoes capillary

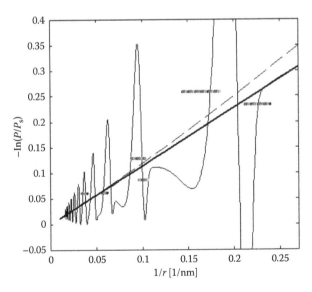

FIGURE 8.5 Relative vapor pressure vs. inverse of pore radius of n-hexadecane in nanoconfinement is plotted on a semi-log scale. Straight line represents the classical Kelvin equation, dashed line represents corrected Kelvin equation, solid symbols represent experimental data (Fisher, L. R. and Israelachvili, J. N. *J. Colloid Interface Sci.* 80: 528, 1981), and oscillating line represents data (Adapted from Chen, Y. et al., *J. Colloid Interface Sci.* 300: 45, 2006. With permission.)

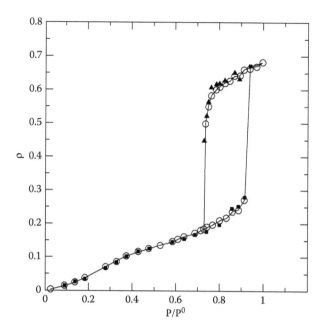

FIGURE 8.6 Adsorption/desorption isotherms of adimensional density (ρ) vs relative pressure (P/P_0) at a typical reduced temperature for an open-ended slit geometry, calculated via grand canonical Monte Carlo (open circles) and molecular dynamics (filled squares, adsorption; filled triangles, desorption). (Adapted from Sarkisov, L. and Monson, P. A., *Langmuir*, 17: 7600, 2001. With permission.)

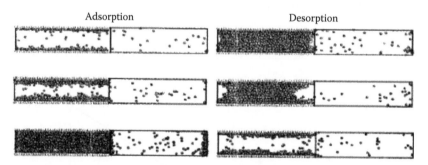

FIGURE 8.7 Computer graphics visualizations of various stages of adsorption and desorption in an open-ended slit geometry. The adsorption column corresponds to a sequence of increasing pressures (from top to bottom); the desorption column corresponds to a sequence of decreasing pressures (from top to bottom). (Adapted from Sarkisov, L. and Monson, P. A., *Langmuir*, 17: 7600, 2001. With permission.)

condensation. In desorption, however, hemispherical menisci develop on the ends of the pore, followed by a transition as these menisci approach each other.

It has been observed that hysteresis invariably occurs with mesoporous materials (pore size between 2 and 50 nm). However, the origin of the hysteresis phenomenon is not completely understood, that is, whether it is mainly due to pore size effects or something else. Some studies related to hysteresis indicate that the phenomenon is usually attributed to thermodynamic or network effect or the combination of the two (Ball and Evans 1989). The thermodynamic effects are related to the metastability of adsorption or desorption (or both) branches of the isotherm. In fact, adsorption or desorption may be delayed and take place at higher or lower pressures, respectively, compared to the pressure of coexistence between the gaslike and liquidlike phases in the pore. In addition, hysteresis may also be caused by pore connectivity (network) effects, which are expected to play an important role in desorption processes. That is, if larger pores have access to the surrounding is only through narrower pores, the former cannot be emptied at the relative pressure corresponding to their capillary evaporation because they are still filled with the condensed adsorbate. So, the larger pores may actually be emptied at the relative pressure corresponding to the capillary evaporation in the smaller connecting pores (or at the relative pressure corresponding to the lower limit of adsorption–desorption hysteresis). Hysteresis loops observed experimentally most likely arise from some combination of thermodynamic and network effects, although the latter are often particularly prominent.

In simulation approaches to phase diagram calculation of confined fluids, the major methodologies include Gibbs ensemble Monte Carlo (GEMC) (McGrother and Gubbins 1999; Panagiotopoulos 1987; Smith and Vortler 1996; Vishnyakov et al. 2001), GCMC (Bock and Schoen 1999, Jiang and Sandler 2005), Gibbs–Duhem Integration (GDI) (Vortler and Smith 2000), and grand canonical transition-matrix Monte Carlo (GC-TMMC) (Errington 2003; Singh and Kwak 2007).

In the GC-TMMC approach, Monte Carlo simulations are conducted in a standard grand canonical ensemble in which the volume (V), chemical potential (μ), and temperature (T) are held constant and the particle number N (density) and energy (U) fluctuate. During a simulation, attempted transitions between the states of different densities are monitored. At regular intervals during a simulation, this information is used to obtain an estimate of the density probability distribution, which is subsequently used to bias the sampling to low probability densities. This ensures that, over time, all densities of interest are sampled adequately. GC-TMMC, among others, recently has been applied to a variety of systems to investigate the vapor–liquid phase equilibria (Singh et al. 2003; Singh and Kofke 2004) and surface tension (Singh and Kwak 2007). Transition matrix Monte Carlo (TMMC) is a sophisticated and intelligent technique that uses the Monte Carlo moves acceptance ratio to fill up the transition matrix. Hence, even rejected moves contribute to the transition matrix, leading to an efficient way to use all the information available from the simulation. Subsequently, detailed balance is used to obtain the macroscopic probabilities. The advantage of TMMC in estimating the vapor–liquid phase equilibria is two-fold. The first is that it can be applied cumulatively, meaning that the existing transition-probability information does not need to be discarded upon the redefinition of the sampling bias. Sampling bias is a technique used to explore the entire configuration space efficiently. In sampling bias, a weighting function (which is inversely proportional of the current estimate of the macrostate probabilities) is introduced that encourages the system to sample all states evenly. In this scheme, weighting function can be updated periodically without discarding the data obtained previously. This aspect of TMMC is helpful because the method relies on an iterative scheme to evolve the sampling bias, which ensures that all particle numbers are sampled sufficiently. In practice, a simulation proceeds with periodic updates of the weighting function until the macrostate probabilities converge. Another important aspect of the TMMC method is the use of the histogram reweighting method to evaluate the phase coexistence value of the chemical potential. This determination is readily performed from the knowledge of the probability distribution from the TMMC calculations.

The result is an efficient self-adaptive method for determining the density probability distribution over a specified range of densities (typically a range that corresponds to the densities of two potentially coexisting phases). Once a probability distribution has been collected at a given value of chemical potential, μ_0, histogram reweighting (Ferrenberg and Swendsen 1988) is used to shift the probability distribution to coexistence value of the chemical potential, μ, using the following relation:

$$\ln \prod (N, \mu) = \ln \prod (N, \mu_0) + \beta(\mu - \mu_0) N \qquad (8.19)$$

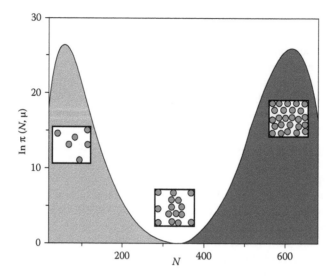

FIGURE 8.8 A typical coexistence probability distribution generated from Monte Carlo simulation and histogram reweighting for a confined model fluid on semi-logarithmic plot.

To determine the coexistence chemical potential, μ, this relation is applied to estimate the chemical potential that produces a coexistence probability distribution as shown in Figure 8.8, in which the area under the peak corresponding to the vapor phase and midpoint of the probability distribution becomes equal to that of liquid peak and the midpoint of the probability distribution. Saturated vapor and liquid densities are related to the first moment of the vapor and liquid peaks respectively, of the coexistence probability distribution, $\Pi(N, \mu)$.

In the close vicinity of the critical point, the characteristic size of the density fluctuations increases; therefore, the simulation cannot be performed near the critical point. Wegner (1972) showed that away from the critical point the difference of coexisting vapor and liquid densities can be written in the following form:

$$\rho_l - \rho_v = C_0\left(1 - \frac{T}{T_c}\right)^{\beta} + C_1\left(1 - \frac{T}{T_c}\right)^{\beta+\Delta} + C_2\left(1 - \frac{T}{T_c}\right)^{\beta+2\Delta} + \cdots \tag{8.20}$$

where ρ_l, ρ_v, and T_C are coexistence liquid and vapor number densities and critical temperature, respectively; Δ is the gap exponent; and C_i are the correction amplitude or coefficients. The parameter β is known as the order parameter critical exponent. However, for the temperatures moderately close to the critical point, $0 < (1-(T/T_c)) < 0.2$, the gap exponent terms that describe behavior *far away* from the critical point were expected to be very small compared to other leading terms (Wegner 1972). Moreover, all the higher terms C_i for $i > 0$ in the Wegner expansion, that is, in Equation 8.20, can be excluded without significant change of the estimated critical temperature, T_c. Therefore, the estimation of T_c is generally based on the knowledge of several liquid–vapor coexistence points in an appropriate temperature range of $0 < (1-(T/T_c)) < 0.2$, using the following simplified expression:

$$\rho_l - \rho_v = C_0\left(1 - \frac{T}{T_C}\right)^{\beta} \tag{8.21}$$

To estimate the critical density, ρ_c, an equation for the diameters, $(\rho_l + \rho_v)/2$, of the coexistence curve is used (Sengers and Levelt-Sengers 1978):

$$\frac{\rho_l + \rho_v}{2} = \rho_c + B_1\left(1 - \frac{T}{T_c}\right)^{\psi} + B_2\left(1 - \frac{T}{T_c}\right) + B_3\left(1 - \frac{T}{T_c}\right)^{\psi + \Delta} + \cdots \tag{8.22}$$

The anomaly in the diameter of the coexistence curve characterized by Ψ is weak and difficult to observe. By combining Equations 8.20 and 8.22, an equation for the coexistence vapor and liquid densities can be obtained as:

$$\rho_{\pm} = \rho_c + B_1\left(1 - \frac{T}{T_c}\right)^{\psi} + B_2\left(1 - \frac{T}{T_c}\right) + B_3\left(1 - \frac{T}{T_c}\right)^{\psi + \Delta} + \cdots$$

$$\pm \frac{1}{2}\left(C_0\left(1 - \frac{T}{T_c}\right)^{\beta} + C_1\left(1 - \frac{T}{T_c}\right)^{\beta + \Delta} + C_2\left(1 - \frac{T}{T_c}\right)^{\beta + 2\Delta} + \cdots\right) \tag{8.23}$$

where ρ_- and ρ_+ represent the vapor, ρ_v, and liquid, ρ_l, phase coexistence densities, respectively. Further it was concluded that, as a result of statistical uncertainties in the simulation data itself, contributions from the higher order terms of Equation 8.23 can be neglected without significant change in the estimation of the critical point (Wegner 1972). More specifically, terms B_1, B_3, and other higher terms in the expression of the diameter, and all the higher terms C_i for $i > 0$ in the Wegner expansion, are excluded. Therefore, Equation 8.23 then simplifies to

$$\rho_{\pm} = \rho_c + B_2\left(1 - \frac{T}{T_c}\right) \pm \frac{1}{2}C_0\left(1 - \frac{T}{T_c}\right)^{\beta} \tag{8.24}$$

which can also be represented in more simplified form as:

$$\frac{\rho_l + \rho_v}{2} = \rho_c + D(T - T_C) \tag{8.25}$$

where D is a fitting parameter. The critical temperature, T_C, estimated from Equation 8.21 is used to calculate the critical density, ρ_c, from the least square fit of Equation 8.25. Equations 8.21 and 8.25 have been used in various recent works (Jana et al. 2009; Singh et al. 2009, 2010a, 2010b) to estimate the critical point of model fluids in confinement. It is worth mentioning that the approach using Equations 8.21 and 8.25 to estimate T_C and ρ_c is merely an approximate means to obtain the critical point data, albeit quite a successful technique for bulk fluids. A more rigorous and accurate technique is Binder's fourth-order cumulant of the order-parameter M, and mixed field finite size scaling technique (Wilding 1995).

In the fourth order cumulant technique, a histogram is recorded $H = H(\mu, \rho)$ at a certain set of temperatures. The idea behind this technique is to record the fourth cumulant of the order parameter M at different temperatures for different system sizes near a guessed critical temperature. The fourth order cumulant, U_L, is given by the following expression:

$$U_L = 1 - \frac{\langle M^4 \rangle_L}{3\langle M^2 \rangle_L^2} \tag{8.26}$$

where M is deviation from the mean density, that is, $\rho - \langle\rho\rangle$. The cumulant at the critical point where is deviation from the mean density would be a universal point (Binder 1981). To evaluate it, U_L is calculated for different temperatures and is plotted against temperature for different sizes, L. The plots of U_L for different L have a common intersection point corresponding to the critical temperature of an infinite system (Binder 1981). To illustrate the technique, a typical example of a model, square-shoulder, square well (SSSW) fluid (Rżysko et al. 2008) in bulk phase, is shown in Figure 8.9. Such an approach for a confined system was recently applied (Rżysko et al. 2010) for the SSSW model fluid confined in a hard slit pore.

A more rigorous approach is the mixed field finite size scaling theory as used by Liu et al. (2010) for the model confined fluids. This approach utilizes an ordering operator M, which is a linear combination of the number of particles N and total configurational energy U

$$M = N - sU \qquad (8.27)$$

where s is the field mixing parameter. At criticality, the probability distribution of the scaling parameter of x, which is defined as $x = A(M - M_c)$ (where A is a nonuniversal parameter and M_c is the critical value of the ordering operator M), has a universal form. Depending on the dimensionality of the system, the vapor–liquid transition belongs to the 2D or 3D universality class of the Ising model. Accordingly, the distribution can have one of two limiting forms: the 2D universal distribution or the 3D universal distribution. The nonuniversal parameter A and the critical value of the ordering operator M_c are chosen so that the zero mean and unit variance of the distribution are observed in the scaling parameter. To obtain the critical parameters, the thermodynamic parameters T and μ and the field mixing parameter s are adjusted to get best fit to the universal distributions. An objective function is used to minimize the deviation of the probability distribution of the ordering operator from the 2D/3D-dimensional universal distribution. A more detailed description of this method may be found in the literature (Liu et al. 2010).

Recent work done using the molecular simulation technique by Jana et al. (2009), Singh et al. (2009, 2010a) and Liu et al. (2010) have been able to clarify, up to a certain extent, most of the unanswered questions in the field of vapor–liquid phase transitions and critical properties under single pore confinement, using various model fluids. A typical vapor–liquid phase diagram of a simple model fluid in an attractive model slit pore is shown in Figure 8.10, for slit widths ranging

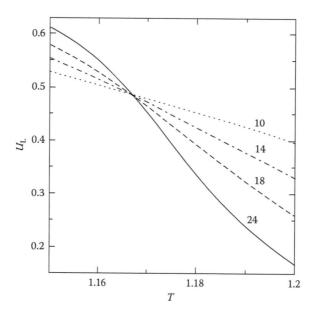

FIGURE 8.9   The cumulants U_L versus the adimensional temperature, T, in the vicinity of the vapor-liquid critical temperature. The numbers in the figure represents the value of L/σ_{ff}. (Adapted from Rżysko, W., et al. *J. Chem. Phys.* 129: 124502, 2008. With permission.)

from 40 to 1 molecular diameter. The following models are used to describe fluid–fluid and fluid–substrate (in this case slit pore) interactions:

$$
u_{ff}(r) = \begin{cases} \infty, & 0 < r < \sigma_{ff} \\ -\varepsilon_{ff}, & \sigma_{ff} \leq r < \lambda_{ff}\sigma_{ff} \\ 0, & \lambda_{ff}\sigma_{ff} \leq r \end{cases}
$$

$$
u_{wf}(z) = \begin{cases} \infty, & 0 < z < \sigma_{ff}/2 \\ -\varepsilon_{wf}, & \sigma_{ff}/2 \leq z < \lambda_{wf}\sigma_{ff} \\ 0, & \lambda_{wf}\sigma_{ff} \leq z \end{cases} \tag{8.28}
$$

where r is the interparticle separation distance, z, is separation distance of the particle from the surface, $\lambda_{ff}\sigma_{ff}$ is the fluid–fluid potential well diameter, ε_{ff} is the depth of the fluid–fluid potential well, σ_{ff} is the diameter of the fluid–fluid hard core, $\lambda_{ff}\sigma_{ff}$ is the fluid–wall potential well diameter, and ε_{wf} is the depth of the fluid–wall potential. All quantities are made adimensional using characteristic energy, ε_{ff} and length scale, σ_{ff}. For example, temperature is scaled by ε_{ff}/k.

In the phase diagram shown in Figure 8.10, the model parameters λ_{ff}, λ_{wf}, and ε_{ff} are fixed at 1.5, 1.0, and 1.0, respectively, and ε_{wf} is varied discretely from 2 to 6; however, for the sake of clarity of the behavior, only a typical case of $\varepsilon_{wf} = 4$ is shown. Solid curves represent the bulk phase coexistence densities. Figure 8.10 suggests that phase envelopes shrink under confinement, compared to bulk. In addition, with a decrease in the pore width, the temperature range in which a stable vapor–liquid transition is observed is significantly reduced. The coexistence envelope shape is characterized by the relation of the order parameter, $\Delta\rho = (\rho_l - \rho_v)/2$, with temperature. Below

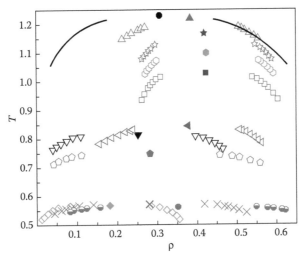

FIGURE 8.10 Temperature–density vapor–liquid coexistence curve for the square-well fluid in the slit pore confinement of pore width, H, varying from 40 to 1 molecular diameter with a wall–fluid interaction, $\varepsilon_{wf} = 4$. Solid curves represent the bulk coexistence densities with the filled black circle as the bulk critical points. Open and filled symbols represent the coexistence densities and critical points, respectively. In the figure from top, symbols triangle, star, hexagon, square, left triangle, inverted triangle, pentagon, diamond, cross, and half-filled circle represent $H = 40, 12, 8, 6, 4, 3, 2, 1.9, 1.25,$ and 1, respectively. Thick cross symbol represents critical point for $H = 1.25$. (Adapted from Singh, S. K., Saha, A. K., and Sing, J. K. *J. Phys. chem. B.*, 114: 4283, 2010a. With permission.)

T_{cp}, $\Delta\rho$ should follow a universal scaling law (Binder and Hohenberg 1974): $\Delta\rho \sim [(T_{cp} - T)/T_{cp}]^{\beta}$. It has been observed (Singh et al. 2010a) that the critical exponent, β, obtained for a different surface field, approaches the 2D Ising value (Chandler 1987) of 0.125 with decreasing H. Moreover, Figure 8.10 indicates that the pore critical temperature monotonically decreases to the quasi–2D region and then becomes approximately constant until the pure 2D region. On the other hand, the pore critical density shows oscillating behavior. Similar behavior is observed with various other model fluids under confinement (Liu et al. 2010; Singh et al. 2009, 2010b).

The pore critical temperature of vapor–liquid equilibrium is one of the most important characteristics of a confined fluid system. Statistical mechanical theories (Evans 1986b; Heffelfinger et al. 1988) of capillary condensation in a single idealized pore suggest that hysteresis arises from the existence of metastable states. The same theory also predicts that "adsorption jump" in adsorption isotherms is associated with capillary condensation and should decrease with increasing temperature, finally vanishing at the pore critical temperature, T_{cp}. The pore critical temperature depends on the pore size and lies below the bulk critical temperature T_{cb}. On the other hand, it is observed that as temperature increases the hysteresis loops shrink in size and eventually vanish at some higher temperature, T_{ch}, (although still below T_{cb}).

The disappearance of the hysteresis loops depends on the average pore size of the materials (Ball and Evans 1989); that is, with decreasing average pore size, the disappearance of loops occurs at a comparatively lower temperature. So, clearly, there are two aspects, disappearance of the adsorption jump and disappearance of hysteresis loop. Moreover, real mesoporous materials always have pore size distribution, to some extent; hence it becomes difficult to clearly identify T_{cp} on the basis of the disappearance of the adsorption jump. It is believed that in real mesoporous materials these two aspects are related to each other; and it has been generally thought that for materials with a pore size distribution and interconnected pores (Ball and Evans 1989; Burgess et al. 1989; Evans et al. 1986a, 1986b), T_{ch} is equal to T_{cp} because of the disappearance of the metastable state or hysteresis loop in pores at T_{cp}. So, for $T > T_{cp}$, adsorption is reversible and for $T < T_{cp}$, it is irreversible, that is, this results in the occurrence of hysteresis loops. Moreover, there is a consensus among some workers (de Keizer et al. 1991; Evans 1990; Maddox et al. 1997; Michalski et al. 1991; Rathousky et al. 1995) that the disappearance of a jump as a function of temperature in real porous materials manifests itself as the shrinking and finally the disappearance of hysteresis loops with increasing temperature. Thus, there is a lack of reliable criteria for experimental identification of T_{cp} and T_{ch} in real porous materials (which have interconnected pores along with a pore size distribution) from adsorption isotherms. In general, the experimental measurements of the pore critical temperatures, T_{cp}, in such systems are nontrivial. Further, with the investigation of mesoporous MCM-41 and MCM-48 materials, in which the pore networking effects or interconnection of pores of different sizes and shapes are negligibly small, it has become possible to quantify the pore critical temperature, T_{cp}, more accurately, as reported for some systems such as hexafluoroethane in MCM-41 (Machin 2003) and nitrogen in MCM-48 (Morishige et al. 2003).

In the investigation of hexafluoroethane in MCM-41 (Machin 2003), which has unconnected pores, it has been pointed out that hysteresis is absent above the temperature T~196 K, designated as the hysteresis critical temperature. A first-order transition, recognized by the decrease jump in the adsorption isotherm, is found to be absent for temperatures greater than 255 K, which is designated as the pore critical temperature for first-order capillary condensation. However, it is not clear why hysteresis is absent at the much lower temperature of 196 K. Moreover, nitrogen in MCM-48 materials, which have interconnected pores of almost the same pore size and geometry, shows behavior similar to that of T_{ch} and T_{cp}, as shown in Figure 8.11. Further in this direction, Morishige and Shikimi (1998) examined the temperature dependence of the slope of the isotherms of Ar, N_2, O_2, C_2H_4, and CO_2 in a single cylindrical pore of radius $r_p = 1.2$ nm of MCM-41 material. This investigation also indicates that in the independent pores, T_{ch} is not equal to T_{cp}. Therefore it becomes clear that, in a single pore, in unconnected pores, or even in interconnected pores of almost the same

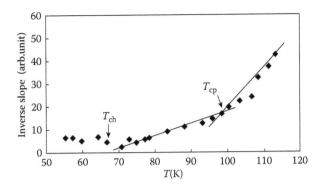

FIGURE 8.11 Temperature dependence of the inverse slope of adsorption step associated with capillary condensation of nitrogen onto MCM-48. Hysteresis and pore critical temperatures, T_{ch} and T_{cp}, respectively, are depicted in the plot. Solid lines are guides for the eye. (Adapted from Morishige, K. and Tateishi, M., *Langmuir* 22: 4165, 2003. With permission.)

pore size and geometry, the pore critical temperature, T_{cp}, is different from the hysteresis critical temperature, T_{ch}. However, the determination of an accurate pore critical temperature, T_{cp}, in real mesoporous materials with interconnected pores of different pore sizes and shape is still an open area of significant research for experimental, simulation, and theoretical approaches.

Nevertheless, it is a well-established fact that T_{cp} is suppressed under confinement and the decrease in T_{cp} increases as pore size decreases (Burgess et al. 1989; de Keizer et al. 1991; Machin 1994; Wong and Chan 1990; Wong et al. 1993). On the other hand, the discovery of mesoporous material of well-defined geometry, such as MCM-41 (Kresge 1992), MCM-48, and SBA-15 (Zhao et al. 1998), has allowed direct experimental measurement of critical points with some quantitative linear trends. For example, Thommes and Findenegg (1994) and Morishige and Shikimi (1998) observed experimentally that the shift in critical temperature has linear dependence on the inverse pore width. However, because of certain limitations with the available experimental techniques, it has not been possible to perform exhaustive investigations with smaller nanopores. For a more systematic investigation of pore critical properties in nanopores, molecular modeling techniques have been employed. For example, the critical temperature and density of argon in cylindrical pores were estimated by Panagiotopoulos (1987), methane in a graphite slit pore by Jiang et al. (1993), and Lennard-Jones fluid in a uniform slit pore by Forsman and Woodward (1999) and Vishnyakov et al. (2001) performed Monte Carlo simulations on a carbon slit pore and obtained similar results as seen experimentally in former investigations, that is, linear dependence of the shift in critical temperature on the inverse pore width. However, the simulations were limited to gaps of a pore width of a few molecular diameters. Further investigations on model fluids (Vortler 2008) have suggested a nonlinear dependence of the shift in critical temperature as a more generic behavior in nanopores.

Recent investigations (Jana et al. 2009; Liu et al. 2010; Singh et al. 2009, 2010a) of various model fluids and confinements have shown more than two different linear regimens in the shift in critical temperature with inverse slit width. A typical illustration is shown in Figure 8.12 with a square-well model fluid in an attractive slit-pore confinement. Five different linear regimes for the shift in critical temperature with inverse slit width are identified for the illustrated model system. In another investigation pertaining to square well fluids, Zhang and Wang (2006) studied the shift in critical temperature in a cylindrical pore for various wall–fluid and fluid–fluid interaction strengths, using DFT calculations, and found nonmonotonic behavior. In a subsequent work, Singh and Kwak (2007) reported a similar observation of the shift in critical temperature for square-well fluids in slit pores; critical temperature first increases with an increase in the wall–fluid interaction strength and subsequently decreases on further increases in the wall–fluid interaction strength.

Investigations of critical density, ρ_{cp}, of fluid confined in attractive slit pores reveals fluctuations with local maxima and minima. Figure 8.13 presents the comparison of variation in ρ_{cp}/ρ_{cb} versus slit width, H, for a typical model fluid, with a wall–fluid interaction, $\varepsilon_{wf} = 4$. Figure 8.13 shows that ρ_{cp}/ρ_{cb} follows a nonmonotonic path contrary to T_{cp}/T_{cb} (Singh et al. 2010a); however, it shows signs of approaching the 3D bulk value, $\rho_{c,R}(3D)$ as H→∞. Noticeably, all critical properties remain unaffected by the strength of surface–fluid interactions for pore width H ≤ 2 molecular diameter (Singh et al. 2010a).

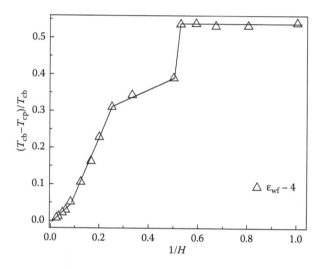

FIGURE 8.12 Shift in the pore critical temperature versus inverse H is shown for a model fluid with a typical wall–fluid interaction, $\varepsilon_{wf} = 4$. Lines serve as a guide to the eye. (Adapted from Singh, S. K., Saha, A. K., and Sing, J. K. *J. Phys. chem. B.*, 114: 4283, 2010a. With permission.)

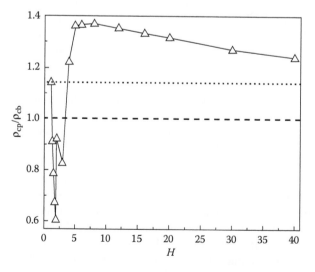

FIGURE 8.13 The dependence of pore critical density (reduced by the corresponding bulk value) on slit width, H, varying between 3D (represented by dash line) to 2D (represented by dotted line) geometry for a typical model fluid. Symbols represent the simulation data. Line serves as a guide to the eye. (Adapted from Singh, S. K., Saha, A. K., and Sing, J. K. *J. Phys. chem. B.*, 114: 4283, 2010a. With permission.)

8.5.2 CROSSOVER FROM THREE-DIMENSIONAL TO TWO-DIMENSIONAL

Crossover behavior in fluid properties is of significant interest from practical as well as fundamental understanding point of view. For example, a lubricant between two surfaces moving relative to each other can undergo an undesirable fluid–solid phase transition when spacing between the moving surfaces is lowered to a certain threshold value, resulting in a drastic increase in effective viscosity. Hence, it becomes increasingly important to understand the 3D to 2D crossover behavior of confined fluids. A common question that arises is, under confinement, when does a fluid start exhibiting 2D–like behavior? Does this departure from 3D to 2D–like depend on the fluid and confining surfaces characteristics?

Fisher and Nakanishi (1981), with the help of scaling arguments, showed that the decrease in critical temperature in larger pores should obey the relation $(T_{cb}-T_{cp})/T_{cb} = kH^{-1/\nu}$, where ν is the critical exponent for the correlation length and k is a proportionality constant and H is the pore width. However, to account for the strongly adsorbed layer on pore walls it is necessary to replace the true pore width H, by a modified pore width, H_{eff}, which accounts for the adsorbed layers that form before capillary condensation. Using an Ising 3D (bulk) correlation length critical exponent (Guida and Zinn-Justin 1998) $\nu(3D) = 0.63$ and Ising 2D correlation length critical exponent (Evans et al. 1986a, 1986b), $\nu(2D) = 1$, we have evaluated the effective layering thickness, t, of the adsorbed layer in the pore by fitting the logarithmic form of the relation, $(T_{cb}-T_{cp})/T_{cb} = k(H-t)^{-1/\nu}$. We observed that for larger pore widths the layering thickness, t, is insignificant, irrespective of the surface attraction and pore shape. However, for smaller pores it is appreciable and cannot be neglected. In some of our current work (Singh et al. 2009, 2010a, 2010b), we investigated the crossover behavior of the various model fluids confined in different attractive slit pores, and crossover from 3D to 2D is observed at around $H\sim 14\text{-}16$ molecular diameters for the studied systems.

Further, in continuation of the crossover studies, we investigated the comparative effect of hard and attractive slit pore surfaces on crossover behavior and also compared the effect of pore shape on this behavior. A typical comparison of crossover behavior is shown in Figure 8.14 with a typical model fluid. In Figure 8.14, the shift in critical temperature, $(T_{cb}-T_{cp})/T_{cb}$, is plotted as a function of effective pore width, H_{eff} or D_{eff} (i.e., $H-t$ or $D-t$) on a log–log scale. This investigation indicates that, for a given model fluid confined in a hard slit-pore, the crossover behavior is significantly reduced

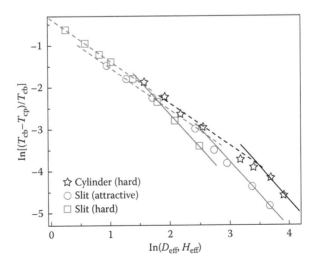

FIGURE 8.14 Crossover from 3D to 2D of a model fluid in slit and cylindrical pore are shown. Correlation length critical exponent, ν, is represented by solid and dashed lines for 3D and 2D regimes, respectively. (Adapted from Singh, S. K. and Singh, J. K., *Fluid. Phase. Equil.* 300: 182, 2011. With permission.)

compared to that of attractive slit pores. For example, for the model fluid shown in Figure 8.14, crossover from 3D to 2D in a hard slit pore is observed at around $H\sim6$ molecular diameters, and in an attractive slit pore at around $H\sim14$ molecular diameters (Singh et al. 2010a). Moreover, the effect of pore shape on crossover behavior is also significant, as depicted in Figure 8.14. It has been shown that for the same model fluid, the crossover from 3D behavior in a hard cylindrical pore is observed at around $H\sim28$ molecular diameters (Singh and Singh 2011), which is significantly larger than for a hard slit pore. A similarly significant difference in crossover behavior is observed when comparing attractive slit and cylindrical pores. This indicates that, for a fluid, pore shape has a significant effect on crossover behavior.

8.6 LIQUID–LIQUID PHASE TRANSITION IN CONFINEMENT

Liquid–liquid transitions are of interest in connection with oil recovery, lubrication, coating technology, and pollution control. The effect of confinement on the solubility of dilute solutes is of great importance in understanding the dispersion of pollutants in soils, in removing trace pollutants from water, and in lubrication. The experimental techniques available for liquid–liquid phase transition studies are mainly nonlinear dielectric effect (NDE) and light transmission measurements (Sliwinska-Bartkowiak et al. 1984). In both experiments, measurements are made along a path of constant composition, starting at a temperature in the one-phase region above the liquid–liquid coexistence line. The temperature is then reduced until a phase transition is observed. This is repeated at different compositions to define the liquid–liquid coexistence curve. The NDE is defined as the change in the medium permittivity ε in a strong external electric field, E:

$$\frac{\Delta\varepsilon}{E^2} = \frac{\varepsilon^E - \varepsilon^0}{E^2} \tag{8.29}$$

where ε^E is the permittivity in the electric field E and ε^0 is the permittivity in the absence of an electric field. The permittivity is measured by the pulse method. NDE is sensitive to fluid inhomogeneities in the vicinity of the critical point and exhibits an anomalous increase related to density fluctuation in the medium. The temperature dependence of NDE, when approaching the critical point along the critical composition line, is described by the following formula (Hoye and Stell 1984; Oxtoby and Metiu 1976):

$$\left(\frac{\Delta\varepsilon}{E^2}\right)_{fl} \approx \left(\frac{\Delta\varepsilon}{E^2}\right) - \left(\frac{\Delta\varepsilon}{E^2}\right)_b \approx At^{-\phi} \tag{8.30}$$

where $(\Delta\varepsilon/E_2)_b$ is the background effect measured at high temperatures and interpolated in the vicinity of the critical temperature, $t = (T–T_c)/T_c$, A is the amplitude, and ϕ is the critical exponent, which is defined as $\gamma-2\beta$, where γ and β are the usual critical exponents for the isothermal compressibility and density difference (order parameter) between the two phases, respectively (Hoye and Stell 1984). Light transmission measurements are used to determine the temperature of the phase separation in the mixtures, that is, the temperature at which a strong turbidity of the system is observed.

Liquid–liquid phase transition in confined mixtures have been well described by Evans and Marconi (1987). If μ_2^b and μ_2^p are the chemical potentials for component 2 of the coexisting phases α and β in the bulk mixture and in the pore, respectively, and if the α phase is the one rich in 2, then

$$\mu_2^b - \mu_2^p = \frac{2\gamma_{\alpha\beta}\cos\theta}{r_p B\rho_{1,\alpha}} \tag{8.31}$$

where $\gamma_{\alpha\beta}$ is the interfacial tension between the two fluid phases, θ is the contact angle between $\alpha\beta$ meniscus and the surface, r_p is the pore radius, $\rho_{l,\alpha}$ is the number density of the α phase at chemical potential μ_2, and $B = [(x_2^b/x_1^b)_\alpha - (x_2^b/x_1^b)_\beta]$, where x is the mole fraction. If the contact angle $\theta < \pi/2$ phase separation in the pore will occur for $\mu_2^p < \mu_2^b$, that is, at smaller mole fractions x_2; this will occur when the surface favors the α phase. The reverse will hold when $\theta > \pi/2$. However, Equation 8.31 is no longer valid for smaller pores and results in very serious error for micropores (pore width <2 nm).

Experimental and theoretical studies of liquid–liquid phase transitions in confinement is studied successfully using NDE and the light transmission measurement (Sliwinska-Bartkowiak et al. 1997) and the Kierlik-Rosinberg-Rosenfeld (Kierlik and Rosinberg 1990) form of DFT, respectively. Using a DFT approach, the effect of pore width (graphitic surface of the pore wall taken as the 10-4-3 potential of Steele (1973) on the phase diagram of liquid–liquid equilibrium of the confined and symmetric ($\sigma_{11} = \sigma_{12} = \sigma$, $\varepsilon_{11} = \varepsilon_{12} = \varepsilon$, and $\varepsilon_{12} = k\varepsilon$ where k is less than unity; the departure of the parameter k from unity is a measure of the nonideality of the mixture) Lennard-Jones fluid for the parameter ratio $\varepsilon_{sf1}/\varepsilon_{sf2} = 1.2$ is illustrated in Figure 8.15 (Sliwinska-Bartkowiak et al. 1997).

The coexistence curve for the bulk and two reduced pore widths, $H = 10.0$ and 20.558, is shown in Figure 8.15. It is clear that confinement causes the region of miscibility to be decreased, and the critical mixing temperature is lowered. Also, a stronger fluid–wall attraction of component 1, ε_{sf1}, leads to a shift in the coexistence curve toward the component 1–rich side. As the pore width is decreased further, the selectivity for component 1 increases and the envelope of miscibility is further decreased. Thus, from Figure 8.15, it becomes clear that confinement leads to a decrease in solubility of component 2 in the component 1–rich phase, and correspondingly a much larger increase in solubility of component 1 in the component 2–rich phase. Also, as the ratio of surface-fluid parameters $\varepsilon_{sf1}/\varepsilon_{sf2}$ increases, the range of miscibility decreases; critical mixing temperature is lowered, and the coexistence curve is further shifted toward the component 1–rich side of the phase diagram. So, the qualitative effect of increasing the $\varepsilon_{sf1}/\varepsilon_{sf2}$ ratio with a fixed confinement is similar to a fixed $\varepsilon_{sf1}/\varepsilon_{sf2}$ ratio and decreased confinement.

Further investigations using molecular simulation techniques for liquid–liquid equilibria in pores have been reported by several investigators with symmetric mixture of model fluids. In

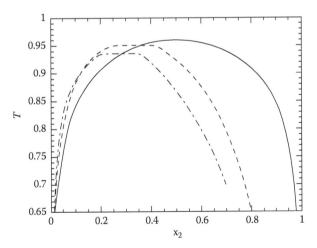

FIGURE 8.15 Liquid–liquid phase diagram of bulk and confined symmetric LJ mixture for $k = 0.8$, $P = 0.1$, $\varepsilon_{sf1}/\varepsilon_{sf2} = 1.2$. Temperature, T, is in adimensional form. The bulk system (solid line) and two pore widths, $H = 10.0$ (dotted-dashed line) and 20.558 (dashed line) are shown. (Adapted from Sliwinska-Bartkowiak, M., Sowers, S. L. and Gubbins, K. E., *Langmuir* 13: 1182, 1997. With permission.)

these studies, liquid–liquid coexistence curves with a homogeneous positive azeotrope (Goźdz et al. 1995; Kierlik et al. 1995) at high pressure and a heterogeneous azeotrope (Fan et al. 1993; Panagiotopoulos 1989; Stapleton and Panagiotopoulos 1990) at low pressure were produced. In these studies of the confined fluid, the two components had the same interaction potential with the pore walls, and the pores were of slit shape; thus the mixture was also symmetric with respect to components 1 and 2 in the pore, and the results show the effects of pure confinement. The main qualitative conclusions of these investigations are that confinement leads to (1) a decreased region of miscibility (i.e., selective adsorption of the dilute component), and (2) lowering of the critical solution/mixing temperature, which is similar to the previously described theoretical (DFT) investigations (Sliwinska-Bartkowiak et al. 1997). The decreased miscibility range is interesting because the two components experience the same fluid–wall interaction. It arises because the effect of the weak unlike pair interaction is reduced as a result of the lower dimensionality associated with decreased pore size. Further, in a related study, Kumar et al. (1994) reported GEMC simulations for a symmetric mixture of LJ chain molecules (same like pair interactions, and weak unlike molecule pair interactions). The liquid–liquid coexistence curves found for this symmetric chain molecule mixture were qualitatively similar to those found for the symmetric LJ mixture mentioned earlier, that is, increased solubility of the dilute component and a lowering of critical solution temperature.

We finish by noting that some experimental results have also been presented, for example, Sliwinska-Bartkowiak et al. (1997) have provided data for a nitrobenzene/n-hexane mixture in a controlled pore glass having a mean pore size of 100 nm. In this work, the results for liquid–liquid coexistence in the pores were obtained using NDE and light transmission measurements and support the behavior observed with model fluids, as explained earlier using molecular simulation (Kumar et al. 1994) and theoretical approaches (Sliwinska-Bartkowiak et al. 1997). However, understanding the liquid–liquid phase transition behavior of various other real systems will require detailed work with unsymmetric mixtures to be performed using molecular simulation, theoretical techniques, and better, yet to be developed, experimental methods. Additionally, the application of this powerful combination of approaches will also allows us to understand phase transitions in smaller nanopores in which the effect of confinement will be more pronounced.

REFERENCES

Altwasser, S., Welker, C., Traa, Y., and Weitkamp, J. 2005. Catalytic cracking of n-octane on small-pore zeolites. *Microporous Mesoporous Mater.* 83: 345.

Aranovich, G. L., and Donohue, M. D. 2003. Adsorption compression: An important new aspect of adsorption behavior and capillarity. *Langmuir* 19: 2722.

Ball, P. C., and Evans, R. 1989. Temperature dependence of gas adsorption on a mesoporous solid: Capillary criticality and hysteresis. *Langmuir* 5: 714.

Bhatia, S. 1998. Adsorption of binary hydrocarbon mixtures in carbon slit pores: A density functional theory study. *Langmuir* 14: 6231.

Binder, K. 1981. Critical properties from Monte Carlo coarse graining and renormalization. *Phys. Rev. Lett.* 47: 693.

Binder, K., and Hohenberg, P. C. 1974. Surface effects on magnetic phase transitions. *Phys. Rev. B.* 9: 2194.

Bock, H., and Schoen, M. 1999. *Physical Rev. E.* Phase behavior of a simple fluid confined between chemically corrugated substrates. 59: 4122.

Burgess, C. G. V., Everett, D. H., and Nuttal, S. 1989. *Pure Appl. Chem.* Adsorption hysteresis in porous materials. 61: 1845.

Carnahan, N. F., and Starling, K. E. 1969. *J. Chem. Phys.* Equation of state for nonattracting rigid spheres. 51: 635.

Casanova, F., Chiang, C. E., Li, C.-P., Roshchin, I. V., Ruminski, A. M., Sailor, M. J., and Schuller, I. K. 2008. *Nanotechnology.* Gas adsorption and capillary condensation in nanoporous alumina films. 19: 315709.

Chandler, D. 1987. *Introduction to Modern Statistical Mechanics.* Oxford University Press, Oxford.

Chen, Y., Wetzel, T., Aranovich, G. L., and Donohue, M. D. 2006. *J. Colloid Interface Sci.* Generalization of Kelvin's equation for compressible liquids in nanoconfinement. 300: 45.

Choudhury, N., and Ghosh, S. K. 2003. *J. Phys. Chem. B*. New weighted density functional theory based on perturbative approach. 107: 7155.

deKeizer, A., Michalski, T., and Findenegg, G. H. 1991. *Pure Appl. Chem*. Fluids in pores: Experimental and computer simulation studies of multilayer adsorption, pore condensation and critical-point shifts. 63: 1495.

Errington, J. R. 2003. *J. Chem. Phys*. Direct calculation of liquid–vapor phase equilibria from transition matrix Monte Carlo simulation. 118: 9915.

Evans, R. 1990. *J. Phys.: Condens. Matter*. Fluids adsorbed in narrow pores: Phase equilibria and structure. 2: 8989.

Evans, R. 1992. Density functionals in the theory of nonuniform fluids. In *Fundamentals of Inhomogeneous Fluids*. Eds. Douglas Henderson. Dekker, New York.

Evans, R., and Marconi, U. M. B. 1987. *J. Chem. Phys*. Phase equilibria and solvation forces for fluids confined between parallel walls. 86: 7138.

Evans, R., Marconi, U. M. B., and Tarazona, P. 1986a. *J. Chem. Phys*. Fluids in narrow pores: Adsorption, capillary condensation, and critical points. 84: 2376.

Evans, R., Marconi, U. M. B., and Tarazona, P. 1986b. *J. Chem. Soc. Faraday Trans*. Capillary condensation and adsorption in cylindrical and slit-like pores. 82: 1763.

Fan, Y., Finn, J. E., and Monson, P. A. 1993. *J. Chem. Phys*. Monte Carlo simulation study of adsorption from a liquid mixture at states near liquid–liquid coexistence. 99: 8238.

Ferrenberg, A. M., and Swendsen, R. H. 1988. *Phys. Rev. Lett*. New Monte Carlo technique for studying phase transitions. 61: 2635.

Fisher, L. R., and Israelachvili, J. N. 1981. *J. Colloid Interface Sci*. Experimental studies on the applicability of the Kelvin equation to highly curved concave menisci. 80: 528.

Fisher, M. E., 1964. *J. Math. Phys*. Correlation Functions and the Critical Region of Simple Fluids. 5: 944.

Fisher, M., and Nakanishi, H. 1981. *J. Chem. Phys*. Scaling theory for the criticality of fluids between plates. 75: 5857.

Forsman, J., and Woodward, C. E. 1999. *Mol. Phys*. Simulations of coexistence between layered phases in planar slits. 96: 189.

Go´zdz, W. T., Gubbins, K. E., and Panagiotopoulos, A. Z. 1995. *Mol. Phys*. Liquid-liquid phase transitions in pores. 84: 825.

Gregg, S. J., and Sing, K. S. W. 1982. *Adsorption, Surface Area and Porosity*. 2nd ed. Academic: London.

Guida, R., and Zinn-Justin, J., 1998. *J. Phys. A: Math. Gen*. Critical exponents of the N-vector model. 31: 8103.

Gupta, S. A., Cochran, H. D., and Cummings, P. T. 1997. *J. Chem. Phys*. Shear behavior of squalane and tetracosane under extreme confinement. II. Confined film structure. 107: 10327.

Hamada, Y., Koga, K., and Tanaka, H. 2007. *J. Chem. Phys*. Phase equilibria and interfacial tension of fluids confined in narrow pores. 127: 084908.

Heffelfinger, G. S., Van, S. F., and Gubbins, K. E. 1988. *J. Chem. Phys*. Adsorption hysteresis in narrow pores. 89: 5202.

Hoye, J. S., and Stell, G. 1984. *J. Chem. Phys*. Kerr effect. I. Field effect on correlation in polarizable fluids. 81: 3200.

Jackson, C. L., and McKenna, G. B. 1990. *J. Chem. Phys*. The melting behavior of organic materials confined in porous solids. 93: 9002.

Jana, S., Singh, J. K., and Kwak, S. K. 2009. *J. Chem. Phys*. Vapor-liquid critical and interfacial properties of square-well fluids in slit pores. 130: 214707.

Jiang, J., and Sandler, S. I. 2005. *Fluid Phase Equil*. Adsorption and phase transitions on nanoporous carbonaceous materials: insights from molecular simulations. 189: 228.

Jiang, S. Y., Rhykerd, C. L., and Gubbins, K. E. 1993. *Mol. Phys*. Layering, freezing transitions, capillary condensation and diffusion of methane in slit carbon pores. 79: 373.

Katsov, K., and Weeks, J. D. 2001. *J. Phys. Chem. B*. On the mean field treatment of attractive interactions in nonuniform simple fluids. 105: 6738.

Kierlik, E., and Rosinberg, M. 1990. *Phys. Rev. A*. Free-energy density functional for the inhomogeneous hard-sphere fluid: Application to interfacial adsorption. 42: 3382.

Kierlik, E., Fan, Y., Monson, P. A., and Rosinberg, M. L. 1995. *J. Chem. Phys*. Liquid–liquid equilibrium in a slit pore: Monte Carlo simulation and mean field density functional theory. 102: 3712.

Kim, S. C., and Lee, S. H. 2004. *J. Phys.: Condens. Matter*. A density functional perturbative approach for simple fluids: The structure of a nonuniform Lennard-Jones fluid at interfaces. 16: 6365.

Koga, K., and Tanaka, H. 2005. *J. Chem. Phys*. Phase diagram of water between hydrophobic surfaces. 122: 104711.

Kresge, C. T., Leonowicz, M. E., Roth, W. J., Vartuli, J. C., and Beck, J. S. 1992. *Nature.* Ordered mesoporous molecular sieves synthesized by a liquid-crystal template mechanism. 359: 710.

Kruk, M., Jaroniec, M., and Sayari, A. 1997. *Langmuir.* Application of large pore MCM-41 molecular sieves to improve pore size analysis using nitrogen adsorption measurements. 13: 6267.

Kumar, S. K., Tang, H., and Szleifer, I. 1994. *Mol. Phys.* Phase transitions in thin films of symmetric binary polymer mixtures. 81: 867.

Lastoskie, C., Gubbins, K. E., and Quirke, N. 1993. *J. Phys. Chem.* Pore size distribution analysis of microporous carbons: a density functional theory approach. 97: 4786.

Liu, Y., Panagiotopoulos, A. Z., and Debenedetti, P. G. 2010. *J. Chem. Phys.* Finite-size scaling study of the vapor-liquid critical properties of confined fluids: Crossover from three dimensions to two dimensions. 132: 144107.

Lutsko, J. F. 2008. *J. Chem. Phys.* Density functional theory of inhomogeneous liquids. II. A fundamental measure approach. 128: 184711.

Machin, W. D. 1994. *Langmuir.* Temperature dependence of hysteresis and the pore size distributions of two mesoporous adsorbents. 10: 1235.

Machin, W. D. 1999 *Langmuir.* Properties of three capillary fluids in the critical region. 15: 169.

Machin, W. D. 2003. *Phys. Chem. Chem. Phys.* Capillary condensation of hexafluoroethane in an ordered mesoporous silica. 5: 203.

Maddox, M. W., Olivier, J. P., and Gubbins, K. E. 1997. *Langmuir.* Characterization of MCM-41 using molecular simulation: heterogeneity effects. 13: 1737.

Matranga, K. R., Myers, A. L., and Glandt, E. D. 1992. *Chem. Eng. Sci.* Storage of natural gas by adsorption on activated carbon. 47: 1569.

McGrother, S. C., and Gubbins, K. E. 1999. *Mol. Phys.* Constant pressure Gibbs ensemble Monte Carlo simulations of adsorption into narrow pores. 97: 955.

Melrose, J. C. 1966. *AIChE J.* Model calculations for capillary condensation. 12: 986.

Michalski, T., Benini, A., Findenegg, G. H. 1991. *Langmuir.* A study of multilayer adsorption and pore condensation of pure fluids in graphite substrates on approaching the bulk critical point. 7: 185.

Miyahara, M., Kanda, H., Yoshioka, T., Okazaki, M. 2000. *Langmuir.* Modeling capillary condensation in cylindrical nanopores: A molecular dynamics study. 16: 4293.

Morishige, K., and Shikimi, M. 1998. *J. Chem. Phys.* Adsorption hysteresis and pore critical temperature in a single cylindrical pore. 108: 7821.

Morishige, K., and Tateishi, M. 2003. *Langmuir.* Accurate relations between pore size and the pressure of capillary condensation and the evaporation of nitrogen in cylindrical pores. 22: 4165.

Morishige, K., Tateishi, N., Fukuma, S. 2003. *J. Phys. Chem. B.* Capillary condensation of nitrogen in MCM-48 and SBA-16. 107: 5177.

Muller, M., MacDowell, L. G., Yethiraj, A. 2003. *J. Chem. Phys.* Short chains at surfaces and interfaces: A quantitative comparison between density-functional theories and Monte Carlo simulations. 118: 2929.

Nicholson, D., and Parsonage, N. G. 1982. *Computer Simulation and the Statistical Mechanics of Adsorption.* Academic, London.

Oxtoby, D. W., and Metiu, H. 1976. *Phys. Rev. Lett.* Magnetic moment of the 6+ isomeric state of ^{134}Te. 36: 1072.

Panagiotopoulos, A. Z. 1987 *Mol. Phys.* Adsorption and capillary condensation of fluids in cylindrical pores by Monte Carlo simulation in the Gibbs ensemble. 62: 701.

Panagiotopoulos, A. Z. 1989. *Int. J. Thermophys.* Exact calculations of fluid-phase equilibria by Monte Carlo simulation in a new statistical ensemble. 10: 447.

Peng, B., and Yu, Y.-X. 2008. *J. Phys. Chem. B.* A density functional theory with a mean-field weight function: Applications to surface tension, adsorption, and phase transition of a lennard-jones fluid in a slit-like pore. 112: 15407.

Rathousky, J., Zukal, A., Franke, O., and Shulz-Ekloff, G. 1995. *J. Chem. Soc., Faraday Trans.* Adsorption on MCM-41 mesoporous molecular sieves. Part 2.—Cyclopentane isotherms and their temperature dependence. 91: 937.

Ravikovitch, P. I., Vishnyakov, A., and Neimark, V. 2001. *Phys. Rev. E.* Density functional theories and molecular simulations of adsorption and phase transitions in nanopores. 64: 011602.

Rivera, J. L., McCabe, C., and Cummings, P. T. 2002. *Nano Lett.* Layering behavior and axial phase equilibria of pure water and water + carbon dioxide inside single wall carbon nanotubes. 2: 1427.

Rżysko, W., Patrykiejew, A., Sokolowski, and S. Pizio, O. 2010. *J. Chem. Phys.* Phase behavior of a two-dimensional and confined in slitlike pores square-shoulder, square-well fluid. 132: 164702.

Rżysko, W., Pizio, O., Patrykiejew, A., and Sokolowski, S. 2008. *J. Chem. Phys.* Phase diagram of a square-shoulder, square-well fluid revisited. 129: 124502.

Sarkisov, L., and Monson, P. A. 2001. *Langmuir.* Modeling of adsorption and desorption in pores of simple geometry using molecular dynamics. 17: 7600.

Sengers, V., and Levelt-Sengers, J. M. H. 1978. *Progress in Liquid Physics.* 4th Ed. Eds. C. A. Croxton. Chichester, Wiley.

Singh, J. K., and Kofke, D. A. 2004. *J. Chem. Phys.* Molecular simulation study of effect of molecular association on vapor-liquid interfacial properties. 121: 9574.

Singh, J. K., and Kwak, S. K. 2007. *J. Chem. Phys.* Surface tension and vapor-liquid phase coexistence of confined square-well fluid. 126: 024702.

Singh, J. K., Kofke, D. A., and Errington, J. R. 2003. *J. Chem. Phys.* Surface tension and vapor–liquid phase coexistence of the square-well fluid. 119: 3405.

Singh, S. K., and Singh, J. K. 2011. *Fluid. Phase. Equil.* Effect of pore morphology on vapour–liquid phase transition and crossover behaviour of critical properties from 3D to 2D. 300: 182.

Singh, S. K., Saha, A. K., and Singh, J. K. 2010a. *J. Phys. Chem. B.* Molecular simulation study of vapor-liquid critical properties of a simple fluid in attractive slit pores: crossover from 3d to 2d. 114: 4283.

Singh, S. K., Singh, J. K., Kwak, S. K., and Deo, G. 2010b. *Chem. Phys. Lett.* Phase transition and crossover behavior of colloidal fluids under confinement. 494: 182.

Singh, S. K., Sinha, A., Deo, G., and Singh, J. K. 2009. *J. Phys. Chem. C.* Vapor-liquid phase coexistence, critical properties, and surface tension of confined alkanes. 113: 7170.

Sliwinska-Bartkowiak, M., Sowers S. L., and Gubbins, K. E. 1997. *Langmuir.* Liquid-liquid phase equilibria in porous materials. 13: 1182.

Sliwinska-Bartkowiak, M., Szurkowski, B., and Hilczer, T. 1984. *Chem. Phys. Lett.* Non-linear dielectric effect in critical and far pre-critical solutions of nitrotoluene. 94: 609.

Smith, W. R., and Vortler, H. L. 1996. *Chem. Phys. Lett.* Monte Carlo simulation of fluid phase equilibria in pore systems: Square-well fluid distributed over a bulk and a slit-pore. 249: 470.

Stapleton, M. R., and Panagiotopoulos, A. Z. 1990. *J. Chem. Phys.* Application of excluded volume map sampling to phase equilibrium calculations in the Gibbs ensemble. 92: 1285.

Steele, W. A. 1973. *Surf. Sci.* The physical interaction of gases with crystalline solids: I. Gas-solid energies and properties of isolated adsorbed atoms. 36: 317.

Sweatman, M. B. 2001. *Phys. Rev. E.* Weighted density functional theory for simple fluids: Supercritical adsorption of a Lennard-Jones fluid in an ideal slit pore. 63: 031102.

Tang, Y. P., and Wu, J.-Z. 2003. *J. Chem. Phys.* A density-functional theory for bulk and inhomogeneous Lennard-Jones fluids from the energy route. 119: 7388.

Thommes, M., and Findenegg, G. H. 1994. *Langmuir.* Pore condensation and critical-point shift of a fluid in controlled-pore glass. 10: 4270.

Vishnyakov, A., Piotrovskaya, E. M., Brodskaya, E. N., Votyakov, E. V., and Tovbin, Y. K. 2001. *Langmuir.* Modeling of adsorption and desorption in pores of simple geometry using molecular dynamics. 17: 4451.

Vortler, H. L. 2008. *Collect. Czech. Chem. Commun.* Simulation of fluid phase equilibria in square-well fluids: from three to two dimensions. 73: 518.

Vortler, H. L., and Smith, W. R. 2000. *J. Chem. Phys.* Computer simulation studies of a square-well fluid in a slit pore. Spreading pressure and vapor–liquid phase equilibria using the virtual-parameter-variation method. 112: 5168.

Weeks, J. D., Chandler, D., and Anderson, H. C. 1971. *J. Chem. Phys.* Role of repulsive forces in determining the equilibrium structure of simple liquids. 54: 5237.

Wegner, F. 1972. *Phys. Rev. B.* Corrections to scaling laws. 5: 4529.

Wilding, N. B. 1995. *Phys. Rev. E.* Critical-point and coexistence-curve properties of the Lennard-Jones fluid: A finite-size scaling study. 52: 602.

Wong, A. P. Y., and Chan, M. H. W. 1990. *Phys. Rev. Lett.* Liquid-vapor critical point of ^{4}He in aerogel. 65: 2567.

Wong, A. P. Y., Kim, S. B., Goldburg, W. I., and Chan, M. H. W. 1993. *Phys. Rev. Lett.* Phase separation, density fluctuation, and critical dynamics of N_2 in aerogel. 70: 954.

Yu, Y.-X., You, F.-Q., Tang, Y., Gao, G.-H., and Li, Y.-G. 2006. *J. Phys. Chem. B.* Structure and adsorption of a hard-core multi-Yukawa fluid confined in a slitlike pore: Grand canonical Monte Carlo simulation and density functional study. 110: 334.

Zangi, R. 2004. *J. Phys. Condens. Matter.* Water confined to a slab geometry: A review of recent computer simulation studies. 16: S5371.

Zhang, X., and Wang, W. 2006. *Phys. Rev. E.* Square-well fluids in confined space with discretely attractive wall-fluid potentials: Critical point shift. 74: 062601.

Zhao, D. Y., Feng, J. L., Huo, Q. S., Melosh, N., Fredrickson, G. H., Chmelka, B. F., and Stucky, G. D. 1998. *Science.* Triblock copolymer syntheses of mesoporous silica with periodic 50 to 300 angstrom pores. 279: 548.

9 Solid–Liquid Phase Transition under Confinement

Sang Kyu Kwak and Jayant K. Singh

CONTENTS

9.1 INTRODUCTION

The confinement of fluid in nanopores is a common occurrence in sensors and devices. Therefore understanding the phase equilibria, and having knowledge of structural and transport properties, of confined fluids is important in the development of new technologies for manufacturing and in the modification of current methods. Interestingly, the phase behavior of fluid in porous material is dramatically different from those of a bulk fluid because of the competition of fluid–fluid and fluid–wall interaction energies. These geometric constraints and the presence of external forces are the primary sources for a range of different phase transitions, such as layering, prewetting, and capillary condensation and a shift in critical properties and melting/freezing properties (Gelb et al. 1999).

Understanding freezing and melting of fluid in nanopores is of practical importance in the fields of lubrication, nanotribology, adhesion, and frost heaving. Bearing this in mind, this chapter deals with the molecular simulation of solid–liquid transition induced by confinement. Our aim is twofold: first, to provide a review of recent molecular simulation techniques for solid-phase transitions under confinement, with relevant case studies, and second, to demonstrate the possibilities and limitations of the simulation methods discussed here, showing recent results for some systems and properties selected from the author's fields of research. The chapter is divided into two sections— one deals with free energy methodologies relevant for the evaluation of phase transition, and the other focuses mainly on the simulation aspect and case studies of fluid–properties of nanoconfined systems and highlights important considerations in the application of molecular simulation techniques.

9.2   SOLID FREE ENERGY CALCULATIONS

Free energy plays a most fundamental role in thermodynamics and its related fields. Its derivatives are directly formulated to provide thermophysical properties, within which the matters are built, and a relative stability can be accurately determined only by it. More importantly, to locate the first order phase transition such as fluid–solid, gas–fluid, and so forth, one must know that the first order derivative of the free energy is discontinuous. Considering its importance in notion and usage, it is obvious to see that much work has been dedicated to develop several methods for calculating the free energy. Recent reports on the methods of the free energy calculation have been published with details (Monson and Kofke 2001, Rickman and LeSar 2002), but our main interest is not to elucidate all methods but to characterize the phenomena of the fluid–solid phase transition under applied confinement (i.e., slitlike, cylindrical, spherical, disordered, etc.) via obtaining the free energy and investigating its derivatives. In particular, Helmholtz free energy is of interest to us because it provides the stability information of the phase under a fixed temperature and density, on which most of our simulation is based. Unfortunately, there has been little work on free energy calculation satisfying the previously mentioned confinement environment except small studies size (Rosenfeld et al. 1996). Since the decoupling of interactions of particles-to-particles and particles-to-confining walls is possible, we briefly introduce a few methods for free energy calculations, which are feasible for the confined system.

It is well known that direct calculation of Helmholtz free energy is a nontrivial problem, which is readily seen from a classical statistical mechanical formula:

$$\frac{f}{k_BT} = -\ln Q(N,V,T)$$ (9.1)

where k_B is Boltzmann's constant, T is the temperature, and Q is the canonical partition function, which depends on the number of particles N, the volume V, and T. To evaluate Q, one must know how to integrate the exponential function of the Hamiltonian over positions and momenta of all atoms in the system (McQuarrie 1976). Although direct and analytical integration is impossible even with the help of simulation, except for the small system, the partition function can be reduced to the configuration integral, of which function depends only on the positions of particles, and in some simple cases, analytical expressions for free energy can emerge.

For a system at a nonideal state, one method is to find a reversible path from the current state of interest to a reference state (Frenkel and Smit 1996). The ideal gas is a good choice for the reference state because the exact free energy can be easily calculated. The usual approach to find this particular path is thermodynamic integration and is readily explained from a thermodynamic relation as follows:

$$\left(\frac{\partial f}{\partial V}\right)_{N,T} = -P$$ (9.2)

where P is the pressure. If double Lagrange transformations are applied to Equation 9.2 with respect to two independent parameters V and N, one can express Helmholtz free energy in a form with the inverse temperature β (i.e., $(\partial(\beta f)/\partial\beta)_{V,N} = E$), where $\beta = 1/k_BT$. From Equation 9.2, it can be seen that the work $-PdV$ is done by the system under fixed N and T; to find the free energy of the state of interest, one must add the form of work energy, which is required to change the state of the system from the reference state. From a simulation of fixed N, V, and T (i.e., canonical ensemble), the pressure can be directly measured through virial formalism (Allen and Tildesley 1987). To make this integration simpler, appropriate equations of state (i.e., van der Waals, Peng-Robinson, Soave-Redlich-Kwong, etc.) can be used depending on the nature and chemical or physical properties of particles constituting the system, that is, one can find the mathematical expression of the pressure, which explicitly depends on T and V (or density).

Although this particular path of the integration works well only for finding the free energy of dilute and dense liquids, one must seek a reference other than the ideal gas if the system of interest is crystalline solid phase. Let us consider a system of a fixed number of indistinguishable particles. First, one can define an interaction potential function in such a way that a coupling constant is introduced to accommodate the linearity between the reference and target systems as follows:

$$U(\lambda) = U_0 + \lambda(U_0 - U_0) \tag{9.3}$$

where U_0 and U are the potential energies between particles in reference and target systems, respectively, and λ is Kirkwood's coupling parameter (1935). When the coupling parameter is introduced to Equation 9.3, the free energy derivative with respect to λ is equal to the expectation value of the potential energy derivative. Thus the free energy of the target system, for which λ is unity, becomes:

$$f_1 = f_0 + \int_{\lambda=0}^{\lambda=1} \left\langle \frac{\partial U(\lambda)}{\partial \lambda} \right\rangle_\lambda d\lambda \tag{9.4}$$

where f_0 is the free energy of the reference system at $\lambda = 0$ and $\langle...\rangle_\lambda$ represents the ensemble average for a system with the potential energy $U(\lambda)$ (Frenkel and Smit 1996, Monson and Kofke 2001). The application of this approach is independent of the type of the potential function. It has been applied to calculate the free energy difference of fluid mixture containing Lennard-Jones and Stockmayer potentials (de Leeuw et al. 1990, Mooij et al. 1992), more fundamentally, the free energy has been successfully calculated for close-packed hard sphere (HS) solids chosen out of arbitrary solids (Frenkel and Ladd 1984). Both continuous and discontinuous potentials work well with this scheme because the Einstein crystal has been chosen as the reference system. To picture the Einstein crystal, one can think of the system in which particles are harmonically bound at their lattice sites in crystal. Continuous increase of λ (i.e., harmonic spring constant in this case) may bring any solids down to a noninteracting (ideal) Einstein crystal state, free of phase transitions. However, an appropriate value of the constant becomes an issue because the accuracy of the free energy through wide-range integration over λ can be compromised. It is rather practical to find the optimal value of λ (i.e., for an interacting Einstein crystal), which Frenkel and Ladd (1984) obtained in the quest for calculating the free energy of the HS solid. They first introduced an intermediate crystal between the noninteracting Einstein crystal and the HS solid and calculated the free energy difference between the two Einstein crystals by driving a virial expression of the mean-square particle displacement $\langle r^2 \rangle_\lambda$, considering nearest neighbor interactions only. The analytical expression was assessed by comparing $\langle r^2 \rangle_\lambda$ from the Monte Carlo simulation (with umbrella sampling), and a small discrepancy, which overestimates the free energy of the interacting Einstein crystal, was found and excluded. Another Monte Carlo simulation between the interacting Einstein crystal (i.e., now with λ_{max} found from the previous simulation) and the HS solid was performed to obtain the free energy difference. With the free energy of the noninteracting Einstein crystal, which was analytically obtained, the free energy of the HS solid is finally revealed. However, in the derivation of the Monte Carlo simulation, the constraint of the fixed center of mass (COM) was not properly treated, especially to the momenta of the system. This would have been problematic if one considers the calculation of the absolute free energy for a system of small size. Later, this method was been reexamined by Polson et al. (2000), who found that the missed COM effect on the momenta constraint is $5/2\ln N/N$ in the calculation of the absolute free energy of three-dimensional HS crystal. When applied to obtain the excess free energy (i.e., the free energy difference between real solid and ideal gas), the leading term of the finite-size correction results in $\ln N/N$. To this end, the original accuracy has been improved and the corrected values were independently supported by other workers such as Vega and Noya (2007), who took the Einstein molecular approach; the method is similar to Frenkel and Ladd's, but the difference is found from the notion of the *carrier* particle (i.e., arbitrarily chosen), movement of which is fixed at its lattice site while others are not under the static lattice condition. This modified method becomes easier in the implementation of the simulation scheme.

Although Frenkel and Ladd's method is currently the most popular for solid free energy calculation, the easiest in terms of concept and implementation is probably calculating the phase space volume, which is specifically restricted by the accessibility of a particle. The fundamental idea comes from an evaluating mechanism of the microstate, enumerated numbers of which reveal the entropy. In this process, the phase space trajectory of the system (i.e., particles) is of great importance and monitoring it is easily accomplished by simulation. Even though the concept of entropy has been well established with the states of the system (McQuarrie 1976), Ma (Ma 1981, Ma and Payne 1981) in fact first actualized the connection between the trajectory and the entropy. The examples studied by Ma and Payne were one-dimensional Ising model with 16 spins and two-dimensional Sherrington-Kirkpatrick spin class at low temperature (Ma and Payne 1981). Their results will reproduce the ones from theoretical approaches but with a small deviation less than approximately 5%. The accuracy was not an issue, but the main drawback in applicability of the method lies in the limited size of the system because of exponential increase of the enumeration of the states. This size limitation has been challenged by Rickman and Srolovitz (1993), who extended Ma's idea by introducing the concept of the density of states, which are directly obtained by the configurations (i.e., trajectory) of the system. Their approach indeed produced the partition function under the canonical ensemble, hence the "absolute" free energy. Compared to Ma's coincidence counting method, this advanced method shows dominant accuracy and efficiency and the system size is larger. Because the internal energy is rather easily obtainable in Hamiltonian operation, the main focus for the free energy calculation narrows to the entropy calculation. For actualizing the entropy concept to the canonical ensemble (i.e., constant-temperature molecular dynamics) simulation, the entropy is now expressed as follows:

$$S = k_B \sum_E P(E) \ln\left(\frac{\Omega(E)}{P(E)}\right) \tag{9.5}$$

where $P(E)$ is the probability, which can be the ratio of the spent time in states of the energy E to the total time of the trajectory, and $\Omega(E)$ is the total number of states with the energy E, also known as the partition function of the microcanonical ensemble. Note that the inclusion of the probability in Equation 9.5 is necessary because the system is in contact with heat reservoir; hence the fluctuation of the energy state is accounted for. Under a constant energy, Equation 9.5 reduces to Boltzmann equation of the entropy $S = k_B \ln \Omega(E)$. In this case, the partition function $\Omega(E)$ can be thought of as the portioned volume in the phase space and easily calculated in some simple systems such as harmonic oscillator and HS. Within the same context of the evaluating mechanism of the microstate, it is realized that the quazi–zero-dimensional system (Rosenfeld et al. 1996), in which a confining volume can hold only one particle, is the essence of the free volume approach, which Velasco et al. (1998) adopted for free energy calculation. They showed a simple expression of the Helmholtz free energy of three-dimensional HS crystal, as follows:

$$F/Nk_B T = -\ln\left(\frac{v(\rho)}{\Lambda^3}\right) \tag{9.6}$$

where $v(\rho)$ is the accessible volume by a particle in a small space confined by neighboring particles and Λ is the de Broglie thermal wavelength. In particular, the accessible volume is generally called the free volume when the surrounding particles are fixed at their lattice sites. Because the shape of the free volume is analogous to that of the Wigner-Seitz (WS) cell and is further molded by the neighboring particles, they adopt the well-known size scaling parameter $(\alpha-1)/\alpha$, where α is the nearest neighbor distance as a function of density. The end products of their works on HS face–centered cubic (FCC) and base-centered cubic (BCC) solids are $v_{FCC}(\rho)=(\alpha-1)^3 8/\sqrt{2}$ and $v_{BCC}(\rho)=(\alpha-1)^3 32/3\sqrt{3}$, which generally produce good results of the free energies over a wide range of densities, compared with the ones from MC simulations, except at the vicinity of melting point. By comparing with an analytical expression obtained by Buehler et al. (1951) for the HS FCC solid, one can find that Velasco's $v_{FCC}(\rho)$ predicts the low value in volume and causes a certain incline (i.e., ~0.04 $k_B T$ at melting) in the free energy. Although the free volume can be easily obtained through simulation with such a high accuracy, it is of great importance to avoid even a small error because the free energy is magnified by Boltzmann's formula.

Moreover, the idea of a direct evaluation of the phase space volume well coincides with the evaluation of the free energy functional from the density functional theory (DFT). The free volume depends on the density, which controls the variation of the nearest neighbor distance, without affecting the density distribution of a particle. Thus, a question may arise in the degree of true reflection of the free volume approach toward the actual system. This issue is not of concern in then DFT. As seen from earlier work by Kirkwood and Monroe (1940, 1941), who attempted to calculate the solid free energy by a DFT approach, the solid was treated as an highly inhomogeneous fluid. It is a reasonable treatment in considering the coexistence of fluid and solid; the structural information (i.e., pair correlation function) of fluid can be used for study of the other phase. The integral equation was solved to attempt to get the periodicity of the one-body density distribution in the crystal, yet an accurate prediction of the phase transition was not successful until appearance of the work done by Ramakrishnan and Yussouff (1977, 1979) and Haymet and Oxtoby (1986), who tried to find the global minimum of the grand potential using a functional Taylor expansion of the free energy functional. A typical expression for the grand potential without applied external fields is:

$$\Omega = F[\rho] - \int \rho(r)\mu dr \tag{9.7}$$

where $F[\rho]$ is the Helmholtz free energy, $\rho(r)$ is the single molecule density distribution, and μ is the chemical potential. Under a homogeneous fluid phase, the second term becomes $N\mu$ because $\rho(r)$ becomes uniform. To obtain the global minimum of the grand potential, one finds the equilibrium molecular density while varying $\rho(r)$ and defining $F[\rho]$. From those pioneering works in search of the solid free energy using the DFT, there have been extensive studies bearing several types of DFTs; the weighted density approximation (WDA) (Curtin and Ashcroft 1985, 1986, Tarazona 1985), the modified WDA (Denton and Ashcroft 1989), the effective liquid approximation (ELA) (Baus 1990), the generalized ELA (Lutsko and Baus 1990a, Lutsko and Baus 1990b, Wang and Gast 1999), the fundamental measure theory (FMT) (Rosenfeld 1989), and so on. Those methods are distinguished mainly by the different interpretations to constituent variables in $F[\rho]$, where a common characteristic is found as a form of the density distribution function, formalism of which was first adopted with a Fourier expansion by Kirkwood and Monroe, but later changed to a Gaussian ansatz with variational parameters (Tarazona 1984).

This density function provides the distribution of the particles centered around their lattice sites (Tarazona 1985). In this approach, the free energy is considered to be the functional $F[\rho(r)]$ of the density distribution $\rho(r)$; thus a general free energy functional can be expressed with the local density approximation as follows:

$$F[\rho] = \int \rho(r)\Psi[\rho(r)]dr \tag{9.8}$$

where $\Psi[\rho(r)]$ is the free energy per particle in a homogeneous system with the density $\rho(r)$. This quantity can be easily obtained from the equation of state, yet Equation 9.8 has shortcomings to be applied to an inhomogeneous system such as a solid or inclusion of walls because of an assumption that $\rho(r)$ should be a smooth function. Thus, it is well applicable only for liquid state or under a weak external field. Considerable efforts have been made to overcome this problem. Later, position-independent weighted density was introduced by Lutsko and Baus (1990a, 1990b), whose work finally produced an accurate prediction of HS solid free energy.

A more versatile version of the DFT has been introduced by Rosenfeld (1989) and Rosenfeld et al. (1996) and is known as the fundamental measure theory (FMT). In this approach, only one integrand is required, and thus the excess free energy functional is expressed as:

$$F_{excess}[\rho] = \beta \int \Phi[\{n_a(r)\}] dr \qquad (9.9)$$

where $\Phi[\{n_a(r)\}]$ is the excess free energy density, in which $[\{n_a(r)\}]$ is a set of the dimensional weighted densities

$$n_a(r) = \int \rho(r') \omega^{(a)}(r - r') dr' \qquad (9.10)$$

where $\omega^{(a)}(r-r')$ is the set of weight functions that are independent of density. This initiative approach of FMT works for predicting inhomogeneous fluid applications but not for HS solid properties until Rosenfeld et al.'s (1996) modifications with empirical corrections were adapted. More accurate treatment on the FMT was later demonstrated by Tarazona (2000), who introduced a new free energy density functional. This approach has been free from the misinterpretation of the nonlocality of the free energy functional with the introduction of the variational parameters, which include lattice size, Gaussian parameter, the anisotropy, and the unit cell occupancy. With this variational minimization on the tensor weighted density, the equation of state and the free energy for the FCC HS solid has been found very accurate and good qualitative behavior of BCC HS solid has been shown.

9.3   SIMULATION OF CONFINEMENT INDUCED FLUID-SOLID PHASE TRANSITIONS

Literature review reveals that there are three common levels of model accuracy exist for the design of these simulations. Here we briefly describe each, noting previous applications and drawing the reader's attention to important considerations when deciding the most appropriate level of accuracy.

9.3.1   MODELS

9.3.1.1   Primitive Models

The simplest possible model represents the surfaces as structureless walls, and the confined molecules are modeled as spheres or chains of spheres depending on the confined fluid to be studied. The molecules and surface may be purely repulsive, or attractive interactions might also be considered; however, the surfaces are typically repulsive. Ayappa and Ghatak (2002, 2007) considered nanoconfined Lennard-Jones spheres and investigated various aspects of solid–liquid transition. Their focus was mainly on understanding the structure of the confined solidlike phases and considering both the effect of surface separation on a generic Lennard-Jones potential and, in a later work, a potential chosen to mimic octamethylcyclotetrasiloxane (OMCTS). Using the generic potential (Fortini and Dijkstra 2006), they observed that opposed to a single solidlike phase, there exists a sequence of solid phases as a function of surface separation (i.e., pore size). Ayyappa and Ghatak (2002), using a parameter set representative of OMCTS, not only demonstrate the same phases as for the generic potential, but also find evidence that even if the layers in contact with the surface are disordered, the central layers in a confined system may be ordered. Other examples include the work of Fortini and Dijkstra (2006), who calculated the complete phase diagram of solid–liquid and various crystal structures for a system of HSs between hard plates.

The use of these relatively simple systems facilitates the calculation of more complex properties, such as free energy, which, coupled with their relatively low computational cost, allow for a more complete investigation, compared to the more detailed methods described in the following section. However, they are not able to provide a reliable or accurate approximation to real systems and thus miss much of the subtle details of what occurs in these systems.

9.3.1.2 United Atom Level Approach

At the next level of accuracy, the confining surfaces are modeled as structured and the confined molecules are described at a united atom (UA) level. Although the surfaces are structured, the structure is typically simplified, most commonly FCC. The particles forming the surface may be allowed to vibrate around their lattice points; however, this is not necessary and the same confined structure will be obtained if the particles are held stationary. This level of molecular model is very common and has led to the identification of important factors determining whether a fluid–solid transition occurs, as well as the nature of the formed solid. Here we provide examples of two important studies utilizing united atom molecular models, noting their specific contributions.

We begin by considering the pioneering work of Cui et al. (2001a, 2001b, 2003), who performed molecular dynamics simulations of dodecane confined between mica sheets. In their work, Cui et al. described the mica sheets using an FCC lattice of Lennard-Jones spheres and the Siepmann, Karaborni, and Smit (SKS) alkane model for the dodecane molecules. In addition to providing insight into confinement-induced freezing, the success of the work of Cui et al. (2001a, 2001b, 2003) resulted in their choice of models being becoming standard in the simulation of nanoconfined alkanes. Despite that, it is worth noting that the SKS model for alkanes may be replaced by any number of UA level models (e.g., Optimized Potentials for Liquid Simulations-United Atom (OPLS-UA)) and the same system behavior obtained.

Cui et al. (2001a, 2001b, 2003) performed constant normal pressure molecular dynamics simulations and made a number of interesting discoveries, two of which we consider here. The first of these is that they successfully showed that dodecane undergoes a phase transition from a disordered fluid state to a layered and herringboned ordered, solidlike state as a function of confinement (i.e., surface separation). The second insight from their work is the role of wall–fluid interaction strength on confinement-induced freezing. Specifically, they showed that in order for a fluid–solid phase transition to occur, the wall–fluid interaction strength must be sufficiently stronger than the fluid–fluid interaction. Interestingly, this is in agreement with the global theory of phase transitions under nanoconfinement developed by Radhakishnan et al. (2000, 2002). It is thus clear that care must be taken to ensure the use of an appropriate wall–fluid potential if the correct behavior is to be observed. Interestingly, if the wall–wall interaction strength is chosen in order to match the experimental surface energy of mica, as was done for the work of Cui et al. (2001a, 2001b, 2003), Lorentz-Berthelot mixing rules result in a wall–fluid interaction, which is approximately 4.5 times the fluid–fluid interaction strength for dodecane.

Other important information obtained using UA level models may be found in the work of Jabbarzadeh et al. (2005, 2006a and 2006b), who presented a series of articles in which they investigated how changing the surface structure of the confining sheets affects the structure of nanoconfined dodecane. Making use of the same models and parameters as Cui et al. (2001a, 2001b, 2003) Jabbarzadeh et al. (2005, 2006a and 2006b) not only confirm the formation of a herringbone ordered structure for FCC surfaces, but also show that when the FCC is replaced with BCC surfaces, the herringbone structure still forms, but rotated by 45°, ensuring that the alignment of the confined structure with the surface is retained. Further to this, they consider the effect of an amorphous surface structure, noting that as the confined surfaces become increasingly amorphous, the induced layered herringbone structure becomes less well defined. This is particularly interesting because, based on the experimental evidence that fluids of different geometry behave similarly, it is often thought that any transition is purely a function of confinement, that is, surface structure does not play a role.

Although the UA approach described earlier has provided valuable insight into the effects of nano-confinement, it could be argued that approximations in the models and simulations may introduce artificialities that affect the results. Specifically, concerns exist that the relatively small system size, together with the shape and periodic nature of the simulation box, stabilized the herringbone ordered structure. In addition to this, the relatively simplistic models, particularly those describing the mica surfaces, might not adequately represent the experimental system. With regard to the role of system size, it is worth noting that in their recent work Jabbarzadeh et al. (2005, 2006a and 2006b) used mica surfaces approximately four times the surface area of those used in the work of Cui et al. (2001a, 2001b, 2003) (51.84 nm^2 versus 12.96 nm^2) and that, in our own work, we have used even larger sheets (80 to 400 nm^2). In both cases, the system exhibited the same behavior (i.e., the formation of a herringbone ordered structure). Thus, it seems that system size did not play a determining role in earlier simulation studies. Docherty and Cummings (2010) and Cummings et al. (2010), however, recently addressed concerns over the use of simplistic models rather than using atomistically detailed simulations.

9.3.1.3 All Atom Approach

Although we describe the highest level as one of the three common levels of accuracy, we note that the use of such accurate models is quite recent, primarily because of computational limitations. However, it seems likely they will continue to increase in popularity over the next few years now that computational resources have made them feasible. These models consist of trying to describe the confined systems, both the surface and the fluid, in a fully atomistic manner, including the incorporation of explicit partial charges where appropriate. As an example, we will use the recent work of Cummings et al. (2010) simulating systems of dodecane and cyclohexane nanoconfined between mica sheets using the atomistic mica model of Heinz et al. (2005) coupled with the OPLS all-atom model for the alkanes. There are many advantages to the use of these models; however, perhaps the most important is that by using a fully flexible model of mica and the confined fluid, together with explicit coulombic interactions, the results may be considered to be truly indicative of what occurs in the real system, opposed to merely being valid for a model mica-alkane.

In the case of dodecane, and Cummings et al. (2010) observed that behavior is qualitatively similar to that seen in the work of Cui et al. (2001a, 2001b, 2003) and Jabbarzadeh et al. (2005, 2006a, 2006b, 2007a and 2007b), that is, the formation of a layered and herringboned ordered solid. However, a few differences are worth noting. First, whereas the UA approach using FCC (or BCC) surfaces results in patches of parallel dodecane molecules aligned perpendicularly, the patches for the atomistic models are aligned at approximately 60° (or 120°). Second, the formation of a stable layered structure up to nine layers thick is seen, compared to the seven layers observed by Cui et al (2001a, 2001b, 2003). The reasons for these differences may be realized by noting the insights provided by the work of Cui et al. (2001a, 2001b, 2003) and Jabbarzadeh et al. (2005, 2006a, 2006b, 2007a and 2007b); that is, the increased number of stable layers is mostly like a result of a stronger wall–fluid interaction, and the 60° alignment is consistent with the hexagonal surface structure of the confining surfaces. This demonstrates that although a good qualitative description may be obtained with the UA approach, in order to capture fully the more subtle details of confinement-induced phenomena, fully atomistic simulations are required.

The important question regarding the role of geometry on the confined fluids can also be addressed using detailed atomistic simulation. It is noted that the experimental observation of fluids ranging from linear to cyclic exhibit the same behavior upon nanoconfinement. Thus, confinement-induced phase transitions are thought to not depend on the fluid being commensurate with the confining structure. Here Docherty and Cummings (2010) investigate whether the shape of the confined molecule affects the formation of a layered and ordered structure by simulating the nanoconfinement of cyclohexane, which, in addition to differing in geometry to dodecane, was involved in early Surface Forces Apparatus (SFA) experiments (Klein 1980).

The results of their simulations show that, in common with dodecane, cyclohexane exhibits an abrupt transition to a layered and ordered structure as a function of confinement. However, it only forms a stable structure up five layers thick, compared to the nine layers observed for dodecane. This value compares well to Perkin et al.'s (2006) experimental value of six stable layers. In terms of the form of the structure (Figures 9.1 to 9.3), although, as expected, the structure is different from that of dodecane, it is still hexagonal in nature; this highlights the importance of the surface structure of the confining material and suggests that the common assumption that this is not the case needs to be reassessed.

In summary, the use of atomistically detailed molecular models provides a level of fidelity and thus confidence that cannot be matched by the other models described here. Thus, if one wishes to state with confidence that one's simulations describe a particular experimental system, there is no substitute. However, the use of such models is not without cost, which, in this case, is the dramatically larger computational time required.

9.3.2 Choice of Ensemble

Another important decision in the simulation of a confined system (and of course in any simulation) is the choice of ensemble in which to perform the simulation. Here we outline some choices that have been successfully applied to these systems.

9.3.2.1 Canonical Ensemble

The canonical ensemble is perhaps the most suitable choice for two types of studies. The first is that performed by Fortini et al. (2006) when trying to determine the confined structure as a function of

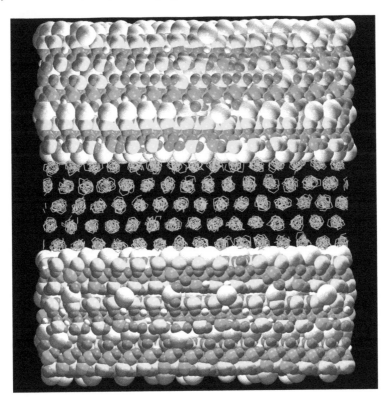

FIGURE 9.1 Side view of an equilibrated all-atom simulation of cyclohexane confined between atomistically detailed mica surfaces. Note the formation of a five-layered hexagonal solidlike structure. (From Cummings, P. T., et al. *AIChE J.* 56: 842, 2010. With permission.)

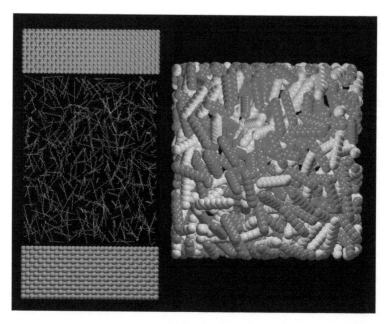

FIGURE 9.2   Side and top views of a typical starting configuration for an isothermal–isobaric simulation of dodecane confined between FCC surfaces. (From Cummings, P. T., et al. *AIChE J.* 56: 842, 2010. With permission.)

FIGURE 9.3   Side and top views of an equilibrated configuration for an isothermal–isobaric simulation of dodecane confined between FCC surfaces. Note the formation of five distinct layers with an intralayer herringbone structure (top view). (From Cummings, P. T., et al. *AIChE J.* 56: 842, 2010. With permission.)

density. A few difficulties should be noted when performing such a study. Primarily, when the confining surfaces are not allowed to move, jamming, particularly of longer chain molecules, may be problematic and require, at best, unfeasibly long simulation times. The second type of simulation for which we recommend the use of a canonical ensemble is the study of phase coexistence under confinement. Although this technique is perhaps most easily applied to the study of vapor–liquid equilibria, it should be possible to simulate a system consisting of a confined solid and fluid phase in equilibrium.

9.3.2.2   Isothermal–Isobaric

Given that the majority of experimental studies take place at atmospheric condition, a natural choice for simulation studies is the use of the isothermal–isobaric ensemble. This ensemble was the method of choice for both Cui et al. (2001a, 2001b, 2003) and Jabbarzadeh et al. (2005, 2006a, 2006b, 2007a and 2007b) in their studies of nanoconfined dodecane. Therefore, we will use dodecane between FCC mica surfaces as an example to further clarify the use of this ensemble.

The system requires placing N dodecane molecules between FCC surfaces (Figure 9.4). The system may be three-dimensionally periodic or may have what is known as a slab geometry (i.e., not replicated in the dimension perpendicular to the confining surfaces). At this stage, a straightforward isothermal–isobaric ensemble simulation with the pressure controlled anisotropically in all three dimensions (only possible for flexible surfaces) may be performed or, commonly, the pressure may be controlled just in the dimension normal to the confining surfaces (suitable for rigid surfaces). In either case, the observed effects of nanoconfinement will be the same. What occurs will depend not only on the imposed temperature and pressure, but also on the number of fluid molecules confined. Specifically, if the number of dodecane molecules is less than the number required to form n_c layers, the fluid will undergo a transition to a layered and ordered solidlike structure. Otherwise it will remain fluidic in nature.

There are a few caveats to this. The dodecane molecules will tend to form an integer number of layers, forming what is known as a herringbone structure (Figure 9.3); however, the transition to this structure is much more rapid (less simulation required) and the final structure contains fewer flaws if the number of dodecane molecules is chosen to correspond as nearly as possible to the number required to form n "perfect" layers. To find the number of molecules in a perfect layer, one option is to assume a reasonable level of packing and use this to predict. Alternatively, because the layers immediately adjacent to the confining surfaces tend to form quite rapidly (as a result of the strong wall–fluid interaction) and are mostly independent of the number of layers formed in the ordered system, running a relatively short simulation allows for the formation of a not perfect, but well-packed first layer from which the number of molecules per layer may be "counted." Of course, these two methods may be combined.

The isothermal–isobaric ensemble appears to provide a reasonable balance between fidelity to experimental studies and computational cost, with its biggest criticism being that the periodic nature of the simulation cell may introduce an artificial stability in the confined structure as well as affect the nature of its ordering (e.g., promote a herringbone structure). With regard to the latter point, it is worth noting that simulations performed using grand canonical molecular dynamics, described below, result in the same herringbone structure, suggesting this concern is probably unfounded.

9.3.2.3   Isothermal–Isobaric Grand Canonical Molecular Dynamics

Another alternative is the grand canonical molecular dynamics technique (Isobaric-Isothermal-GCMD) (Steinhardt et al. 1983). This method may be thought of as taking the isothermal–isobaric simulation cell described earlier (Figure 9.4) and, placing a large amount of bulk fluid on two opposite sides of the confining sheets such that molecules are free to enter, and leave the slit from the bulk fluid (i.e., the confined fluid and surrounding bulk is in equilibrium). The system is three-dimensionally periodic, with the box length in the lateral direction used to maintain the pressure. This method has two key advantages. First, the confining surfaces remain stationary (i.e., the

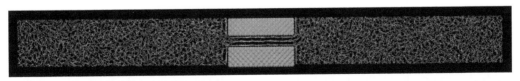

FIGURE 9.4   Side view of an equilibrated NPT-GCMD simulation of united atom dodecane confined between FCC Lennard-Jones mica surfaces with a fixed surface separation of three dodecane. (From Cummings, P. T., et al. *AIChE J.* 56: 842, 2010. With permission.)

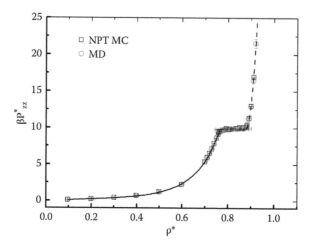

FIGURE 9.5 Comparison of the equation of state (reduced axial pressure versus reduced numerical density) of Square-Well molecules of $\lambda_{ff} = 1.5$ confined in cylindrical hard pore with diameter, $D/\sigma_{ff} = 2.2$, obtained by isobaric–isothermal Monte Carlo (NPT MC) and molecular dynamic (MD) simulations. Here, squares indicate NPT MC result and circles the MD result. The solid line indicates an analytical fit of the result at the fluid branch, and the dash line is the second order polynomial fit to the solid branch. Error bars are the standard deviation of five independent runs. (From Huang, H. C., *J. Chem. Phys.*, 132, 224504, 2010. With permission. Copyright 2010, American Institute of Physics.)

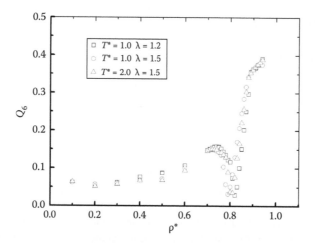

FIGURE 9.6 The bond order parameter Q_6 as the function of density, $\rho^* = \rho\sigma_{ff}^3$, for the SW molecules confined in a cylindrical hard pore of $D = 2.2\sigma_{ff}$ at different temperature and potential range. (From Huang, H. C., *J. Chem. Phys.*, 132, 224504, 2010. With permission. Copyright 2010, American Institute of Physics.)

pore width is constant), which eliminates confinement rate as a source of error. Second, because molecules are free to enter and leave the pore, the number of confined molecules is not artificially fixed. Typically the results of this technique are consistent with isothermal–isobaric simulations. For example, for the UA n-dodecane/FCC Lennard-Jones (LJ) system, a herringbone ordered structure with the same structure and orientation relative to the confining surfaces is obtained. The drawback to this method is that it requires a significantly larger number of fluid molecules, resulting in increased computational cost. Given this, the technique is probably best used sparingly as a check of the computationally cheaper isothermal–isobaric method.

9.3.3 ANALYSIS

Having dealt with methods to perform simulations of confined fluid–solid transitions, we turn to the determination of whether a transition has occurred, as well as measures of the structure of the confined system. The global order of the system indicates whether a system is fluidlike or solidlike and is characterized by the bond order parameters introduced by Steinhardt et al. (1983). Here a bond is not a chemical bond but a vector joining two neighboring atoms whose distance is less than a cutoff radius. In order to calculate bond order parameters, a spherical harmonic associated with every bond r_{ij} in the system needs to be determined. The quantity is defined as:

$$Q_{lm} = Y_{lm}(\theta(r_{ij}), \phi(r_{ij}))$$

(9.11)

where Y_{lm} is the spherical harmonic and θ and ϕ are the polar and azimuthal angles, respectively, of the bond with respect to a fixed coordinate frame that can be arbitrary chosen. The global bond order parameter Q_l, which is rotationally invariant is then defined as follows:

$$Q_l \equiv \left(\frac{4\pi}{2l+1} \sum_{m=-l}^{l} |\bar{Q}_{lm}|^2 \right)^{1/2}$$

(9.12)

where $\bar{Q}_{lm}$ is the average of Q_{lm}'s over all bonds in the system. The presence of any order under confinement can be identified by the use of Q_6 as the order parameter. Q_6 is very sensitive to the global structure, which increases significantly when order appears and vanishes for the isotropic fluid.

To illustrate this, we take an example from our recent work (Huang et al. 2010), in which we considered the simple case of a square-well fluid confined in an HS cylindrical pore and investigated the fluid–solid transition and structural changes under confinement. Using NPT-based molecular dynamics simulation, P_{zz}, the spreading pressure (defined as pressure along the axis of the cylindrical tube) is calculated at different densities. Separate NVT (i.e. constant volume)-based Monte Carlo simulations were also conducted to verify the results. Figure 9.7 presents a typical spreading pressure versus density isotherm. Phase equilibria data are estimated from the P_{zz} equality criteria (which is not a strict condition). Results from both types of simulations are in agreement. The discontinuity in the equation of state, which is observed by both methods, is indicative of a fluid–solid transition. This comparison indicates that both methods are sufficiently accurate to determine the equation of state of both fluid and solid branches. However, for determining the fluid–solid coexistence regime, molecular dynamics (MD) is better than NPT MC, because MD has a much smaller relaxation time, which enables better accuracy in determining the coexistence density regime and coexistence pressure, whereas NPT MC generally cannot pinpoint these properties with such accuracy. However, it is noted that precise calculation of phase equilibria is not feasible from this approach, for which a separate free energy calculation is necessary, which is described briefly later.

Even though the equation of state, as shown in Figure 9.5, illustrates a coexistence region, it is appropriate to look at the structure of states near the coexistence region to verify the liquid or solid state behavior. As stated earlier, precise solid–liquid equilibria calculation under confinement from pressure–density isotherms is not possible; thus investigation of the structural order parameter is necessary as a check. Figure 9.6 illustrates a bulk bond order parameter Q_6 as a function of density in a pore of size $D = 2.2 \sigma_{ff}$. Q_6 for this system is found to be nonzero for dilute fluid because of a confinement effect. The system with a random set of N_b bonds, has a Q_l value of $1/\text{sqrt}(N_b)$, but not zero (Rintoul and Torquato, 1996), as within the tight confinement, because the direction of bonds joining each pair of atoms is not stochastic but has limited orientations. An interesting phenomenon has been observed with the increase of fluid density. Q_6 increases to reach a value around 0.15 at a reduced density around 0.75. Similar behavior is found for the two-dimensional bond orientation parameter, which is described by Equation 9.13.

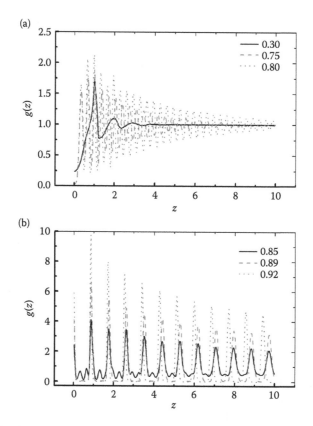

FIGURE 9.7 Axial pair distribution function g(z) for SW molecules confined in a cylindrical pore of $D = 2.2\sigma_{ff}$ at $kT / \varepsilon_{ff} = 1.0$ and $\lambda_{ff} = 1.2$, at different average densities, $\rho^* = 0.30$ (dilute fluid), 0.75 (dense fluid), 0.80 (around freezing density), 0.85 (within fluid–solid coexistence), 0.89 (around melting density), and 0.92 (solid). Note that z is normalized by σ_{ff}. (From Huang, H. C., *J. Chem. Phys.*, 132, 224504, 2010. With permission. Copyright 2010, American Institute of Physics.)

$$\Psi_6 = \left| \frac{1}{M} \sum_{i=1}^{M} \frac{1}{N_i} \sum_{j=1}^{N_i} \exp(i6\theta_{ij}) \right| \tag{9.13}$$

where M is the number of particles in the two-dimensional layer (in the previous case it would be cylindrical layer) and N_i is the number of nearest neighbors of particles i. The value of Ψ_6 is equal to 1 for a perfect hexagonal plane, but far from 1 for a disordered phase.

We should not take agreement between the global bond order and local order to be the general behavior for confined solids. As shown by Coasne et al. (2007) the bulk bond-order parameter is not a suitable parameter to study the structural parameters because it is sensitive to the structure of the surface. For a crystal phase, the $\bar{Q}_{lm}$ components add up coherently. However, for example, for disordered porous materials, Q_6 would provide erroneous information about the structure because a large degree of disorder of the porous materials would prevent coherency of the orientational order of the confined phase. Hence, a more appropriate parameter for the study is a two-dimensional hexagonal order parameter (in-plane two-dimensional orientational order parameter). Other than order parameter calculations, radial distribution functions can also provide necessary structural information for confined fluids. For example, Figure 9.7 illustrates the axial distribution function $g(z)$, for the aforementioned confined square-well fluid case. Corresponding molecular snapshots are shown in Figure 9.8.

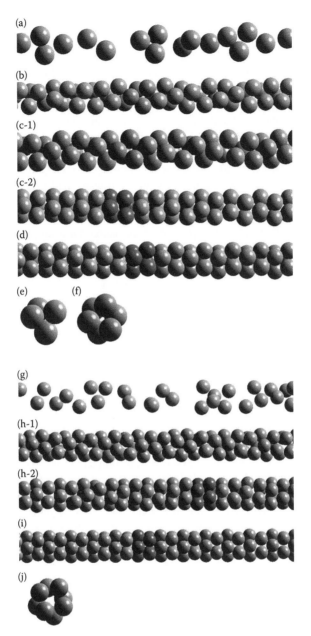

FIGURE 9.8 Snapshots for SW molecules confined in narrow cylindrical pores of $D = 2.2$ (a to d), and $D = 2.5$ (g to i) at different average densities: a, $\rho = 0.3$ (dilute fluid); b, $\rho = 0.75$ (dense fluid, spiral packing of tetrahedron); c, $\rho = 0.85$ (fluid–solid coexistence); d, $\rho = 0.92$ (solid, triangular packing); e, tetrahedral packing; f, triangular packing; g, $\rho = 0.3$ (dilute fluid); h, $\rho = 0.85$ (fluid–solid coexistence); and i, $\rho = 0.95$ (solid, square packing); j, square packing. Two different structures are found in the corresponding coexistence densities in $D = 2.2$ (c-1 and c-2) and $D = 2.5$ (h-1 and h-2). (From Huang, H. C., *J. Chem. Phys.*, 132, 224504, 2010. With permission. Copyright 2010, American Institute of Physics.)

At a reduced density of $\rho = 0.30$ ($\sigma_{ff} = 1$ and asterisk is dropped in the rest of the chapter for the sake of convenience), $g(z)$ shows a peak at $z = 1.0$ and a valley at $z = 1.2$, which are the hard core diameter and the extent of the potential well, respectively, and it quickly decays to 1.0 as z increases, indicating the system does not have a long-range order at this density. For $\rho = 0.75$ and 0.80, peaks appear at intervals of approximately 0.33, with the largest peak occurring at $z = 1.0$ before dropping smoothly to g(r) = 1.0. In contrast to ρ = 0.30, g(r) at z = 0 is nearly zero. Interestingly, the position of the peaks matches the axial distance of the peaks of tetrahedrons. This supports the evidence from bond order parameters, which suggests there is a partially ordered phase of spiral packing of tetrahedra around this density range. For densities of 0.85 (within coexistence densities), 0.89 (around melting point), and 0.92 (solid phase), a sharp peak appears at z = 0 and is followed by a series of peaks periodically spaced at intervals of about 0.87, with the largest peak being the first of these. This phenomenon indicates that the tetrahedral packing has transformed to triangular packing. Note that for the density within coexistence, there exist two small peaks between the sharp peak at z = 0 and z = 0.87, which indicates that both tetrahedral and triangular packing coexist in the system. Similar behavior is observed for the case of diameter, D = 2.5; however, the twisted helical structure disappears at ρ = 0.95, where the structure is transformed to a square shape consisting of four particles sharing the same z against another nearest square layer by rotating by an angle of π/4 (Figure 9.10). In addition to the in-plane positional correlation function, the in-plane bond-orientational pair correlation function, $G_6(r)$, which measures for each two-dimensional plane the correlation between the local bond order parameter, $\Psi_6(r)$ at two positions separated by a distance r:

$$G_6(r) = \langle \psi_6^*(0)\psi_6(r)\rangle. \tag{9.14}$$

Similar to the positional pair correlation function, the G_6 function would possess a short-range orientation order for liquid phases and long-range orientation order for crystal layers. Hence, analysis of the in-plane two-dimensional pair correlation function and orientation correlation function jointly can provide the transition between liquid and solid. Nevertheless, precise calculation of coexistence liquid–solid phases can be done only by evaluating the chemical potential of the individual phases, or equality, of the grand potential $\Omega = F - \mu N$ (Fortini and Dijkstra 2006).

Finally, for this section, we note that, in addition to inducing a fluid–solid transition, confinement can change stable crystal structures and may lead to a variety of different and new structures, which are not seen in bulk solids, as shown by Fortini and Dijkstra (2006) for HSs confined between two hard plates. The authors determine for the first time the complete phase diagram of solid–liquid and various crystal structures of the confined HS system by calculating the free energy using a thermodynamic integration technique. The coexisting solid phase consists of crystalline layers with either triangular (T) or square (S) symmetry (similar to that observed for square-well fluids in hard cylindrical pores). A sequence of crystal structures, from nT→(n + 1)S → (n + 1)T ..., is observed by the authors with increase in plate separation. Here, n is the number of crystal layers. This behavior is in agreement with experiments on colloids (Fontecha et al. 2005).

9.4   SUMMARY

We finish by noting that the study of nanoconfined systems is of vital importance to the fledgling development of nanoscale devices. Direct experimental observation of what occurs in these systems is difficult and expensive, when it is possible at all. This may be due to the speed at which a particular phenomenon occurs (e.g., wetting) or simply the lack of a tool with sufficient resolution to observe what is occurring (e.g., confinement-induced freezing). This weakness provides an exciting opportunity to apply any of a wide range of increasingly powerful molecular simulation techniques to the study of nanoconfined systems. With their inherent molecular resolution, such methods are a synergetic complement to current experimental studies, providing direct insight into the underlying mechanisms of nanoconfinement-induced behavior.

9.5   ACKNOWLEDGEMENTS

We are grateful for support extended by the Department of Science and Technology, India.

REFERENCES

Allen, M. P., Tildesley, D. J. 1987. *Computer simulation of liquids*. Oxford University Press, USA.

Ayappa, K. G., and Ghatak, C. The structure of frozen phases in slit nanopores: A grand canonical Monte Carlo study. 2002. *J. Chem. Phys.* 117: 5373.

Ayappa, K. G., and Mishra, R. K. Freezing of fluids confined between mica surfaces. 2007. *J. Chem. Phys. B* 111: 14299.

Baus, M. The generalized effective liquid approximation for the freezing of hard spheres. 1990. *J. Phys. Condens. Matter* 2: 2111.

Buehler, R. J., Wentorf, R. H., Hirschfelder, J. O., and Curtiss, C. F. The free wolume for rigid sphere molecules. 1951. *J. Chem. Phys.* 19: 61.

Coasne, B., Jain, S.K., Naamar, L., and Gubbins, K.E. Freezing of argon in ordered and disordered porous carbon. 2007. *Phys. Rev. B* 76: 085416.

Cui, S. T., McCabe, C., Cummings, P. T. and Cochran, H.D. Structural transition and solid-like behavior of alkane films confined in nano-spacing. 2001. *Fluid. Phase. Equil.* 183: 381.

Cui, S. T., Cummings, P. T. and Cochran, H. D. Molecular simulation of the transition from liquidlike to solid-like behavior in complex fluids confined to nanoscale gaps. 2001. *J. Chem. Phys.* 114: 7189.

Cui, S. T., McCabe, C., Cummings, P. T., and Cochran, H. D. Molecular dynamics study of the nano-rheology of n-dodecane confined between planar surfaces. 2003. *J. Chem. Phys.* 118: 8941.

Cummings, P. T., Docherty, H., Iacovella, C. R. and Singh, J. K. Phase transitions in nanoconfined fluids: The evidence from simulation and theory. 2010. *AIChE J.* 56: 842 .

Curtin, W.A., and Ashcroft, N.W. Weighted-density-functional theory of inhomogeneous liquids and the freezing transition. 1985. *Phys. Rev. A Atom. Mol. Opt. Phys.* 32: 2909.

Curtin, W.A., and Ashcroft, N.W. Density-functional theory and freezing of simple liquids. 1986. *Phys. Rev. Lett.* 56: 2775.

de Leeuw, S.W., Smit, B., and Williams, C.P. Molecular-dynamics studies of polar nonpolar fluid mixtures .1. Mixtures of Lennard-Jones and stockmayer fluids. 1990. *J. Chem. Phys.* 93: 2704.

Denton, A.R., and Ashcroft, N.W. Modified weighted-density-functional theory of nonuniform classical liquids. 1989. *Phys. Rev. A Atom. Mol. Opt. Phys.* 39: 4701.

Docherty, H., and Cummings, P.T. Direct evidence for fluid-solid transition of nanoconfined fluids. 2010. *Soft Matter* 6: 1640.

Fontecha, A.B., Schope, H.J., Konig, H., Palberg, T., Messina, R., and Lowen, H. A comparative study on the phase behaviour of highly charged colloidal spheres in a confining wedge geometry. 2005. *J. Phys. Condens. Matter* 17: S2779.

Fortini, A., and Dijkstra, M. Phase behaviour of hard spheres confined between parallel hard plates: Manipulation of colloidal crystal structures by confinement. 2006. *J. Phys. Condens. Matter* 18: L371.

Fortini, A., Hynninen, A.P., and Dijkstra, M. Gas-liquid phase separation in oppositely charged colloids: Stability and interfacial tension. 2006. *J. Chem. Phys.* 125: 094502.

Frenkel, D., and Ladd, A.J.C. New Monte-Carlo method to compute the free-energy of arbitrary solids - Application to the Fcc and Hcp phases of hard-spheres. 1984. *J. Chem. Phys.* 81: 3188.

Frenkel, D., and Smit, B. 1996. *Understanding Molecular Simulation: From algorithms to applications*. Academic Press, London.

Gelb, L.D., Gubbins, K.E., Radhakrishnan, R., and Sliwinska-Bartkowiak, M. Phase separation in confined systems. 1999. *Rep. Prog. Phys.* 62: 1573.

Haymet, A.D.J., and Oxtoby, D.W. A molecular theory for freezing - comparison of theories, and results for hard-spheres. 1986. *J. Chem. Phys.* 84: 1769.

Heinz, H., Koerner, H., Anderson, K.L., Vaia, R.A., and Farmer, B.L. Force field for mica-type silicates and dynamics of octadecylammonium chains grafted to montmorillonite. 2005. *Chem. Mater.* 17: 5658.

Huang, H.C., Chen, W.W., Singh, J.K., and Kwak, S.K. Direct determination of fluid-solid coexistence of square-well fluids confined in narrow cylindrical hard pores. 2010. *J. Chem. Phys.* 132: 224504.

Jabbarzadeh, A., Harrowell, P., and Tanner, R.I. Very low friction state of a dodecane film confined between mica surfaces. 2005. *Phys. Rev. Lett.* 94: 126103.

Jabbarzadeh, A., Harrowell, P., and Tanner, R.I. Low friction lubrication between amorphous walls: Unraveling the contributions of surface roughness and in-plane disorder. 2006. *J. Chem. Phys.* 125: 034703.

Jabbarzadeh, A., Harrowell, P., and Tanner, R.I. Crystal bridge formation marks the transition to rigidity in a thin lubrication film. 2006. *Phys. Rev. Lett.* 96: 206102.

Jabbarzadeh, A., Harrowell, P., and Tanner, R.I. Crystal bridges, tetratic order, and elusive equilibria: The role of structure in lubrication films. 2007. *J. Phys. Chem. B* 111: 11354.

Jabbarzadeh, A., Harrowell, P., and Tanner, I. The structural origin of the complex rheology in thin dodecane films: Three routes to low friction. 2007. *Tribol. Int.* 40: 1574.

Kirkwood, J.G. Statistical mechanics of fluid mixtures. 1935. *J. Chem. Phys.* 3: 300.

Kirkwood, J.G., and Monroe, E. On the theory of fusion. 1940. *J. Chem. Phys.* 8: 845.

Kirkwood, J.G., and Monroe, E. Statistical mechanics of fusion. 1941. *J. Chem. Phys.* 9: 514.

Klein, J. Forces between mica surfaces bearing layers of adsorbed polystyrene in cyclohexane. 1980. *Nature* 288: 248.

Lutsko, J.F., and Baus, M. Nonperturbative density-functional theories of classical nonuniform systems. 1990. *Phys. Rev. A Atom. Mol. Opt. Phys.* 41: 6647.

Lutsko, J.F. and Baus, M. Can the thermodynamic properties of a solid be mapped onto those of a liquid. 1990. *Phys. Rev. Lett.* 64: 761.

Ma, S.K. Calculation of entropy from data of motion. 1981. *J. Stat. phys.* 26: 221.

Ma, S.K., and Payne, M. Entropy of a spin-glass model with long-range interaction. 1981. *Phys. Rev. B* 24: 3984.

McQuarrie, D.A. 1976. *Statistical Mechanics.* Harper & Row, New York.

Monson, P. A., and Kofke, D. A. Solid-fluid equilibrium: Insights from simple molecular models. 2001. *Adv. Chem. Phys.* 115: 113.

Mooij, G.C.A.M., de Leeuw, S.W., Smit, B., and Williams, C.P. Molecular-dynamics studies of polar nonpolar fluid mixtures .2. Mixtures of stockmayer and polarizable Lennard-Jones fluids. 1992. *J. Chem. Phys.* 97: 5113.

Perkin, S., Chai, L., Kampf, N., Raviv, U., Briscoe, W., Dunlop, I., et al. Forces between mica surfaces, prepared in different ways, across aqueous and nonaqueous liquids confined to molecularly thin films. 2006. *Langmuir* 22: 6142.

Polson, J.M., Trizac, E., Pronk, S., and Frenkel, D. Finite-size corrections to the free energies of crystalline solids. 2000. *J. Chem. Phys.* 112: 5339.

Radhakrishnan, R., Gubbins, K.E., and Sliwinska-Bartkowiak, M. Effect of the fluid-wall interaction on freezing of confined fluids: Toward the development of a global phase diagram. 2000. *J. Chem. Phys.* 112: 11048.

Radhakrishnan, R., Gubbins, K.E., and Sliwinska-Bartkowiak, M. Global phase diagrams for freezing in porous media. 2002. *J. Chem. Phys.* 116: 1147.

Ramakrishnan, T.V., and Yussouff, M. Theory of the liquid-solid transition. 1977. *Solid State Comm.* 21: 389.

Ramakrishnan, T.V., and Yussouff, M. First-principles order-parameter theory of freezing. 1979. *Phys. Rev. B Condens Matter and Mater. Phys.*19: 2775.

Rickman, J.M., and LeSar, R. Free-energy calculations in materials research. 2002. *Annu. Rev. Mater. Res.* 32: 195.

Rickman, J.M., and Srolovitz, D.J. Efficient determination of thermodynamic properties from a single simulation. 1993. *J. Chem. Phys.* 99: 7993.

Rintoul, M.D., and Torquato, S. Computer simulations of dense hard-sphere systems. 1996. *J. Chem. Phys.* 105: 9258.

Rosenfeld, Y. Free-energy model for the inhomogeneous hard-sphere fluid mixture and density-functional theory of freezing. 1989. *Phys. Rev. Lett.* 63: 980.

Rosenfeld, Y., Schmidt, M., Lowen, H., and Tarazona, P. Dimensional crossover and the freezing transition in density functional theory. 1996. *J. Phys. Condens. Matter* 8: L577.

Steinhardt, P.J., Nelson, D.R., and Ronchetti, M. (1983) Bond-orientational order in liquids and glasses. *Phys. Rev. B* 28: 784.

Tarazona, P. Density functional for hard sphere crystals: A fundamental measure approach. 2000. *Phys. Rev. Lett.* 84: 694.

Tarazona, P. A Density functional theory of melting. 1984. *Mol. Phys.* 52: 81.

Tarazona, P. Free-energy density functional for hard-spheres. 1985. *Phys. Rev. A Atom. Mol. Opt. Phys.* 31: 2672.

Caballero, J.B., Noya, E.G., and Vega, C. Complete phase behavior of the symmetrical colloidal electrolyte. 2007. *J. Chem. Phys.* 127: 154113.

Mederos, L., Velasco, E., and Navascues, G. Phase diagram of colloidal systems. 1998. *Langmuir* 14: 5652.

Wang, D.C., and Gast, A.P. Crystallization of power-law fluids: A modified weighted density approximation model with a solid reference state. 1999. *J. Chem. Phys.* 110: 2522.

10 Computing Transport in Materials

Mario Pinto, Venkata Gopala Rao Palla, and Ajay Nandgaonkar

CONTENTS

10.1 INTRODUCTION

To use a new material for a specific application, it is vital to understand and characterize its *transport properties* such as thermal transport, mass transport, and electrical transport. With the advent of large and fast massively parallel computers, it is now possible to devise new molecular modeling (MM) methods that can reliably compute transport properties for complex materials from the bottom up. In this chapter, we provide an overview of some commonly used techniques to compute transport coefficients for materials, using classical molecular dynamics (MD). We apply MD to two interesting physical situations as illustrative examples.

This chapter is divided into three parts. In Section 10.2, we discuss the interesting problem of heat transfer in novel materials called nanofluids, which are suspensions of nanoparticles in liquids. Here, the central question is to understand the heat transfer across the interface between a nanoparticle and the surrounding base fluid. We believe that understanding heat transfer across the interface provides crucial insights into the observed enhanced thermal conductivities of nanoparticle suspensions in polar liquids (Choi 2009). We provide an overview of the computation of thermal conductivity for inhomogeneous systems using MD simulations, followed by a discussion on the heat transfer due to radiative heating.

Then in Section 10.3, we address the problem of computing mass transport coefficients in porous materials called zeolites. Zeolites are materials with a wide range of applications, such as petrochemical separation, water purification, and catalysis. Understanding and predictably computing mass transport coefficients for a variety of molecules in these molecular sieves instruct optimal use of the appropriate zeolite for an application. We describe the methodology used to compute diffusivities of *n*-alkanes in a particular zeolite, the silicalite. We conclude this section with computation of diffusivities for methane and butane with different force fields and other simulation parameters.

The choice of these two problems is guided by the following considerations:

- While each physical problem is interesting in itself, the need to study it is essentially driven by an application.
- The two physical problems are completely different from each other. However, they are dealt with through the same tool (i.e., LAMMPS), thereby adding perspective.
- Both problems are active pursuits of researchers today.

We conclude the chapter with a brief discussion of the benchmarks (Plimpton 1995) of one of the many open source tools for performing MD simulations: the LAMMPS package (http://lammps.sandia.gov/bench.html). We aim to provide the reader with a glimpse of what it means to run a large-scale molecular dynamic simulation on a parallel computer in Section 10.4.

10.2   THERMAL TRANSPORT IN NANOFLUIDS

10.2.1   Introduction to Nanofluids

Technological developments in areas such as electronics, automobiles, and nuclear energy demand better thermal management. Conventional methods of enhancing the heat removal rate are reaching their limits. There is an immediate need to look for innovative means to achieve enhanced heat removal capacity, and the low thermal conductivity of conventional heat transfer fluids is proving to be a serious limitation. Given this scenario, nanofluids have emerged as an attractive solution to meet these challenges (Yu et al. 2007).

Nanofluids, a term coined by Choi (Choi 1995, 2009), refers to a new class of nanotechnology-based heat transfer fluids, engineered by dispersing and stably suspending nanoparticles with typical length of the order of 1 to 50 nm in traditional heat transfer fluids. Solids have thermal conductivity approximately two orders of magnitude higher than liquids. Hence, by dispersing solids in liquids, one would expect the effective thermal conductivity of the mixture to increase. This is not a new idea. Maxwell, in his theoretical work, proposed a model to predict the effective thermal conductivity of such a solid suspension (Maxwell 1873). However, attempts in the past to disperse micron-sized particles have lead to problems such as sedimentation, clogging, abrasion, and increased pressure drops. In nanofluids, interactions between particles and base fluid have lead to a range of remarkable thermal properties, while overcoming these difficulties. For example, oil dispersed with carbon nanotubes (CNTs) shows a 150% enhancement in thermal conductivity at low concentrations (Choi et al. 2001). Other interesting thermal properties include nonlinear increase of thermal conductivity with particle concentration, little or modest increase in pressure drop as compared with the base fluid, size-dependent single-phase convective heat transfer, and critical heat flux enhancement. These characteristics have spurred explorations on the applicability of nanofluids in areas such as electronic cooling, transportation, nuclear system cooling, defense, space, and biomedicine (Wong and Leon 2010).

Two methods are commonly used to prepare nanofluids: the single-step process and the two-step process. In the two-step method, nanoparticles are first prepared, followed by dispersing them in the base fluid. In the single-step method, preparation and dispersion of nanoparticles happens simultaneously. The important characteristics of a nanofluid are a good mono-sized dispersion and a stable suspension. Commonly used base fluids are water and organic solvents such as ethylene glycol; common particle materials are metals such as gold, silver, or copper; metal oxides such as CuO or Al2O3; and CNTs. There has been a steady growth of literature dealing with the synthesis and characterization of nanofluids since the early 2000s (e.g., Das et al. 2007 and Yu et al. 2007).

10.2.2 Computing Thermal Conductivity Using Molecular Dynamics

The favorable thermal properties of nanofluids and the possibility of their application in a large number of areas have lead to exploration of different methods to predict properties of nanofluids and mechanisms that enhance their thermal performance. Classical MD is an attractive option for such studies. Within MD, transport coefficients can be calculated using either *equilibrium molecular dynamics* (EMD) or *nonequilibrium molecular dynamics* (NEMD).

In EMD, transport coefficients are calculated using the well-known statistical mechanical Green–Kubo formalism (Allen and Tildesley 1989; Fernández et al. 2004; Rapaport 1995). The fluctuation–dissipation theorem relates equilibrium microscopic fluctuations to externally imposed gradients, thereby enabling the computation of transport coefficients. The heat conduction capacity of a material is characterized by the thermal conductivity κ, as defined by Fourier's law of heat conduction. Within the linear-response Green–Kubo formalism, the thermal conductivity of a material is given by the integral of the time autocorrelation function of the microscopic energy flux **J** (Kaburaki 2005):

$$\kappa = \frac{V}{3k_B T^2} \int_0^\infty \langle \mathbf{J}(0).\mathbf{J}(t) \rangle \mathrm{dt} \tag{10.1}$$

where the energy flux **J** is defined as:

$$\mathbf{J} = \frac{1}{V}\left[\sum_i e_i v_i + \sum_{i<j} (f_{ij}.v_j)r_{ij} \right] = \frac{1}{V}\left[\sum_i e_i v_i + \frac{1}{2}\sum_{i<j} (f_{ij}.(v_i + v_j))r_{ij} \right] \tag{10.2}$$

$$e_i = \frac{1}{2}m_i v_i^2 + \sum_{i<j} \varphi(r_{ij}) \quad \text{and} \quad f_{ij} = -\frac{\partial \varphi(r_{ij})}{\partial r_{ij}} \tag{10.3}$$

The $\langle \ \rangle$ denotes the ensemble average, V is the volume, T is the temperature, e_i and v_i are the total energy and velocity of the ith particle, f_{ij} is the force between particles i and j, and φ is the pair potential function. For a two-phase system, the mean partial enthalpy also comes into the equation of **J** (Eapen et al. 2007). Long runs are necessary to achieve reliable values, because a single value of **J** is obtained in one time step. One is also faced with the problem of integrating a correlation function that may have a long time tail (Allen and Tildesley 1989; Rapaport 1995).

An alternate method is the NEMD methodology, wherein transport coefficients are calculated as a ratio of a flux to an appropriate driving force, extrapolated to the limit of zero driving force. The method involves the simultaneous addition and removal of heat into and out of the system, which is achieved by velocity rescaling, thereby setting up a thermal gradient across the system. Thermal conductivity can then be inferred using Fourier's law. As against EMD, a different simulation setup is required to calculate the thermal conductivity along each direction, a consideration vital for anisotropic systems. Vogelsang et al. (1987) provide a comparison between the EMD and NEMD methods as applied to a Lennard-Jones (LJ) liquid.

A special case of NEMD is the relatively new reverse nonequilibrium molecular dynamics method (RNEMD) (Muller-Plathe and Bordat 2004; Muller-Plathe and Reith 1999). The essential difference between NEMD and RNEMD arises from the way the steady-state fluxes are maintained in the system. RNEMD involves an exchange of particle momentum from the designated cold and hot regions, and it has been applied to calculate the thermal conductivity of both simple and molecular fluids (Muller-Plathe 1997; Zhang et al. 2005). Recently, Eapen et al. (2006, 2007) used the Green–Kubo formulation to calculate the thermal conductivity of the Pt–Xe nanofluid. Keblinski et al. (2002) used the NEMD methodology to calculate the thermal conductivity of eight LJ particles

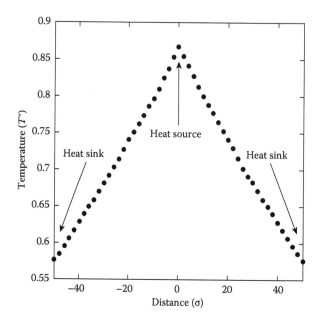

FIGURE 10.1   Temperature profile obtained from heat source–sink simulation of the LJ fluid. The slope of the temperature profile allows the determination of κ for the LJ liquid. Positions of heat source and sink are as indicated in the figure. A straight line in the temperature profile is obtained after sufficient equilibration and a large number of averaging steps in the MD simulation.

dispersed in an LJ fluid and concluded that the enhancement measured in the simulations was in accordance with Maxwell's theory (Maxwell 1873). In other words, these computations did not lead to dramatic increases in the thermal conductivity as found in the experiments. One of the reasons for not finding such enhancements in the κ values is the small ratio of the thermal conductivities of the LJ solid to an LJ fluid. A high $\kappa_{solid}/\kappa_{fluid}$ is a crucial attribute in the observation of significant enhancements. Further, classical descriptions such as the LJ solids cannot capture the high thermal conductivities of metals because electrons play an important role in the thermal transport.

The LAMMPS package (http://lammps.sandia.gov/bench.html) has provisions for calculating thermal conductivity using all the methods described so far. Due to its flexibility of application and ease of use, especially for complex systems, we used the NEMD method for all the results reported here. Although we do not exhaustively deal with computing thermal conductivity for inhomogeneous systems, we provide an overview of the methodology using the LJ system as an example. The system consists of an LJ fluid with $T^* = 0.70$ and $\rho^* = 0.84$. Fluid atoms interacted with each other via the LJ potential, with $\varepsilon = 1.0$ and $\sigma = 1.0$. In other words, we measured everything in the units of $\varepsilon = \sigma = 1$. The simulation box size was 104 σ × 50 σ × 50 σ and contained 223,200 atoms. The time step used was 0.005 τ. The heat source and sink consisted of 2 σ-wide regions located at the center and edges of the box. The system was first equilibrated at the desired state point. Next, heat was added into the source region and simultaneously removed from the sink region at a rate of 200 ε/τ, which induced a temperature gradient in the fluid. A steady-state temperature profile was established in 150,000 time steps, followed by another 150,000 time steps to collect the data. The resultant temperature profile is shown in Figure 10.1. The thermal conductivity value was then readily calculated using Fourier's law. The value of κ for this specific computation was 6.68 in LJ units.

10.2.3   Predicting Thermal Properties of Nanofluids: Radiative Heating

After the survey of a range of MD methods to compute thermal transport, we focused on one specific way to address the thermal transport in nanofluids vis-à-vis the radiative heating simulation.

The parameters that control the thermal transport in a nanofluid are particle material, size, shape, and volume fraction of the nanoparticles, base fluid material, additives, acidity, and the operating temperatures. Experiments (Choi et al. 2001) seem to suggest that enhancements in thermal conductivity are well above those given by Maxwell's relations (1873). This has led to many mechanisms and models being proposed to explain the unusual enhancement of thermal conductivity: Brownian motion, layering at solid–liquid interface, ballistic phonon transport, and clustering (Keblinski et al. 2002). Models for convection and boiling heat transfer have also been proposed. Keblinski et al. (2008) have shown that most reported enhancements lie in the so-called Hashin and Shritkman limits. These investigations seem to point to the fact that particle clustering and particle–fluid interfacial effects are the major parameters that govern the thermal conductivity enhancement. However, no quantitative agreement among simulations, theory, and experiments has been found to date, and a comprehensive understanding remains elusive (Murshed 2009). The nanometer dimension of particles in nanofluids leads to a large density of interfaces. Consequently, the thermal coupling at the liquid–solid interface becomes an important parameter that controls the thermal conductivity of the nanofluid (Patel et al. 2005; Shenogina et al. 2009; Xue et al. 2003). This coupling is quantified by the interfacial thermal resistance (R_K), defined as:

$$R_K = \frac{\Delta T}{J} \tag{10.4}$$

where ΔT is the temperature jump at the interface and $\mathbf{J}$ is the heat flux (Barrat and Chiaruttini 2003). Another convenient way of parameterizing the interfacial thermal coupling is the *Kapitza length lk*, defined as:

$$l_K = R_K \kappa \tag{10.5}$$

where κ is the thermal conductivity of the fluid. The Kapitza length is the length of the fluid column such that the temperature drop would be identical to the one occurring at the interface (Kapitza 1941).

A clean way to probe the interfacial resistance at the nanoparticle–fluid interface is by studying radiative heating using molecular dynamical simulations in which the particles are irradiated using radiation and the heat transfer pathway is clearly from the particle to the fluid. Merabia et al. studied the effect of surface curvature on the critical heat flux using a nanoparticle–fluid system (Merabia, Keblinski, et al. 2009). A similar setup can be used to probe the effect of particle–fluid interaction on the temperature jump at the interface and consequently the interfacial resistance. We describe this computation briefly. The system consisted of an LJ solid comprising 555 atoms surrounded by an LJ fluid made up of approximately 42,000 atoms. The LJ potential is used to describe all interactions. In addition, the solid atoms interacted via Finitely Extensible Nonlinear Elastic (FENE) bonds (or springs). The solid–fluid interaction is given by:

$$\varphi(\alpha\beta) = 4\varepsilon\left[\left(\frac{\sigma}{r}\right)^{12} - c_{\alpha\beta}\left(\frac{\sigma}{r}\right)^{6}\right] \tag{10.6}$$

such that, by changing $c_{\alpha\beta}$, the interaction between the solid and the fluid can be changed from hydrophobic (for low $c_{\alpha\beta}$ values) to hydrophilic (for high $c_{\alpha\beta}$ values). Heating of the particle and cooling of the fluid at a certain distance from the particle is modeled by scaling the velocities of the particle and fluid atoms. The time step used was $0.005\ \tau$.

The system was first equilibrated at $T^* = 0.75$ and $P^* = 0.015$. Then the temperature of the particle was maintained at $T^* = 1.5$. Fluid at a distance greater than $16\ \sigma$ was maintained at $T^* = 0.75$.

A steady-state temperature profile evolved over 100,000 time steps, following which temperature data were gathered over the next 100,000 time steps and binned in shells of thickness 0.15σ. This procedure was repeated for values of $c_{\alpha\beta}$ ranging from 0.50 to 2.00 in steps of 0.25. The resultant temperature profile plot is as shown in Figure 10.2. From the figure, it is clear that solid–fluid coupling has a strong influence on the interfacial resistance. A similar study for plane surfaces was carried out by Barrat and Chiaruttini (2003) and Keblinski et al. (2003).

Recently, Merabia, Shenogin, et al. (2009) studied a gold–water system with 484 gold atoms and 10,000 water molecules. Gold–gold and gold–oxygen interactions were modeled using the LJ potential and the SPC/E water model was used to describe water. The choice of this model was guided by the correct reproduction of thermal properties for water with reasonable computation cost. A time step of 1 fs was used. The system was first equilibrated at a temperature of 450 K and a pressure of 80 bar. This was followed by heating of the particle with a constant flux. Fluid at a distance greater than 33Å from the particle center was maintained at 450 K. The temperature profile for the gold–water system at heating rates of 350, 525, and 700 nW, reproduced by us, is shown in Figure 10.3.

Based on the temperature drops at the interface, the interfacial resistance can be calculated using Equation 10.3. From Figure 10.2, it is clear that a large variation in the temperature jump at the interface can be obtained by varying the solid–fluid interaction. For a given particle material, shape, and size, the enhancement in thermal conductivity depends on the interfacial resistance value: the less the resistance, the great the enhancement. Parameters such as charge, acidity, or the surfactant used to keep the nanoparticles from agglomerating modify the interfacial resistance value. Therefore, the thermal response of the nanofluid can be fine-tuned by adjusting these parameters. To accurately predict the thermal response of a nanofluid, new methods to reliably compute, calibrate, and validate the interfacial resistance are needed. One of the most important parameters for an MD scheme to be successful in capturing the physics and complexity of this system are the right interatomic potentials. Devising new interatomic potentials and force fields and choosing the right force field for a particular problem is central to the success of the simulation. For a comprehensive overview of a range of force fields (or potentials) the reader is requested to refer to the force field

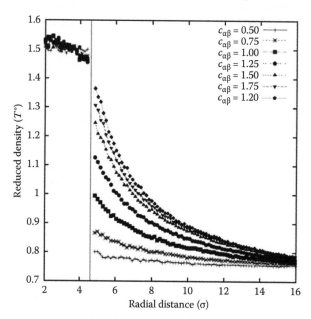

FIGURE 10.2 Temperature profile at the particle–fluid interface for different values of $c_{\alpha\beta}$. The position of the particle surface is indicated by the vertical line. The discontinuity at the interface denotes the value of the interfacial resistance. Wetting interactions between the particle and the fluid lead to a smaller interfacial resistance.

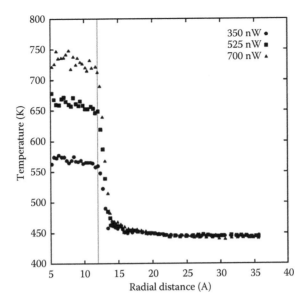

FIGURE 10.3 Temperature profiles across the gold nanoparticle–water interface for three different heating rates of 350 nW (circle), 525 nW (squares), and 700 nW (triangles).

article in *Wikipedia*: http://en.wikipedia.org/wiki/Force_field_(chemistry). New thermal conductivity models that appropriately take into account the interfacial resistance are also needed to get a handle on accurate predictions, as well as a deeper understanding of thermal conductivity in these new materials. Such new methods and accurate models will enable the design of new nanomaterials that are created for specific uses. Thermal transport in nanofluids continues to be an active area of research.

10.3 DIFFUSION OF *N*-ALKANES IN ZEOLITES

10.3.1 ZEOLITES

Zeolites are microporous, aluminosilicate minerals commonly used as commercial adsorbents. The term *zeolite* was coined in 1756 by the Swedish mineralogist Axel Fredrik Cronstedt, who observed that upon rapidly heating the material stilbite, it produced large amounts of steam from water that had been adsorbed by the material. Different zeolite structures have similar chemical composition; however, they differ from each other in their pore sizes and pore distributions. Given that the pore sizes are in the range of few Angstroms, zeolites are well suited for applications like catalysis because the amount of surface area per gram can be of the order of hundreds of square meters. Zeolites are widely used in petrochemical industries for processes like cracking, isomerization, and alkylation. Further, zeolites are useful for water purification, detergents, agriculture, and medicine. We show a schematic representation of the pore topology of a particular zeolite, silicalite, in Figure 10.4. Clear straight pores can be seen along the Y-direction, whereas the X-direction has a zigzag channel.

When zeolites are used as catalysts, the effectiveness of the catalytic action on the reaction inside the pores depends on the following parameters:

- The adsorption of the reactants
- The diffusion of the reactants to the active sites
- The chemical conversion at the active site
- The diffusion of products away from the reaction site
- The desorption of the products from the zeolite surface

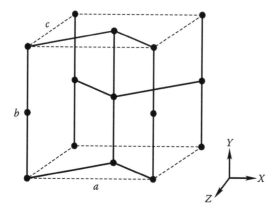

FIGURE 10.4   Schematic diagram of networks of pores in a silicalite. Solid dots represent junctions, and the lines represent pores. Crystallographic axes are represented as a, b, and c.

Therefore, the choice of a zeolite for a particular application requires knowledge of the kinetics and the thermodynamics of the sorbate molecules. Among these factors mentioned, we focus on diffusivity in this section.

Experimentally, the diffusivity of molecules in a zeolite is obtained using techniques like quasi-elastic neutron scattering (QENS) or pulse field gradient nuclear-magnetic resonance (PFG-NMR) techniques (Jobic 2000; Jobic and Theodorou 2006; Jobic et al. 2006). Given the wide range of zeolites (more than 500 different types) and the extraordinary range of molecules, it is impossible and prohibitively expensive to measure the diffusivities of all the molecules in all zeolites. However, modern computational methods and molecular simulations would soon make it possible to have accurate estimates of the diffusivity values that act as decision support mechanisms in deciding the choice of a zeolite for a specific application.

10.3.2   Modeling Zeolites

In a molecular simulation, a zeolite is described by its geometry and the interactions among atoms (namely, the force fields). This atomic or molecular level information then needs to be translated into measurable macroscopic quantities so that computations can be validated against experiments. Statistical-modeling techniques, such as the classical Monte Carlo method, can be used to accurately compute the static properties of zeolites, provided the force fields assigned to the system are accurate enough and are based on experimental data. Dynamic properties, such as thermal conductivity or mass diffusivity, are most readily computed using classical MD.

We shall briefly discuss one example of computing diffusivity of n-alkanes in silicalite as an illustrative example. As we shall show below, the diffusivity values computed using classical MD for n-alkanes are in good agreement with experiments for smaller chains. The values start deviating from the experimental values as we go to longer and longer chains (Jobic and Theodorou 2006; Leroy et al. 2004). There is a range of factors that contribute to the accuracy of the simulation:

- The force fields used to describe the system: The so-called anisotropic united atom (AUA) potential gives better results than does the united atom (UA) potential (Leroy and Rousseau 2004).
- The effect of flexibility of the zeolite on the diffusivity: A flexible zeolite yields better results with smaller molecules and lower loading (number of atoms per unit cell of the sorbate molecules).
- The length of the simulation: Longer simulation times yield better results for the diffusivity values; however, this means dealing with large systems and thus longer computations increasing the computational cost immensely.

For a detailed comparison between QENS experiments and molecular simulations (Jobic and Theodorou 2006) and for an excellent review of molecular simulations, the reader is requested to refer to articles by Smit et al. (2000) and Smit and Krishna (2001, 2003).

As an illustration, we have reproduced the results for the diffusivity of methane in a silicalite by using the UA potentials (Leroy et al. 2004). We then used a better and a more detailed force field polymer consistent force field (PCFF) to compute the diffusivity of butane in silicalite. We see that the PCFF description of the problem yields results closer to experiments.

10.3.3 COMPUTING DIFFUSIVITY USING CLASSICAL MOLECULAR DYNAMICS

Within the linear response approximation, the rate of transport (mass, momentum, or energy) through a system is proportional to the gradient (of concentration, velocity, and temperature), with the transport coefficient being the proportionality constant. This proportionality constant can be computed using equilibrium description of the system through the so-called fluctuation dissipation theorems. One such equation, relating equilibrium fluctuations to the diffusion constant, is Einstein's well-known equation:

$$\left\langle [x(t) - x(0)]^2 \right\rangle = 2Dt \tag{10.7}$$

The relation between the means-square displacement (MSD) to the diffusion constant is valid in the physical limit of the observation time being larger compared with the mean-collision time (Haile 1991). We used Equation 10.6 to compute diffusivities through MD simulations. We computed diffusivity, D, along each direction using the projected Einstein equations:

$$D_\alpha = \frac{1}{2} \lim_{t \to \infty} \frac{d}{dt} \left\langle |r_\alpha(t) - r_\alpha(0)|^2 \right\rangle \tag{10.8}$$

where α can be the x, y, or z component. Self-diffusivity can then be computed as:

$$D = \frac{1}{3}(D_x + D_y + D_z) = \frac{1}{6} \lim_{t \to \infty} \frac{d}{dt} \left\langle |r(t) - r(0)|^2 \right\rangle \tag{10.9}$$

While Equation 10.7 is valid for an isotropic medium, we used it to compute D for sorbate molecules in a zeolite, with an assumption that the simulations would run long enough for the molecules to explore a range of pores (Fritzsche et al. 1992), though this is an approximation. Alternatively, diffusivity can also be computed using Green–Kubo formalism. In the present case, it is equivalent to computing diffusivity using Equation 10.7 (Frenkel and Smit 1996). Another commonly used technique is the method of *shifted origins* for trajectories of sorbate molecules, thereby improving statistics (Rapaport 1995). This modifies the right side of Equation 10.7 as:

$$\left\langle |r_\alpha(t) - r_\alpha(0)|^2 \right\rangle = \frac{1}{N_{\text{origin}} N_{\text{mol}}} \sum_{i=1}^{N_{\text{origin}}} \sum_{j=1}^{N_{\text{mol}}} \left(r_{\alpha,j}(t_i + t) - r_{\alpha,j}(t_i) \right)^2 \tag{10.10}$$

We chose to model the zeolite with only the oxygen atoms in it at the appropriate positions in the zeolite geometry. As described before, methane was modeled using the UA potential, whereas we chose to model butane using the PCFF description (Hill and Sauer 1994). The interaction parameters of the UA force field are shown in Table 10.1.

TABLE 10.1

Interaction Parameters Between Groups for the UA Potential

	ε (kcal/mol)	σ (A)
O–O	0.18586	3.00
O–CH$_4$	0.19168	3.60
CH$_4$–CH$_4$	0.29420	3.73

We performed simulations on a system of $2 \times 2 \times 2$ unit cells of the zeolite. A typical MD simulation involved the following steps:

1. Prepare the system: After defining the geometry of the zeolite, the desired number of sorbate molecules (the loading) were placed within the silicalite using the configuration-biased Monte Carlo (CBMC) technique (Frenkel and Smit 1996; Smit et al. 2000). We used another open source software called TOWHEE (http://towhee.sourceforge.net) to prepare the initial state (Martin and Siepmann 1999).
2. The sorbate molecules were assigned velocities based on the Gaussian distribution to get the target temperature (say 300 K).
3. Equilibration: We chose a time-step of 5 fs to equilibrate the system in the microcanonical ensemble through velocity scaling at 300 K for a total duration of 2 ns. We further equilibrated the system for 4 ns in the canonical ensemble, vis-à-vis the system was in contact with a heat bath of the desired temperature $\beta = 1/k_B T$.
4. Sampling and Analysis: We continued the canonical ensemble simulation for another 10 to 20 ns, while the trajectories of the sorbate molecules were written out every 100 fs. The analysis of these trajectories enabled us to compute the mean square displacements and therefore the diffusivity.

We show the average means-square displacements (MSD) for methane molecules in Figure 10.5. Notice that the plot is on a log-scale. After an initial ballistic motion, the MSD becomes linear, thereby indicating the diffusive regime. The slope of this line is twice the diffusion constant (see Equation 10.7). We repeated this procedure in all three directions and averaged to get the final diffusion constant values. Note that it is better to compute the quantity, D_α, in each direction first and then average rather than doing it the other way around. The diffusivity values for methane (Leroy and Rousseau 2004) in a silicalite for the loading values of 4, 8, and 12 molecules per unit cell are shown in Table 10.2. Other explorations of the system such as the effect of the MD time step, length of the runs, use of other force fields, and detailed comparisons with experiments will be published elsewhere (Nandgaonkar unpublished).

We conclude this section by briefly discussing the effect of a force field on the diffusivity value for butane. We computed D for *n*-butane in silicalite for a loading of two molecules per unit cell. We use the PCFF description for the interatomic potentials and compared them against the AUA or UA description (Leroy et al. 2004). The results are summarized in Table 10.3. The PCFF force field is clearly closer to the experimental values than the UA or AUA description noted by Leroy et al. (2004). Choosing the right force fields to describe the problem is the cornerstone of an MM scheme. This is illustrated by the significant improvement in results from a more comprehensive force field like the PCFF over the UA or AUA force fields. Getting Universal force fields that would suit almost all situations in MM still remains the elusive goal; however, newer and ever-improving unifying schemes are emerging. We believe that such unification is the key to successful and true multiscale modeling.

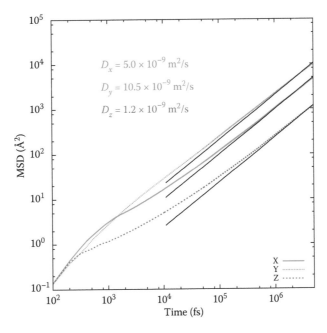

FIGURE 10.5 MSD's of methane at a loading of four molecules per unit cell along x (solid red), y (dotted magenta), and z (dashed blue) directions as a function of time on a log–log plot. The straight line fits to the curves for longer times denote that MSDs are proportional to the diffusivity values along each direction.

TABLE 10.2

Diffusivity of Methane in Silicalite at Different Loadings

Loading	Molecules/Unit Cell	Molecules/Unit Cell	Molecules/Unit Cell
D_x	9.2	5.1	2.7
D_y	14.3	8.9	4.6
D_z	1.9	1.1	0.65
D	8.5	5.0	2.7

Loadings are in the units of molecules per unit cell. Diffusivity values are reported in the units of 10^{-9} m²/s. As expected, the diffusivity decreases with increased loading.

TABLE 10.3

Comparison of Diffusivity Values for Butane in Silicalite Using UA and PCFF Force Fields

Symbol	UA Force Field	PCFF Force Field	Experimental
D_x	1.7	1.4	—
D_y	7.9	2.0	—
D_z	0.54	0.33	—
D	3.4	1.24	0.98

All values reported are for a loading of 2 molecules per unit cell. Diffusivity values are reported in the units of 10^{-9} m²/s.

The flexibility of the zeolite is another parameter that plays a role for longer chain molecules at low loading values. Investigations for getting D close to the experimental value are underway for longer chain molecules and shall be published elsewhere (Nandgaonkar unpublished).

10.4   LAMMPS BENCHMARKS

Large-scale atomic and molecular massively parallel simulator (LAMMPS) is an open source software package with potentials for soft materials (e.g., biomolecules and polymers), solid-state materials (e.g., metals and semiconductors), and coarse grained or mesoscopic systems. It is one of the most widely used MD tools for modeling atoms or, more generically, as a parallel particle simulator at the atomic, mesoscopic, or continuum scale. LAMMPS runs on single processors or in parallel, using the message-passing techniques with a spatial decomposition of the simulation domain (Plimpton 2005). The code is designed to be easy to modify or extend with new functionality, and one of the authors (MP) has contributed a new thermal fix to the software. More information about LAMMPS can be found at its website http://lammps.sandia.gov.

Among the various benchmarks reported on the website of LAMMPS, we used two benchmarks for evaluating the performance of the simulator on a large-scale parallel machine—*Eka*, the largest commercially available high-performance computing system (*Eka*: http://www.crlindia.com):

1. The LJ atomic fluid: We conducted an MD simulation for 100 time steps of 32,000 atoms interacting with each other via the well-known 6-12 LJ potential (Lennard-Jones 1924):

$$V_{LJ}(r_{ij}) = 4 \in_{ij} \left[\left(\frac{\sigma_{ij}}{r_{ij}} \right)^{12} - \left(\frac{\sigma_{ij}}{r_{ij}} \right)^{6} \right] \tag{10.11}$$

We worked at the reduced density $\rho^* = 0.8442$ such that the system of atoms was a liquid. A force cutoff of 2.5 σ was used. The system had 55 neighbors per atom within the force cutoff, and we performed Constant number, Volume and Energy (NVE) time integration to simulate the system, signifying the simulation of an isolated system.

2. The rhodopsin protein problem: An all-atom rhodopsin protein was set in a solvated lipid bilayer described via the Chemistry at Harvard Molecular Mechanics (CHARMM) force field. Long-range coulomb interactions were described via the particle–particle mesh. SHAKE constraints were applied to the system for the definitions of the force field and constraints. Further, the model consisted of counter ions with a reduced amount of water; the effect was to have a total system with 32,000 atoms that was simulated for 100 time steps. This simulation was performed at a constant pressure and temperature with an LJ force cutoff of 10.0 Angstroms. In this problem, the total number of neighbors per atom was 440 within this force cutoff. More information about the benchmark problem can be found at http://lammps.sandia.gov/bench.html#rhodo

The main difference between the two systems is that the latter is a charged system in which all particles talk to each other via long-range coulomb interactions, whereas the former is a neutral system with short-range interactions. The long-range component of energy is computed in the momentum space via a Fast Fourier transform (FFT).

For each of these problems, performance was measured for two cases:

1. Fixed-size scaling: The 32,000-atom system was distributed over different processors, and the total run time for 100 MD steps was measured.
2. Scaled-size problem: The problem was run on P processors, and the total number of atoms was equal to P times the number of atoms per processor or core.

TABLE 10.4

List and Specifications of Various Clusters on Which LAMMPS Performance Is Compared

Nickname	Machine	Processors	Site	CPU	Bandwidth (Mb/sec)	Latency (μsec)
Eka	HP	14,440	Computational Research Laboratories Ltd., (CRL) Pune	3 GHz dual Xeons	2000	2
Spirit	HP	512	Sandia National Laboratories (SNL)	3.4 GHz dual Xeons	230	9
HPCx	IBM	512	Daresbury	1.7 GHz Power4+	1450	6
Blue Gene Light	IBM	65,636	Lawrence Livermore National Laboratory (LLNL)	700 MHz Power PC	150	3
Red Storm	Cray	10,000	SNL	2.0 GHz Opteron	1000	7
Intel Xeon	Dell	8	SNL	2.66 GHz		

The bandwidth and latencies defined are at the MPI level, i.e., the ones seen by a code-like LAMMPS in actual run time. For further details see the LAMMPS website.

Thus a scaled 64-processor simulation has 2,048,000 atoms; and a 2048-processor run has 65,536,000 atoms.

Performance and scaling of the simulator program were measured via parallel efficiency. *Parallel efficiency* refers to the ratio of the ideal run time to the actual run time. For example, if the perfect speed-up would have given a run time of 10 seconds and the actual run time was 12 seconds, then the parallel efficiency is 10/12, or 83.3%. Care must be taken while measuring parallel efficiencies so that there are no other jobs running on the cluster that would share the CPU time with the benchmark runs.

On a parallel computer like *Eka*, there are two components to a computation: the actual computation on each of the processors and the communication between the processors. We examined the results for parallel efficiency on a number of parallel computers in order to extract the common behavior of the simulator. Comparative specifications of different computers are given in Table 10.4.

10.4.1 PERFORMANCE RESULTS: FIXED SIZE

For a *fixed-size* problem, the parallel efficiency drops dramatically with an increasing number of cores, because the communication overheads far supersede the actual computation—an observation valid for both the LJ liquid and the rhodopsin protein problems. The parallel efficiency drops to below 10%. The communication time contributes to more than 80% of the total run time. Typically, one would resort to a fixed-size scaling *only* when the problem size is such that it cannot fit on one core. It is the scaled-size problems that present an interesting aspect for large-scale computations.

10.4.2 PERFORMANCE RESULTS: SCALED SIZE

We plotted the parallel efficiency for the scaled-size LJ liquid problem in Figure 10.6. The plot is from 1 to 1024 processors. In contrast to the fixed-size problem, the scaled-size problem shows near 100% parallel efficiency up to 1024 cores, with a dip at 512 cores. Because the problem is scaled with each new added processor, only boundary information needs to be communicated to nearest processors in the topology of the parallel computer. The percentage of communication time for the largest system is of the order of 12%, thereby providing excellent scaling. Such good scaling enables simulations of large system sizes with millions of atoms without an appreciable loss in performance.

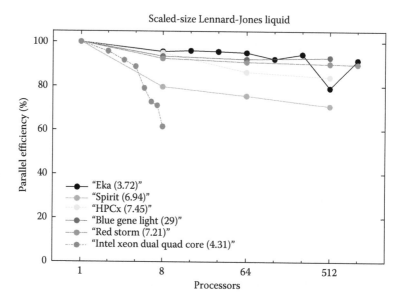

FIGURE 10.6 Parallel efficiency of the scaled LJ liquids as a function of the number of processors on various parallel computers. The numbers in brackets in the legend represent single processor times for each computer. Data for machines apart from *Eka* were obtained from the LAMMPS website.

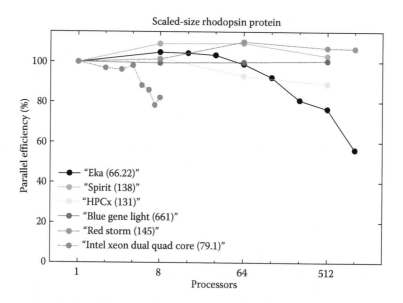

FIGURE 10.7 Parallel efficiency of the scaled rhodopsin protein problem as a function of the number of processors on various parallel computers. The numbers in brackets in the legend represent single processor times for each computer. Data for machines apart from *Eka* were obtained from the LAMMPS website.

The scaling, however, is different when long-range interactions come into play. Such is the case with the rhodopsin protein problem. A similar plot is made for the scaled-size rhodopsin protein problem in Figure 10.7. Interestingly, a super-linear behavior is seen up to 32 cores, thereby indicating that the simulator works faster and better on a parallel computer. This performance subsequently drops to about 60% parallel efficiency for 1024 cores on *Eka*. The performance further degrades to about 40% parallel efficiency for 2048 processors (not shown in the figure). For the largest problem handled with 65,536,000 atoms on 2048 cores, the communication time is still negligible (barely 1%). The additional computations, which involve transformation to the momentum space to treat long-range coulomb forces, take the bulk of computation time. FFT consumes approximately 77% of the computation time, and the scaling of a parallel FFT is the essential bottleneck.

While it is important to look at scaling and parallel efficiencies of the codes that are designed to run on large-scale parallel machines, another crucial factor is also the net turnaround clock time. With faster and better processors, novel interconnects, and smarter topologies connecting a number of processors to each other, the clock times for simulations have been consistently decreasing. New parallel algorithms (e.g., FFT), coupled with this rapid change in chip technology, are sure to open new doors for large-scale high-performance computing in the not-so-distant future.

10.5 ACKNOWLEDGMENTS

We thank Murali Sastry and Sushrut Bhanushali for their insights into the experimental measurements of thermal properties of nanofluids. Kunj Tandon and Mahesh Mynam are acknowledged for many illuminating discussions on the zeolite problem. We thank D. G. Kanhere and the editors, especially Beena Rai, who carefully read the manuscript, gave us feedback, and helped us improve and contextualize this article for the purpose of this book.

REFERENCES

Allen, M. P. and Tildesley, D. J. 1989. *Computer Simulation of Liquids*. Clarendon Press, Oxford.
Barrat, J. L., and Chiaruttini, F. 2003. "Kapitza resistance at the liquid-solid interface", Mol. Phys., Vol. 101, issue 11 pp 1605–1610.
Choi, S. U. S. 1995. "Enhancing thermal conductivity of fluids with nanoparticles", *Developments and Applications of Non-Newtonian Flows*. Ed. D. A. Singer and H. P. Wang. ASME, New York, Vol. 231, pp 99–105.
Choi, S. U. S. 2009. "Nanofluids: from vision to reality through research", J. Heat Transf., Vol. 131 Article ID: 033106
Choi, S. U. S., Zhang, Z.G., Yu, W., Lockwood, F. E., and Grulke, E. A. 2001. "Anomalous thermal conductivity enhancement in nano-tube suspensions". Appl. Phys. Lett., Vol. 79, pp 2252–2254.
Das, S. K., Choi, S. U. S., Yu, W., and Pradeep, T. 2007. *Nanofluids—Science and Technology*. John Wiley & Sons, Inc., Hoboken, NJ, USA, 2008, 397 pages, ISBN 978-0-470-07473-2.
Eapen J., Li, J., and Yip, S. 2006. "Probing transport mechanisms in nanofluids by molecular dynamics simulations", *Proceedings of the 7th ISHMT-ASME Heat and Mass Transfer Conference*, Guwahati, India.
Eapen, J., Li, J., and Yip, S. 2007. "Mechanism of thermal transport in dilute nanocolloids", *Phys. Rev. Lett.* Vol. 98: Article ID 028302.
Evans, W., Fish, J., and Keblinski, P. 2006. "Role of Brownian motion hydrodynamics on nanofluid thermal conductivity", *Appl. Phys. Lett.* Vol. 88 Article ID 093116.
Fernández, G. A., Vrabec, J., and Hasse, H. 2004. "A molecular simulation study of shear and bulk viscosity and thermal conductivity of simple real fluids", *Fluid Phase Equilibr.* Vol. 221 pp 157–163.
Frenkel, D., and Smit, B. 1996. *Understanding Molecular Simulation: From Algorithms to Applications*. Academic Press, San Diego, California.
Fritzsche, S., Haberlandt, R., Karger, J., Pfeifer, H., and Heinzinger, K. 1992. "An MD simulation on the applicability of the diffusion equation for molecules adsorbed in a zeolite". *Chem. Phys. Lett.* 198:3-4 pp 283–287.
Haile, J. M. 1991. *Molecular Dynamics Simulation*. Wiley Interscience, New York.
Hill, J. R., and Sauer, "Molecular mechanics potential for silica and zeolite catalysts based on ab initio calculations. 1. Dense and microporous silica" J., *J. Phys. Chem.* 98:4 pp 1238–1244.
Jobic, H. 2000. "Diffusion of linear and branched alkanes in ZSM-5. A quasi-elastic neutron scattering study" *J. Mol. Catal.* Vol. 158 pp 135–142.
Jobic, H., and Theodorou, D. N. 2006. "Diffusion of Long n-Alkanes in Silicalite. A Comparison between Neutron Scattering Experiments and Hierarchical Simulation Results" *J. Phys. Chem. Lett. B* Vol. 110 issue 5: pp 1964–1967.

Jobic, H., Schmidt, W., Krause C. B., and Karger, J. 2006. "PFG NMR and QENS diffusion study of n-alkane homologues in MFI-type zeolites" *Microporous and Mesoporous Materials*, Vol. 90, pp 299–306.

Kaburaki, H. 2005. "Thermal Transport Process by the Molecular Dynamics Method", *Handbook of Materials Modeling*. Ed. S. Yip. Springer, Netherlands, 763.

Kapitza, P. L. 1941. "The study of heat transfer in helium II" *J. Phys. USSR* Vol. 4 pp 181–210.

Keblinski, P., Parasher, R., and Eapen, J. 2008. *J. Nanopart. Res.* 10: 1089.

Keblinski, P., Philpot, S. R., Choi, S. U. S., and Eastman, J. A. 2002. "Mechanisms of heat flow in suspensions of nano-sized particles" *Int. J. Heat Mass Tran.* Vol. 45 No. 4, pp 855–863.

Lennard-Jones, J. E. 1924. "*On the Determination of Molecular Fields. II. From the Equation of State of a Gas*", *Proc. R. Soc. Lond. A* Vol. 106 No. 738 pp 463–477.

Leroy, F., and Rousseau, B. 2004. "Self-diffusion of n-alkanes in MFI type zeolite using molecular dynamics simulations with an anisotropic united atom (AUA) forcefield" *Mol. Simul.* Vol. 30 pp 617–620.

Leroy, F., Rousseau, B., and Fuchs, A. H. 2004. "Self-diffusion of n-alkanes in silicalite using molecular dynamics simulations. A comparison between rigid and flexible frameworks", *Phys. Chem. Chem. Phys.* Vol. 6 pp 775–783.

Martin, M. G., and Siepmann, J. I. 1999. "Novel configurational-bias Monte Carlo method for branched molecules. Transferable potentials for phase equilibria. 2. United-atom description of branched alkanes", *J. Phys. Chem. B* Vol. 103 pp 4508–4517.

Maxwell, J. C. 1873. *Treatise on Electricity and Magnetism,* Clarendon Press, Oxford.

Merabia, S., Shenogin, S. L., Keblinski, P., and Barrat, J.-L. 2009. "Heat transfer from nanoparticles: A corresponding state analysis", *Proc. Natl. Acad. Sci. USA,* Vol. 106 pp 15113–15118.

Merabia, S., Keblinski, P., Joly, L., Lewis, L. J., and Barrat, J.-L. 2009. "Critical heat flux around strongly heated nanoparticles", *Phys. Rev. E* Vol. 79: Article ID 021404.

Muller-Plathe, F. 1997. "A Simple Nonequilibrium MD Method For Calculating The Thermal Conductivity", *J. Chem. Phys.* Vol. 106, pp 6082–6085.

Muller-Plathe, F., and Bordat, P. 2004. "Reverse Non-equilibrium Molecular Dynamics", *Lecture Notes in Physics*, Vol. 640, pp 310–326.

Muller-Plathe, F., and Reith, D. 1999. "Cause and effect reversed in non-equilibrium molecular dynamics: an easy route to transport coefficients", *Comput. Theor. Polym.* S Vol. 9: pp 203–209.

Murshed, S. M. S. 2009. "Correction and comment on 'thermal conductance of nanofluids: is the controversy over?'"*J. Nanopart. Res.* Vol. 11, pp 511–512.

Nandgaonkar, A., Mario, P., and Venkata Gopala Rao, P., unpublished.

Patel, H. A., Garde, S., and Keblinski, P. 2005. "Thermal Resistance of Nanoscopic Liquid-Liquid Interfaces: Dependence on Chemistry and Molecular Architecture", *Nano Lett.* Vol. 5 No. 11 pp 2225–2231.

Plimpton, S. J. 1995. "Fast Parallel Algorithms for Short-Range Molecular Dynamics", *J. Comp. Phys.* Vol. 117 pp 1–19.

Rapaport, D. C. 1995. *The Art of Molecular Dynamics Simulation,* Cambridge University Press, Cambridge.

Shenogina, N., Godawat, R., Keblinski, P., and Garde, S. 2009. "How Wetting and Adhesion Affect Thermal Conductance of a Range of Hydrophobic to Hydrophilic Aqueous Interfaces", *Phys. Rev. Lett.* Vol. 102, issue 15, pp 156101–156104.

Smit, B., and Krishna, R. 2001. "Monte Carlo simulations in zeolites", *Curr. Opin. Solid State Mater. Sci.* Vol. 5, issue 5, pp 455–461.

Smit, B., and Krishna, R. 2003. "Molecular simulations in zeolitic process design", *Chem. Eng. Sci.* Vol. 58 pp 557–568.

Smit, B., and Maeson, Theo, L. M. 2000. "Molecular Simulations of Zeolites: Adsorption, Diffusion, and Shape Selectivity", *Chem. Rev.* Vol. 108, No. 10 pp 4125–4184.

Vogelsang, R., Hoheisel, C., and Ciccotti, G. 1987. "Thermal conductivity of the Lennard Jones liquid by molecular dynamics calculations", *J. Chem. Phys.* Vol. 86, No. 11 pp 6371–6375.

Wong, K. V., and Leon, O. D. 2010. "Applications of Nanofluids: Current and Future", *Adv. Mech. Eng.* Vol. 2010 Article ID 519659.

Xue, L., Keblinski, P., Phillpot, S. R., Choi, S. U. S., and Eastman, J. A. 2003. "Two regimes of thermal resistance at the liquid-solid interface", *J. Chem. Phys.* Vol. 118, pp 337–339.

Yu, W., France, D. M., Choi, S. U. S., and Routbort, J. L. 2007. "Review and Assessment of Nanofluid Technology for Transportation and other Applications", Energy Systems Division, Argonnel National Laboratory, ANL/ESD/07-9 Available at http://www.transportation.anl.gov/pdfs/MM/401.pdf

Zhang, M., Lussetti, E., de Souza, L. E. S., and Muller-Plathe, F. 2005. "Thermal Conductivity Of Molecular Liquids By RNEMD", *J. Phys. Chem. B,* Vol. 109 pp 15060–15067.

11 Simulation of Crystals with Chemical Disorder at Lattice Sites

Ricardo Grau-Crespo and Umesh V. Waghmare

CONTENTS

11.1 REPRESENTATIONS AND MODELS OF SITE-DISORDERED SOLIDS

Many materials of technological importance, from super-alloys with high strength at elevated temperatures to catalysts for oxidation of harmful gases, involve solutions of two or more crystalline materials with inherent randomness in occupation of lattice sites with their constituent atoms—site disorder. Thus, site disorder is a kind of disorder that results from *nonperiodic* occupation of lattice sites in a crystal structure. It is different from amorphous disorder in that it does not destroy the long-range periodicity of the lattice sites, except possibly with small local atomic displacements with respect to lattice sites. Computational modeling of these solids is challenging because periodic boundary conditions cannot be applied in the same straightforward way as in perfect crystals. The models employed can be classified in three broad groups (Figure 11.1).

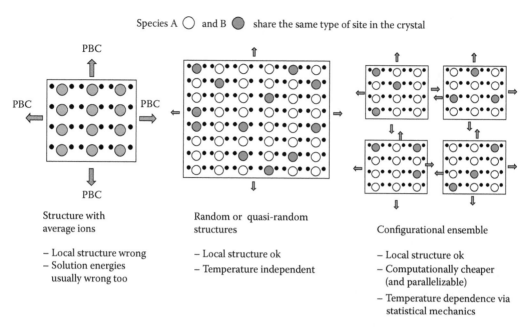

Species A ○ and B ● share the same type of site in the crystal

Structure with
average ions

– Local structure wrong
– Solution energies
  usually wrong too

Random or quasi-random
structures

– Local structure ok
– Temperature independent

Configurational ensemble

– Local structure ok
– Computationally cheaper
  (and parallelizable)
– Temperature dependence via
  statistical mechanics

FIGURE 11.1   Three types of computational representations of site-disordered solids.

The first group comprises all methods in which a sort of average atom is defined, thus allowing recovering the perfect periodicity of the crystal. In the context of classical calculations, based on analytical interatomic potentials, this is done by making each site experience a potential that is the mean or weighted average of all possible configurations corresponding to disordered atomic positions. This approach is implemented, for example, in the General Utility Lattice Program (GULP) (Gale 1997; Gale and Rohl 2003) and can sometimes be useful for preliminary simulations of very disordered (random) systems. An equivalent method in the world of quantum-mechanical simulations is the virtual crystal approximation (VCA), in which the potential felt by electrons is the one generated by average atoms, that is, average of potentials of atoms that can occupy a given site and its periodic images (Bellaiche and Vanderbilt 2000; Tripathi et al. 2010). The main drawback of this kind of method is that the local structure around each particular ion in a real material is very poorly represented by the geometry around these average ions.

In the second group of methods for treating site-disordered solids, a large periodic supercell is employed, with a more or less random distribution of ions at the sites. This type of representation is computationally more expensive than the "average ion" models, but it provides a better description of the local geometries found in the real system. It is assumed that (1) distribution of atoms on lattice sites is random, and (2) the supercell is of sufficient size to include a large number of possible local arrangements of ions, amounting to spatial averaging over configurations. A useful variation of this model is the special quasi-random structure, in which the ion positions in the supercell are chosen to mimic as closely as possible the most relevant near-neighbor pair and multisite correlations of a random substitutional alloy (Zunger et al. 1990). Special quasi-random structures are particularly useful in evaluation of the electronic structure and related properties such as magnetic moments of site-disordered solids. Their main limitation comes from the inflexibility of the ion distribution in the structure, which is fixed to mimic a disordered (a truly random) solid. However, we often desire to investigate varying degrees of disorder; for example, short-range ordering can be present depending on the temperature used in the synthesis (Burton et al. 2006), for which we need a more flexible representation.

The third type of method is the multiconfigurational supercell approach, which is the focus of the present chapter. Within this approach, an infinite site-disordered solid is modeled with a set of configurations with various site occupancies in a supercell representing a piece of the solid. Each configuration corresponds to a particular arrangement of the atoms within the supercell and has an associated probability of occurrence. The idea behind the method is that listing all possible configurations (in principle) and their probabilities can provide a good description of the distribution of the ions and their level of disorder, at least within the range marked by the supercell size. In order to attribute experimental meaning to this kind of representation, we can image the case of a site-disordered (but otherwise perfect) surface that is being studied with an electron microscope capable of taking atomic resolution photographs of the surface. If we take a very large number of photographs at randomly selected sections of the surface, we have all the information required to describe any distribution pattern with ordering range shorter than the size of the photographs. The probability of a given configuration can then be defined as its frequency of appearance in the limit of a very large number of images. In the limit of infinite size of the system, this scheme becomes exact.

From a theoretical point of view, the central challenge in the multiconfigurational supercell approach is to calculate the probabilities of occurrence of the configurations, under the assumption of configurational equilibrium. A typical approximation consists of assigning energy to each configuration and then applying a formalism based on Boltzmann-Gibbs statistical mechanics in such a way that those configurations with lower energies have higher probabilities and the dispersion of the distribution is controlled by the temperature. At low temperatures only the most stable configurations occur, whereas at high temperatures more configurations are accessible to the ions, leading to higher degree of disorder.

We should note that it not trivial to define an energy characterizing the stability of each configuration, because, strictly speaking, the energy contribution from a given ionic configuration in a cell depends on the configuration of the ions in the neighboring cells. Only for large supercells, where *intracell* interactions are much more important than *intercell* interactions, the energy of a given configuration can be considered a function of the arrangement of the ions within the supercell. In what follows we will consider that this is the case, that is, the energy of a given configuration is independent of the distribution of cations next to the cell boundary. Then, we could as well assume that the distribution of ions in the region adjacent to the supercell is identical to that in the supercell, and we can simply calculate the configuration energy using periodic boundary conditions, which is straightforward using modern computer programs for solid state simulations.

11.2 THREE TYPES OF METHODS FOR ENERGY EVALUATIONS

Evaluation of the energy of a periodic crystal with integer site occupancies is a typical task in computational physics and chemistry, and the different methods and approximations involved have been widely discussed elsewhere (Catlow 1997). Here, we present a classification of the alternatives, in increasing order of sophistication and computational cost based on the information used in and that remains accessible from a simulation.

- *Type I methods:* In this case, the energy is a function of only the site occupancies. Electronic and structural (geometric) degrees of freedom are not included explicitly, but are implicit within a model that yields the energy from nearest-neighbor (NN), next-nearest-neighbor (NNN) configurations, or longer-distance effective pair interactions or including terms for clusters of more than two ions. These types of methods include Ising-type models of alloys and cluster expansion methods and have been used extensively in determination of phase diagrams of alloys. We note that an Ising cluster expansion can be derived from energies obtained with structural relaxation (local distortions), but they do not provide access to these details for an arbitrary configuration.
- *Type II methods:* These methods include geometric relaxations explicitly, but electronic effects only implicitly. The energy is evaluated in this case via a classical interatomic

potential function or force field, which is a function of the ionic coordinates. These are quite efficient computationally and can be used in studies of symmetry-breaking structural phase transitions such as those in a ferroelectric or shape memory alloy. In such a case, the structural degrees of freedom are directly relevant to the problem and cannot be integrated out.

- *Type III methods:* These methods include explicit geometric and electronic relaxation for each configuration. This category comprises all quantum-mechanical methods, including those based on the density functional theory (DFT) and its extensions, Hartree–Fock (HF) and post-HF approaches, hybrid DFT-HF, and semiempirical methods such as tight-binding and so forth. Computationally, these methods are quite expensive and can be used only with relatively small supercells. They have to be used in problems that involve electronic phase transitions (such as magnetic or metal–insulator transition) and those in which site-specific chemistry is relevant, for example, in catalysts.

In the evaluation of energies of a large number of configurations to investigate the thermodynamics of disorder, type I methods are still the most commonly employed. Not only are they are computationally cheaper, they are also easier to integrate with sampling algorithms (e.g., Metropolis–Monte Carlo) within a single computer program.

In contrast, energy evaluations using type II and III methods are more expensive and typically require a specialized program that deals with only one geometric configuration at a time. Configurational sampling in these cases requires multiple calls to the quantum-mechanical or interatomic potential code from an external program and has a very long running time.

However, there are good reasons to move toward more sophisticated (types II and III) methods to evaluate energies in a multi-configurational simulation. First, such calculations provide not only energies but many other properties as well (e.g., local geometries and cell parameters, and in the case of type III methods, electronic structure information). Any property that can be obtained for each configuration can be averaged over the ensemble to obtain effective values for the disordered solid. Second, methods of type II and III also provide access to vibrational properties of the solid and its response to external pressure, thereby allowing an integration of configurational and vibrational degrees of freedom in the construction of complex phase diagrams (as a function of composition, temperature, and pressure). Finally, if interactions in the system are long-range, energy evaluations in terms of simple type I methods might not provide good precision or would require a large number of terms and parameters (van de Walle and Ceder 2002). The main disadvantage of type II and III methods, although evaluating a large number of configurations, is their computational cost, but with the developments in computer hardware and efficient algorithms, they are becoming much more affordable.

11.3 CONFIGURATIONAL THERMODYNAMICS FROM CANONICAL ENSEMBLES

We will now discuss in detail the statistical formulation of the configurational equilibrium in a site-disordered binary system. For the sake of simplicity, in this initial formulation all configurations are constrained to have the same compositions, and we will ignore vibrational and pressure effects in the thermodynamics; the corresponding generalizations are introduced later in this chapter.

The extent of occurrence of each configuration (labeled with an index k) is described in this approximation by a Boltzmann-like probability that is calculated from the energy E_k of the configuration and the temperature T:

$$P_k = \frac{1}{Z}\exp(-E_k/k_\mathrm{B}T) \qquad (11.1)$$

where k_B is Boltzmann's constant (it is formally equivalent to use the gas constant R instead, and expressing the molar energies of supercells, but we follow here the usual notation in statistical mechanics in terms of k_B):

$$Z = \sum_{k=1}^{K} \exp(-E_k/k_B T) \qquad (11.2)$$

where Z is the partition function and K is the total number of configurations with the given composition in the supercell. For a binary system, K can be calculated as a number of combinations:

$$K = \frac{N!}{(N-n)!n!} \qquad (11.3)$$

where N is the number of exchangeable sites in the supercell and n is number of ions of one of the species (it can also be a vacancy) that can occupy these sites. The molar concentrations are then $x = n/N$ for the first species and $1-x$ for the second species.

The definition of the configurational ensemble and the associated probabilities allows us to obtain the effective value in the disordered solid of any quantity that can be theoretically obtained for each ordered configuration. If A_k is the value of the given magnitude for configuration k, then the effective value in the disordered solid is:

$$A = \sum_{k=1}^{K} P_k A_k \qquad (11.4)$$

In this way, it is possible to obtain effective values even for quantities such as the cell parameters, which are not strictly defined but still have experimental meaning in a solid with nonperiodic distribution of ions on lattice sites. Equation 11.4 also allows us to obtain the effective energy of the solid from the configurational energies E_k, as:

$$E = \sum_{k=1}^{K} P_k E_k \qquad (11.5)$$

In evaluating the thermodynamic stability of a disordered solid at a given temperature, not only the energy but also the configurational multiplicity of the system should be taken into account, which is done by introducing the (configurational) free energy:

$$F = -k_B T \ln Z = -k_B T \sum_{k=1}^{K} \exp(-E_k/k_B T) \qquad (11.6)$$

The difference per temperature unit between the average energy and the free energy defines the configurational entropy, which can also be expressed in terms of the probabilities P_k:

$$S = \frac{E-F}{T} = -k_B \sum_{k=1}^{K} P_k \ln P_k \qquad (11.7)$$

There are two limiting cases here:

1. *Perfect order:* This occurs when one configuration (say $k = 1$) is much more stable than the rest; that is, it is separated from the other configurations by an energy difference much

larger than $k_B T$. In this case $P_1 = 1$, while $P_k = 0$ for $k \neq 1$, and the system will have zero configurational entropy. The configurational free energy simply corresponds to the energy of the most stable configuration.

2. *Perfect disorder:* This occurs when the energies of all the configurations are very similar (again in comparison with $k_B T$) or formally in the limit $T \to \infty$. In this case, all configurations have the same probability $P_k = 1/K$ and the configurational entropy reaches its maximum possible value:

$$S_{max} = k_B \ln K = k_B \ln \frac{N!}{[N(1-x)]![Nx]!} \tag{11.8}$$

which, in the limit of an infinitely large supercell ($N \to \infty$ at constant x), using Stirling's formula, converts to the well-known expression

$$S_{ideal} = -k_B N (x \ln x + (1-x) \ln(1-x)) \tag{11.9}$$

Intermediate to these two limiting cases, there is a continuum of situations with varying degrees of ordering, which can be described within the same formalism, leading to temperature-dependent entropy values given by Equation 11.7. It is clear from the previous equations that any finite supercell is unable to describe exactly the perfect disorder limit. In order to correct for this, it is convenient to rewrite the free energy (Equation 11.6) as

$$F = -k_B T \ln K - k_B T \ln \left(\frac{1}{K} \sum_{k=1}^{K} \exp(-E_k / k_B T) \right) \tag{11.10}$$

as suggested by Becker et al. (2000). The first term represents the entropy contribution in the limit of perfect disorder; the second term contains the energy contribution plus the correction to the entropy contribution resulting from partial ordering. The first term can then be adjusted to its correct value, that is:

$$F = N k_B T (x \ln x + (1-x) \ln(1-x)) - k_B T \ln \left(\frac{1}{K} \sum_{k=1}^{K} \exp(-E_k / k_B T) \right) \tag{11.11}$$

which is equivalent to amending the temperature-dependent entropy by a term $\Delta S_{corr} = S_{ideal} - S_{max}$, thus guaranteeing the correct behavior in the limit of perfect disorder. However, this adjustment also breaks down in the description of the perfect order limit, by introducing a spurious configurational entropy contribution to the free energy (which should be zero in this limit). Therefore, this correction should be applied only to simulations of systems with near-perfect disorder.

11.4 INCLUDING VIBRATIONAL AND FINITE PRESSURE EFFECTS

In order to consider vibrational contributions to the thermodynamics of a disordered solid within the multi-configurational formalism, we write the total partition function of the system as:

$$Z = \sum_{k=1}^{K} \sum_{v} \exp(-E_{k,v} / k_B T) \tag{11.12}$$

where v is an index (or, strictly speaking, a collection of indices) that characterizes each vibrational state (with energy $E_{k,v}$) of configuration k. In terms of the vibrational partition function $Z_k^{(vib)}$ and the vibrational free energy $F_k^{(vib)}$ of each configuration, this becomes:

$$Z = \sum_{k=1}^{K} Z_k^{(vib)} = \sum_{k=1}^{K} \exp(-F_k^{(vib)}/k_B T) \tag{11.13}$$

which, by comparison with Equation 11.6, indicates that the formalism including vibrational contributions is equivalent to the one introduced in the previous section, except that now the vibrational free energy $F_k^{(vib)}$ of each configuration should be used instead of the energy E_k to define the probabilities of Equation 11.1. Using types II and III methods, in which the energy depends explicitly on ionic coordinates, vibrational free energies can be evaluated from the vibrational frequencies of each configuration, by invoking the harmonic approximation (Dove 1993). This, of course, adds considerably to the cost of the simulations, especially if the equilibrium geometry of each configuration is obtained by minimizing the free energy and not just the energy, but can become affordable if methods of type II are being used (Benny et al. 2009).

Analogously, if we want to introduce the effect of a finite external pressure p, we should use the Gibbs free energy:

$$G_k^{(vib)} = F_k^{(vib)} + pV_k = H_k - TS_k^{(vib)} \tag{11.14}$$

to calculate the configurational probabilities, where V_k, H_k, and $S_k^{(vib)}$ are the supercell volume, the enthalpy, and the vibrational entropy, respectively, for the particular configuration. In this case, the effective thermodynamic potentials for the disordered solid are:

$$H = \sum_{k=1}^{K} P_k H_k \tag{11.15}$$

$$G = -k_B T \sum_{k=1}^{K} \exp(-G_k^{(vib)}/k_B T) \tag{11.16}$$

and

$$S = \frac{H-G}{T} = \sum_{k=1}^{K} P_k S_k^{(vib)} - k_B \sum_{k=1}^{K} P_k \ln P_k \tag{11.17}$$

where in the last expression for the entropy, the first term is the vibrational contribution and the second term is the configurational contribution. The correction defined by Equation 11.11 can be analogously applied here to treat highly disordered systems.

11.5 ACCESSING THE CONFIGURATIONAL SPACE

Computing energies and other properties of all possible configurations of ions in a mixed solid can be a rather demanding task, even for relatively small cells. Let us consider the case of a body-centered cubic (BCC) binary alloy, with a $2 \times 2 \times 2$ supercell, which has only 16 exchangeable sites. The number of configurations as a function of the substitution fraction x increases very quickly and reaches a maximum at $x = 0.5$, when there is a total of 12,870 configurations (Figure 11.2). This number is tractable with type I or even type II methods, but becomes too expensive for type III methods. In a $3 \times 3 \times 3$ supercell, with 54 exchange sites, the maximum is approximately 2×10^{15} configurations, which becomes very expensive even for type I methods.

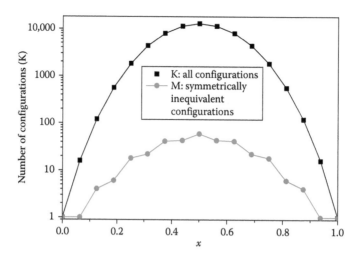

FIGURE 11.2   Number of configurations (K) in a model of a binary alloy with BCC structure using a $2 \times 2 \times 2$ supercell, compared with the number of symmetrically inequivalent configurations.

It is therefore clearly necessary to find strategies to reduce the number of configurations to be evaluated. We will discuss here three possible routes: (1) taking advantage of the crystal symmetry, (2) random sampling, and (3) importance sampling using Metropolis–Monte Carlo algorithms.

11.5.1   Taking Advantage of the Crystal Symmetry

If we are dealing with relatively small supercells, for example, when doing quantum-mechanical calculations for each configuration, it is possible to reduce the number of configurations by taking advantage of the crystal symmetry of the lattice (Grau-Crespo et al. 2007). Within this approach, two configurations are considered equivalent when they are related by a symmetry (an isometric) operation, for example, a reflection (Figure 11.3). A list of all possible isometric transformations is provided by the group of symmetry operators in the parent structure (the original structure without any substitutions). They include the symmetry operators in the space group of the crystal unit cell (scaled in an appropriate way to account for the cell multiplicity of the supercell), the supercell internal translational operators, and the combinations between them. It is then possible (at least for small systems) to start with all possible configurations through explicit enumeration and reduce to those that are symmetrically inequivalent for energy or properties evaluation. We also need to keep track of the degeneracy of each independent configuration, that is, how many times it repeats in the whole configurational space (this is similar to the number of members in the star of k-points in the Brillouin zone in the representation and group theory of crystals). This algorithm is implemented in the Site Occupancy Disorder (SOD) program (Grau-Crespo et al. 2007).

It is necessary to slightly adapt the equations for configurational statistics to operate in the reduced space of inequivalent configurations. If E_m is the energy and Ω_m is the degeneracy of the independent configuration m ($m = 1,..., M$), its contribution to energy or other properties needs to be weighted by a probability:

$$P_m = \frac{\Omega_m}{Z} \exp(-E_m / k_B T) \tag{11.18}$$

which means that if we want to compare the stability of two independent configurations in energetic terms, we should not use their energies E_m but instead the value $E_m - k_B T \ln \Omega_m$. Average values can be obtained using an expression analogous to Equation 11.4:

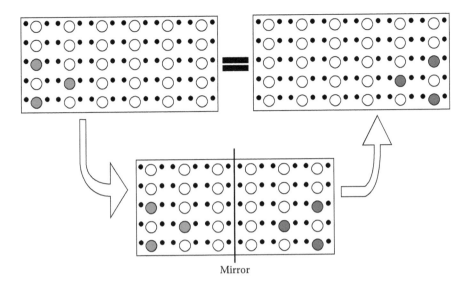

FIGURE 11.3 Two configurations are identical if they are related by an isometric transformation, for example, a reflection.

$$A = \sum_{m=1}^{M} P_m A_m \tag{11.19}$$

for scalar properties, that is, if A_m is the same for all the Ω_m equivalent configurations that the inequivalent configuration m represents. For example, if we are modeling a cubic system, we cannot obtain the average cell parameter a from the cell parameters a_m of the inequivalent configurations, because this result could be different from the direct average of the b_m or c_m values, breaking the cubic symmetry. We therefore, need to find first a related magnitude that is invariant in the subspace of equivalent configurations, that is, the volume V_m in the given example. We can then define the average cell parameter of the cubic system as:

$$a = \left(\sum_{m=1}^{M} P_m V_m \right)^{1/3} \tag{11.20}$$

Otherwise, one needs to explicitly symmetrize the property obtained using Equation 11.19 with all the symmetry operations used in obtaining the reduced set of configuration (e.g., electric dipole or polarization in a ferroelectric).

Symmetry reduction is practical only when working with relatively small supercells. This is typically the case when properties other than the energy are being evaluated, using types II and III methods. When evaluating the thermodynamic functions of the solution, the enthalpy tends to converge very quickly with supercell size, but the convergence of the entropy, which depends on configuration counting, is much slower. Therefore, for a complete thermodynamic characterization of solid solutions, it is generally necessary to consider supercells much larger than those tractable by symmetry-reduction methods.

11.5.2 Random sampling

A relatively straightforward alternative to deal with a large number of configurations occurring in larger supercells is to generate them randomly and increase the size of the sample until the thermodynamic magnitudes of interest have converged. The evaluation of the energy (or enthalpy) from a random sample of size $K' < K$ can be done using an expression analogous to Equation 11.5:

$$E = \frac{\displaystyle\sum_{k=1}^{K'} E_k \exp(-E_k/k_B T)}{\displaystyle\sum_{k=1}^{K'} \exp(-E_k/k_B T)}$$

which generally converges very quickly with the sample size, even for $K' \ll K$ (Todorov et al. 2004). However, the evaluation of free energies (and entropies) from the random sample should not be done from the analogous to Equation 11.6 because the convergence would be very slow. As pointed out by Todorov et al. (2004), it is convenient to use here the separation given in Equation. 11.10, but substituting K by K' only in the second term, that is:

$$F = -k_B T \ln K - k_B T \ln\left(\frac{1}{K'}\sum_{k=1}^{K'} \exp(-E_k/k_B T)\right) \tag{11.21}$$

The first term remains constant with the sample size because it only depends on the concentration and supercell size (of course, the correction defined by Equation 11.11 can also be applied here), and only the second term should be converged with respect to sample size. Even after this correction, free energies and entropies tend to show slower convergence with sample size than the energy. Therefore it is always important to check convergence with respect to supercell size for the thermodynamic potential of interest. We note that the differences in free energies of two macroscopic states, however, converge much faster.

11.5.3   Importance Sampling with Metropolis–Monte Carlo Algorithm

In situations in which the configurational space is too large, random sampling becomes inefficient and still requires the evaluation of too many configurations to achieve convergence. It is often the case that a vast number of the configurations that are produced by random sampling contribute relatively little to the equilibrium averages because they have high energies. Importance sampling is a method that attempts to emphasize the evaluation of "important" configurations (those with low energies in this case) and avoid the evaluation of nonimportant ones by generating a chain or a set of configuration with a distribution of occurrence that is proportional to the probability in Equation 11.1 (Landau and Binder 2009).

The classic algorithm for the generation of configurations for importance sampling was first introduced by Metropolis et al. (1953). The averages are performed only over a sequence of accepted configurations, which accumulate at the bottom of the energy spectrum. This is guaranteed by the way configurations are selected: a configuration with energy E_i is generated from a previously accepted one (with energy E_{i-1}) by a small modification and is accepted to become part of the ensemble with a probability:

$$p = \begin{cases} 1 & \text{if } E_i < E_{i-1} \\[2mm] \exp\left(\dfrac{E_i - E_{i-1}}{k_B T}\right) & \text{if } E_i > E_{i-1} \end{cases} \tag{11.22}$$

This is typically achieved by generating at each step a random number r between 0 and 1. If $r < p$, the configuration is accepted; otherwise it is left out. Another modification in configuration is then introduced from the previous accepted configuration, and the procedure is repeated. With an ensemble generated in this way, averages are calculated as simple arithmetic averages:

$$A = \frac{1}{N_{\text{steps}}} \sum_{i=1}^{N_{\text{steps}}} A_i \tag{11.23}$$

because the distribution of probabilities given by energies and the temperature is already introduced during the sampling.

This approach is particularly efficient when type I methods are being employed, allowing an easy integration of configurational sampling, energy evaluation, and statistical processing within a single computer program. A detailed description of these methods can be found elsewhere (Landau and Binder 2009).

11.5.4 Combined SOD and Monte Carlo Approach

The SOD approach, presented in Section 11.5.1, yields exact statistical thermodynamic analysis of a system of given size. There are two steps in this analysis in which an SOD-based simulation may become unfeasible. First, the generation of symmetry inequivalent configurations in the reduced subspace can become computationally untractable for large system sizes. Second, a limitation in the application of the SOD approach often arises from the computational intensity of evaluation of energies and other properties of a rather large number of configurations even in the reduced space. In the latter, SOD and the Monte Carlo method, summarized in Section 5.3, can be combined by using importance sampling (1) in the reduced set of configurations obtained by SOD scheme and (2) with a free energy function defined as $E_m - k_B ln\Omega_m$ instead of the energy E_m.

11.6 CONFIGURATIONAL THERMODYNAMICS FROM GRAND-CANONICAL ENSEMBLES

We now present one more generalization of the formalism of configurational statistical mechanics in a supercell. It is very useful for certain applications to extend the ensemble to configurations with changes in composition, that is, to make it grand-canonical instead of canonical (Grau-Crespo et al. 2009). In a binary system, the composition is characterized by a single index n, which is the number of substitutions made in the parent structure. In the grand-canonical ensemble for this system, each configuration is then labeled by two indices, n and m (we will assume here that we are working only with the symmetrically inequivalent configurations obtained with the SOD technique), and its probability is given by:

$$P_m^{(n)} = \frac{\Omega_m^{(n)}}{\Xi} \exp\left(-\frac{E_m^{(n)} - n\mu}{k_B T}\right) \tag{11.24}$$

where $E_m^{(n)}$ and $\Omega_m^{(n)}$ are the energy and the degeneracy of the given configuration

$$\Xi = \sum_{n=1}^{N} \sum_{m=1}^{M_n} \Omega_m^{(n)} \exp\left(-\frac{E_m^{(n)} - n\mu}{k_B T}\right) \tag{11.25}$$

is the grand-canonical partition function, and μ is the chemical potential of the new species with respect to the original one.

In practical calculations, two different approaches can be used to choose the value of the chemical potential. The first approach is used when the ion exchange in the solid is in equilibrium with a known environment, for example, a gas phase, for which we can calculate the chemical potentials of the species. This approach was followed by Grau-Crespo et al. (2009) to calculate the distribution of hydrogen vacancies in a metal hydride (MgH_2) in equilibrium with a gas of H_2 molecules. The chemical potential of a molecule in a gas is a known function of partial pressure and temperature, and the equilibrium implies that the chemical potential of the species in the solid and in the gas are the same. The application to MgH_2 is described in more detail in Section 11.7.3.

The second approach to fix the value of the chemical potential is by establishing its connection with the overall composition of the solid. The substitution fraction is related to the chemical potential by the equation

$$x \equiv \frac{<n>}{N} = \frac{1}{N}\sum_n n \sum_m P_m^{(n)}(\mu) \tag{11.26}$$

where we have emphasized the dependence of the probabilities on the μ value. This equation allows the calculation of the μ value required to achieve an arbitrary composition x, and it can be proved that a unique function $\mu(x)$ exists and is easy to obtain numerically (Haabgood et al. 2011). The basis for this is that μ is a Lagrange multiplier for imposing a constraint on the number of particles in the thermodynamic simulation.

Besides allowing the equilibration with external phases, the grand-canonical procedure has the advantage of decreasing the supercell size required for certain types of problems. For example, low concentrations of substitutions require very large cells in the canonical formulation, whereas in the grand-canonical approach they can be represented using the relevant proportion of cells with $n = 0$ and cells with $n > 0$. This sort of amounts to interpolation to low concentrations x, between the results obtained with vanishing and a nonzero composition. Furthermore, the grand-canonical procedure allows a more exhaustive consideration of the configurational space for a given finite supercell and therefore, more accurate determination of the thermodynamic potentials. To illustrate this last point, consider a binary system with $x = 0.5$. In a canonical approach, the maximum entropy that can be obtained with a small supercell of $N = 16$ sites (e.g., a $2 \times 2 \times 2$ cell for a BCC structure) is given by Equation 11.8, and is approximately $0.591\ k_B$ per site, but the ideal entropy for this composition is $k_B \ln 2 \approx 0.693\ k_B$ per site, according to Equation 11.9. In the grand-canonical ensemble there are 2^N configurations for this composition, and therefore the maximum entropy agrees exactly with the ideal value. It fact, it can be demonstrated that the configurational entropy in the perfect disorder limit is always exact for the grand-canonical ensemble, for any composition x and supercell size N, but the general demonstration for $x \neq 0.5$ is a bit too long to be included here.

Despite these advantages, there are two important aspects of the grand-canonical approach one should keep in mind. First, it obviously requires a much higher number of configurations to be evaluated than the canonical method for a given cell size, which is an important practical limitation. The second limitation is more fundamental: when the size mismatch between the exchangeable ions is considerable, cells with different compositions will have significantly different cell parameters. The elastic stress required to combine these cells in the solid is not included in the formalism described earlier, but it can have a significant effect on the probability of occurrence of configurations. Therefore, the application of the grand-canonical formalism as described earlier is restricted to either a very low concentration of substitutions or cases in which there is negligible size mismatch between the exchangeable species. To include the effects of change in lattice size with x, it would be necessary to modify the formalism described above.

11.7 EXAMPLES

We describe now some recent applications of the methodologies described in this chapter. For convenience, we discuss examples from our own work. The selection therefore, does not attempt to represent the wide variety of research that is currently done in the field of site-disordered alloys and compounds, using similar or very different methodologies.

11.7.1 Using Symmetry-Reduced Ensembles to Identify Favorable Cation Distributions in Oxides

We start with the simplest possible use of the multi-configurational representation of the ionic distribution in a mixed solid: finding the most stable configurations. We use the iron oxide γ-Fe_2O_3 (maghemite) as a first example.

Maghemite is the second most stable polymorph of iron (III) oxide. Its magnetism, chemical stability, and low cost have led to its wide application as magnetic pigment in electronic recording media since the late 1940s (Dronskowski 2009). Maghemite nanoparticles are also widely used in biomedicine, because their high magnetic moment allows manipulation with external fields and they are biocompatible and potentially nontoxic to humans (Levy et al. 2008; Pankhurst et al. 2003). Despite the compositional simplicity, its precise structure has been the subject of debate for decades. Like magnetite (Fe_3O_4), maghemite exhibits a spinel crystal structure, but whereas the former contains both Fe^{2+} and Fe^{3+} cations, in maghemite all the iron cations are in trivalent state and the charge neutrality of the cell is guaranteed by the presence of cation vacancies. The debate about the maghemite structure has focused on the degree of ordering of these vacancies in the solid.

The unit cell of magnetite can be represented as $(Fe^{3+})_8[Fe^{2.5+}]_{16}O_{32}$, where the brackets () and [] designate tetrahedral and octahedral sites, respectively. The maghemite structure can be obtained by creating 8/3 vacancies out of the 24 Fe sites in the cubic unit cell of magnetite. These vacancies are known to be located in the octahedral sites (Waychunas 1991) and therefore, the structure of maghemite can be approximated as a cubic unit cell with composition $(Fe^{3+})_8[Fe^{3+}_{5/6} \square_{1/6}]_{16}O_{32}$. If the cation vacancies were randomly distributed over the octahedral sites, as was initially assumed, the space group would be Fd3m, like in magnetite. However, there is evidence for a higher degree of ordering. Braun (1952), for example, noticed that maghemite exhibits the same superstructure as lithium ferrite ($LiFe_5O_8$), which is also a spinel with unit cell composition $(Fe^{3+})_8[Fe^{3+}_{3/4}Li^{1+}_{1/4}]_{16}O_{32}$, and suggested this was due to similar ordering in both compounds. In the space group $P4_332$ of lithium ferrite, there are two types of octahedral sites, one with multiplicity 12 in the unit cell and one with multiplicity four, which is the one occupied by lithium. In maghemite, the same symmetry exists if the Fe vacancies are constrained to these Wyckoff 4b sites, instead of being distributed over *all* the 16 octahedral sites. It should be noted, however, that some level of disorder persists in this structure, because the 4b sites have fractional (1/3) iron occupancies. Finally, there is also evidence of a fully ordered structure, exhibiting a tetragonal cell with space group $P4_12_12$ and $c/a \approx 3$ (spinel cubic cell tripled along the *c* axis) (Grau-Crespo et al. 2010; Jorgensen et al. 2007).

A computational investigation of the energetics of vacancy ordering in maghemite was presented by Grau-Crespo et al. (2010). A $1 \times 1 \times 3$ supercell of the cubic structure was used to obtain the spectrum of energies of all the ordered configurations that contribute to the partially disordered $P4_332$ cubic structure. The energies were evaluated using long-range Coulomb contributions and classical interatomic potentials to describe short-range interactions (parameters derived by Lewis and Catlow [1985]). The core-shell model of Dick and Overhauser (1958) was employed to account for the polarizability of the anions. The calculations were performed with the GULP code (Gale 1997, 2005; Gale and Rohl 2003). Although not as sophisticated as quantum-mechanical calculations, this methodology allows for accurate calculations of ion relaxations and configuration energies.

The total number of combinations of the four iron ions on the so-called L sites of the supercell (Figure 11.4a) is $12!/(4! \times 8!) = 495$, but only 29 of these are inequivalent, as determined using the SOD program (Grau-Crespo et al. 2007). The calculated energies for these 29 configurations are shown in Figure 11.4b. Only one of these configurations has the space group $P4_12_12$, found by Shmakov et al. (1995) for fully ordered maghemite. This configuration is indeed the most stable one, with a significant energetic separation from the second most stable configuration (32 kJ/mol). The energy range covered by the configurational spectrum is quite wide (~850 kJ/mol), indicating that full disorder is very unlikely. The distinctive feature of the most stable configuration ($P4_12_12$) is the

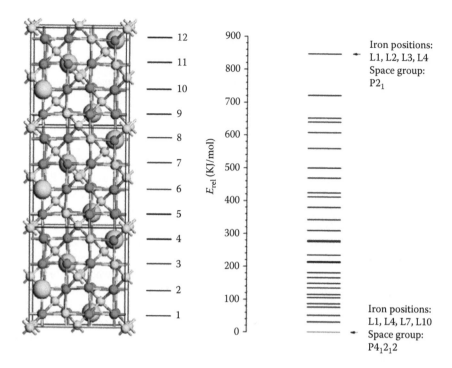

FIGURE 11.4 (a) The exchangeable sites in the maghemite tetragonal cell: four Fe ions and eight vacancies are distributed over these "L" sites. (b) The calculated configurational spectrum.

maximum possible homogeneity of iron cations and vacancies over the L sites. This configuration is the only one in which vacancies never occupy three consecutive layers; there are always two layers containing vacancies separated by a layer without vacancies, which instead contains Fe^{3+} cations in the L sites (e.g., positions L1 - L4 - L7 - L10) and the $P4_12_12$ configuration is therefore the one that minimizes the electrostatic repulsion between these cations.

In order to interpret the energy differences in the configurational spectrum in terms of the degree of vacancy ordering in the solid, we can calculate the probability of occurrence of each independent configuration, using Equation 11.18. Figure 11.5 shows the probabilities of the most stable configuration ($P4_12_12$) and of the second most stable configuration (with space group $C222_1$) as a function of temperature. At 500 K, a typical synthesis temperature for maghemite (Shmakov et al. 1995) the cumulative probabilities of all the configurations excluding the most stable $P4_12_12$ is less than 0.1%. This contribution increases slowly with temperature, but at 800 K this cumulative probability, which measures the expected level of vacancy disorder, is still less than 2%. At temperatures above 700 to 800 K, maghemite transforms irreversibly to hematite (α-Fe_2O_3) and considering higher temperatures is therefore irrelevant. It thus seems clear that perfect crystals of maghemite in configurational equilibrium should have a fully ordered distribution of cation vacancies. Further analysis of the cation distribution in this oxide can be found in the work of Grau-Crespo et al. (2010).

11.7.2 CE$_{1-x}$ZRO$_2$ SOLID SOLUTIONS: CONFIGURATIONAL AVERAGES IN THE BULK AND THE SURFACE

We discuss now some applications of the concept of configurational average, using the $Ce_{1-x}ZrO_2$ solid solution as a case study. This material is used as a support for the noble metals in the catalyst employed for reduction of harmful emissions from car exhausts. A computational study of this solid solution was presented by Grau-Crespo (2011).

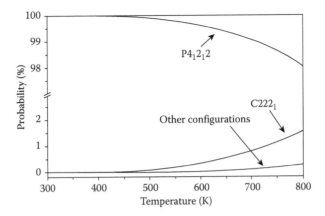

FIGURE 11.5 Probabilities of the two most stable configurations in the maghemite supercell as a function of temperature.

A supercell with 36 atoms was used there to model the bulk system, in particular the cerium-rich part of the solid solution ($0 < x < 0.5$ in $Ce_{1-x}Zr_xO_2$), which exhibits cubic symmetry (Cabanas et al. 2001; Lee et al. 2008). In this case, all calculations were performed using quantum-mechanical calculations, based on the density functional theory (DFT), as implemented in the VASP (Vienna Ab Initio Simulation Package) code (Kresse and Furthmüller 1996a,b). From the calculations, it was immediately clear that the lowest energy configurations were those in which all the zirconium ions are grouped together, indicating a tendency to ex-solution. The tendency to ex-solution within bulk phases can be quantified by calculating the enthalpy of mixing:

$$\Delta H_{mix} = H[Ce_{1-x}Zr_xO_2] - (1-x)H[CeO_2] - xH[c\text{-}ZrO_2] \qquad (11.27)$$

where $H[CeO_2]$ and $H[C–ZrO_2]$ are the DFT energies per formula unit of ceria and cubic zirconia, respectively, and $H[Ce_{1-x}Zr_xO_2]$ is the effective energy of the solid solution, calculated as a configurational average. The resulting enthalpy of mixing is strongly positive, in agreement with recent calorimetric measurement (Lee et al. 2008) (Figure 11.6).

Assuming a regular solid solution model (Prieto 2009; Ruiz-Hernandez et al. 2010), the enthalpy of mixing at low zirconium content was fitted with a polynomial of the form:

$$\Delta H_{mix} = Wx(1-x)$$

which gives $W = 38$ kJ/mol. This result is intermediate between the value of 28 kJ/mol obtained by Du et al. (1994) from fitting a regular solution model to experimental solubility data, and the value of 51 kJ/mol obtained by Lee et al. (2008) from fitting directly to calorimetric measurements. The positive values of the enthalpy of mixing suggest that cation ordering is not a stabilizing factor in ceria–zirconia solid solutions, at least for the compositions examined here, and confirm that the zirconium ions have an energetic preference to segregate or form a separate zirconium-rich phase. The origin of this tendency is due to the difference between the ionic radii of the cations ($r[Ce^{4+}] = 0.97$ Å and $r[Zr^{4+}] = 0.84$ Å, for eight-fold coordination, according to Shannon (1976). It should be noted that real samples, where homogeneity at the atomic level can be achieved using special synthesis methods (Cabanas et al. 2000, 2001) might not experience this trend unless subjected to temperatures high enough to overcome the cation diffusion barriers.

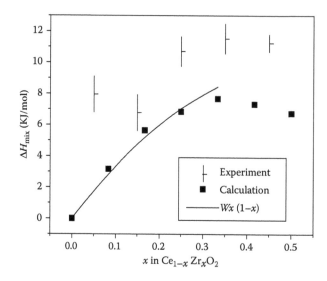

FIGURE 11.6 Calculated enthalpies of mixing for $Ce_{1-x}Zr_xO_2$ in comparison with experimental results (From Lee et al. *J. Mater. Res.,* 23: 1105, 2008). The curved line represents the fitting of a regular-solution quadratic polynomial to the calculated values for low zirconium concentrations.

To describe the thermodynamic stability of the solid solution at any finite temperature, entropies and free energies of mixing should also be calculated. It was found that, even assuming ideal configurational entropy, the resulting free energy of mixing is positive except for very small values of x. Furthermore, because zirconium-rich phases are known to be monoclinic (Garvie 1970) at the temperatures of interest here, the mixing free energy should be calculated with respect to the more stable monoclinic zirconia phase (m-ZrO_2), which makes the mixed phase even less stable with respect to phase separation. In order to estimate the solubility limit of zirconium in CeO_2, the mixing free energy function

$$\Delta G_{mix}(x,T) = Wx(1-x) + \Delta H_t x + RT[x\ln x + (1-x)\ln(1-x)] \tag{11.28}$$

was considered, in which the enthalpy of the monoclinic-cubic zirconia phase transformation $\Delta H_t = 8.8$ kJ/mol (Navrotsky et al. 2005) was introduced. The use of the ideal entropy is justified because at very low zirconium content the disorder should be nearly perfect. This analytical function allows the interpolation to x values smaller than those directly obtainable with the simulation supercell, and its minimum with respect to x at a given temperature provides an estimation of the solubility limit. Figure 11.7 shows that the maximum equilibrium solubility of zirconium from monoclinic zirconia into the ceria structure is approximately 0.4 mol% at 973 K, and increases to 2 mol% at 1373 K. Thus, although ceria–zirconia solid solutions in the whole range of compositions can be synthesized under adequate conditions (Cabanas et al. 2000), these results taken together with previous experimental evidence clearly show that these solid solutions are metastable with respect to phase separation into cerium-rich and zirconium-rich phases. This phase separation can actually occur in a close-coupled catalytic converter, where temperatures of up to 1373 K could lead to rearrangement of the cations in the solid solution.

Simulations of the distribution of cations near the (111) surface of the solid were performed in the same study, by using the periodic slab model shown in Figure 11.8. The number of configurations in the slab was reduced by only including those keeping the inversion symmetry of the cell and then selecting the symmetrically inequivalent ones. The equilibrium zirconium content of a particular cation layer parallel to the (111) surface depends both on the overall zirconium content of the slab and on the temperature and can be calculated by taking the configurational average,

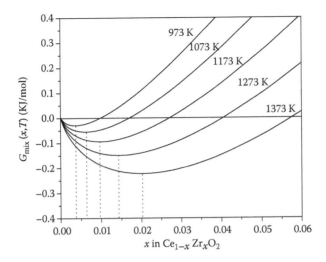

FIGURE 11.7 Free energies of mixing for low zirconium concentrations, as obtained from Equation 11.5. The vertical dotted lines mark the solubility limit of zirconium in CeO_2 at the particular temperature.

$$c_l = \frac{\sum_m f_{ml}\Omega_m \exp(-E_m/k_B T)}{\sum_m \Omega_m \exp(-E_m/k_B T)} \qquad (11.29)$$

where f_{ml} is the fraction of sites occupied by zirconium in the layer l for configuration m. The results are shown in Figure 11.8 for temperatures between 800 and 1600 K. The most obvious feature of the cation distribution is the low concentration of zirconium at the top (111) layer. Even for the 50:50 solid solution, at the highest temperature considered (1600 K), the equilibrium zirconium content of the surface is only approximately 10%. The dependence of the calculated concentrations on temperature is relatively weak, especially at the top layer, but it is clear that increasing temperatures lead to more homogeneity in the composition of the interior of the slab, by equalizing the zirconium content in the second and third layers. Thus, according to these results, the redistribution of cations at high temperatures should occur with significant cerium-enrichment of the (111) surface of ceria–zirconia, regardless of the overall composition of the solid solution. These conclusions are discussed in detail, in comparison with the experimental evidence, in the work by Grau-Crespo et al. (2010).

11.7.3 Hydrogen Vacancies in Pure and Doped MgH_2: Application of the Grand-Canonical Extension

This example is brought to illustrate the grand-canonical extension formalism and is about the distribution of hydrogen vacancies in magnesium hydride (MgH_2). This is the parent compound of an important group of hydrogen storage materials, which exhibits high volumetric and gravimetric storage density. However, the pure Mg–H system exhibits very slow kinetics of dehydrogenation and hydrogenation (Johnson et al. 2005) and also thermodynamic limitations, which limit its practical applications. The kinetic behavior can be interpreted in terms of a model of phase nucleation and growth, where which particle surfaces consist mainly of hydride (β phase) during hydrogenation and metallic Mg (α phase) during dehydrogenation (Rudman 1979; Sastri et al. 1998). The rate-limiting step for hydrogen absorption has been suggested to be the diffusion through the

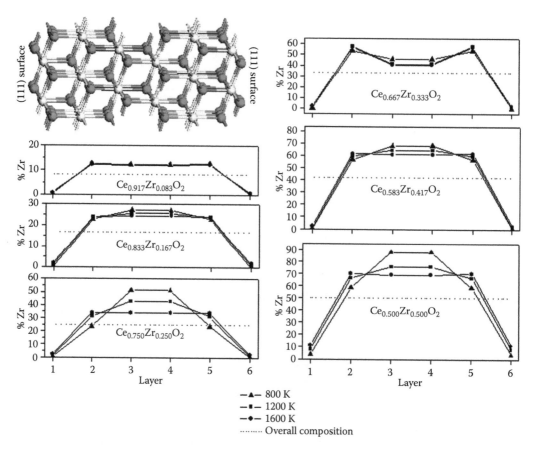

FIGURE 11.8 Calculated equilibrium concentrations of zirconium as a function of the distance to the (111) surface in the $Ce_{1-x}Zr_xO_2$ solid solution. Because of the slab construction, layers 1, 2, and 3 are equivalent to layers 6, 5, and 4, respectively.

external hydride phase layer (Friedlmeier and Groll 1997; Vigeholm et al. 1983). This diffusion is most likely mediated by defects (H vacancies) in the MgH_2 structure, which is why these defects are of interest.

A theoretical investigation of the microscopic configurations and related thermodynamics of hydrogen vacancies in the hydride β phase, near the transformation point to the α phase, was reported by Grau-Crespo et al. (2009). DFT calculations were performed with the VASP code (Kresse and Furthmuller 1996a, 1996b) to obtain the energies and relaxed geometries for different defect distributions in a $2 \times 2 \times 2$ supercell of the rutilelike MgH_2 structure, with composition $Mg_{16}H_{32-n}$, where n is the number of hydrogen vacancies. Only the symmetrically inequivalent configurations were calculated, as determined using the SOD program (Grau-Crespo et al. 2007).

The calculated DFT energies (Figure 11.9) showed significant variations in stability among the different vacancy distributions for each composition. The results reveal that the most stable configurations with multiple vacancies per cell involve the formation of vacancy clusters. In the case of $n = 2$, for example, the lowest energy configuration consists of two vacancies located at the shared edge of one MgH_6 octahedron, corresponding to the shortest H–H distance in the perfect structure (2.48 Å). The second most stable configuration also has two vacancies in the same octahedron but with a different orientation of the pair, in this case involving the shared corner of the octahedron corresponding to the next shortest H–H distance in the structure (2.75 Å). The other configurations with composition $n = 2$ are much higher in energy. For $n = 3$, a smaller energy gap exists between

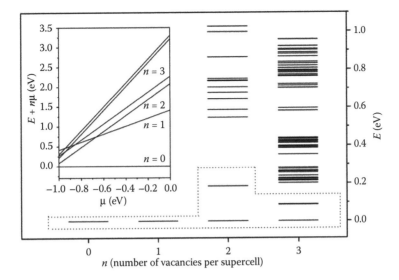

FIGURE 11.9   Configurational energies (relative to the lowest energy at each composition) in the $2 \times 2 \times 2$ supercell of magnesium hydride with composition $Mg_{16}H_{32-n}$. The dotted lines enclose the configurations for which a separate statistical treatment, including vibrational contributions, was performed. The inset shows the variation in the stability of the selected configurations with the hydrogen chemical potential.

the configurations, but again the lowest energy one exhibits vacancy aggregation: two vacancies in the shared edge and one in the shared corner of the same octahedron. The vacancy formation energy

$$\text{VFE} = \frac{1}{n}\left(E[Mg_{16}H_{32-n}] + \frac{n}{2}E[H_2] - E[Mg_{16}H_{32}] \right) \tag{11.30}$$

is always lower for a vacancy in a cluster (1.04 and 1.07 eV for the most stable di-vacancy and tri-vacancy configurations, respectively) than for an isolated vacancy (1.41 eV). The energetic preference for vacancy aggregation was therefore clear from these results.

However, energetic preference of clusters versus monovacancies does not necessarily mean more abundance of the former. To illustrate these, grand-canonical probabilities were calculated using Equation. 11.24. The chemical potential of hydrogen in the hydride is assumed to be identical to the potential per atom in the gas phase (equilibrium condition):

$$\mu = \frac{1}{2}g_{H_2}(T, p_{H_2}) = \frac{1}{2}\left(E[H_2] + E_{ZP}[H_2] + \Delta g_{H_2}(T, p_{H_2})\right) \tag{11.31}$$

where $E[H_2]$ is the DFT energy of an isolated hydrogen molecule, $E_{ZP}[H_2] = 0.273$ eV is the zero-point energy of the molecule, calculated from the experimental vibrational frequency (the difference from using the theoretical value is insignificant), and Δg_{H_2} is the increase, from 0 K to temperature T, of the free energy per molecule in a gas at pressure P_{H_2}, as obtained from thermodynamic tables (Fukai 2005). The molar fraction δ of vacancies at any temperature and hydrogen partial pressure can then be calculated from the grand-canonical probabilities.

The calculations for the configurational equilibrium of vacancies in MgH_2 were performed in two steps. First, all configurations for $n = 0, 1, 2$, and 3 were included in the analysis, and the probabilities from Equation 11.2 were evaluated using the DFT energies. In the second step, apart from the $n = 0$ and 1 configurations, only the most stable, cluster-forming, configurations were considered for $n > 1$ (two for $n = 2$ and three for $n = 3$, as shown in Figure 11.9), and the probabilities in this case were evaluated by adding the vibrational free energy contribution to the energy of each configuration. The vibrational contribution at each temperature was obtained from the phonon frequencies calculated from the structure optimized at 0 K; that is, no explicit free energy minimization was performed. Test calculations confirmed that the ensemble truncation does not affect the predicted concentrations of each species, because at low vacancy concentrations the contribution of configurations with $n \geq 1$ to the partition function is negligible. Pressure effects were only included via the pressure dependence of the chemical potential in Equation 11.31, because zero external pressure was assumed in all DFT calculations.

The pressure-composition isotherms at 600 to 800 K are shown in Figure 11.10. The horizontal portions (labeled $\alpha + \beta$) of each isotherm correspond to an equilibrium mixture of α and β phases. The values of equilibrium pressure at each temperature were taken from experimental data (Stampfer et al. 1960). The β part of each isotherm was calculated from DFT results using the method described above. Higher concentrations of vacancies are predicted for lower hydrogen partial pressures and higher temperatures, as expected. The intersection of this β line with the horizontal part of the isotherm determines the maximum fraction of vacancies in the hydride at that particular temperature, and this fraction is predicted to be very small, varying between 10^{-8} and 10^{-6} for temperatures in the range 600 to 800 K. This result agrees with experimental observations by Stampfer et al. (1960), who found that the hydride in equilibrium with the α phase has an H/Mg ratio of 1.99 ± 0.01 within the investigated range of temperatures (713 to 833 K).

The predicted concentrations of each vacancy species at the transition point (Table 11.1) clearly reveal that the relative abundance of the different species is not well correlated with the VFEs. Di-vacancies, for example, although energetically favorable with respect to isolated vacancies, are less abundant under these conditions by one or two orders of magnitude. Tri-vacancy clusters are even rarer, despite the fact that their formation energies are relatively low. In the context of a canonical formulation, with a fixed number of vacancies in a much larger supercell, this effect would arise from the much higher degeneracy of configurations with isolated vacancies compared to those with clusters. Equivalently, in the grand-canonical treatment, with much smaller cells, this is a consequence of introducing the H_2 chemical potential in the energy balance: weakly negative values of μ stabilize configurations with smaller values of n, as seen in the inset of Figure 11.9.

The results of this study suggested that the slow kinetics of hydrogen diffusion in pure magnesium hydride, which is one of the factors limiting its practical applications in hydrogen storage, is

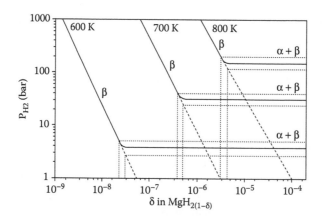

FIGURE 11.10 Pressure-composition isotherms for MgH_2 at pressures near and above the β-α transformation point. The dashed lines represent the hypothetical extensions of the isotherms in the absence of a phase transition. The dotted lines represent the error margins of the experimental determination of the transition pressure (From Stampfer, J. F., Holley, C. E., and Suttle, J. F., *J. Am. Chem. Soc.*, 82: 3504, 1960) and their propagation to the theoretically predicted concentrations.

TABLE 11.1

Vacancy Formation Energies (VFE) and Predicted Fractions of Vacancies at the Phase Transition Point for Each Vacancy Species in the Hydride (VnH Represents the Most Stable Cluster of n Vacancies)

Species	VFE* (eV/ Vacancy)	Fraction of Vacancies at Transition Pressure		
		600 K	700 K	800 K
V_{2H}	1.41	3×10^{-8}	5×10^{-7}	4×10^{-6}
V_{2H}	1.04	7×10^{-10}	3×10^{-8}	4×10^{-7}
V_{3H}	1.07	1×10^{-14}	5×10^{-12}	2×10^{-10}

related to the low concentration of vacancies at the conditions of hydrogen release. More details can be found in the work by Grau-Crespo et al. (2009).

11.7.4 SITE DISORDER AND ITS EFFECTS ON STRUCTURAL AND MAGNETIC TRANSITIONS

Relaxor ferroelectrics form a class of ferroelectrics in which the nature of site-disorder between heterovalent cations is believed crucial to their spectacular dielectric response, which is of great importance to technological applications. For example, short-range order or disorder in scandium and niobium ions at the B site of pervoskite structure of $Pb(ScNb)_{0.5}O_3$ or that in Mg and Nb in $Pb_3(MgNb_2)O_9$ is associated with their glassy dielectric response that is an order magnitude stronger than that of conventional ferroelectric materials, such as $BaTiO_3$. In conventional ferroelectrics, an inversion symmetry breaking structural transition essentially drives a large dielectric response and anomaly as a function of temperature. Thus, simulations aimed at understanding these interesting properties of relaxor ferroelectrics necessarily require (1) a realistic representation of disorder or short-range cationic order and (2) structural degrees of freedom that cannot be integrated out merely by minimization of energy through local distortions (in the language developed here, vibrational free energy cannot be obtained within the harmonic approximation as discussed in Section 11.4) and need to be included on the same footing as the chemical site-disorder (Burton et al. 2006 and references therein). A computationally viable strategy then involves use of (1) a simplified Ising cluster expansion (type I method) to capture the effects of disorder or the short-range order on inter-atomic force constants and local fields (type II methods) and (2) using Monte Carlo or molecular dynamic simulations to determine temperature-dependent structural and vibrational properties and a possible structural phase transition. Only recently, a scheme has been developed that essentially determines a free energy function relevant to a structural phase transition (Kumar and Waghmare 2010) resulting in a generalized Landau theory derived from first-principles.

Another problem of current importance is the magnetism in transition metal-doped semiconductors (e.g., Co, Mn–doped ZnO). Magnetism in $Zn_{1-x}TM_xO$, where TM is a transition metal, has been a rather controversial materials issue, because different methods or conditions of synthesis yield different results for magnetism in it. Indeed, disorder or short-range ordering of TM (also known as clustering on the cation lattice in this field) is known to be one of the important reasons for the synthesis-dependent properties of dilute magnetic semiconductors. In simulations and design of such materials (Sluiter et al. 2005), one uses model II for representation of disorder with a large supercell and Ising cluster expansion approach to generate one configuration (or many configurations, if affordable) of realization of disorder in model II. For each of these configurations, the spin Hamiltonian with distance-dependent and site-dependent exchange coupling derived from spin-dependent density functional theory calculations is simulated using Monte Carlo simulations as a function of temperature. Although the predictions of such modeling and simulations agree only qualitatively with experiments, they are quite useful in understanding trends and cost-effective screening of materials for design by experiment.

11.8 CONCLUSIONS

We have demonstrated how ionic or atomic disorder on a lattice is omnipresent and important to the structure and properties of materials through some recent examples ranging from catalysis, to energy storage, to multifunctional smart materials. We have reviewed strategies for modeling and simulations to capture the physics and chemistry of disorder in influencing various properties of materials. These typically involve access to configurational information at different levels, such as electronic properties and atomic displacements, and have varied computational cost. Although Ising cluster expansions have been used extensively in determination of phase diagrams of alloys and similar problems, we have emphasized here the methods that attempt essentially an exact statistical thermodynamic analysis using the SOD technique with a relatively smaller system, but having access to as much information and properties as possible. Such an approach is becoming quite practical in understanding and design of disordered materials, thanks to advances in computers and algorithms. There are many other interesting current problems (e.g., gas adsorption and sensors, B- and C-doped graphene) in which these methodologies can be quite effective. Needless to say, one usually has to use a combination of the techniques presented here (and Ising cluster expansion) to be able to solve challenging materials problems.

11.9 ACKNOWLEDGMENTS

We are grateful to our coauthors in the case studies described here: Dr. Said Hamad, Professor Nora de Leeuw, Professor Richard Catlow, Ms. Asmaa Ai-Baitai, Mr. Kyle Smith, Professor Tim Fisher, Dr. B. P. Burton, Dr. E. J. Cockayne, Professor Marcel Sluiter, and Professor Y. Kawazoe. Umesh V. Waghmare thanks the Asian Office of Aerospace R&D (AOARD) for its support through grant numbers FA2386-10-1-4150 and FA2386-10-1-4062.

REFERENCES

Becker, U., Fernandez-Gonzalez, A., Prieto, M., Harrison, R. and Putnis, A. 2000. Direct calculation of thermodynamic properties of the barite/celestite solid solution from molecular principles. *Phys. Chem. Miner.* 27: 291.

Bellaiche, L. and Vanderbilt, D. 2000. Virtual crystal approximation revisited: Application to dielectric and piezoelectric properties of perovskites. *Phys. Rev. B.* 61: 7877.

Benny, S., Grau-Crespo, R., and de Leeuw, N. H. 2009. A theoretical investigation of α-Fe_2O_3–Cr_2O_3 solid solutions. *Phys. Chem. Chem. Phys.* 11: 808.

Braun, P. B. 1952. A superstructure in spinels. *Nature* 170: 1123.

Burton, B. P., Tinte, S., Cockayne, E. J., and Waghmare U. V. 2006. First-principles based simulations of relaxor ferroelectrics. *Phase Trans.* 79: 91.

Cabanas, A., Darr, J. A., Lester, E. and Poliakoff, M. 2000. A continuous and clean one-step synthesis of nanoparticulate $Ce_{1-x}Zr_xO_2$ solid solutions in near-critical water. *Chem. Commun.* 901.

Cabanas, A., Darr, J. A., Lester, E. and Poliakoff, M. 2001. Continuous hydrothermal synthesis of inorganic materials in a near-critical water flow reactor; the one-step synthesis of nano-particulate $Ce_{1-x}Zr_xO_2$ ($x = 0 - 1$) solid solutions. *J. Mater. Chem.* 11: 561.

Catlow, C. R. A. 1997. *Computer Modelling in Inorganic Crystallography.* Academic Press Limited, London.

Dick, B. G., and Overhauser, A.W. 1958. Theory of the dielectric constant of alkali halide crystals. *Phys. Rev.* 112: 90.

Dove, M. T. 1993. *Introduction to Lattice Dynamics.* Cambridge University Press, Cambridge.

Dronskowski, R. 2001. The little maghemite story: A classic functional material. *Adv. Func. Mater.* 11: 27.

Du, Y., Yashima, M., Koura, T., Kakihana, M. and Yoshimura, M. 1994. Thermodynamic evaluation of the ZrO_2–CeO_2 system. *Scripta Metall. Mater.* 31: 327.

Friedlmeier, G., and Groll, M. 1997. Experimental analysis and modelling of the hydriding kinetics of Ni-doped and pure Mg. *J. Alloys Comp.* 550: 253.

Fukai, Y. 2005. *The Metal–Hydrogen System,* 2nd ed. Springer, Berlin.

Gale, J. D. 1997. GULP: A computer program for the symmetry-adapted simulation of solids. *J. Chem. Soc. Faraday Trans.* 93: 629.

Gale, J. D. 2005. GULP: Capabilities and prospects. *Z. Kristallogr.* 220: 552.

Gale, J. D., and Rohl, A. L. 2003. The General Utility Lattice Program (GULP). *Mol. Sim.* 29: 291.

Garvie, R. C. 1970. Oxides of Rare Earths, Titanium, Zirconium, Hafnium, Niobium, and Tantalum. In *High temperature oxides. Part II.* Ed. M. A. Alper, Academic Press, San Diego, CA.

Grau-Crespo, R., et al. 2007. Symmetry-adapted configurational modelling of fractional site occupancy in solids. *J. Phys. Condens. Matter* 19: 256201.

Grau-Crespo, R., et al. 2009. Thermodynamics of hydrogen vacancies in MgH2 from first-principles calculations and grand-canonical statistical mechanics. *Phys. Rev. B* 80: 174117.

Grau-Crespo, R., et al. 2010. Vacancy ordering and electronic structure of γ-Fe2O3 (maghemite): A theoretical investigation. *J. Phys. Condens. Matter* 22: 255401.

Grau-Crespo, R., et al. 2011. Phase separation and surface segregation in ceria-zirconia solid solutions. *Proc. Royal Soc. A Math. Phys. Eng Sci.* DOI:10.1098/rspa.2010.0512

Haabgood, M., Grau-Crespo, R. and Price, S. L. 2011. Substitutional and orientational disorder in organic crystals: a symmetry-adapted ensemble model. *Phys. Chem. Chem. Phys.* 13: 9590.

Johnson, S. R., et al. 2005. Chemical activation of MgH_2: A new route to superior hydrogen storage materials. *Chem. Comm.* 2823.

Jorgensen, J. E., Mosegaard, L., Thomsen, L. E., Jensen, T. R. and Hanson, J. C. 2007. Formation of gamma-Fe_2O_3 nanoparticles and vacancy ordering: An in situ X-ray powder diffraction study. *J. Solid State Chem.* 180: 180.

Kresse, G., and Furthmuller, J. 1996a. Efficiency of ab-initio total energy calculations for metals and semiconductors using a plane-wave basis set. *Comp. Mater. Sci.* 6: 15.

Kresse, G., and Furthmuller, J. 1996b. Efficient iterative schemes for ab initio total-energy calculations using a plane-wave basis set. *Phys. Rev. B* 54: 11169.

Kumar, A., and Waghmare, U. V. 2010. First-principles free energies and Ginzburg-Landau theory of domains and ferroelectric phase transitions in $BaTiO_3$. *Phys. Rev. B* 82: 05411.

Landau, D. P., and Binder, K. 2009. *A Guide to Monte-Carlo Simulations in Statistical Physics.* University Press, Cambridge.

Lee, T. A., Stanek, C. R., McClellan, K. J., Mitchell, J. N. and Navrotsky, A. 2008. Enthalpy of formation of the cubic fluorite phase in the ceria–zirconia system. *J. Mater. Res.* 23: 1105.

Levy, M., Wilhelm, C., Siaugue, J. M., Horner, O., Bacri, J. C. and Gazeau, F. 2008. Magnetically induced hyperthermia: size-dependent heating power of gamma-Fe_2O_3 nanoparticles. *J. Phys. - Condens. Mat.* 20: 204133.

Lewis, G. V., and Catlow, C. R. A. 1985. Potential models for ionic oxides. *J. Phys. C Solid State Phys.* 18: 1149.

Metropolis, N., Rosenbluth, A. W., Rosenbluth, M. N., Teller, A. H. and Teller, E. 1953. Equation of state calculations by fast computing machines. *J. Chem. Phys.* 21: 1087.

Navrotsky, A., Benoist, L., and Lefebvre, H. 2005. Direct calorimetric measurement of enthalpies of phase transitions at 2000 degrees–2400 degrees C in yttria and zirconia. *J. Am. Ceramic Soc.* 88: 2942.

Pankhurst, Q. A., Connolly, J., Jones, S. K. and Dobson, J. 2003. Applications of magnetic nanoparticles in biomedicine. *J. Phys. D.-Appl. Phys.* 36: R167.

Prieto, M. 2009. Thermodynamics of solid solution-aqueous solution systems. *Thermodynam. Kinet. Water-Rock Interact.* 70: 47.

Rudman, P. S. 1979. Hydrogen-diffusion-rate-limited hydriding and dehydriding kinetics. *J. Appl. Phys.* 50: 7195.

Ruiz-Hernandez, S. E., et al. 2010. Thermochemistry of strontium incorporation in aragonite from atomistic simulations. *Geochim. Cosmochim. Acta* 74: 1320.

Sastri, M. V. C., Viswanathan, B., and Achyuthlal Babu, R. S. 1998. *Kinetics of Metal Hydride Formation and Decomposition in Metal Hydrides: Fundamentals and Applications*. Ed. M.V.C. Sastri. Springer-Verlag, New York.

Shannon, R. D. 1976. Revised effective ionic radii and systematic studies of interatomic distances in halides and chalcogenides. *Acta Crystallogr.* A32: 751.

Shmakov, A. N., Kryukova, G. N., Tsybulya, S. V., Chuvilin, A. L. and Solovyeva, L. P. 1995. Vacancy ordering in gamma-Fe_2O_3 - synchrotron X-ray-powder diffraction and high-resolution electron-microscopy studies. *J. Appl. Crystallogr.* 28: 141.

Sluiter, M. H. F., Kawazoe, Y., Sharma, P., Inoue, A., Raju, A. R., Rout, C., and Waghmare, U. V. 2005. First-principles based design and experimental evidence for a ZnO-based ferromagnet at room temperature, *Phys. Rev. Lett.* 94: 187204.

Stampfer, J. F., Holley, C. E., and Suttle, J. F. 1960. The magnesium hydride system. *J. Am. Chem. Soc.* 82: 3504.

Todorov, I. T., Allan, N. L., Lavrentiev, M. Y., Freeman, C. L., Mohn, C. E. and Purton, J. A. 2004. Simulation of mineral solid solutions at zero and high pressure using lattice statics, lattice dynamics and Monte Carlo methods. *J. Phys. Condens. Matter* 16: S2751.

Tripathi, S., Kumar, A., Waghmare, U. V., and Pandey, D. 2010. Effect of $NaNbO_3$ substitution on the quantum paraelectric behavior of $CaTiO_3$, *Phys. Rev. B* 81: 212101.

van de Walle, A., and Ceder, G. 2002. The effect of lattice vibrations on substitutional alloy thermodynamics. *Rev. Mod. Phys.* 74: 11.

Vigeholm, B., Kjoller, J., Larsen, B. and Pedersen, A. S. 1983. Formation and decomposition of magnesium hydride. *J. Less Common Metals* 89: 135.

Waychunas, G. A. 1991. Crystal chemistry of oxides and oxyhydroxides. *Rev. Mineral. Geochem.* 25: 11.

Zunger, A., Wei, S. H., Ferreira, L. G. and Bernard, J. E. 1990. Special quasirandom structures. *Phys. Rev. Lett.* 65: 353.

12 Design of Compound Semiconductor Alloys Using Molecular Simulations

Jhumpa Adhikari

CONTENTS

12.1 INTRODUCTION

Compound semiconductors are composed of elements of either group III and group V (known as III-V semiconductors) or group II and group VI (known as II-VI semiconductors) as compared to elemental semiconductors which belong to group IV. Examples of III-V semiconductors are (B, Al, Ga, In) – (N, P, As, Sb) and their mixtures, whereas (Zn, Cd, Hg) – (O, S, Se, Te) and their mixtures are examples of II-VI semiconductors. Compound semiconductors are substitutional solid solutions, generally of the zinc blende (or wurtzite) crystal structure. The zinc blende (or wurtzite) structure consists of two inter-connected face centred cubic, fcc, (or hexagonal close packed, hcp) sublattices; one of which is occupied by group III (or group II) elements and the other by group V (or group VI) elements. These alloys have found use in or have the potential to be used in the fabrication of novel opto-electronic devices. Wu in a 2009 review article has stated that the market of opto-electronic devices has reached $20 billion a year and is expected to expand further (Wu 2009).

The many semiconductor device applications are due to the nano-structured semiconducting heterostructures present in these alloys, which account for novel electronic and optical properties. Applications of group III – nitrides are in multi-colour light emitting devices (LEDs), lasers, photoelectrodes, high electron mobility transistors (HEMT; also known as heterostructure field effect transistors, HFET) (Wu 2009). For example, InGaN forms the active layer in high brightness blue LEDs (Bellotti et al. 2007). AlN, GaN and InN are used in the manufacture of short wavelength electro-luminescent devices, such as LEDs and laser diodes (LDs) in the visible and ultra-violet region (Benkabou et al. 2000). BN is potentially useful due to its extreme hardness and high thermal conductivity (Moon et al. 2003).

InGaAs forms the active region for infra-red lasers used in spectroscopy and medicine; as also in optical fibre applications for the telecommunications industry (Titantah et al. 2007, 2008). GaAs thin films are widely used in photonics, microelectronics and spinotronics. GaAs based devices include LEDs, lasers, infrared detectors, solar cells, cellular phones and wireless communication devices, direct broadcasting systems, global positioning systems, fibre optic drivers and receivers, collision avoidance radars (Murdick et al. 2005).

Polycrystalline CdS/CdSe thin films have been used in the manufacture of solar cells, which have achieved efficiencies over 15% (Britt and Ferekides 1993). ZnTe/ZnSe have been used to fabricate room temperature operating blue and blue-green LDs (Itoh et al. 2000). CdSe quantum dots are used as the active component in devices such as single-photon emitters, LEDs, photovoltaic (PV) cells and lasers (Hendry et al. 2006).

This chapter deals with molecular simulation tools to determine the structural and thermodynamic properties of compound semiconductor alloys. Macroscopic phenomena (such as melting, growth of alloys, defect formation, miscibility) can be understood in terms of the atomistic mechanism using molecular simulation techniques. Experimentation has provided us with knowledge of many of the properties of these alloys. However, the cost of development of these alloys into marketable devices using only experimental research is prohibitively high. Using computer simulations in conjunction with experiments lowers the costs involved in designing these alloys as simulations can be used to reduce the alternatives to the point where only the useful alloys can be subjected to experiments. The best method to model materials is to use *ab initio* first principles simulations, by solving quantum mechanical equations of motions. However, the *ab initio* methods are computationally expensive. This has led to development of empirical, environment dependent potential models as discussed in the following section. Classical molecular simulation methods, both molecular dynamics and Monte Carlo, have been used in conjunction with the empirical potential models to characterize these novel alloys. These simulations and their results have been discussed in subsequent sections.

12.2　POTENTIAL MODELS

All matter consists of atoms which interact with each other to result in observable physical phenomenon. The classical interatomic potential models described below approximate the complex interparticle interactions present in these alloys. Compound semiconductors are predominantly covalent solids with some ionic characteristics. Covalent solids are tetrahedrally bonded open structures. Extensive research has been conducted to develop empirical environment dependent interatomic potential models for elemental semiconductors such as silicon as also for compound semiconductors. For silicon, the available potential models include the Stillinger-Webber potential (Stillinger and Weber 1985), Biswas- Haman potential (Biswas and Hamann 1986), PTHT (Pearson, Takai, Halicogulu, and Tiller) potential (Pearson et al. 1984) and Tersoff potential models (Balamane et al. 1992). The Tersoff model has been used to describe interatomic interactions in silicon (Tersoff 1986; Tersoff 1988a), carbon (Tersoff 1988b) and germanium; and their mixtures with silicon, e.g. SiGe and SiC (Tersoff 1989). The parameters of the Tersoff model have been determined for compound semiconductor alloys by other authors (Powell et al. 2007). The other extensively used potential model describing the interatomic interactions in these alloys is the Valence Force Field (VFF) model (Keating 1966). Both these models have been described in detail in the following sub-sections.

12.2.1　VFF Model

The VFF model had been first proposed for covalent solids by P. N. Keating in 1966 (Keating 1966). This model, hence, is also referred to as the Keating model. The VFF model gives a measure of the strain energy which arises when two compounds with different lattice constants are mixed. This strain energy is the sum of two-body bond stretching interactions and three-body bond bending interactions. For a cubic zinc blende crystal, the strain energy (as defined by Ho and Stringfellow in their seminal 1996 paper (Ho and Stringfellow 1996)) is as follows:-

$$E_m = \frac{3}{8}\sum_{i=1}^{4}\alpha_i \frac{(d_i^2 - d_{i0}^2)^2}{d_{i0}^2} + \frac{6}{8}\sum_{i=1}^{4}\sum_{j=i+1}^{4}\frac{\beta_i + \beta_j}{2}\left(\frac{\left(\overrightarrow{d_i d_j} + d_{i0}d_{j0}/3\right)^2}{d_{i0}d_{j0}}\right) \tag{12.1}$$

where d_i is the distance between the central atom and a corner atom in the tetrahedron (as shown in Figure 12.1), d_{i0} is the equilibrium bond length in the binary compound, α_i is the bond stretching force constant and β_i is the bond bending force constant.

The two parameters in the potential model, α and β, for compound semiconductors alloys were determined by Martin in 1970 indirectly from experiments at room temperature and are shown in Table 12.1 (Martin 1970).

Modifications introduced by Takayama and co-workers (Takayama et al. 2000, 2001a, 2001b) to model the wurtzite crystal structure are included in Equation 12.2 which measures the local potential energy centred around the j atom.

$$U_j = \sum_{i=1}^{3}\frac{3\alpha_{ij}}{8d_{ij}^{e2}}\left(r_{ji}^2 - d_{ji}^e\right)^2 + \frac{3\alpha_{i4}}{8d_{i4}^{c\,2}}\left(r_{j4}^2 - d_{j4}^c\right)^2$$

$$+2\sum_{i=1}^{3}\left\{\sum_{k>i}^{3}\frac{3\beta_{ijk}}{8d_{ji}^e d_{jk}^e}\left(r_{ji}r_{jk} - d_{ji}^e d_{jk}^e \cos\theta_{ijk}^e\right)^2\right\} \tag{12.2}$$

$$+2\sum_{i=1}^{3}\left\{\frac{3\beta_{ij4}}{8d_{ji}^e d_{j4}^c}\left(r_{ji}r_{j4} - d_{ji}^e d_{j4}^c \cos\theta_{ij4}^c\right)^2\right\}$$

In Equation 12.2, the angles between the bonds are not perfect $109.47°$ but the two types of angles in the wurtzite structure are the angle around the j atom involving neighbours at different z-coordinate values, θ_{ij4}^c, and the other angle is with neighbours having the same z-coordinates, θ_{ijk}^e. The distance, d_{j4}^c, is the bond length between the atoms with the same x- and y- coordinates and the equatorial distance, d_{ji}^e, is the bond length between the atoms with different x- or y- coordinates or both.

The VFF model has generally been used as the molecular model to study the solution thermodynamics of ternary and quaternary alloys in compound semiconductor alloys. Energy minimization techniques have been used to determine the interaction parameter in the regular solution theory and then, the binodal and spinodal curves have been calculated. Kim et al., have

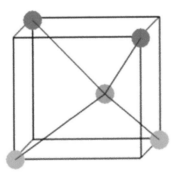

FIGURE 12.1   Zinc Blende Tetrahedron.

TABLE 12.1

VFF Model Parameters (Martin 1970; Takayama et al. 2001a)

System	a_0(Å)	α (N/m)	β (N/m)
InP	5.868	43.04	6.24
GaP	5.451	47.32	10.44
AlP	5.466	47.29	9.08
InSb	6.47	29.61	4.77
GaSb	6.09	33.16	7.22
AlSb	6.1355	35.35	6.77
InAs	6.0584	35.18	5.50
GaAs	5.653	41.19	8.95
AlAs	5.6605	43.05	9.86
InN	4.98	63.58	8.049
GaN	4.52	81.09	12.16
AlN	4.38	86.53	14.30
ZnS	5.41	44.92	4.78
ZnSe	5.67	35.24	4.23
ZnTe	6.10	31.35	4.45
CdSe	6.08	31.20	3.10
CdTe	6.48	29.02	2.43

Source: From Martin R. M., *Phys. Rev. B*, 1: 4005, 1970; Takayama, T., et al. *J. Cryst. Growth*, 222: 29, 2001a.

generalized the VFF model (Equation 12.1) by using cosine of the ideal bond angle and this model is generally referred to as the KVFF model (Kim et al. 2002). Further modifications have been introduced to the KVFF model by Biswas and co-workers to predict the formation energies and the binodal and spinodal curves by taking into account the bond-length/bond-angle interactions (Biswas et al. 2008).

12.2.2 Tersoff Model

In the late 1980s, J. Tersoff of the IBM Thomas Watson Research Center developed a new empirical potential to model covalent systems such as silicon (Tersoff 1986, 1988a), germanium, carbon (Tersoff 1988b) and their mixtures (Tersoff 1989). The Tersoff model has a longer range than the VFF model and this range can be adjusted. Tersoff model potential function is based on the fact that the strength of the bond between neighbouring atoms depends on the local environment. Unlike the VFF model, which is limited to modelling crystals with tetrahedral structure allowing only small distortions, the Tersoff model allows for more significant deformations. However, the potential function was found by Tersoff to be not long-ranged enough to describe silicon melts successfully. The potential was found by Tersoff to predict the properties of silicon with various polymorphous forms with good accuracy and was also found to have transferability. This model was extended to III-V systems such as InGaAlAs by other authors by calculating the various fitting parameters. The growth of the InGaAs alloy has been studied by P.A. Ashu et al. (1995) using the molecular dynamics approach in 1995 and modifications to the potential were introduced by M. Nakamura et al. in 2000 (Nakamura et al. 2000). Nakamura et al. pointed out that the bonds in InGaAs alloy have some ionic character and introduced an additional Coulombic term. Tersoff applied his form of the potential function to SiC in the rocksalt (NaCl) structure, which has a considerable ionic character and observed the lattice constant and cohesive energy to be a reasonable approximation (slight underestimation) whereas for the cubic structure, the results were in excellent agreement as compared to experimental values (Tersoff 1989). The III-V alloys crystallize in the cubic zinc-blende structure and the hexagonal wurtzite structures, which have similar tetragonal nearest-neighbour arrangements. As Tersoff has noted, in the short-range of the potential, the difference in energy between the two structures is minor. Thus, in most studies, the cubic zinc-blende structure is considered and the improvement in accuracy provided by the introduction of the Coulombic term does not merit the increase in computational speed that the additional calculations will cause.

The functional form of the Tersoff potential model is motivated by the structural chemistry of covalent systems. The primary basis for this model is that the strength of a bond (i.e. bond order) depends on the local environment (i.e. coordination number). Most predominantly covalent systems have open structures, as an atom with fewer neighbours will form stronger bonds as compared to a close-packed structure where an atom has many neighbours. Tersoff states that the energy is modelled as a sum of pair-like interactions, where, however the coefficient of the attractive term in the pair-like potential (which plays the role of bond order) depends on the local environment, giving a many-body potential. The empirical interatomic potential function for multi-component systems as proposed by Tersoff is as shown below.

$$E = \sum_i E_i = \frac{1}{2}\sum_{i \neq j} V_{ij} \tag{12.3}$$

$$V_{ij} = f_C(r_{ij})[f_R(r_{ij}) + b_{ij}f_A(r_{ij})] \tag{12.4}$$

$$f_R(r_{ij}) = A_{ij}\exp(-\lambda_{ij}r_{ij}) \tag{12.5}$$

$$f_A(r_{ij}) = -B_{ij}\exp(-\mu_{ij}r_{ij}) \tag{12.6}$$

$$f_C(r_{ij}) = \begin{cases} 1.0, & r_{ij} < R_{ij}; \\ \frac{1}{2} + \frac{1}{2}\cos\left[\frac{\pi(r_{ij} - R_{ij})}{(S_{ij} - R_{ij})}\right], & R_{ij} < r_{ij} < S_{ij}; \\ 0.0, & r_{ij} > S_{ij} \end{cases} \tag{12.7}$$

$$b_{ij} = \left(1 + \beta_{ij}^{n_{ij}}\zeta_{ij}^{n_{ij}}\right)^{-1/2n_{ij}} \tag{12.8}$$

$$\zeta_{ij} = \sum_{k \neq i,j} f_C(r_{ij})g(\theta_{ijk}) \tag{12.9}$$

$$g(\theta_{ijk}) = 1 + \frac{c_i^2}{d_i^2} - \frac{c_i^2}{\left[d_i^2 + (h_i - \cos\theta_{ijk})^2\right]} \tag{12.10}$$

In Equation 12.3, E is the total energy of the system, E_i is the site energy and V_{ij} is the bond energy. The repulsive pair potential is represented by f_R and is defined as in Equation 12.5. f_A represents the attractive pair potential associated with bonding, which is defined in Equation 12.6. Equation 12.7 shows the smooth cut-off function, f_C. Generally, the cutoff is defined such that only the first shell

TABLE 12.2

Parameters of the Potential for the InGaAs System (Ashu et al. 1995)

Parameter	Units	As-As	Ga-Ga	In-In	Ga-As	In-As	In-Ga
A_{ij}	(eV)	1571.86	993.88	2975.54	2579.46	2246.55	1214.917
B_{ij}	(eV)	546.431	136.123	360.61	317.21	417.665	177.22
λ_{ij}	(Å^{-1})	2.38413	2.50842	2.6159	2.82805	2.53034	2.5621
μ_{ij}	(Å^{-1})	1.72872	1.49082	1.68117	1.72303	1.67123	1.58600
R_{ij}	(Å)	3.4	3.4	3.4	3.4	3.4	3.4
S_{ij}	(Å)	3.6	3.6	3.6	3.6	3.6	3.6
β_{ij}		0.007488	0.23586	2.10871	0.35719	0.38712	0.70524
n_{ij}		0.60879	3.47290	3.40223	6.31747	6.33190	3.43739
c_i		5.27313	0.076297	0.084215	1.22630	1.30678	0.080256
d_i		0.75102	19.7964	19.2626	0.79040	91.4432	195.2950
h_i		0.15292	7.14591	7.39228	−0.51848	−0.56983	7.26910

Source: From Ashu, P. A., et al. *J. Cryst. Growth*, 150: 176, 1995.

of nearest neighbours is included. The bond order parameter b_{ij} is defined as per Equation 12.8 and describes the effect of neighbouring atoms (represented by index k) on the bond-formation energy. Equation 12.9 counts the number of bonds to atom i except the ij bond and gives the effective coordination number ζ_{ij}. The parameter n_{ij} quantifies the measure by which the closer neighbours are preferred over more distant ones in the competition to form bonds. The bond angle between bonds ij and ik is given by θ_{ijk} and the interatomic distance by r_{ij}. Here, A_{ij}, B_{ij}, λ_{ij}, μ_{ij}, R_{ij}, S_{ij}, β_{ij}, n_{ij}, c_i, d_i and h_i are fitting parameters defined in Table 12.2 as taken from Ashu et al.(1995).

12.3   MOLECULAR SIMULATION METHODS FOR DESIGN OF SEMICONDUCTING ALLOYS

The simulation methodology first involves the choice of a suitable empirical interatomic potential model to study the alloy system, which could be either of the potentials discussed above. Having selected the interatomic potential, the simulation methodology for prediction of the structural and thermophysical properties is discussed in detail below. Knowledge of the properties predicted by molecular simulations will enable design of alloys for specific applications and require experiments only for particular compositions. The use of molecular simulations, thus, reduces the cost of development of these alloys into marketable devices. *A priori* knowledge of the predicted lattice constants and bond lengths (and their changes with temperature and composition) help in the selection of substrate with the least lattice mismatch, to be used for the growth of alloys with minimal defects. Energy gap also varies with composition and temperature, and, thus, the use of the device fabricated will also change. Many of the devices use heterostructures so that properties such as thermal expansivity are important. Elastic constants, specific heats, cohesive energies, radial distribution functions (structural information) and other simulation predicted properties that can be matched against experimental data, where available, are used for validation of the selected potential (with its parameter values) and the simulation methodology used. Once validated, molecular simulations can be used to characterize new alloys and predict properties of existing alloys at conditions for which literature data is not available.

For the ternary (pseudo-binary) alloys, the energy band gap of the ternary alloy range between those of the two binary alloys. By mixing the two binary compounds, the properties of the resulting ternary alloy can be tuned to intermediate values. While the energy band gap is tuned with composition, the lattice constant also changes, causing a lattice mismatch with the substrate on which the alloy is grown and the quality of the crystal grown may suffer. Prediction of properties using molecular simulations will allow the selection of the substrate with matching lattice constant or one with the least mismatch reducing the need for experimentation with the material itself and hence, lowering costs. In the InGaN alloy, for example, the wide band gap and the ability to operate at room temperature, permits the alloy to be the paramount detector material for monitoring and imaging applications in the telecommunications industry. The validation of the method is generally done by comparison of results predicted by molecular simulation and experimental results where available. There are many properties and phenomena involving compound semiconductor alloys, which are difficult to obtain from experiments. An example of such a case is the microphase separation that influences the important optoelectronic properties of these alloys. The optoelectronic properties, which allow the alloys to be used in the fabrication of many devices, are affected greatly even by the occurrence of small quantities of component 1-rich or component 2-rich domains in the ternary alloys. An understanding of the microstructure of these alloys will help in the design of materials customized for a specific application.

As noted previously, under certain conditions, these alloys show a tendency to phase segregate. This is difficult to quantify experimentally and conducting experiments to determine the miscibility characteristics can also be exorbitantly expensive. The existence of even minor microphases can have a disproportionate effect on the opto-electronic properties of these semiconducting alloys. The miscibility diagrams (temperature versus composition) have been predicted theoretically using the energy minimization techniques in conjunction with a molecular model (generally, the VFF model) and assuming a mixing model (generally, the regular solution model). However, the advantage to using molecular simulations is that the assumption of the mixing model can be avoided and hence, a more accurate picture of the mixing effects is expected within the confines of the molecular model assumed. The simulation predicted miscibility diagram can provide guidelines for the selection of operating conditions for the Metal-Organic Chemical Vapour Deposition (MOCVD) reactors to enable the synthesis of the alloy of desired equilibrium composition. And, knowing the operating conditions can limit the experiments to alloys with the potential to be used in the manufacture of practical devices.

The synthesis and growth of the compound semiconductors alloys have also been modelled using molecular dynamics simulations. Molecular dynamics techniques have been performed to study the formation and propagation of defects in these semiconducting alloys.

Discussion on the use of molecular dynamics and Monte Carlo methods is given in detail in the sub-sections that follow.

12.3.1 MOLECULAR DYNAMICS

Classical molecular dynamics (MD) simulations (Allen and Tildesley 1987; Frenkel and Smit 2002) have been used extensively with the empirical Tersoff model to predict structural, dynamic and thermodynamic properties of elemental as well as compound semiconductors. This method involves the integration of Newton's equations of motion and has been used in the canonical (N, V, T) ensemble, where the number of atoms, N, simulation cell volume, V, and the temperature, T, are held constant. Tersoff used MD with his proposed potential model to calculate the cohesive energies, elastic constants, defect formation energies, lattice constants, bulk modulus values, nearest neighbour bond lengths and radial distribution functions for the diamond and graphite forms of carbon, amorphous carbon, silicon, germanium, and SiGeC alloys (Tersoff 1986, 1988a, 1988b, 1989). The simulation results were found to be in excellent agreement with experimental data and it was concluded that the then newly developed empirical Tersoff potential could accurately describe solid phase elemental semiconductors and their alloys.

The parameters in the Tersoff model have been calculated for the compound semiconductor alloys by fitting to a database of experimental and theoretical values available for cohesive energies, lattice parameters, bond lengths, elastic constants, etc. (Ashu et al. 1995; Powell et al. 2007). Then, MD simulations have been used to predict properties including cohesive energies, elastic constants, bulk moduli, equilibrium lattice parameters, bond lengths, thermal expansion coefficients, isochoric heat capacities, radial distribution functions, mean square displacements, diffusivity, Debye temperatures and melting points for a variety of binary alloys. In 1998, Benkabou et al., conducted MD simulations for the zinc blende GaN alloy in the canonical ensemble with a system size of 216 atoms ($3 \times 3 \times 3$ unit cells) and a time step of 1.86 fs in the temperature range from 300 K to 900 K (Benkabou et al. 1998). Benkabou et al., also performed MD studies for BAs (Benkabou et al. 1999); and GaN, AlN and InN (Benkabou et al. 2000). Kanoun and co-workers conducted MD to predict the properties of cubic CdTe and ZnTe in the canonical ensemble with 216 atoms at a fixed density of 5.86 gm/cc and a time step of 4.74 fs (Kanoun et al. 2000, 2003). El-Mellhouhi et al., calculated the properties of BP in the zinc blende crystal structure using MD (El-Mellouhi et al. 2002). Moon and Hwang investigated the reliability of the Tersoff model to predict the physical properties of GaN system by conducting MD in the canonical ensemble with a time step of 2.1 fs in a 216 atom system(Moon and Hwang 2003). Moon and co-workers also predicted the structural properties of cubic BN using MD as in previous study, however, with a time step of 0.2 fs (Moon et al. 2003). The properties of cubic AlN have been calculated by Goumri-Said and co-workers using the MD technique with a time step of 1.37 fs (Souraya Goumri-Said et al. 2004).

The summary of the structural, thermodynamic and transport properties determined by various authors (as discussed above) has been shown in Tables 12.3, 12.4 and 12.5, respectively. The lattice constant values calculated by different authors for GaN have been observed to be in good agreement. The same has been found to be true for the AlN alloy also. However, the room temperature linear thermal expansion coefficient values for GaN vary widely. Unlike GaN, the expansivities for the AlN alloy, as determined by different studies, are in good agreement and the values for the other binary alloys are of similar order. The isochoric heat capacity, C_V, values show that these alloys follow the Dulong-Petit law for solids whereby at high temperatures, $C_v \rightarrow 3R$. It should also be noted that the diffusion coefficients for the binary nitrides have similar values at room temperature.

TABLE 12.3
Structural Properties as Predicted Using MD Simulations

System	Structural Property		References
	Lattice Constant (Å)	Linear Thermal Expansion Coefficient (K^{-1})	
GaN	$a(T) = 4.489 + 6.0 \times 10^{-5}T$	$\alpha(T = 300\,K) = 1.32 \times 10^{-5}$	Benkabou et al. 2000
	$a(T) = 3.609 + 2.0141 \times 10^{-5}T$ $+ 3.5182 \times 10^{-9}T^2$	$\alpha(T) = 5.174 \times 10^{-6}$ $+ 1.466 \times 10^{-9}T$	Moon, W.H. and H.J. Hwang, 2003
	$a(T = 300K) = 4.528$	$\alpha(T = 300K) = 5.613 \times 10^{-6}$	
	$a(T = 300K) = 4.538$	$\alpha(T = 300K) = 5.497 \times 10^{-6}$	Sun et al. 2005
	$a(T = 300K) = 4.54$	$\alpha(T = 300K) = 4.59 \times 10^{-5}$	Benkabou et al. 1998
AlN	$a(T) = 4.384 + 3.0 \times 10^{-5}T$	$\alpha(T = 300K) = 6.8 \times 10^{-6}$	Benkabou et al. 2000
	$a(T) = 4.3733 + 2.73 \times 10^{-5}T - 6.66 \times 10^{-11}T^2$ $a(T = 300K) = 4.379$	$\alpha(T = 300K) = 6.22 \times 10^{-6}$	Souraya Goumri-Said et al., 2004
BAs	$a(T) = 4.777 + 2.306 \times 10^{-5}T - 5.55 \times 10^{-9}T^2$ $a(T = 300K) = 4.777$	$\alpha(T = 300K) = 4.1 \times 10^{-6}$	Benkabou et al. 1999
BN	$a(T) = 3.609 + 2.0141 \times 10^{-5}T + 3.5182 \times 10^{-9}T^2$ $a(T = 300K) = 3.609$	$\alpha(T = 300K) = 6.166 \times 10^{-6}$	Moon et al. 2003
BP	$a(T = 300K) = 4.5536$	—	El-Mellouhi et al. 2002
CdTe	$a(T) = 6.497 + 3.469 \times 10^{-5}T + 7.0 \times 10^{-9}T^2$ $a(T = 300K) = 6.487$	$\alpha(T = 300K) = 5.34 \times 10^{-6}$	Kanoun et al. 2000
ZnTe	$a(T) = 6.1116 + 5.3 \times 10^{-5}T + 2.236 \times 10^{-8}T^2$ $a(T = 300K) = 6.1169$	$\alpha(T = 300K) = 8.7 \times 10^{-6}$	Kanoun et al. 2003

TABLE 12.4

Thermodynamic Properties as Predicted by MD Simulations

System	Isochoric Heat Capacity (J/mol K)	Debye Temperature (K)	Melting Point (K)	References
GaN	25.031	—	—	Benkabou et al. 2000
	25.856	823	2332 ± 300	Moon et al. 2003b
	25.156	—	—	Benkabou et al. 1998
AlN	25.031	—	—	Benkabou et al. 2000
	25.184	911.6	2387 ± 300	Goumri-Said et al. 2004
BAs	25.156	—	—	Benkabou et al. 1999
BN	25.108	1710	—	Moon et al. 2003a
BP	24.110	1362.49	3638.02 ± 300	El-Mellouhi et al. 2002
CdTe	25.357	—	—	Kanoun et al. 2000
ZnTe	25.58	—	—	Kanoun et al. 2003

TABLE 12.5

Diffusivity Values at Room Temperature (300 K) as Predicted by MD Simulations

System	Diffusivity (cm²/s)	References
GaN	0.414×10^{-8}	Benkabou et al. 2000
	0.66×10^{-8}	Benkabou et al. 1998
AlN	0.493×10^{-8}	Benkabou et al. 2000
InN	0.454×10^{-8}	Benkabou et al. 2000

MD has also been used to study the growth process of the compound semiconductor alloys as well as the formation and dynamics of defects. In 1995, Ashu and co-workers (Ashu et al. 1995) conducted MD simulations with up to 1700 atoms of (100) InGaAs/GaAs strained heterostructure using the Tersoff model to observe the detailed atomic structure at the interface and to measure the critical thickness for the formation of interface misfit dislocations. The authors also simulated the threading dislocation dynamics in the InGaAs overlayer. Nakamura et al. (2000) conducted MD simulations of molecular beam epitaxy (MBE) growth for the III-V InAs/GaAs system using a new potential model which added the Coulombic interaction term to the Tersoff model in a simulation cell of size $2.4232 \times 2.142 \times 2.142$ nm³. The substrate temperature has been fixed at 400° C and two atomic layers from the bottom are fixed to reproduce the rigidity of the substrate. The authors have used a time step of 4 fs and a maximum of 125,000 time steps have been considered in the study, which revealed that lattice mismatch causes severe distortion in the arrangement of atoms in the first layer during the initial growth stages of InAs on GaAs substrate. Nord and co-workers (Nord et al. 2002) conducted a MD study of the ion-irradiation induced amorphization process of silicon, germanium and GaAs, and found that the irradiated material contains high levels of defects. Gartner and his group (Gärtner 2006, 2010; Gärtner and Clauß 2010) have published a series of three papers on ion implantation damage in AlGaAs, where the velocity form of the Verlet algorithm has been used to integrate Newton's equations of motion with a maximum time step of 0.25 fs. The simulation cell has dimensions of $10 \times 10 \times 10$ unit cells with a total of 8000 atoms and a composition ranging between $x = 0$ and $x = 1$. In the first paper, Gartner has investigated the generation of point defects and calculated the critical energy (displacement energy) for the generation of one stable Frenkel pair. The second paper by the same author is a study of generation of different kinds of point defects in crystalline AlGaAs due to recoils in energy range of 50-400 eV (above displacement energies). Gartner and Clauß, in the most recent paper in the series, have conducted computer simulations of damage production in AlGaAs with 200 keV Ar at 20 K for a wide range of compositions. The modelling of the production process for cubic BN using a parallel version of Soermer-Verlet MD algorithm, has been conducted by Sibona and co-workers in 2003 (Sibona et al. 2003).

Sun et al. (2005) have used MD with a pair potential energy function known as the Buckingham potential model to predict the phase transition, lattice constant, thermal expansivity, isothermal bulk modulus and heat capacity for GaN in the isothermal isobaric (N, P, T) ensemble for a temperature range of 300 K to 3000 K and for pressures from 0 to 65 GPa. In the isothermal-isobaric ensemble, the number of atoms, N, pressure, P, and temperature, T, are held constant while allowing the volume, V, and internal energy, E, to vary.

12.3.2 Monte Carlo

Metropolis Monte Carlo (MC) simulations (Allen and Tildesley 1987; Frenkel and Smit 2002) have been used to predict the structural and thermodynamic properties of mixtures of elemental semiconductors as also compound semiconductors. MC simulations have been conducted using both the VFF and Tersoff potential models to describe the interatomic interactions. The structural properties determined include lattice constants, thermal expansion coefficients and bond lengths. The temperature versus composition miscibility diagram of ternary alloys at a given pressure, and the miscibility envelope for quaternary alloys at given temperature and pressure conditions have been determined using the transition matrix Monte Carlo (TMMC) method.

Kelires and his co-workers (Salvador et al. 2000; Tzoumanekas and Kelires 2002) have determined the lattice parameter and bond length variations with composition in silicon – germanium – carbon and silicon – germanium alloys, respectively, using MC methods in conjunction with the Tersoff potential. The authors used MC simulations as they were unable to attain equilibrium using MD due to extremely slow diffusion in the bulk phase of the alloys. The MC simulations have been conducted in the semigrand isothermal isobaric (N, P, T, ξ_2) ensemble where the total number of atoms, $N(N = N_1 + N_2)$, pressure, P, temperature, T, and fugacity coefficient of component 2, ξ_2, are held constant; while varying the internal energy, volume and composition (N_1, N_2) of the system. Simulations in this ensemble require identity change moves besides the atom displacement trials and volume change trials. The conclusions reached by the authors indicated that a fraction of carbon incorporated substitutionally into the silicon-germanium matrix contributes considerably to the change of lattice parameter and that the bowing effect is independent of the germanium and carbon contents. The nearest neighbour bond lengths have been found to vary with composition such that the values approach the Pauling limit, and also the variation depends on the type of bond. The analysis of next nearest neighbour bond lengths has been done analogously.

Adhikari and Kumar (2007) have validated the use of the Monte Carlo simulations with the Tersoff model for III-V binary alloys by comparing the simulation predicted values for lattice constants, thermal expansion coefficients and bond lengths with experimental data available for the GaAs binary alloy. It was found that good agreement exists between the experimental data and simulation results as seen in Table 12.6.

Hence, this method was then used to predict the properties for the InAs alloy, where literature data is not extensively available. An interesting experimental observation of these materials is that at low temperatures, the thermal expansion coefficients are negative as also decrease with increasing temperature. The simulations were able to capture this behaviour qualitatively, though unable to correctly reproduce the quantitative data, as seen in Figure 12.2a and 12.2b. The system size

---

TABLE 12.6

Comparison of property data obtained from simulation with literature data at 300 K. Number in parenthesis indicates error in the last decimal place

Property	Simulation	Literature
GaAs lattice constant (Å)	5.67137 (8)	5.65325 (2)
Bond length Ga-As (Å)	2.45890 (4)	2.44793
GaAs linear thermal expansion coefficient (K^{-1})	9.53E-6	5.73E-6
InAs lattice constant (Å)	6.074 (35)	6.058
Bond length In-As (Å)	2.63455 (4)	2.623
InAs linear thermal expansion coefficient (K^{-1})	8.31E-6	—

Source: From Adhikari, J. and Kumar, A. *Mol. Simul.* 33: 623, 2007. With permission.

---

considered in this study has been 64 atoms as larger system sizes investigated showed insignificant finite size effects.

For the ternary InGaAs alloy, a comparison has been done between the simulation results using the VFF model as the interatomic potential and those predicted by using the Tersoff model (Adhikari 2008). The simulations have been conducted in the semigrand isothermal isobaric ensemble as described by Kelires and his co-workers (Salvador et al. 2000; Tzoumanekas and Kelires 2002). However, the trial swap of positions of In and Ga atoms has been used in place of the identity change move used by the Kelires group, as the composition of the system at any given temperature and pressure conditions has been maintained constant. It was concluded that the VFF model describes accurately the room temperature structural properties as observed from experiments. However, the Tersoff model (and its parameters) simulation results follow the Vegard's Law for the lattice parameters and virtual crystal approximation for the bond lengths, which are different from the known experimental observations. The VFF model simulations, however, fail at temperatures above 300 K where an unphysical contraction of the crystal with increase in temperature has been predicted. The Tersoff model however is able to predict the variation of lattice parameters and bond lengths with increasing temperature correctly. It was concluded that the parameters of the Tersoff model need to be adjusted to improve the predictions at room temperature. The VFF model parameters are determined from experiments at room temperature and are not transferable to higher temperatures for structural property calculations.

The seminal work in the modelling of the miscibility behaviour of compound semiconductor alloys was done by authors Ho and Stringfellow (1996, 997). This work studied the solution thermodynamics of the InGaN alloy. Various authors have since refined the molecular model used by the authors (VFF model) and have developed the phase diagrams for other ternary and quaternary compound semiconductor alloys using the energy minimization approach of Ho and Stringfellow (Saito and Arakawa 1999; Takayama et al. 2000, 2001a, 2001b). Authors Adhikari and Kofke (2004a) first used the molecular simulation approach to predict the miscibility diagram and local composition using the Valence Force Field (VFF) model for the InGaN alloy. Structural properties, such as the lattice constant and bond lengths, have also been calculated by Adhikari and Kofke. The solution thermodynamics of the zinc blende structured InGaAlN alloys have also been studied using Monte Carlo simulations by the same authors (Adhikari and Kofke 2004b).

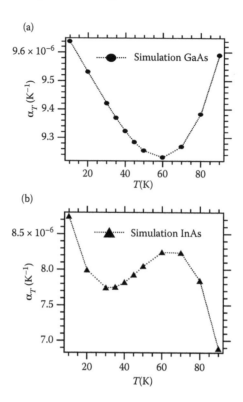

FIGURE 12.2 (a) Plot of linear thermal expansion coefficient of GaAs as a function of temperature in the low temperature range. Results shown are for N = 64 atoms system size. (b) Plot of linear thermal expansion coefficient of InAs as a function of temperature in the low temperature range. Results shown are for N = 64 atoms system size. (From Adhikari, J. and A. Kumar, *Mol. Sim.*, **33**(8): 623, 2007. Reproduced with permission from Taylor & Francis (http://www.informaworld.com).)

In the TMMC method, Monte Carlo (MC) simulations are conducted in the isothermal isobaric semigrand ensemble which involves atom displacement, volume change and identity change trials. Since, no atom insertion or deletion trials are involved, this method can be used to model solid solutions. Multicanonical weighting is done to promote a more uniform sampling of compositions. Each TMMC simulation covered the entire composition range from $x = 0$ to $x = 1$. The transition probability for every identity change move, i.e., $N_1 \rightarrow N_1 - 1$, $N_2 \rightarrow N_2 + 1$; or $N_1 \rightarrow N_1$, $N_2 \rightarrow N_2$; or $N_1 \rightarrow N_1 + 1$, $N_2 \rightarrow N_2 - 1$ (irrespective of its acceptance or rejection) during the simulation, has been noted and detailed balance has been used to calculate the state probabilities, $\Pi(S)$, and hence, the free energies have been determined as $\beta G = -ln\Pi(S)$. The free energy change of mixing ($\beta \Delta G_{mix} = \beta(G - xG_1 - (1 - x)G_2)$) as a function of composition, x, has been obtained (as shown in Figure 12.3(a) for the InGaAs system (Adhikari 2009)).

The coexistence compositions have been calculated from the change in free energy due to mixing curves by drawing common tangents and the miscibility diagram has been constructed (as shown in Figure 12.3b for the InGaAs system (Adhikari 2009)). The simulation predicted miscibility diagram for InGaN alloy indicated a higher upper critical solution temperature (UCST) of 1550 K than the 1505 K value calculated by Ho and Stringfellow's energy minimization method (Adhikari and Kofke 2004a). Also, the predicted phase diagram has been found to be asymmetric as there was no assumption of random mixing as is true in regular solution theory. The extension of this work to other group III (Al, Ga, In)-nitride ternary alloys showed that the UCST increased from 195 K for the AlGaN alloy to 2700 K for the InAlN alloy as the lattice mismatch between the binary alloys increased (Adhikari and Kofke 2004). The excess entropy and excess enthalpy values have

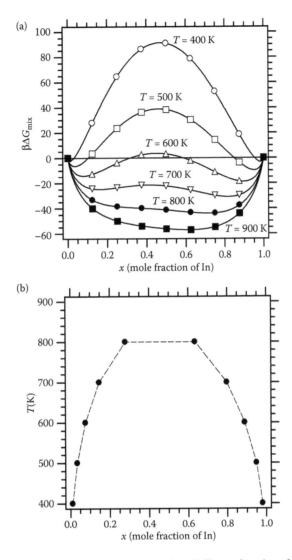

FIGURE 12.3 (a) Plot of the free energy of mixing (in units of kT) as a function of composition at different temperatures using the TMMC method in conjunction with the VFF model. The system size is N = 216 atoms. Error bars are smaller than the markers where the markers have been shown. (b) T-x diagram as predicted by the VFF model using the TMMC method. The UCST is approximately 850 K (not shown in the figure).The line is a guide to the eye. (From Adhikari, J, *Mol. Phys.*, **107**(16): 1641, 2009. Reproduced with permission from Taylor & Francis Ltd. (http://www.informaworld.com).)

been calculated and the excess entropy has been been found to be almost zero whereas the excess enthalpy was observed to be almost independent of temperature, and hence, the regular solution theory has been revealed as a good approximation for the mixing behaviour of the ternary group III-nitrides. The TMMC method has been used to develop the miscibility diagram for InGaAs using both the VFF model and the Tersoff model (Adhikari 2009). The Tersoff model is a more realistic description of the interatomic interactions in a predominantly covalent solid and the miscibility diagram predicted by this model is expected to give a more accurate picture of the mixing characteristics (Figures 12.4a and 12.4b (Adhikari 2009)). At high temperatures, the Tersoff model predicts the existence of a new phase (Figure 12.4b), which was not observed in the VFF phase diagram (Figure 12.4a). It has been found that the Tersoff model simulation predictions indicate a

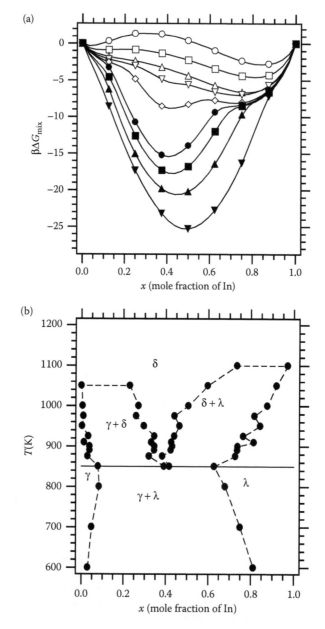

FIGURE 12.4   (a) Plot of the free energy of mixing (in units of kT) as a function of composition at different temperatures using the TMMC method in conjunction with the Tersoff model. The system size is N = 64 atoms. The temperature increases as the curves are observed from top to bottom of the figure. The open circle shaped markers represent the data at 600 K, open squares at 700 K, open triangles at 800 K, open inverted triangles at 850 K, open diamond markers at 900 K, shaded circles at 950 K, shaded squares at 1000 K, shaded triangles at 1050 K and shaded inverted triangles at 1100 K. Error bars are smaller than the markers where the markers have been shown. (b) T-x diagram as predicted by the Tersoff model using the TMMC method. The line is a guide to the eye. The "eutectic" temperature is approximately 850 K and is indicated as a solid line. The three single stable phases are indicated as GaAs–rich phase (γ), the InAs-rich phase (λ) and the third phase (δ). The three two-phase regions are marked as γ + λ, γ + δ and γ + λ. (From Adhikari, J, *Mol. Phys.*, **107**(16): 1641, 2009. Reproduced with permission from Taylor & Francis Ltd. (http://www.informaworld.com).)

complex phase diagram with three two-phase regions as compared to the VFF model which shows only one two-phase region. In Figure 12.4b, the "δ" phase is anticipated to be an amorphous region as the experimental melting point for the InGaAs alloy is known to be at higher temperatures. It was noted that the "eutectic" temperature predicted in Figure 12.4b is the approximately the same as the UCST observed in Figure 12.3b. It was concluded that experiments have to be conducted to determine if such a "δ" phase exists and also re-parameterization of the Tersoff model is necessary at higher temperatures, especially the values of the parameters, R and S, in the cut-off function as longer range interactions are expected at higher temperatures.

12.4 CONCLUSIONS

Compound semiconductors alloys are used in the fabrication of optoelectronic devices. The cost of development of these ternary and quaternary alloys into marketable devices using only experimental research is very high. Using computer simulations in conjunction with experiments lower the costs involved in researching these alloys as simulations can be used to reduce the alternatives to the point where only the useful alloys can be subjected to experiments. Classical molecular simulation methods, both MD and MC have been successfully used in combination with empirical potential energy functions, generally either the Tersoff model or the VFF model, to predict the thermophysical properties as also to model the growth processes involved in the MBE and MOCVD reactors. Molecular simulation methods, thus, find extensive use in the design and characterization of the material and in modelling the growth process, for these semiconducting alloys.

REFERENCES

Adhikari, J. 2008, Molecular simulation study of the structural properties in InxGa1-xAs alloys: Comparison between Valence Force Field and Tersoff potential models. *Comp. Mater. Sc.*, **43**(4): pp 616–622.
Adhikari, J. 2009, Miscibility of InxGa1-xAs alloys : a study using atomistic simulations. *Mol. Phys.*, **107**(16): pp. 1641–1648.
Adhikari, J. and D.A. Kofke 2004a, Molecular simulation study of miscibility in InxGa1-xN ternary alloys. *J. Appl. Phys.*, **95**(8): pp. 4500–4502.
Adhikari, J. and D.A. Kofke 2004b, Molecular simulation study of miscibility of ternary and quaternary InGaAlN alloys. *J. Appl. Phys.*, 2004. **95**(11): pp. 6129–6137.
Adhikari, J. and A. Kumar 2007, Study of structural and thermodynamic properties of GaAs and InAs using Monte Carlo simulations. *Mol. Sim.*, **33**(8): pp. 623–628.
Allen, M.P. and D.J. Tildesley 1987, *Computer Simulation of Liquids.*, Oxford: Clarendon Press.
Ashu, P.A., J.H. Jefferson, A.G. Cullis, W.E. Hagston and C.R. Whitehouse 1995, Molecular dynamics simulation of (100)InGaAs/GaAs strained-layer relaxation processes. *J. Cryst. Growth*, **150**(1–4): pp. 176–179.
Balamane, H., T. Halicioglu, and W.A. Tiller 1992, Comparative study of silicon empirical interatomic potentials. *Phys. Rev. B*, **46**(4): pp. 2250–2279.
Bellotti, E., F. Bertazzi, and M. Goano 2007, Alloy scattering in AlGaN and InGaN: A numerical study. *J. Appl. Phys.*, **101**(12): pp. 123706 (1–8).
Benkabou, F., H. Aourag, Pierre J. Becker, and M. Certier 2000, Molecular Dynamics Study of Zinc-Blende GaN, AlN and InN. *Mol. Sim.*, **23**(4): pp. 327–341.
Benkabou, F., P. Becker, M. Certier and H. Aourag 1998, Structural and Dynamical Properties of Zincblende GaN. *physica status solidi* (b), **209**(2): pp. 223–233.
Benkabou, F., Chelahi Chikr.Z, H.Aourag, Pierre J. Becker and M. Certier 1999 , Atomistic study of zinc-blende BAs from molecular dynamics. *Phys. Lett. A*, 1999. **252**(1–2): pp. 71–76
Biswas, K., A. Franceschetti, and S. Lany 2008, Generalized valence-force-field model of (Ga,In)(N,P) ternary alloys. *Phys. Rev. B*, **78**(8): pp. 085212 (1–10).
Biswas, R. and D.R. Hamann 1986, Simulated annealing of silicon atom clusters in Langevin molecular dynamics. *Phys. Rev. B*, **34**(2): pp. 895–901.
Britt, J. and C. Ferekides 1993, Thin film CdS/CdTe solar cell with 15.8% efficiency. *Appl. Phys. Lett.*, **62**(22): pp. 2851–2852.

El-Mellouhi, F., W. Sekkal, and A. Zaoui 2002, A modified Tersoff potential for the study of finite temperature properties of BP. *Physica A*, **311**(1–2): pp. 130–136.

Frenkel, D. and B. Smit 2002, *Understanding Molecular Simulation: From Algorithms to Applications*. 2nd ed., San Diego: Academic Press.

Gärtner, K. 2010, MD simulation of ion implantation damage in AlGaAs: II. Generation of point defects. Nucl. Instrum. *Methods Phys. Res. Sect. B-Beam Interact. Mater. Atoms*, **268**(2): pp. 149–154.

Gärtner, K. 2006, MD simulation of ion implantation damage in AlGaAs: I. Displacement energies. Nucl. Instrum. *Methods Phys. Res. Sect. B-Beam Interact. Mater. Atoms*, **252**(2): pp. 190–196.

Gärtner, K. and T. Clauß 2010, MD simulation of ion implantation damage in AlGaAs: III. Defect accumulation and amorphization. *Nucl. Instrum. Methods Phys. Res. Sect. B-Beam Interact. Mater. Atoms*, **268**(2): pp. 155–164.

Goumri-Said, S., M.B. Kanoun, A.E. Merad, G. Merad and H. Aourag 2004, Prediction of structural and thermodynamic properties of zinc-blende AlN: molecular dynamics simulation. *Chem. Phys.*, **302**: pp. 135–141.

Hendry, E., M. Koeberg, F. Wang, H. Zhang, C. de Mello Donega', D. Vanmaekelbergh and M. Bonn 2006, Direct Observation of Electron-to-Hole Energy Transfer in CdSe Quantum Dots. *Phys. Rev. Lett.*, **96**: pp. 057408 (1–4).

Ho, I.H. and G.B. Stringfellow 1996, Solid phase immiscibility in GaInN. *Appl. Phys. Lett.*, **69**: pp. 2701–2703.

Ho, I.H. and G.B. Stringfellow 1997, Solubility of nitrogen in binary III-V systems. *J. Crys. Growth*, **178**(1–2): pp. 1–7.

Itoh, S., K. Nakano, and A. Ishibashi 2000, Current status and future prospects of ZnSe-based light-emitting devices. *J. Cryst. Growth*, **214–215**: pp. 1029–1034

Kanoun, M.B., A.E. Merad, H. Aourag, J. Cibert and G. Merad 2003, Molecular-dynamics simulations of structural and thermodynamic properties of ZnTe using a three-body potential. *Solid State Sci.*, **5**(9): pp. 1211–1216.

Kanoun, M.B., W. Sekkal, H.Aourag, and G. Merad 2000, Molecular-dynamics study of the structural, elastic and thermodynamic properties of cadmium telluride. *Phys. Lett. A*, **272**(1–2): pp. 113–118.

Keating, P.N. 1966, Effect of Invariance Requirements on the Elastic Strain Energy of Crystals with Application to the Diamond Structure. *Phys. Rev.*, **145**(2): pp. 637–645.

Kim, K., P.R.C. Kent and A. Zunger 2002, Atomistic description of the electronic structure of InxGa1-xAs alloys and InAs/GaAs superlattices. *Phys. Rev. B*, **66**(4): pp. 045208 (1–15).

Martin, R.M. 1970, Elastic Properties of ZnS Structure Semiconductors. *Phys. Rev. B*, **1**(10): pp. 4005–4011.

Moon, W.H. and H.J. Hwang 2003, Structural and thermodynamic properties of GaN: a molecular dynamics simulation. *Phys. Lett. A*, **315**(3–4): p. 319–324.

Moon, W.H., M.S. Son, and H.J. Hwang 2003, Molecular-dynamics simulation of structural properties of cubic boron nitride. *Physica B*, **336**(3–4): pp. 329–334.

Murdick, D.A., X.W. Zhou, and H.N.G. Wadley 2005, Assessment of interatomic potentials for molecular dynamics simulations of GaAs deposition. *Phys. Rev. B*, **72**(20): pp. 205340 (1–14).

Nakamura, M., H. Fujioka, K. Ono, M. Takeuchi, T. Mitsui and M. Oshima 2000, Molecular dynamics simulation of III-V compound semiconductor growth with MBE. *J. Cryst. Growth*, **209**: pp. 232–236.

Nord, J., K. Nordlund, and J. Keinonen 2002, Molecular dynamics simulation of ion-beam-amorphization of Si, Ge and GaAs. *Nucl. Instrum. Methods Phys. Res. Sect. B-Beam Interact. Mater. Atoms*, **193**(1–4): pp. 294–298.

Pearson, E., T. Takai, T. Halicioglu and W.A. Tiller 1984, Computer modeling of Si and SiC surfaces and surface processes relevant to crystal growth from the vapor. *J. Cryst. Growth*, **70**(1–2): pp. 33–40

Powell, D., M.A. Migliorato, and A.G. Cullis 2007, Optimized Tersoff potential parameters for tetrahedrally bonded III-V semiconductors. *Phys. Rev. B*, **75**: p. 115202 (1–9).

Saito, T. and Y. Arakawa 1999, Atomic structure and phase stability of $In_xGa_{1-x}N$ random alloys calculated using a valence-force-field method. *Phys. Rev. B*, **60**(3): pp. 1701–1706.

Salvador, D.D., M. Petrovich, M. Berti, F. Romanato, E. Napolitani, A. Drigo, J. Stangl, S. Zerlauth, M. Mühlberger, F. Schäffler, G. Bauer and P. C. Kelires 2000, Lattice parameter of Si1-x-yGexCy alloys. *Phys. Rev. B*, **61**(19): pp. 13005–13013.

Sibona, G.J., S. Schreiber, R. H. W. Hoppe, B. Stritzker and A. Revnic 2003, Numerical simulation of the production processes of layered materials. *Mater. Sci. Semicond. Process*, **6**(1–3): pp. 71–76.

Stillinger, F.H. and T.A. Weber 1985, Computer simulation of local order in condensed phases of silicon. *Phys. Rev. B*, **31**(8): pp. 5262–5271.

Sun, X., Q. Chen, Y. Chu and C. Wang 2005, Structural and thermodynamic properties of GaN at high pressures and high temperatures. *Physica B*, **368**(1–4): pp. 243–250.

Takayama, T., M. Yuri, K. Itoh, T. Baba and J. S. Harris, Jr. 2000, Theoretical analysis of unstable two-phase region and microscopic structure in wurtzite and zinc-blende InGaN using modified valence force field model. *J. Appl. Phys*, **88**(2): pp. 1104–1110.

Takayama, T., M. Yuri, K. Itoh, T. Baba and J.S. Harris Jr. 2001a, Analysis of phase-separation region in wurtzite group III-nitride quaternary material system using modified valence force field model. *J. Cryst. Growth* **222**: pp. 29–37.

Takayama, T., M. Yuri, K. Itoh and J. S. Harris, Jr. 2001b, Theoretical predictions of unstable two-phase regions in wurtzite group-III-nitride-based ternary and quaternary material systems using modified valence force field model. *J. Appl. Phys.*, **90**(5): pp. 2358–2369.

Tersoff, J. 1986, New empirical model for the structural properties of silicon. *Phys. Rev. Lett.*, **56**(6): pp. 632–635.

Tersoff, J. 1988a, Empirical interatomic potential for silicon with improved elastic properties. *Phys. Rev. B*, **38**(14): pp. 9902–9905.

Tersoff, J. 1988b, Empirical Interatomic Potential for Carbon, with Applications to Amorphous Carbon. *Phys. Rev. Lett.*, **61**(25): pp. 2879–2882.

Tersoff, J. 1989, Modeling of solid state chemistry: Interatomic potentials for multicomponent systems. *Phys. Rev. B*, **39**(8): pp. 5566–5568.

Titantah, J.T., D. Lamoen, M. Schowalter and A. Rosenauer 2008, Size effects and strain state of Ga1−xInxAs/GaAs multiple quantum wells: Monte Carlo study. *Phys. Rev. B*, **78**(16): pp. 165326 (1–7).

Titantah, J.T., D. Lamoen, M. Schowalter and A. Rosenauer 2007, Bond length variation in Ga1−xInxAs crystals from the Tersoff potential. *J. Appl. Phys.*, **101**(12): pp. 123508 (1–4).

Tzoumanekas, C. and P.C. Kelires 2002, Theory of bond-length variations in relaxed, strained, and amorphous silicon-germanium alloys. *Phys. Rev. B*, **66**(19): pp. 195209 (1–11).

Wu, J. 2009, When group-III nitrides go infrared: New properties and perspectives. *J. Appl. Phys.*, **106**(1): pp. 011101 (1–28).

13 Structural Properties of Cement Clinker Compound by First Principles Calculations

Ryoji Sakurada, Abhishek Kumar Singh,
and Yoshiyuki Kawazoe

CONTENTS

13.1 INTRODUCTION

The cement clinker manufactured by heating raw materials such as limestone, clay, bauxite, and other trace materials together in a rotary cement kiln at high temperature ($1400°$–$1600°C$) contains four major compounds: tricalcium silicate ($3CaO \cdot SiO_2$, alite), dicalcium silicate ($2CaO \cdot SiO_2$, belite), tricalcium aluminate ($3CaO \cdot Al_2O_3$), and tetracalcium aluminoferrite ($4CaO \cdot Al_2O_3 \cdot Fe_2O_3$). Calcium originated from limestone, silicon, and aluminum from clay and iron oxide from bauxite fuse together to produce a clinker consisting of these four major compounds. These four major compounds are expressed in a shorthand notation of cement chemistry as $C_3S = 3CaO \cdot SiO_2$, $C_2S = 2CaO \cdot SiO_2$, $C_3A = 3CaO \cdot Al_2O_3$, $C_4AF = 4CaO \cdot Al_2O_3 \cdot Fe_2O_3$, where $C = CaO$, $S = SiO_2$, $A = Al_2O_3$, $F = Fe_2O_3$, $Cs = CaSO_4$, and $H = H_2O$.

The cooling process before grinding of the clinker causes phase changes in these compounds because of temperature. For instance, five polymorphs of belite (C_2S) have been known to form (α-form, α_H'-form, α_L'-form, β-form, and γ-form) during the cooling process of the cement clinker (Figure 13.1). Slower cooling in the temperature range of $400°$ to $500°C$ may cause the transformation from β-C_2S to γ-C_2S, which reacts very slowly with water comparison to β-C_2S. The crystal structure also transforms with phase change: hexagonal in α-form, orthorhombic in α_H'-form and α_L'-form, monoclinic in β-form, and orthorhombic in γ-form. Symmetry of the crystal decreases from hexagonal to orthorhombic.

The difference in hydraulic activity of dicalcium silicates arises as a result of the difference in stability of inherent crystal structure. The usual form of dicalcium silicate in Ordinary Portland Cement (OPC) is β-C_2S, which reacts more slowly with water and results in the lowest rate of heat

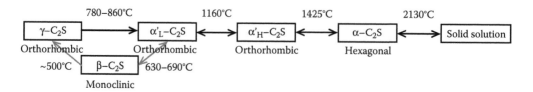

FIGURE 13.1   Phase change of Belite C$_2$S.

evolution compared to the other major compounds (C$_3$S, C$_3$A, C$_4$AF). β-C$_2$S contributes to strength development in a later stage of hydration (after 28 days). The hydraulic activity of β-C$_2$S and γ-C$_2$S has been analyzed by X-ray crystallographic measurements (Jost et al. 1977). This study reveals that the Ca–Ca bond length has a strong relationship with the hydraulic activity. However, little effort has been made to theoretically understand the relationship.

First-principles *(ab-initio)* calculation can be a powerful tool to analyze and predict the hydraulic activity of cement clinker compounds at atomic level. These calculations employ no other statistical assumptions and phenomena-based calculation model. These calculations directly solve either Schrödinger's equation or a Schrödinger-like equation at the quantum level.

The first-principles calculations have been used extensively to investigate the electronic and band structure of not only bulk materials but also nanoscale electronic devices: magnetism in metal-doped silicon nanotubes (Singh et al. 2004a, 2004b), silicon nanowires (Singh et al. 2005), and magnetic behavior of Mn clusters (Briere et al. 2002). However, there is very little information available on the usage of *ab initio* calculation in the study of crystal structure of the cement clinker compounds. There is a report of calculation of Young's moduli of Portlandite and Forshagite minerals as calcium-silicate-hydroxide (C-S-H) by using density functional theory (Laugesen 2003).

We reported earlier the first-principles study on the crystal structure and hydraulic activity of β- and γ-form belites (Sakurada et al. 2009a, 2009b). In this chapter, we present our DFT simulation results on the study of hydraulic properties of cement phases. Ca–Ca, Ca–O, and Si–O bond lengths are chosen as a yardstick for making reliable prediction on the hydraulic activity of β- and γ-C$_2$S phases. Moreover, to find a change in crystal structure of β-C$_2$S on substitution of the strontium atom or barium atom for a calcium atom, a periodic boundary condition has been applied on a large supercell (504 atoms, a × 3, b × 3, c × 2) to remove any spurious interactions.

13.2   HYDRAULIC REACTIONS OF CEMENT CLINKER COMPOUND

The chemical reaction of cement compounds with water is very complex because the hydration of each compound (C$_3$S, C$_2$S, C$_3$A, and C$_4$AF) is closely related with the reactions among the rest of the compounds. The OPC is a mixture of four major compounds, as mentioned in Section 13.1. Therefore, at first, it is important to understand the hydration of pure cement compounds in order to have complete understanding of the hydration mechanism of OPC.

Typical hydration reactions of the alite (C$_3$S), β-belite (C$_2$S), aluminate phase with the existence of gypsum (C$_3$A) and the ferric phase (C$_4$AF) are expressed as follows:

Alite phase

$$2C_3S + 6H \rightarrow C_3S_2H_3 + 3CH \tag{13.1}$$

β-Belite phase

$$2\beta\text{-}C_2S + 4H \rightarrow C_3S_2H_3 + CH \tag{13.2}$$

Aluminate phase with gypsum

$$C_3A + 3CsH_2 + 26H \rightarrow C_3ACs_3H_{32} \tag{13.3}$$

$$C_3ACs_3H_{32} + 2C_3A + 4H \rightarrow 3C_3ACsH_{12} \tag{13.4}$$

Ferric phase

$$4\,C_4AF + 2CH + 10H \rightarrow C_3AH_6 + C_3FH_6 \tag{13.5}$$

Equations 13.1 to 13.5 are part of the whole hydration reaction of a relatively complex mixture of cement phases. Following are the hydration reactions of β-belite in which several kinds of calcium silicate hydrates (xC-yS-zH) have been found (Babushkin et al. 1985).

$$\beta\text{-}C_2S + 1.17H \rightarrow C_2SH_{1.17} \tag{13.6}$$

$$2\beta\text{-}C_2S + 4H \rightarrow C_3S_2H_3 + CH \tag{13.7}$$

$$6\beta\text{-}C_2S + 7H \rightarrow 2C_4S_3H_{1.5} + 4CH \tag{13.8}$$

$$6\beta\text{-}C_2S + 7H \rightarrow C_6S_6H + 6CH \tag{13.9}$$

$$6\beta\text{-}C_2S + 12.5H \rightarrow C_5S_6H_{5.5} + 7CH \tag{13.10}$$

$$3\beta\text{-}C_2S + 6.5H \rightarrow C_2S_3H_{2.5} + 4CH \tag{13.11}$$

$$2\beta\text{-}C_2S + 5H \rightarrow CS_2H_2 + 3CH \tag{13.12}$$

In addition to the four major compounds comprising primary mineral components, calcium, silicon, aluminum, and iron, OPC also contains following trace impurities (Uchikawa 1993):

- Magnesium, sulfur, phosphorus, sodium, potassium, and titanium in maximum concentration of 2.5%
- Chlorine, zinc, chromium, manganese, strontium, barium, and fluorine in maximum concentration of 900 ppm
- Other small quantities of impurities in concentration lower than 28 ppm

Trace impurities in cement influence the burning ability of clinkers, clinkering temperature, mineral compositions of clinkers, hydraulic activity, stabilizing activity, and strength development (Benarchid et al. 2005). For instance, sodium as a dopant in the C_3A phase decreases its hydraulic activity. Li et al. (1998) demonstrated through their quantum chemical simulations that the decreased hydraulic activity in Na-doped C_3A is due to the increased energy of the lowest energy unoccupied molecular orbital (LUMO), which in turn lowers the Fermi energy of the electronic states. Fierence and Tirlocq (1983) reported that the stabilizing elements boron, vanadium, phosphorus, arsenic, and chromium affect the stability and hydration rate of dicalcium silicates. From the kinetic constants of heat liberation, hydration reactivity sequences during the early stage are observed to be: $V = P > B > Cr > As$ for low concentration (0.5%) and $Cr > P > B > As > V$ for high concentration (3%). Interestingly, they also showed that the chemical stabilization ability is strongly related to the charge versus ionic radius ratio (C/R) concerning the charge surface potential: $V(C/R8.5) = P(C/R14.3) > B(C/R13) > Cr(C/R11.5) > As(C/R10.9)$ for low concentration (0.5%) and $Cr(C/R11.5) > P(C/R14.3) > B(C/R13) > As(C/R10.9) > V(C/R8.5)$ for high concentration (3%).

Gao et al. (1992) studied the phase transformation tendency of belite as a function of the composition of minor impurities. They synthesized belite doped with Li^+, K^+, Na^+, Sr^{2+}, Mn^{2+}, Mg^{2+}, Fe^{3+}, Al^{3+}, Ti^{4+}, Mn^{4+}, P^{5+}, and S^{6+} ions by sintering at 1500°C for three hours and slower cooling. The formation of β-phase was found to be more sensitive to belite doped by the multivalent ions such as Fe^{3+}, Al^{3+}, and S^{6+} compared to the univalent and divalent ions. The amount of the β-phase formed depends on the kinds of the ions and their composition in belite. The β-phase was obtained in all of Si^{4+}-substituted samples but not in the Ca^{2+}-substituted samples. This result indicates that the substitution of Si^{4+} is preferred to stabilize the β-C_2S phase rather than the substitution of Ca^{2+}.

13.3   COMPUTATIONAL MODEL

13.3.1   First-Principles Calculation and Its Application to Engineering Problem

First-principles molecular dynamics (Atkins and Friedman 2005) is based on the density functional theory (DFT) and the pseudopotentials representing ion–electron interaction. DFT has been applied for calculation of the electronic structure of many-body systems. The total energy of electron system $E[\rho(r)]$ is expressed as a function of electronic charge density $\rho(r)$ at a particular point r (Equation 13.13). This energy functional $E[\rho(r)]$ gives the minimum value of energy in ground-state for the real electronic charge density. The electronic charge density at the exact ground-state $\rho(r)$ is given by Equation 13.14.

$$E[\rho(r)] = T[\rho(r)] + \int V_{ext}(r)\rho(r)dr + \frac{1}{2}\iint \frac{\rho(r)\rho(r')}{|r-r'|}drdr' + E_{xc}[\rho(r)] \tag{13.13}$$

$$\rho(r) = \sum_{i=1}^{n}|\psi_i(r)|^2 \tag{13.14}$$

where r indicates coordinate vector in the real space and $\psi_i(r)$ denotes the one-electron spatial orbital for $i = 1, 2, 3, ...n$. The terms on right-hand side of Equation 13.13 are the kinetic energy of a system of noninteracting electrons, the potential energy resulting from electron–nucleus attraction, the Coulomb interaction energy between the total charge distribution and the exchange-correlation energy of the system, respectively.

By applying the variation principle to the total energy functional $E[\rho(r)]$ of the electronic charge density $\rho(r)$ in Equations 13.13 to 13.14, the Kohn-Sham equation in a similar form to that of Shrödinger equation can be easily obtained for the one-electron orbital (Equations 13.15 to 13.16)

$$\left(-\frac{h^2}{2m}\nabla^2 + V_{eff}(r)\right)\psi_i(r) = \varepsilon_i\psi_i(r) \tag{13.15}$$

$$V_{eff}(r) = V_{ext}(r) + \int \frac{\rho(r')}{|r-r'|}dr' + \mu_{xc}[r] \tag{13.16}$$

where h is Planck's constant ($\hbar = h/2\pi$), m is mass of electron, ε_i are the Kohn-Sham orbital energies, and $V_{eff}(r)$ is effective potential involving the external potential, interaction potential between electrons, and the exchange-correlation potential $\mu_{xc}[r]$, which is the functional derivative of the exchange-correlation energy $E_{xc}[\rho(r)]$:

$$\mu_{xc}[r] = \frac{\delta E_{xc}[\rho(r)]}{\delta\rho(r)} \tag{13.17}$$

The solutions for Equation 13.15 can be easily found by solving the Shrödinger equation for noninteracting particles moving under an effective potential $V_{eff}(r)$. The *ab-initio* calculations were carried out using Vienna *ab-initio* simulation package, VASP (Kresse and Furthmüller 1996). The calculations were performed using a plane wave method employing the ultrasoft pseudo-potentials for calcium, silicon, and oxygen, and the generalized gradient approximation (GGA) for the exchange-correlation potential. K-space is sampled by Γ-point for calculations.

13.3.2 COMPUTATIONAL MODEL OF BELITE

The structure of β-C$_2$S has been studied by X-ray diffraction analysis (Jost et al. 1977). This crystal structure belongs to monoclinic space group $P2_{1/n}$ (C_{2h}^5) with lattice constants of $a_m = 5.502$Å, $b_m = 6.745$ Å, $c_m = 9.297$ Å, and the monoclinic angle $= 94.59°$. The atom coordinates of the monoclinic cell are tabulated in Table 13.1. The crystal structure of γ-C$_2$S is orthorhombic, and belongs to P_{bnm} space group with lattice constants of $a_o = 5.081$Å, $b_o = 11.224$Å, $c_o = 6.778$Å (Czaya 1971, Udagawa et al. 1980). The atom coordinates are listed in Table 13.2. The structures of β-C$_2$S and γ-C$_2$S are illustrated in Figure 13.2 and Figure 13.3, respectively.

13.4 RESULTS AND DISCUSSIONS

13.4.1 CRYSTAL STRUCTURE OF BELITE

The optimized polyhedral model of β-belite without substitution of any foreign atom is illustrated in Figure 13.4. The crystal structural unit is made up of two kinds of CaO polyhedra (Ca(1)O = 7 Ca–O bonds, Ca(2)O = 8 Ca–O bonds) and SiO$_4$ tetrahedron. The Ca(1)O polyhedron with 7 Ca–O bonds forms a distorted pentagonal bi-pyramid, and the Ca(2)O polyhedron with 8 Ca–O bonds takes shape of a distorted anti-cube as show in Figure 13.5.

TABLE 13.1

Atomic Coordinates for β-C$_2$S

Atom	x	y	z
Ca(1)	0.2738	0.3428	0.5694
Ca(2)	0.2798	0.9976	0.2981
Si	0.2324	0.7814	0.5817
O(1)	0.2864	0.0135	0.5599
O(2)	0.0202	0.7492	0.6919
O(3)	0.4859	0.6682	0.6381
O(4)	0.1558	0.6710	0.4264

TABLE 13.2

Atomic Coordinates for γ-C$_2$S

Atom	x	y	z
Ca(1)	0.0000	0.0000	0.0000
Ca(2)	−0.0099	0.2809	0.2500
Si	0.4275	0.0966	0.2500
O(1)	−0.2543	0.0937	0.2500
O(2)	0.2974	−0.0384	0.2500
O(3)	0.2985	0.1624	0.0575

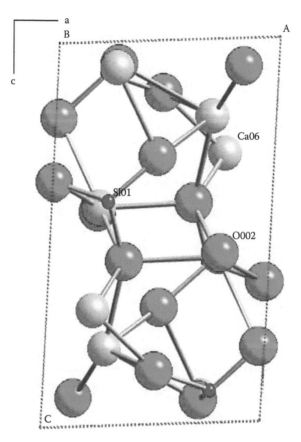

FIGURE 13.2   Crystal structure of β-C$_2$S.

Si–O bond lengths within 3 Å to the neighboring O atoms for all the independent Si atoms in supercell model as obtained from the *ab initio* calculations are tabulated in Table 13.3. The calculated mean distances of Si–O are 1.64 Å and 1.65 Å in β-C$_2$S and γ-C$_2$S, respectively. Si–O mean distance in β-C$_2$S is 0.01Å shorter than that of γ-C$_2$S. The corresponding experimental values obtained from X-ray diffraction measurements are 1.63 Å and 1.65 Å in β-C$_2$S and γ-C$_2$S, respectively (Jost et al. 1977). The silicon atom in SiO$_4$ tetrahedron exhibits shorter distances with the oxygen atoms, almost half of the Ca–O distance (as discussed later). This means that the silicon atom forming the SiO$_4$ tetrahedron lies in a more stable position than the calcium atom in Ca–O polyhedron.

The mechanism of Si–O chemical bonding was analyzed for covalent and ionic bond orders in β- and γ-C$_2$S. Each bond order was evaluated by an overlapping population calculated by the DV-Xα molecular orbital method (Xiuji et al. 1994). There are slight differences in computed covalent and ionic bond orders obtained for both dicalcium silicates. However, Xiuji et al. (1994) demonstrated that the differences in hydraulic activity between β- and γ-C$_2$S do not arise from the difference of Si–O chemical bonding. Further investigation will be required to establish the relationship between these differences in the crystal structure to hydraulic activity of belite.

The computed Ca–O bond lengths within 3 Å to the neighboring O atoms for all the independent calcium atoms of CaO polyhedra are listed in Table 13.3. The computed Ca–O bond lengths, 2.46 Å in β-C$_2$S and 2.38 Å, in γ-C$_2$S are comparable to those obtained from X-ray diffraction analysis (Jost et al. 1977). However, as reported in the literature (Jost et al. 1977), the exact correlation between the Ca–O bond length and the hydraulic activity of cement clinker compounds is not well established. Moreover, the time-of-flight powder diffraction experiments (Mori et al. 2006) provide an interesting result that the charge state of calcium atoms in CaO polyhedron depends on the structure

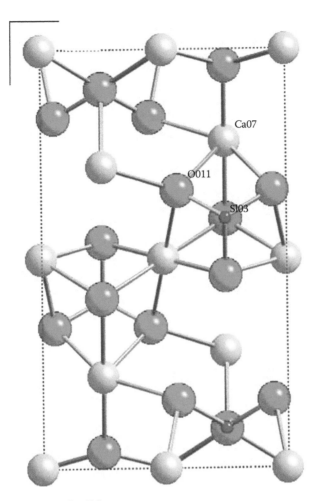

Ca07

O011

S103

FIGURE 13.3 Crystal structure of γ-C$_2$S.

of CaO polyhedron. The charge state of Ca(1) with seven Ca–O bonds and Ca(2) with eight Ca–O bonds were 1.87$^+$ and 2$^+$, respectively.

The longer bond should be advantageous to split the Ca–O bond, resulting into dissolution of calcium atoms into pure water. This hypothesis causes a contradiction to hydraulic activity between β-C$_2$S and C$_3$S, as tabulated in Table 13.3. Jost et al. (1977) pointed out that the differences in Ca–O bond length between the clinker compounds exhibit very little effect on its ability to react faster or slower with water; rather the connection of CaO polyhedra by common faces influences hydraulic activity. There is a connection of CaO polyhedra by common faces in β-C$_2$S. In contrast there is a connection of polyhedra by common edges and corners in γ-C$_2$S. This means that only Ca–O bond lengths provide less information on the hydraulic activity of the belites.

Therefore, whether the mean Ca–Ca bond lengths have a correlation with the hydraulic activity of the cement clinker compounds was investigated. The Ca–Ca bond lengths within 4 Å to the neighboring calcium atoms for all the independent calcium atoms within a supercell are computed. The calculated values of Ca–Ca bond length are listed in Table 13.3 and compared with experimental results reported by Jost et al. (1977).

The mean Ca–Ca distances obtained from this work are 3.56 and 3.75 Å in β-C$_2$S and γ-C$_2$S, respectively. These results agree well with the X-ray diffraction analysis (Jost et al. (1977)): 3.58 Å in β-C$_2$S, 3.75 Å in γ-C$_2$S, 3.40 Å in CaO and 3.47 Å in highly hydraulic active alite (C$_3$S). β-C$_2$S

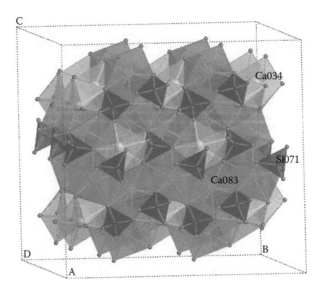

FIGURE 13.4 Converged geometry of β-C$_2$S without substitution.

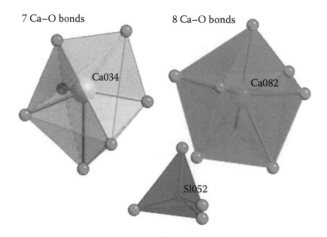

FIGURE 13.5 CaO polyhedra and SiO$_4$ tetrahedron.

exhibits a closer arrangement of calcium atoms than that in γ-C$_2$S. The shortening of Ca–Ca distance causes instability of the structure because of repulsion between cations. Based on these results and experimental Ca–Ca distances, Ca–Ca bond length becomes an important index for assessing the hydraulic activity of the cement clinker compounds.

13.4.2 EFFECT OF TRACE IMPURITIES

The crystal structure of β-C$_2$S obtained by replacing calcium atom by strontium and barium atoms was analyzed using first-principle calculations. There are two nonequivalent calcium atoms present in β-C$_2$S: the Ca(1) atom, having a sevenfold coordination of oxygen atoms, and the Ca(2) atom, having an eightfold coordination of oxygen atoms. To carry out these calculations, a large supercell consisting of 504 atoms (a × 3, b × 3, c × 2) was employed (Figure 13.6). The computed total energy and Ca–Ca bond lengths within 4 Å are tabulated in Table 13.4.

The total energies of the supercell obtained after replacement of Ca(1) and Ca(2) atoms with Sr atom are –3643.314 eV and –3643.273 eV, respectively. This clearly indicates that the substitution of

TABLE 13.3

Interatomic Distances of β-C₂S and γ-C₂S

Cement Clinker Compounds			Ca–Ca Distance, Mean Distance <4 Å (this Work)	Ca–Ca Distance, Mean Distance <4 Å (Jost Reference)	Ca–Ca Distance, Shortest Distance <4 Å (this Work)	Ca–Ca Distance, Shortest Distance <4 Å (Jost Reference)
C	CaO	Lime	–	3.40	–	3.40
C₃S	Ca₃SiO₅	Alite	–	3.47	–	3.16
β-C₂S	β-Ca₂SiO₄	Belite	3.56	3.58	3.01	3.43
γ-C₂S	γ-Ca₂SiO₄	Belite	3.75	3.75	3.39	3.38
Cement Clinker Compounds			Si–O Distance, Mean Distance <3 Å (this Work)	Si–O Distance, Mean Distance <3 Å (Jost Reference)	Ca–O Distance, Mean Distance <3 Å (this Work)	Ca–O Distance, Mean Distance <3 Å (Jost Reference)
C	CaO	Lime	–	–	–	2.40
C₃S	Ca₃SiO₅	Alite	–	–	–	2.43
β-C₂S	β-Ca₂SiO₄	Belite	1.62	1.63	2.46	2.50
γ-C₂S	γ-Ca₂SiO₄	Belite	1.65	–	2.38	2.37

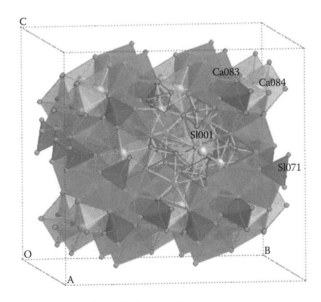

FIGURE 13.6 Converged geometry of one Sr-doped β-C₂S.

the strontium atom for the Ca(1) atom is energetically more favored. For the Ba-doped β-C₂S, the total energies are −3642.578 eV and −3642.269 eV for substitution of the Ca(1) atom and Ca(2) atom, respectively. Once again the results are similar to that obtained for Sr-doped structure. The structure obtained by substituting a Ca(2) atom is less stable than that of the Ca(1) atom. It is important to note that the substitution of one strontium or barium atom for a calcium atom makes the crystal structure of β-C₂S unstable compared to that of pure β-C₂S.

As mentioned previously in Section 13.1, shorter Ca–Ca interatomic distance results in higher hydraulic activity of calcium silicates. The computed distances within 4 Å between the neighboring calcium atoms in the doped supercell are shown in Table 13.4. Ca–Ca bond lengths are 3.5585 Å and

TABLE 13.4

Total Energy and Interatomic Distances of β-C$_2$S Replaced Ca Atom by One Sr Atom or One Ba Atom

Dopants	Total Energy (eV)			Ca–Ca Mean Distance <4 Å		
Ca	Without Substitution	Sr	Ba	Without Substitution	Sr	Ba
Ca(1)	−3644.023	−3643.314	−3642.578	3.5589	3.5585	3.5584
Ca(2)	−3644.023	−3643.273	−3642.269	3.5589	3.5589	3.5588

3.5589 Å for doped β-C$_2$S, with each Ca(1) and Ca(2) atom replaced with one Sr atom, respectively. Corresponding values for Ba-doped β-C$_2$S are 3.5584 Å and 3.5588 Å for the replacement of Ca(1) and Ca(2) atoms, respectively. It is interesting to note that the Ca–Ca bond length in undoped β-C$_2$S is 3.5589 Å. The doping with strontium and barium atoms indeed leads to shorter Ca–Ca distances. As discussed earlier, the strontium or barium atom substitution at the Ca(1) site is energetically more favorable than at the Ca(2) site, which would naturally lead to shorter Ca–Ca distances and improve the hydraulic activity of β-C$_2$S. Furthermore, Ba-doped β-C$_2$S shows shorter Ca–Ca bond length than Sr-doped β-C$_2$S, implying a better hydraulic ability for the former.

The differences observed in the two modified β-C$_2$S crystal structures obtained by substituting only one calcium atom with a strontium atom or two calcium atoms with two strontium atoms is analyzed in this section. A strontium atom can substitute two nonequivalent calcium atoms: Ca(1) atom having sevenfold oxygen coordination and Ca(2) atom having eight-fold coordination. Four symmetrically nonequivalent structures are considered to carry out this study. The structures of 2Sr-1, 2Sr-2, 2Sr-3, and 2Sr-4 are schematically illustrated in Figure 13.7.

- 2Sr-1: In this configuration, two of the nearest neighbor Ca(1) sites are occupied by strontium atoms. Two Sr(1)O polyhedra having sevenfold oxygen coordination are independent and apart by 4.661 Å. Furthermore, in this configuration, one Ca(1)O polyhedron, three Ca(2)O polyhedra and two SiO$_4$ tetrahedra directly bridge two Sr(1)O polyhedra.
- 2Sr-2: This configuration is obtained by replacing calcium atoms with strontium atoms at neighboring Ca(1) and Ca(2) sites of β-C$_2$S crystal. Sr(1)O polyhedron is connected with Sr(2)O polyhedron having eightfold oxygen coordination via edge to edge bond. Moreover, one Ca(1)O polyhedron combines with an Sr(1)O polyhedron through edge to edge bond and with Sr(2)O polyhedron through face to face bond.
- 2Sr-3: The doping sites are similar to those of 2Sr-1. However, in this configuration two Sr(1)O polyhedra connect via edge to edge bond. These polyhedra are connected with one Ca(1)O polyhedron and three Ca(2)O polyhedra.
- 2Sr-4: The doping sites are similar to those of 2Sr-2. However, this configuration is obtained by replacing calcium atoms with strontium atoms at neighboring Ca(1) and Ca(2) sites of a β-C$_2$S crystal. In this configuration, the Sr(1)O polyhedron connects to the Sr(2) polyhedron via face to face bond. Sr(1)O and Sr(2) polyhedra are connected with two CaO polyhedra.

The computed total energies of these four models are summarized in Table 13.5. The 2Sr-1 is the lowest in energy followed by configurations 2Sr-3, 2Sr-4, and 2Sr-2. Once again there is a clear preference for substitution at Ca(1) site. The energy of 2Sr-2 is 0.031 eV higher than that for 2Sr-1. The total energy of 2Sr-4 is also 0.06 eV higher than that of 2Sr-3. Both configurations 2Sr-1 and 2Sr-2 have strontium atoms substituted at the Ca(1) site. Even the presence of an Sr atom at Ca(1) did not alter the preference for the other strontium atom, because it always prefers to go to the Ca(1) site. A 2Sr-2 forming an edge to edge bond between the Sr(1)O polyhedron and Sr(2)O polyhedron lies in the most unstable state, as indicated by the total energy of the crystal.

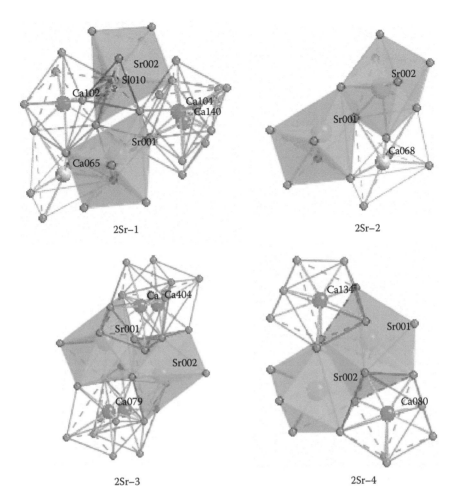

FIGURE 13.7   Computational models of β-C₂S substituted strontium atoms for calcium atoms.

The Ca–Ca distances corroborate nicely the energy differences. The distances within 4 Å to the neighboring calcium atoms for all the independent calcium atoms were calculated for each supercell (Table 13.6). The Ca–Ca interatomic mean distances in 2Sr-1 and 2Sr-3 are 3.5579 Å and 3.5579 Å, respectively, whereas in 2Sr-2 and 2Sr-4 they are 3.5588Å and 3.5582 Å, respectively. It is evident from the results presented in Tables 13.5 and 13.6 that the mean Ca–Ca distance in one strontium atom–doped β-C₂S is longer than that of two strontium atoms–doped β-C₂S. Once again the lowest energy structure has the shortest Ca–Ca distance and therefore should exhibit high hydraulic activity, which is in agreement with the experiments.

We also computed the volumetric change observed in substituted polyhedra arising as a result of the substitution of a calcium atom by a strontium atom, using the relationship ((Ca–O)–(Sr–O))/(Ca–O), where X–O is interatomic mean distances of Ca–O and Sr–O bond lengths. The computed interatomic distance change rate for the 2Sr-1 crystal is found to be –3.610% for the Sr(1)O polyhedron and –3.529% for the other Sr(1)O polyhedron, respectively. The minus value of the change rate means a stretch expansion of SrO polyhedra because of the presence of the Sr atom in the CaO polyhedra. The computed interatomic distance change rate for the 2Sr-2 crystal is observed to be –3.598% in the Sr(1)O and –2.249% in the Sr(2)O polyhedron, respectively. The expansion rate of 2Sr-2 is smaller than that of 2Sr-1. A similar tendency is also observed for 2Sr-3 and 2Sr-4 configurations. It should be noted that the SiO₄ tetrahedra connecting directly to the SrO polyhedra also shrank by 0.1% after Sr doping in every configuration.

TABLE 13.5

Total Energy of β-C$_2$S Replaced Two Ca atoms by Two Sr Atoms

Computational Model	Without Substitution	2Sr–1 Sr$_1$ = 7 Coord. Sr$_2$ = 7 Coord.	2Sr–2 Sr$_1$ = 7 Coord. Sr$_2$ = 8 Coord.	2Sr–3 Sr$_1$ = 7 Coord. Sr$_2$ = 7 Coord.	2Sr–4 Sr$_1$ = 7 Coord. Sr$_2$ = 8 Coord.
Total energy, eV	−3644.023	−3642.616	−3642.585	−3642.613	−3642.607

TABLE 13.6

Interatomic Distances of β-C$_2$S Replaced Two Ca Atoms by Two Sr Atoms

Computational Model	Without Substitution	2Sr–1 Sr(1) = 7 Coord. Sr(2) = 7 Coord.	2Sr–2 Sr(1) = 7 Coord. Sr(2) = 8 Coord.	2Sr–3 Sr(1) = 7 Coord. Sr(2) = 7 Coord.	2Sr–4 Sr(1) = 7 Coord. Sr(2) = 8 Coord.
Ca–Ca bond length (<4 Å), Å	3.5589	3.5579	3.5588	3.5579	3.5582
Bond length change rate (%)	0	0.028	0.005	0.030	0.020
Ca–O bond length (<3 Å), Å	Before substitution	2.493	2.493	2.493	2.493
Sr(1)–O bond length (<3 Å), Å	After substitution	2.583	2.583	2.584	2.591
Ca–O bond length (<3 Å), Å	Before substitution	2.494	2.484	2.493	2.484
Sr(2)–O bond length (<3 Å), Å	After substitution	2.582	2.540	2.584	2.557
Bond length change rate (%)	Sr(1)O polyhedron	−3.610	−3.598	−3.627	−3.925
Bond length change rate (%)	Sr(2)O polyhedron	−3.529	−2.249	−3.650	−2.944
Si–O bond length (<3 Å) change rate (%)	SiO$_4$ tetrahedron	0.125	0.088	0.094	0.090

These computed volumetric change rates led to an interesting observation that higher expansion of SrO polyhedra cause shrinkage in Ca–Ca interatomic distances. Shorter Ca–Ca interatomic distance thus lead to favorable to hydraulic activity of cement clinker compounds, which is in line with the experimental observations (Jost et al. 1977).

13.5 CONCLUSIONS

Crystallographic analysis of dicalcium silicates are conducted using DFT simulations to predict the hydraulic activity of pure β-Belite, as well as those doped with strontium or barium atoms. The results of this analysis are summarized as follows:

- Ca–Ca bond lengths obtained from the DFT calculations are 3.56Å and 3.75Å in β-C$_2$S and γ-C$_2$S, respectively, which agree with the X-ray diffraction analysis. The β-C$_2$S phase exhibits close arrangement of Ca atoms compared to the γ-C$_2$S phase.

- Ca–Ca bond lengths in β-C$_2$S crystals doped with one strontium or barium atom are shorter than that observed in pure β-C$_2$S.
- β-C$_2$S phase forming an edge to edge bond between Sr(1)O polyhedron having sevenfold oxygen coordination and Sr(2)O polyhedron having eightfold oxygen coordination lies in a more unstable state as represented by its total energy.
- Higher expansion of SrO polyhedra as a result of doping of an Sr atom into a CaO polyhedron causes shorter Ca–Ca interatomic distance in the configuration 2Sr-1 in which two SrO polyhedra are independent and 2Sr-3, which forms an edge to edge bond between two Sr(1)O polyhedra. Shorter Ca–Ca interatomic distance is likely to be favorable for the hydraulic activity of cement clinker compounds, which also compares well with the experimental observations.
- Ca–Ca bond lengths computed though DFT can be used as an index to predict the hydraulic activity of β-C$_2$S and other cement clinker compounds.

13.6   ACKNOWLEDGMENT

The authors gratefully acknowledge the supercomputing resources from the Center for Computational Materials Sciences of the Institute for Materials Research, Tohoku University. The authors also express thanks to Dr. Masami Uzawa of Taiheiyo Cement Co., Ltd. for giving useful suggestions on cement clinker properties.

REFERENCES

Atkins, P., and Friedman, R. 2005. *Molecular Quantum Mechanics.* Oxford University Press, New York.

Babushkin, V. I., Matveyev, G. M., and Mchedlov-Petrossyan, O. P. 1985. *Thermodynamics of Silicates.* Springer-Verlag, New York.

Benarchid, M.Y., Diouri, A., Boukhari, A., Aride, J., and Elkhadiri, I. 2005. *Materials Chemistry and Physics* 94: 190.

Briere, T. M., Sluiter, M. H. F., Kumar, V., and Kawazoe, Y. 2002. *Phys. Rev. B* 66: 064412.

Czaya, R. 1971. *Acta Crystallogr. B* 27: 848.

Fierence, P., and Tirlocq, J. 1983. *Cement Concrete Res.* 13: 267.

Gao, C. L., Nojiri, T., and Nakao, K. 1992. *Cement Concrete Res.* 22: 743.

Jost, K. H., Ziemer, B., and Seydel, R. 1977. *Acta Crystallographica B* 33: 1696.

Kresse, G., Furthmüller, J. 1996. *Phys. Rev. B* 54: 11169.

Laugesen, J. L. 2003. *Nanotechnology In Construction.* The Royal Society of Chemistry Press, UK:185.

Li, B., Zhang, W., Zhou, M., and Chen, X. 1998. *Proc. 4th Int. Symp. Cement Concrete* 3: 45.

Mori, K., Kiyanagi, R., Yoneyama, M., Iwase, K., Sato, T., and Itoh, K., et al. 2006. *J. Solid State Chem.* 179: 3286.

Sakurada, R., Singh, A. K., Uzawa, M., and Kawazoe, Y. 2009a. *Proc. 34th Conf. Our World in Concrete Struct.* 28: 297.

Sakurada, R., Singh, A. K., Uzawa, M., and Kawazoe, Y. 2009b. *Proc. 4th Asian Particle Technol. Symp. (APT-2009)* 691.

Singh, A. K., Kumar, V., and Kawazoe, Y. 2004a. *J. Mater. Chem.* 14: 555.

Singh, A. K., Kumar, V., and Kawazoe, Y. 2004b. *Phys. Rev. B* 69: 233406.

Singh, A. K., Kumar, V., Note, R., and Kawazoe, Y. 2005. *Nano Lett.* 5: 2302.

Uchikawa, H. 1993. In *Proceedings of 9th International Congress on the Chemistry of Cement/Special Lecture. Cement & Concrete* 566: 49.

Udagawa, S., Urabe, K., Natsume, M., and Yano, T. 1980. *Cement Concrete Res.* 10: 139.

Xiuji, F., Xinmin, M., and Congxi, T. 1994. *Cement Concrete Res.* 24: 1311.

14 First Principles Modeling of the Atomic and Electronic Properties of Palladium Clusters Adsorbed on TiO$_2$ Rutile (110) Surfaces

Palanichamy Murugan, Vijay Kumar, and Yoshiyuki Kawazoe

CONTENTS

14.1 INTRODUCTION

In recent years, many studies have been devoted to understand the atomic and electronic structure of TiO$_2$ because it has wide applications, for example, (1) as an alternative to indium tin oxide in the field of transparent conductors, (2) in dye-sensitized solar cells, (3) photocatalysis, (4) water splitter, (5) in paints, and (6) support in heterogeneous catalysis in automobile industry for pollution control (Czekaj et al. 2009; Diebold 2003; Feng et al. 2008; Henrich and Cox 1994; Kaden et al. 2009, 2010; Menzies et al. 2004; Sellidj and Koel 1993; Zeng et al. 2008). In the latter application, noble or near noble metal nanoclusters supported on a TiO$_2$ surface are used to oxidize (reduce) hazardous CO (NO) gas that is present in the exhaust of an automobile. In this process the role of metal clusters as well as support are crucial. The combined system, first, oxidizes CO gas where an additional oxygen atom is generally obtained from either the process of NO reduction or the environment. Second, the system is capable of releasing the oxidized CO gas into the environment from the active sites of catalyst to provide further oxidation of CO to continue the heterogeneous catalyst. If the CO$_2$ is not released in such a catalyst, poisoning can occur (Wang et al. 2011). Much attention has been focused on nanoclusters of platinum, palladium, and gold and their alloys as suitable candidates for supported clusters for such reactions. In an actual catalyst there is a distribution of cluster sizes, but it is believed that among these different-sized clusters, small clusters may be catalytically more active. Therefore, understanding the interaction of small clusters of such elements and their alloys as well as their interaction with CO molecules has attracted much interest.

Titanium dioxide (TiO$_2$) substrate is well known to have two polymorphic phases: rutile and anatase. The former phase is structurally more stable compared to anatase (Hanaor and Sorrell 2011) and is preferred as support in catalysis. Surfaces of the rutile phase of TiO$_2$ have been well studied, and it has been found that (110) surface is more stable than (100) and (001) surfaces (Harrison et al. 1999; Ramamoorthy et al. 1994b). Accordingly, several experimental (Kaden et al. 2010; Varazo et al. 2004) and theoretical studies (Bredow and Pacchioni 1999; Bredow et al. 2004; Murugan et al. 2005, 2006; Ramamoorthy et al. 1994a; Sanz et al. 2000) have been done on (110) surface. Theoretical studies often consider slab geometry to represent surfaces. For TiO$_2$, such studies have shown oscillations in the physical and chemical properties (Bredow et al. 2004; Murugan et al. 2006), such as band gap, selected Ti–O bonds, and the local dipole moment, depending on the number of layers in the slab to be even or odd and the thickness. As an example, the energy band gap for an even number of layers is calculated to be higher than the value for bulk, and the gap for an odd number of layers is less than the bulk value. Thus, the band gap oscillates around the bulk value with respect to increasing the thickness of the surface slab, and finally the oscillation converges to the bulk value. Therefore, studies of different numbers of layers offers a way to obtain the correct band gap of bulk slabs. Note that each layer of TiO$_2$ surface slab consists of three sublayers (O–Ti–O), as shown in Figure 14.1, and that are stacked along the z-axis, taken to be perpendicular to the surface. The variation in the properties of TiO$_2$ slab as a function of the thickness also offers a possibility to

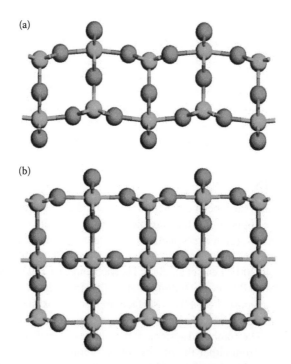

FIGURE 14.1 Side view of the ball and stick model for the optimized (a) 2L and (b) 3L (110) surface slabs of rutile TiO$_2$. Note that each layer of the slab comprises three sublayers (O–Ti–O). Red and green balls correspond to O and Ti atoms, respectively. For the 2L slab, the layers are slightly curved. Channel between bridging oxygen atoms can be seen. There are two types of Ti atoms on the surface having five-fold and six-fold coordination of oxygen.

tailor the properties of thin TiO$_2$ layers, for example, for photocatalysis as well as for the interaction of clusters on TiO$_2$ surfaces.

Unsupported clusters of noble and near noble metals that are interesting from the point of view of catalysis have been shown to have quite different structures and properties compared to bulk, and the difference is more pronounced when the nanoparticles are small. A striking example is of gold nanoparticles. In bulk form, gold is known to be the most noble and catalytically inactive, but small gold nanoclusters have been shown (Gruene et al. 2008) to become catalytically active and are a potential catalyst (Yoon et al. 2005) for the oxidation process in fuel cell applications as well as for the conversion of CO into CO$_2$. Small gold clusters with fewer than 16 atoms have planar structures and beyond that they adopt interesting atomic structures, including fullerene cages (Kumar 2009) that are yet to be fully explored. Doping of gold clusters could lead to new structures that may have interesting magnetic (Yadav and Kumar 2010), optical, and catalytic properties with diverse applications, including biology and sensor applications. Another interesting aspect is that a lower mean coordination in nanoparticles compared with bulk can lead to the development of magnetism in nanoparticles of nonmagnetic bulk metals. Nanoparticles of rhodium, palladium, and platinum (Bae et al. 2004, 2005; Kumar and Kawazoe 2002, 2008) have been shown to possess nonzero magnetic moments even though in bulk they are nonmagnetic. It is also important to understand the magnetic properties as well as the modifications in the atomic and electronic structures of such nanoparticles after they are deposited on an oxide substrate. In the present study, we present results of the properties of Pd$_n$ ($n = 1$–13) nanoparticles on rutile TiO$_2$ (110) surface using plane wave–based first-principles calculations. There are experimental results (Kaden et al. 2009) on this system because of the environmental importance in the conversion of CO into CO$_2$. Further, we shall briefly discuss the results of studies on Pd clusters supported on MgO and Al$_2$O$_3$ that are also often used as support for catalysts and their behavior compared to the results on TiO$_2$.

Recently, a few groups (Kaden et al. 2009, 2010; Lai et al. 1998; Xu et al. 1997) have performed experiments on size-selected Pd clusters on TiO$_2$ (110) support. In one such study (Kaden et al. 2009, 2010), Pd$_n$ clusters with $n = 1, 2, 4, 7, 10, 16, 20,$ and 25 were deposited by soft landing on the TiO$_2$ (110) surface, and the reactivity of CO as well as the conversion of CO into CO$_2$ has been studied. It is worthy of note here that oxide surfaces may often have oxygen vacancies that could act as traps for clusters as well as for molecules. Here, we consider stoichiometric surfaces without such defects. Experiments have shown that a palladium atom supported on the TiO$_2$ (110) surface does not show much reactivity with CO, but with increasing size, the reactivity increases (Kaden et al. 2009). Further, small clusters with n up to 10 have been suggested to have a layer structure and that beyond this size the second layer of Pd begins depositing, leading to formation of three-dimensional structures as well as an increase in the reactivity. Note that in earlier studies (Pan et al. 1988; Pick et al. 1979; Ruckman et al. 1986, 1992) conducted in a different context, enhancement of hydrogen uptake by Pd deposited on an Nb(110) or Ta(110) surface was obtained when a second layer of palladium began depositing on Nb(110). This has been understood in terms of the changes in the atomic and electronic structure of Pd at low coverages. For submonolayer coverage of Pd on Nb(110), palladium atoms are deposited in body-centered cubic (BCC) structure with a slight elongation in the interatomic distances compared with bulk palladium. In this configuration, the palladium layer acts like that of a noble metal with a nearly full $4d$ band (Kumar and Bennemann 1983) and low density of states (DOS) at the Fermi energy, E_F. When a second layer begins depositing, the Pd layer structure transforms locally to the face-centered cubic (FCC) (111) type with bond lengths similar to those on a Pd(111) surface and wider $4d$ band. This creates more holes in the $4d$ bands as well as a shift in the center of the Pd $4d$ band that leads to a higher DOS at the E_F, which makes it possible to dissociate H$_2$ molecules leading to a large H uptake. Therefore, changes in the Pd–Pd interatomic distance are very important in the reactivity, and similar changes in the atomic and electronic structure of the deposited palladium nanoclusters could take place when the fraction of deposited palladium is such that a second layer starts forming, leading to higher reactivity of Pd nanoparticles.

14.2 COMPUTATIONAL METHOD

For understanding the adsorption properties of Pd_n clusters on a rutile TiO_2 (110) surface, we constructed a slab supercell with the surface area of $4c \times 2\sqrt{2}a$, a and c being the lattice parameters for bulk TiO_2. We modeled the surface with two different thicknesses of the slab, namely with two layers (2L) and with three layers (3L) of TiO_2. This can also help to understand the variation in the properties of the clusters as a function of the slab thickness. A large surface supercell is needed for the deposition of Pd clusters so that there is sufficient separation between the cluster and its periodical images to make the interaction between the cluster and the images sufficiently weak. For a Pd monomer adsorption, we constructed the supercell with the dimension of $2c \times \sqrt{2}a$, which is sufficient to reduce interactions between the monomer and its periodical images. Along the z-axis, the periodic supercell is constructed by taking approximately 10 to 12 Å vacuum space. All atoms in the supercell are relaxed completely without any symmetry constraint using first-principles calculations as implemented in the Vienna *Ab Initio* Simulation Package (VASP). We use ultrasoft pseudo-potentials (Kresse and Hafner 1994; Vanderbilt 1990) to describe the electron–ion interactions present in the system. The exchange-correlation energy has been calculated using the generalized gradient approximation (Perdew et al. 1992). The cutoff energy for the plane wave expansion has been taken to be 450 eV. The ionic relaxation process is repeated until the force on each ion is converged within 10 meV/Å. The Brillouin zone of the supercell is sampled by $2 \times 2 \times 1$ **k**-points. However, for supercells with a smaller surface area, we used $4 \times 4 \times 1$ **k**-points. In the case of supported clusters, we performed calculations using the spin-polarized exchange-correlation functional because Pd nanoclusters are known to have magnetic moments (Kumar and Kawazoe 2002). Note that the isolated Pd clusters were studied earlier by the previously mentioned method, and we have taken the lowest energy isomers for Pd_n clusters from Kumar and Kawazoe (2002). Furthermore, we have studied some planar structures that may form as found in experiments. These calculations have been done in a large cubic unit cell with only gamma point for the Brillouin zone integrations. These are required to calculate the adsorption energy of Pd clusters on TiO_2 support.

14.3 RESULTS AND DISCUSSIONS

14.3.1 MODELING OF PRISTINE TiO_2 (110) SURFACE

To understand the adsorption properties of palladium nanoclusters on (110) surface of rutile TiO_2, we studied the optimized lattice constants a and c of bulk TiO_2 in rutile phase and used these results to construct the surface slabs. To obtain the bulk lattice constants, a tetragonal $2a \times 2a \times 2c$ supercell was taken and the Brillouin zone was sampled by Monkhorst-Pack mesh of $4 \times 4 \times 6$ **k**-points The calculated lattice parameters, $a = 4.66$ Å and $c = 2.98$ Å, agree well with the experimental values of 4.594 Å and 2.959 Å, respectively. The optimized short and long Ti-O bond distances from our calculations are 1.97 Å and 2.00 Å, respectively.

Surface slabs were modeled from the optimized bulk lattice constants. The atomic structure of 2L and 3L surface slabs was relaxed along the z-direction, and the resultant optimized structures are shown in Figure 14.1. It can be seen that the surface slabs are reconstructed as a result of relaxation of ions and the type of reconstruction for 2L and 3L surface slabs is quite different. In fact, further calculations on thicker slabs showed that the reconstruction can be classified based on an odd or even number of layers in a surface slab (Murugan et al. 2006). For the 3L (odd number of layers) surface slab, the central layer is almost similar to bulk, whereas in the surface layers (top and bottom), half of the titanium atoms have a reduced coordination number of five and the remaining half of titanium atoms have the same coordination as in bulk (coordination number = 6). Such Ti atoms are referred to Ti(5c) and Ti(6c), respectively. The oxygen atoms within the surface plane are referred to as O(p) and on bridge sites protruding away from the surface as O(b). The Ti(5c) atoms move toward the central layer (toward inside the surface along z-direction) and the separation between Ti(5c) atoms and the plane of O(p) atoms in the surface is –0.30 Å, whereas such separation is negligible for Ti(6c) in the surface layer (0.03Å). For the 2L surface slab, both the top and bottom layers are buckled in the same direction because the bridging oxygen atoms do not have mirror symmetry in a central plane of the slab as in the case of 3L slab. In 2L slab, also, the undercoordinated Ti(5c) atoms are relaxed inward of the surface, leading to the curving of the two layers in 2L slab in the same direction, as shown in Figure 14.1. The separations of Ti(5c) and Ti(6c) atoms from the surface O(p) atoms are –0.39 Å and 0.21 Å, respectively, along the z-axis. The variation in displacement of Ti(5c) and Ti(6c) is also observed for higher number of layers in a surface slab but in a smaller magnitude (Murugan et al. 2006). We also calculated the band gap of 2L and 3L surface slabs and the values are 2.29 eV and 1.27 eV, respectively. Note that the calculated band gap for bulk TiO₂ is ~1.85 eV (Murugan et al. 2005, 2006) and it lies in between the values for 2L and 3L slabs. It has been found that the value of the band gap shows an oscillatory behavior as the number of layers is increased in the surface slab and for more than seven layers in the slab, the band gap tends to converge to the bulk value. Because TiO₂ is a good photocatalyst, it may be possible to tailor its catalytic behavior by appropriately choosing the number of layers in a sample or depositing on a substrate.

14.3.2 ADSORPTION OF A MONOMER OF Pd

To understand the adsorption behavior of a palladium monomer on rutile TiO₂ (110) 2L surface slab, we calculated adsorption on five different sites, namely, the Pd atom at: (1) the bridging site of two bridging oxygen atoms (O(b)) (Figure 14.2a), (2) the bridging sites of two planar oxygen atoms (Figure 14.2b) in which case the Pd atom tilts to form bonds with Ti(5c) and O(b), (3) the bridging sites of two Ti(5c) atoms (Figure 14.2c), (4) on top of one of the planar oxygen atoms (Figure 14.2d) that tilts toward neighboring Ti(6c), and (5) on top of one of Ti(5c) atoms (Figure 14.2e). The first two sites and the last site are attempted because Pd adsorption on these sites saturates the bonding coordination of the undercoordinated Ti(5c) or O(b) atoms. Another two cases, namely, (2) and (4) are studied to understand the bonding stability of O(p) atoms with Ti atoms. The adsorption energy is calculated from:

$$E_{ad} = E(Pd_n) + E(TiO_2) - E(Pd_n/TiO_2)$$

where $E(Pd_n)$, $E(TiO_2)$, and $E(Pd_n/TiO_2)$ are the total energies of the isolated Pd_n cluster, clean TiO₂ surface, and supported cluster, respectively. The calculated adsorption energies for all the five cases are 1.39 eV, 1.29 eV, 1.26 eV, 1.21 eV, and 1.09 eV, respectively. From these results we conclude that the palladium monomer prefers to occupy the bridging sites of two O(b) atoms on 2L slab. Palladium adsorption on these sites increases the bonding coordination of undercoordinated O(b) atoms. This finding is in agreement with those of earlier studies (Bredow and Pacchioni 1999; Sanz et al. 2000).

It is also observed that the O(b)–Pd bond distance (~2.14 Å) in the most stable configuration (Figure 14.2a) is shorter than in other cases (Figure 14.2b and d), in which the bond distance is little longer than 2.3 Å. In general, the shortest optimized bond distances of Pd–O(p) and Pd–Ti(5c) for the cases in Figures 14.2b to d are 2.3 Å and 2.5 Å, respectively. However, the Pd–Ti(5c) bond distance in the last case (Figure 14.2e) is shorter (~2.35 Å), which could lead to the possibility of the formation of Pd–Ti clustering, but it is energetically less stable according to our first-principles calculations. It is also noted that for all the five cases, the spin magnetic moment of the system is nearly zero. Similar calculations have been repeated for a Pd monomer deposited on bridging site of O(b) atoms of a 3L surface slab. As expected, the adsorption energy in this case is increased to 1.54 eV because the 3L surface slab has a lower band gap compared with 2L slab.

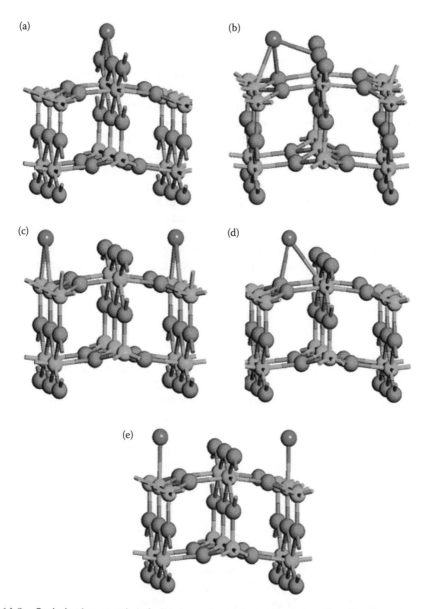

FIGURE 14.2 Optimized geometries of a Pd atom adsorbed on various sites (described in main text) of a 2L slab of TiO_2 (110) surface. Blue, red, and green balls correspond to Pd, O, and Ti atoms, respectively. In some cases two Pd atoms are seen. This is due to periodic boundary conditions.

Sanz et al. (2000) also studied adsorption of one and two palladium atoms on TiO_2 (110) surface using density functional theory (DFT) calculations with 3L, 4L, and 5L slabs. They obtained the highest binding energy for the Ti(5c) site, but the palladium atom is tilted toward the protruding oxygen atoms (O(b)). Their binding energy value of 1.90 eV is similar to our calculations, and they also obtained variations as a function of the number of layers. The site with the highest binding energy also depends on the number of layers included in the calculations, though variation in the binding energy is small, suggesting high mobility of the palladium atom on this surface. It is interesting to note that in experiments, single palladium atom has been difficult to observe. But dimers, trimers, and so on were observed aligned along the [001] direction.

14.3.3   ADSORPTION OF Pd$_2$

Similar to the monomer adsorption case, a dimer of palladium atoms is placed on a 2L surface slab on four different sites, namely: (1) the dimer across the channel that is formed by two rows of bridging oxygen atoms (Figure 14.3a), (2) dimer at the connecting site between O(b) and Ti(5c) atoms (Figure 14.3b), (3) the dimer at the top of O(p) atoms (Figure 14.3c) that tilt toward the neighboring Ti(5c) atoms, and (4) the dimer at the top of O(b) atoms. The calculated adsorption energies for the four cases are 1.93 eV, 1.70 eV, 1.62 eV, and 1.51 eV, respectively. The first configuration has the highest stability among those we studied and agrees with the scanning tunnelling microscope (STM) experiments (Xu et al. 1997). The binding energy per atom with respect to free palladium atoms is $(1.93 + 1.26)/2 = 1.58$ eV/atom, which is 0.19 eV/atom higher than the value for one palladium atom discussed earlier. Here, 1.26 eV is the total binding energy of a free Pd$_2$ dimer. Therefore, it is energetically preferable for two palladium atoms to be neighbors, and aside from high mobility, this could be another reason why a single palladium atom was not observed in the STM experiments. The Pd–Pd bond distance in all cases of different adsorption sites is increased compared to the isolated dimer distance of 2.48 Å. The adsorution energy is found to be highly dependent on the Pd–Pd bond strength. After adsorption of Pd$_2$ on 2L surface slab, Pd–Pd bond distance for the previously discussed four cases is increased to 2.63 Å, 2.71 Å, 2.71 Å, and 2.86 Å, respectively. The dimer bond distance for the last configuration is increased due to the bridging oxygen atoms which are separated by 2.99 Å. Therefore, an increase in the Pd–Pd bond distance leads to a decrease in the adsorption energy of Pd$_2$ on the TiO$_2$ (110) surface. Our result differs from the one reported by Bredow and Pacchioni (1999), who used an embedded cluster model and concluded that palladium dimers adsorbed on a TiO$_2$ (110) surface lose most of the Pd–Pd interaction because of the relatively strong bond with the substrate. However, our result agrees with the calculations of Sanz et al. (2000).

It is also important to compare the adsorption properties of palladium monomer and dimer on the TiO$_2$ surface. In general, both the monomer and dimer of Pd are adsorbed on a surface site that increases the bonding coordination of the undercoordinated surface atoms. But for dimer the adsorption is more favorable on the surface sites that also keep interaction between palladium atoms optimal and the bond length remains close to the value for the isolated palladium dimer. Note that the bond strength of isolated Pd dimer is 0.63 eV/atom. The small value of the binding energy of an isolated Pd$_2$ is due to the closed shell behavior of 4d electrons, because the electronic configuration

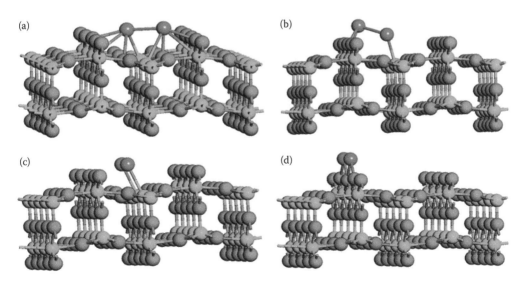

FIGURE 14.3   Pd$_2$ adsorption on various sites of a 2L rutile TiO$_2$ (110) surface slab.

of a palladium atom is $4d^{10}5s^0$. We also calculated the spin magnetic moment of an isolated Pd_2, which is 2 μ_B, but this moment is quenched to zero when a dimer is supported on the TiO_2 (110) surface. We also deposited a Pd_2 cluster at the stable adsorption site of the 3L surface slab, and the adsorption energy increased to 2.41 eV.

14.3.4 ADSORPTION OF Pd₄ CLUSTER

Initially, we studied the atomic, electronic, and magnetic properties of isolated Pd_4 (planar isomer) by first-principles calculations. The bond distance, the binding energy, the highest occupied molecular orbital–lowest unoccupied molecular orbital (HOMO–LUMO) gap, and the total magnetic moments are 2.49 Å (equal bond separation), 1.46 eV/atom, 0.04 eV, and 2 μ_B, respectively. With the intuition from the adsorption of a monomer and a dimer on the TiO_2 (110) surface, a planar Pd_4 cluster is deposited on the channel that is formed by the bridging oxygen atoms. The optimized atomic structure of Pd_4 on 3L surface is shown in Figure 14.4. The adsorption energy for this cluster

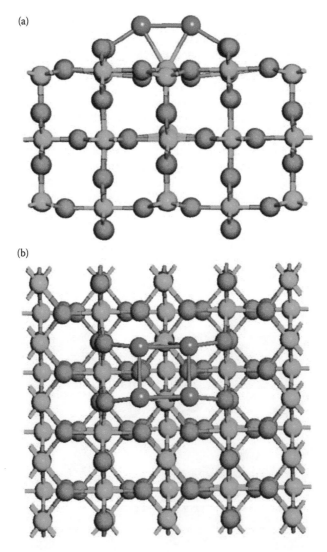

FIGURE 14.4   (a) Side view and (b) aerial view of square Pd_4 isomer adsorbed on a 3L TiO_2 (110) surface slab.

on 2L and 3L surface slabs is calculated to be 2.10 eV and 3.94 eV, respectively. The adsorption energy of this cluster on a 3L surface slab is much higher than that on a 2L surface slab because of the smaller band gap in the former and weaker interaction with Ti(5c) atoms. Note that the increase in the adsorption energy of Pd_4 from Pd_2 cluster is not a multiple of two. This is because there is a significant increase in the binding energy between Pd atoms in free cluster. The observed bond distances of Pd–O and Pd–Ti(5c) are 2.22 Å and 2.85 Å, respectively. Also, the Pd–Pd bond distance is increased slightly. The bond distance of Pd–Pd along channel and across the channel is 2.50 Å and 2.53 Å, respectively.

14.3.5 ADSORPTION OF Pd_6 CLUSTER

Earlier calculations (Kumar and Kawazoe 2002) predicted that the lowest energy geometry of Pd_6 is an octahedron. We optimized an octahedral isomer of Pd_6 from first-principles calculations. The calculated bond distance, the binding energy, the HOMO–LUMO gap, and the total magnetic moments are 2.66 Å, 1.92 eV/atom, 0.09 eV, and 2 μ_B, respectively. As the size of Pd clusters grows, it is more favorable to deposit clusters in the channel which could provide an increase in the interaction between atoms in the cluster and the surface, particularly the undercoordinated surface atoms. Also the atomic structure of the deposited clusters could be different from the one for a free (unsupported) cluster. In this case, instead of attempting various adsorption sites on the surface, it is also interesting to understand the deposition of cluster with different orientations on the support. This could be important because of the size of the channel formed by bridging oxygen atoms. For an octahedral cluster, we considered three orientations (Figure 14.5), namely, vertex, edge, and face toward the surface. In the vertex orientation, the Pd_6 cluster interacts with the support surface mainly through one Pd atom. Similarly, two and three atoms in the cluster are bonded with support for edge and face orientations, respectively.

All three orientations of the Pd_6 cluster are deposited on the 2L surface slab. The calculated adsorption energies for vertex, edge, and face orientations are 1.50 eV, 1.18 eV, and 2.18 eV, respectively. The octahedral cluster of Pd_6 prefers to be deposited on the face orientation which allows interaction of more Pd atoms in the cluster on the surface. It is also noted that the adsorption energy of this cluster is slightly increased with respect to the adsorption of a Pd_4 cluster. In this case, three palladium atoms in the cluster are mainly bonded with the surface and the remaining three atoms

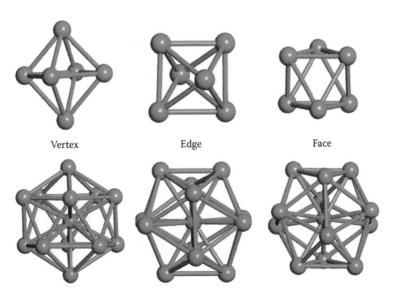

Vertex   Edge   Face

FIGURE 14.5   Stable isomers of Pd_6 (octahedron) and Pd_{13} (icosahedron) are shown in three orientations.

interact weakly with the surface. The total magnetic moment for the three cases is unaltered compared to the isolated cluster, which has the total magnetic moment of 2 μ_B.

As discussed earlier, the adsorption and physical properties of TiO_2 bulk surfaces generally lie in between the properties of 2L and 3L surface slabs. Therefore, we extended this study to further calculate deposition of Pd_6 cluster on a 3L surface slab by first-principles calculations; the optimized structure is shown in Figure 14.6, along with the structure of the cluster on a 2L surface slab. The adsorption energy of this cluster on a 3L surface slab is 3.42 eV, which is much higher than the adsorption energy of the same cluster on a 2L surface slab (2.18 eV). This increase in the adsorption energy could be expected because the band gap of 3L slab is much smaller than that of a 2L surface slab. Often, metallic surfaces are more reactive and therefore enhancement in the adsorption energy could be expected. Also, some effects are likely to be observed because of the curvature of the 2L slab. Furthermore, our results show that certain adsorption sites are more preferable on oxide surfaces. The shortest bond distance of Pd–Ti(5c) for the 3L case is 2.48 Å, whereas for the 2L case, this distance is increased to 2.82 Å, showing quite significant changes in the adsorption behavior as a function of the thickness of the TiO_2 slab. From our results of Pd_6 on 2L and 3L slabs, we conclude that the adsorption energy of a three-dimensional octahedral isomer of Pd_6 cluster on bulk TiO_2 surface lies in the range of 2.18 eV to 3.42 eV.

Experimentally, palladium clusters with fewer than 10 atoms have been found to have single layer structure (Kaden et al. 2009). To check the adsorption behavior of different isomers of Pd_6 cluster, we carried out calculations on the adsorption properties of a planar Pd_6 cluster and the optimized structure is shown in Figure 14.6(c) and (d). Note that the isolated planar isomer is energetically less stable by 2.1 eV compared to an octahedral Pd_6 cluster. Planar isomer has a small HOMO–LUMO gap of about 0.04 eV, and the total magnetic moment is 2 μ_B. This cluster is deposited on the channel of 3L surface slab. The adsorption energy of this cluster on 3L surface slab is 4.81 eV, which is

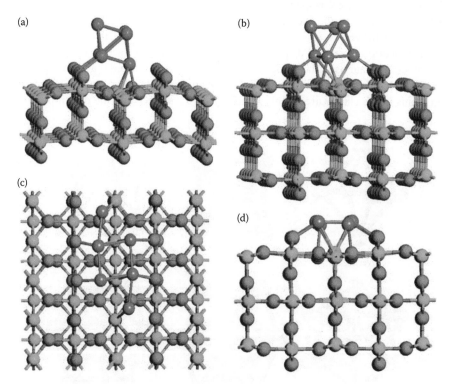

FIGURE 14.6 Face oriented octahedral Pd_6 isomer on (a) a 2L surface slab and (b) a 3L surface slab. This is the preferable orientation of adsorption of this cluster on surface. (c) and (d) show, respectively, the aerial and side views of a planar isomer of Pd_6 on a 3L TiO_2 (110) surface slab.

much higher than that of an octahedral isomer. The increase in the adsorption energy of this cluster is a consequence of the fact that all Pd atoms interact strongly with the surface. In the optimized structure, Pd–O(b), Ti–Pd, and Pd–Pd bond distances are approximately 2.2 Å, 2.6 Å, and 2.6 Å, respectively. We conclude that the planar cluster is adsorbed more strongly on the substrate compared with an octahedral isomer. However, our results show that among these two isomers of Pd$_6$, octahedral isomer on TiO$_2$ (110) surface has lower energy. It is possible that there is another planar atomic arrangement of Pd$_6$ on TiO$_2$ (110) that may have lower energy than the octahedral isomer. Further work in this direction is in progress.

14.3.6 ADSORPTION OF Pd$_{13}$ CLUSTER

For Pd$_{13}$ cluster, an icosahedral isomer has been found (Kumar and Kawazoe 2002) to be the most stable configuration. The optimized cluster has the binding energy of 2.29 eV/atom, and mean bond distance of 2.75 Å, and total magnetic moment of 8 μ_B. Similar to an octahedral cluster, all the faces in an icosahedron are triangular and this cluster can also be deposited with three orientations on the surface, as mentioned earlier (Figure 14.5). First-principles calculations with these orientations of this cluster on a 2L surface slab have been carried out. The adsorption energies of vertex, edge, and face orientations are 2.23 eV, 2.43 eV, and 2.07 eV, respectively. These results show that an icosa-hedral isomer of the Pd$_{13}$ cluster with edge orientation prefers to be deposited on a 2L surface slab, and the optimized atomic structure is shown in Figure 14.7. The face orientation of this cluster is difficult to accommodate in the channel. Hence, such a configuration has lower adsorption energy. In all three cases, the total magnetic moment is reduced to 6 μ_B from the value of 8 μ_B for an isolated icosahedral cluster.

Similar calculations were extended to understand the adsorption properties of this cluster on a 3L surface slab, and the optimized structure is also shown in Figure 14.7. The adsorption energy is increased to 2.69 eV compared with the 2L case as has also been found for other cluster sizes, but the difference is less compared with the case of Pd$_4$ or Pd$_6$. The increase in the adsorption energy in the 3L case also arises from the shift of the undercoordinated O(b) and Ti(5c) atoms toward the cluster. Compared to a clean surface, the displacement of neighboring O(b) and Ti(5c) atoms in 2L surface slab toward the cluster are found to be 0.03 Å and 0.11 Å, respectively. On the other hand, the shifts of O(b) and Ti(5c) atoms in 3L surface slab are 0.15 Å, and 0.43 Å, respectively. Hence, the total magnetic moment of Pd$_{13}$ is quenched similar to the reduction of the magnetic moment in the adsorption of Co$_4$ cluster on oxide surface (Giordano et al. 2001). It would be of

(a)   (b)

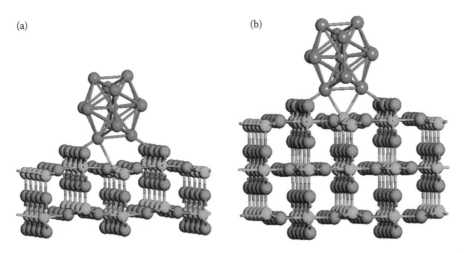

FIGURE 14.7 Icosahedral Pd$_{13}$ cluster adsorbed on (a) 2L and (b) 3L surface slabs. The edge orientation of this cluster on the surface is energetically more favorable.

further interest to study other isomers with bilayer structure if they would have lower energy when supported on a TiO_2 (110) surface.

14.3.7 ELECTRONIC STRUCTURE

Understanding the electronic structure of the supported clusters is very important for reactivity. Figure 14.8 shows the total and partial DOS of Pd_2 cluster deposited on stable sites (Figure 14.2a) on 2L and 3L surface slabs. The energy gap between the valence band (VB), which is mainly composed of O-2p states (refer Figure 14.8) and the conduction band (CB), which is formed by Ti-3d states for 2L surface slab is 2.29 eV, and the energy gap of 3L slab is 1.27 eV (Murugan et al. 2005, 2006). The band width for the 3L case is much larger compared with that of the 2L case. Also seen in Figure 14.8, most of the Pd-4d states are located in between the gap of 2L surface slab, whereas in the case of 3L slab, some of the Pd-states overlap with O-2p states, and this increases the adsorption energy of the Pd_2 cluster on the 3L surface slab. It is also noted that the up- and the down-spin channels of Pd_2 adsorbed on the surface are equally filled, which is indicative that the total spin moment of an isolated Pd_2 cluster (2 μ_B) is quenched to zero upon adsorption.

The DOS for planar and octahedral Pd_6 isomers adsorbed on 3L surface slab are shown in Figure 14.9. Both isomers as isolated clusters have a small HOMO–LUMO gap (< 0.1 eV), which means that these clusters can strongly interact with the substrate. The adsorption energy of these isomers is 4.81 eV and 3.42 eV, respectively. The DOS (Figure 14.9) curve shows that the Pd *4d* states of these isomers strongly overlap with the O *2p* states and to some extent with Ti *3d* states also. As the planar isomer has low dimensionality, Pd *4d* states hybridize well with those of the substrate and are distributed over a range from about –3.5 eV to 0.75 eV. It supports our finding of large adsorption energy of planar Pd_6 isomer compared to the octahedral isomer. It is also noted that the up- and the down-spin channels are unequally occupied in octahedral the Pd_6 isomer on TiO_2 (110) surface, and this gives rise to the total magnetic moment of 2 μ_B. However, the strong interaction between the planar cluster and the substrate results in quenching of the magnetic moment to zero μ_B.

We also compare the DOSs of Pd_n clusters with n = 1, 4, and 13, which are adsorbed on 3L rutile TiO_2 (110) surface slab (Figure 14.10). As the size of the cluster increases, palladium states are distributed widely. For example, in the case of a palladium monomer adsorbed on the surface, most of

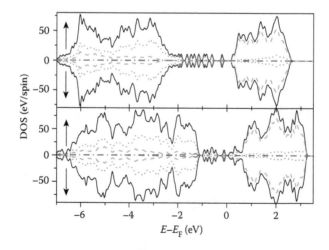

FIGURE 14.8 Comparison of the total and the partial density of states (DOS) of Pd_2 adsorbed on 2L (top panel) and 3L (bottom panel) slabs of TiO_2(110) surface. Black, dashed green, dotted red, and dash-dotted blue lines correspond to total DOS, Ti *3d*, O *2p*, and Pd *4d* states, respectively. Up and down arrows in diagram correspond to up- and down-spin channels, respectively.

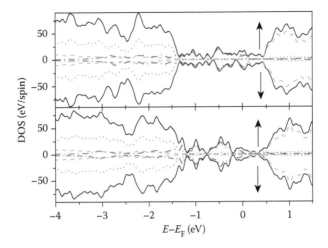

FIGURE 14.9 Comparison between the total and the partial DOS of octahedral (top panel) and planar (bottom panel) isomers of Pd$_6$ on 3L slab of TiO$_2$ (110) surface. Black, dashed green, dotted red, and dash-dotted blue lines correspond to total DOS, Ti $3d$, O $2p$, Pd $4d$ states, respectively. Up and down arrows correspond to up- and down-spin channels, respectively.

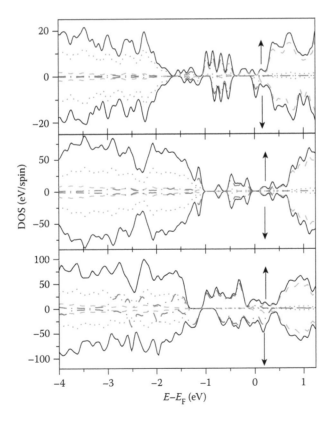

FIGURE 14.10 Comparison between the total and the partial DOS of Pd monomer (top panel), Pd$_4$ (middle panel), and Pd$_{13}$ (bottom panel) clusters on 3L slab of TiO$_2$ (110) surface. Black, dashed green, dotted red, and dash-dotted blue lines correspond to total DOS, Ti $3d$, O $2p$, Pd $4d$ states, respectively. Up and down arrows show up- and down-spin channels, respectively. Note a change in scale in the top panel because the unit cell for 1 Pd is smaller.

the Pd *4d* states are distributed within the gap. Thus, this restricts the hybridization of O *2p* or Ti *3d* with Pd *4d* states and weak adsorption energy. For the case of Pd$_4$ adsorption on 3L surface slab, Pd *4d* states are distributed from –3.0 eV to 0.5 eV. Hence, Pd *4d* states overlap with O *2p* states (closer to the top of the VB) and Ti *3d* (bottom of the CB), which increases the adsorption energy of the cluster on the substrate to 3.94 eV. In the DOS corresponding to the adsorption of an icosahedral Pd$_{13}$ cluster on 3L surface, the Pd *4d* states are distributed widely from –4.5 eV to 0.5 eV. This allows overlap of the Pd *4d* states with O *2p* states and Ti *3d* states, and the surface behaves like the one of a metal. Because only two Pd atoms in this cluster actively interact with the surface, the adsorption energy of this cluster with 3L surface is only 2.69 eV, which is quite comparable to the adsorption energy of Pd dimer (2.45 eV) with the 3L surface. Hence, the magnetic moment of Pd$_{13}$ cluster is slightly altered and is reduced to 6 μ_B from its free cluster value of 8 μ_B. Note that for Pd$_2$, the magnetic moment also reduces from 2 to 0 μ_B upon adsorption on TiO$_2$. In the case of the planar Pd$_4$ cluster, the magnetic moment is completely quenched to zero μ_B as this cluster interacts strongly with the substrate.

14.4   COMPARISON WITH OTHER OXIDE SURFACES

Sterrer et al. (2007) have studied palladium monomer, dimer, and trimer on MgO (001) surface using STM. They obtained mostly palladium monomers on this surface and predominantly one type of adsorption site. DFT calculations (Moseler et al. 2002) have shown Pd atom bonded on top of oxygen atoms to be most favorable. Under certain conditions of bias, Pd dimers were also formed and interpreted to be adsorbed on top of neighboring oxygen sites with a small displacement. Pd timer was also formed, and it was concluded that it does not lie flat on the surface but the increased height suggested perpendicularly oriented Pd$_3$ on MgO (001) surface. Therefore, the behavior of Pd deposited on the MgO (001) surface is quite different from that deposited on TiO$_2$ (110) surface, and oxygen vacancies play an important role for the adsorption of Pd clusters. Al$_2$O$_3$ is another oxide material often used as support in catalysis. It has been shown that thin layers of Al$_2$O$_3$ can be deposited on NiAl (110) surface and this can be used as template to deposit well-arranged and well-separated clusters (Stierle et al. 2004). Therefore, the atomic structure of the support as well as chemical bonding has an effect on the atomic structure of the clusters, and this could be used to design novel catalysts so that the catalytic behavior of the clusters could be varied significantly and tailored for use to specific reaction or improve the reaction rate.

14.5   CONCLUSION

First-principles calculations have been carried out to understand the adsorption behavior and electronic properties of Pd$_n$ (n = 1–13) clusters on rutile TiO$_2$ (110) surface slabs. Such supported clusters are used as heterogeneous catalyst for the oxidation of CO and reduction of NO gases. For small clusters, we studied various adsorption sites on a 2L surface slab to understand the best sites and for large cluster; various orientations of clusters on the surface were also considered. Further studies have also been carried out on palladium clusters on a 3L surface slab. The adsorption behavior on a bulk surface is expected to be between the behaviors for 2L and 3L slabs. Overall, it has been found that the clusters are deposited on undercoordinated sites on the surface so that it increases the bonding coordination number of the undercoordinated O(b) and Ti(5c) atoms on the TiO$_2$ (110) surface. For monomer adsorption, palladium atoms prefer to be on bridging sites of O(b) atoms. As the size of the cluster is increased, it is accommodated in the channels between the protruding oxygen atoms. These results agree with the available experimental data. The adsorption energy of Pd$_n$ clusters increases until n = 4, and then it decreases for three-dimensional isomers because of the limitation of channel size. Compared to 2L surface, there is more metallicity as well as less curvature in the channel of the 3L surface slab, increasing the adsorption energy of the clusters. For Pd$_6$ we find that a planar isomer has lower energy compared to an octahedral isomer on the surface.

This agrees with the experimental findings. Because palladium clusters are magnetic, our calculations show that the magnetic moments are generally partially quenched when clusters are deposited on the TiO_2 (110) substrate, and in the cases of stronger interaction of the clusters with the surface they are completely quenched. Further, we find that Pd clusters on different supports could behave differently because of the different atomic structure of the support surface. The adsorption behavior has been further explained with the help of the electronic structure of the Pd cluster supported on the TiO_2 surface. Further work on other two-dimensional and three-dimensional structures of different-sized Pd clusters as well as the adsorption behavior of CO is in progress.

14.6 ACKNOWLEDGMENTS

We are grateful to the staff of the Center for Computational Materials Science of the Institute for Materials Research, Tohoku University, for allowing the use of Hitachi SR11000 supercomputing system and for their excellent support. Part of the calculations were performed at the HPC facility of CECRI.

REFERENCES

Bae, Y. C., Osanai, H., Kumar, V., and Kawazoe, Y. 2004. Nonicosahedral growth and magnetic behavior of rhodium clusters. *Phys. Rev. B* 70: 195413.

Bae, Y. C., Kumar, V., Osanai, H., and Kawazoe, Y. 2005. Cubic magic clusters of rhodium stabilized with eight-center bonding: Magnetism and growth. *Phys. Rev. B* 72: 125427.

Bredow, T., and Pacchioni, G. 1999. A quantum-chemical study of Pd atoms and dimers supported on TiO_2 (110) and their interaction with CO. *Surface Sci.* 426: 106.

Bredow, T., Giordano, L., Cinquini, F., and Pacchioni, G. 2004. Electronic properties of rutile TiO_2 ultrathin films: Odd-even oscillations with the number of layers. *Phys. Rev. B* 70: 035419.

Czekaj, I., Wambach, J., and Krocher, O. 2009. Modelling catalyst surfaces using DFT cluster calculations. *Int. J. Mol. Sci.* 10: 4310.

Diebold, U. 2003. The surface science of titanium dioxide. *Surface Sci. Rep.* 48: 53.

Feng, X. J., Shankar, K., Varghese, O. K., Paulose, M., Latempa, T. J., and Grimes, C. A. 2008. Vertically aligned single crystal TiO_2 nanowire arrays grown directly on transparent conducting oxide coated glass: Synthesis details and applications. *Nano Lett.* 8: 3781.

Giordano, L., Pacchioni, G., Ferrari, A. M., Illas, F., and Rosch, N. 2001. Electronic structure and magnetic moments of Co-4 and Ni-4 clusters supported on the MgO(001) surface. *Surface Sci.* 473: 213.

Gruene, P., Rayner, D. M., Redlich, B., van der Meer, A. F. G., Lyon, J. T., Meijer, G., and Fielicke, A. 2008. Structures of neutral Au-7, Au-19, and Au-20 clusters in the gas phase. *Science* 321: 674.

Hanaor, D. A. H., and Sorrell, C. C. 2011. Review of the anatase to rutile phase transformation. *J. Mater. Sci.* 46: 855.

Harrison, N. M., Wang, X. G., Muscat, J., and Scheffler, M. 1999. The influence of soft vibrational modes on our understanding of oxide surface structure. *Faraday Discuss.* 114: 305.

Henrich, V. E., and Cox, P. A. 1994. *The Surface Science of Metal Oxides.* Cambridge: Cambridge University Press.

Kaden, W. E., Wu, T. P., Kunkel, W. A., and Anderson, S. L. 2009. Electronic structure controls reactivity of size-selected Pd clusters adsorbed on TiO_2 surfaces. *Science* 326: 826.

Kaden, W. E., Kunkel, W. A., Kane, M. D., Roberts, F. S., and Anderson, S. L. 2010. Size-dependent oxygen activation efficiency over Pd-n/TiO_2(110) for the CO oxidation reaction. *J. Am Chemical Soc.* 132: 13097.

Kresse, G., and Hafner, J. 1994. Norm-conserving and ultrasoft pseudopotentials for first-row and transition-elements. *J. Phys. Condens. Matter* 6: 8245.

Kumar, V., and Bennemann, K. H. 1983. Electronic structure of transition-metal-transition-metal interfaces: Pd on Nb(110). *Phys. Rev. B* 28: 3138.

Kumar, V., and Kawazoe, Y. 2002. Icosahedral growth, magnetic behavior, and adsorbate-induced metal-nonmetal transition in palladium clusters. *Phys. Rev. B* 66: 144413.

Kumar, V., and Kawazoe, Y. 2008. Evolution of atomic and electronic structure of Pt clusters: Planar, layered, pyramidal, cage, cubic, and octahedral growth. *Phys. Rev. B* 77: 205418.

Kumar, V. 2009. Coating of a layer of Au on Al13: The findings of icosahedral Al@Al12Au20− and Al12Au202 − fullerenes using ab initio pseudopotential calculations. *Phys. Rev. B* 79: 085423.

Lai, X., St Clair, T. P., Valden, M., and Goodman, D. W. 1998. Scanning tunneling microscopy studies of metal clusters supported on TiO$_2$ (110): Morphology and electronic structure. *Progr. Surface Sci.* 59: 25.

Menzies, D., Dai, Q., Cheng, Y. B., Simon, G., and Spiccia, L. 2004. Microwave calcination of thin TiO$_2$ films on transparent conducting oxide glass substrates. *J. Mater. Sci.* 39: 6361.

Moseler, M., Hakkinen, H., and Landman, U. 2002. Supported magnetic nanoclusters: Soft landing of Pd clusters on a MgO surface. *Phys. Rev. Lett.* 89: 176103.

Murugan, P., Kumar, V., and Kawazoe, Y. 2005. Ab initio study of magnetism in palladium clusters supported on (110) surface of TiO$_2$ rutile. *Int. J. Mod. Phys. B* 19: 2544.

Murugan, P., Kumar, V., and Kawazoe, Y. 2006. Thickness dependence of the atomic and electronic structures of TiO2 rutile (110) slabs and the effects on the electronic and magnetic properties of supported clusters of Pd and Rh. *Phys. Rev. B* 73: 075401.

Pan, X. H., Johnson, P. D., Weinert, M., Watson, R. E., Davenport, J. W., Fernando, G. W., and Hulbert, S. L. 1988. Localized states at metal-metal interfaces: An inverse photoemission-study of Pd/Nb(110). *Phys. Rev. B* 38: 7850.

Perdew, J. P., Chevary, J. A., Vosko, S. H., Jackson, K. A., Pederson, M. R., Singh, D. J., and Fiolhais, C. 1992. Atoms, molecules, solids, and surfaces: Applications of the generalized gradient approximation for exchange and correlation. *Phys. Rev. B* 46: 6671.

Pick, M. A., Davenport, J. W., Strongin, M., and Dienes, G. J. 1979. Enhancement of hydrogen uptake rates for Nb and Ta by thin surface overlayers. *Phys. Rev. Lett.* 43: 286.

Ramamoorthy, M., Kingsmith, R. D., and Vanderbilt, D. 1994a. Defects on TiO$_2$ (110) Surfaces. *Physical Review B* 49: 7709.

Ramamoorthy, M., Vanderbilt, D., and Kingsmith, R. D. 1994b. 1st-Principles calculations of the energetics of stoichiometric TiO$_2$ surfaces. *Phys. Rev. B* 49: 16721.

Ruckman, M.W., Murgai, V., and Strongin, M. 1986. Morphology and structural phase transitions of Pd monolayers on Ta(110). *Phys. Rev. B* 34: 6759.

Ruckman, M. W., Jiang, L. Q., and Strongin, M. 1992. Study of palladium monolayer and multilayer on Ta(110) between 300-K and 1500-K. *J. Vacuum Sci. Technol. A Vacuum Surfaces Films* 10: 2551.

Sanz, J. F., Hernandez, N. C., and Marquez, A. 2000. A first principles study of Pd deposition on the TiO$_2$(110) surface. *Theor. Chem. Accounts* 104: 317.

Sellidj, A., and Koel, B. E. 1993. Growth-mechanism and structure of ultrathin Pd films vapor-deposited on Ta(110). *Surface Sci.* 281: 223.

Sterrer, M., Risse, T., Giordano, L., Heyde, M., Nilius, N., Rust, H. P., Pacchioni, G., and Freund, H. J. 2007. Palladium monomers, dimers, and trimers on the MgO(001) surface viewed individually. *Angewandte Chem. Int. Ed.* 46: 8703.

Stierle, A., Renner, F., Streitel, R., Dosch, H., Drube, W., and Cowie, B. C. 2004. X-ray diffraction study of the ultrathin Al$_2$O$_3$ layer on NiAl(110). *Science* 303: 1652.

Vanderbilt, D. 1990. Soft self-consistent pseudopotentials in a generalized eigenvalue formalism. *Phys. Rev. B* 41: 7892.

Varazo, K., Parsons, F. W., Ma, S., and Chen, D. A. 2004. Methanol chemistry on Cu and oxygen-covered Cu nanoclusters supported on TiO$_2$(110). *J. Phys. Chem. B* 108: 18274.

Wang, Y. Q., Wei, Z. D., Gao, B., Qi, X. Q., Li, L., Zhang, Q., and Xia, M. R. 2011. The electrochemical oxidation of methanol on a Pt/TNTs/Ti electrode enhanced by illumination. *J. Power Sources* 196: 1132.

Xu, C., Lai, X., Zajac, G. W., and Goodman, D. W. 1997. Scanning tunneling microscopy studies of the TiO$_2$(110) surface: Structure and the nucleation growth of Pd. *Phys. Rev.* 56: 13464.

Yadav, B. D., and Kumar, V. 2010. Gd@Au15: A magic magnetic gold cluster for cancer therapy and bioimaging. *Appl. Phys. Lett.* 97: 133701.

Yoon, B., Hakkinen, H., Landman, U., Worz, A. S., Antonietti, J. M., Abbet, S., Judai, K., and Heiz, U. 2005. Charging effects on bonding and catalyzed oxidation of CO on Au-8 clusters on MgO. *Science* 307: 403.

Zeng, H. B., Liu, P. S., Cai, W. P., Yang, S. K., and Xu, X. X. 2008. Controllable Pt/ZnO porous nanocages with improved photocatalytic activity. *Journal of Physical Chemistry C* 112: 19620.

Index

Soluble salt minerals, 110–111, 127–128
    adsorption states of amine, 125–126
    concentrated alkali halide solutions, 111–114
    flotation of alkali halide salts in saturated solutions, 126–127
    interfacial phenomena of alkali chloride salt crystals, 118–125
Solute–oxygen distribution function (RDF), 111
Solvent/dispersing medium, computation of interaction energy in, 34
Split-valence basis sets, 16
Starch, 51
    selectivity toward hematite, 51–52
Static energy minimization, 33
Statistical ensembles, 10–12
Sulfide and nonsulfide mineral flotation, 66
Sulfur surface, dextrin adsorption at, 148–150
Super-hydrophilic surface, 235–237
Super-hydrophobic surface, 219, 235–237
Super-oleophobic surface, 237
Supersaturation, 171–173
Surface activity, 29
Surface docking approach, 175–176
Surface energy, calculation of, 69–71
Surface polarity and contact angle, 234
Surface(s)
    defined, 28
    types of, 69
Surfactant design framework, 30
Surfactant–metal complexes, 46, 48
Surfactant molecules, 33
    structure of common, 29
Surfactants, 28
    adsorption at interfaces, 28–30
      factors governing, 29–30
    framework for rational design of, 31–32
    interactions affecting the efficacy of, 29

T

Tail, 28
Talc, interaction between amphipathic surfactants and, 134–139
Talc basal plane surface, water dipole moment distribution at, 131
Talc edge surface, water dipole moment distribution at, 131, 132
Talc surfaces; *see also* Talc
    adsorption of cationic surfactant at, 134–136
    adsorption of dextrin at, 136–139
    interfacial water structure at, 129–132
Tersoff model, 330–332
Tetrahydrofuran (THF), 180
Thermal conductivity, using molecular dynamics to compute, 289–290
Thermal roughening, 172
Three body potential, 75–76
TiO_2 (titanium dioxide), 360–361
TiO_2 rutile (110) surface, 359–362; *see also* Pd
    modeling of pristine, 362–363
TiO_2 rutile (110) surface slab
    Pd_2 adsorption on various sites of 2L, 365
    square Pd_4 isomer adsorbed on 3L, 366

Titanium dioxide, *see* TiO_2
Tolbutamide, 161, 164
Total partition function, 308–309
Trajectory, 8
Transition matrix Monte Carlo (TMMC), 253, 336, 338–340; *see also* GC-TMMC approach
Tri-n-butyltin (TBT), 238

U

United atom (UA) level approach, 275–276
United atom (UA) potentials/force fields, 296, 297
    vs. anisotropic united atom (AUA), 294
    interactions parameters between groups for, 294–295
Upper critical solution temperature (UCST), 338, 339

V

Vacancy formation energies (VFE), 322, 323
Vacuum, computation of interaction energy in, 34
Valence cross-terms, 4
Valence Force Field (VFF) model, 328–330, 337
    parameters, 329, 330
Valence interactions, 3
van der Walls interactions, 4
Vapor–liquid phase transition and critical point, 250–260
    crossover from three- to two-dimensional, 261–262
Verlet algorithm, 9
Verlet neighbor list, 7–8
Verlet velocity algorithm, 10
Volumetric adsorption apparatus, 244, 245

W

Water
    adsorption energy for
      on hydroxylated quartz surfaces, 84
      on low energy pure wollastonite surfaces, 93
      on quartz surfaces, 81
    surface energy on low energy pure wollastonite surfaces, 93
Water diffusion coefficients, 123
Water surface residence time, 124
Wenzel model, 221–222, 236
Wettability, 35
Wetting of liquid drop
    on flat substrate, 220
    on textured surface, 221
Wetting temperature, 225
Wollastonite
    calculation of adsorption energy, 71–72
    potential parameters for, 73, 76
    side view of relaxed structure of, 90
    surface energy of reconstructed, 91
Wollastonite flotation system, quartz and, 100–102
Wollastonite surface(s)
    adsorption of water on pure, 91, 93
    dissolution of calcium from, 72–73, 99–100
    hydroxylation of, 93, 95, 100–102
    methanoic acid adsorption on pure, 96
      side view of, 97
      top view of, 97
    methylamine adsorption on, 96, 98
      side view of, 98